Student Solutions Manual

Laurel Technical Services

Elementary Algebra for College Students Early Graphing

Allen R. Angel

PRENTICE HALL, Upper Saddle River, NJ 07458

Executive Editor: Karin Wagner
Supplement Editor: Kate Marks
Special Projects Manager: Barbara A. Murray
Production Editor: Meaghan Forbes
Supplement Cover Manager: Paul Gourhan
Supplement Cover Designer: Liz Nemeth
Manufacturing Buyer: Alan Fischer

Upper Saddle River, NJ 07458

Printed in the United States of America

10 9 8 7 6 5 4 3 2 1

ISBN 0-13-040181-1

Prentice-Hall International (UK) Limited, London
Prentice-Hall of Australia Pty. Limited, Sydney
Prentice-Hall Canada, Inc., Toronto
Prentice-Hall Hispanoamericana, S.A., Mexico
Prentice-Hall of India Private Limited, New Delhi
Prentice-Hall (Singapore) Pte. Ltd.
Prentice-Hall of Japan, Inc., Tokyo
Editora Prentice-Hall do Brazil, Ltda., Rio de Janeiro

Table of Contents

Chapter 1

Exercise Set 1.1

11. To prepare properly for this class, you need to do all the homework carefully and preview the new material that is to be covered in class.

13. At least 2 hours of study and homework time for each hour of class time is generally recommended.

15. **a.** You need to do the homework in order to practice what was presented in class.

b. When you miss class, you miss important information, therefore it is important that you attend class regularly.

17. Answers will vary.

Exercise Set 1.2

1. Understand, translate, calculate, check, state answer

3. Substitute other numbers so that the method becomes clear.

5. Rank the data. The median is the value in the middle.

7. The mean is greater since it takes the value of 30 into account.

9. A mean average of 80 corresponds to a total of 800 points for the 10 quizzes. Jerome's mean average of 79 corresponds to a total of 790 points for the 10 quizzes. Thus, he actually missed a B by 10 points.

11. **a.** $\frac{78+97+59+74+74}{5}=\frac{382}{5}=76.4$

The mean grade is 76.4.

b. 59, 74, 74, 78, 97
The middle value is 74.
The median grade is 74.

13. **a.** $\frac{109.62+62.73+83.79+74.74+121.63}{5}=\frac{452.51}{5}=90.502$

The mean bill is \$90.502.

b. \$62.73, \$74.74, \$83.79, \$109.62, \$121.63
The middle value is \$83.79.
The median bill is \$83.79.

15. **a.** $\frac{24,990+26,260+25,872+26,290+27,100+26,284}{6}=\frac{156,796}{6}\approx 26,132.67$

The mean tuition and fees is about \$26,132.67.

b. \$24,990, \$25,872, \$26,260, \$26,284, \$26,290, \$27,100
The middle values are \$26,260 and \$26,284.

$\frac{26,260+26,284}{2}=\frac{52,544}{2}=26,272$

The median truition and fees is \$26,272.

17. Barbara's earnings = 5% of sales
Barbara's earnings = 0.05(9400)
= 470
Her week's earnings were $470.

19. New balance = old balance – (number of disks) (cost of each disk)
New balance = 312.60 – 5(17.11)
= 312.60 – 85.55
= 227.05
Her new balance is $227.05.

21. a. weekly cost = (days per week)(hours per day) (cost per hour)
weekly cost = (5)(8)(1.50)
= 60
Her weekly cost for parking is $60.

b. savings = cost paying hourly – cost paying weekly
savings = 60 – 35
= 25
She would save $25.

23. a. time to use energy

$$= \frac{\text{kJ in hamburger}}{\text{kJ / min running}}$$

$$= \frac{1550}{80}$$

$$= 19.375$$

It takes 19.375 minutes to run away the energy from a hamburger.

b. time to use energy

$$= \frac{\text{kJ in milkshake}}{\text{kJ / min walking}}$$

$$= \frac{2200}{25}$$

$$= 88$$

It takes 88 minutes to walk away the energy from a chocolate milkshake.

c. time to use energy

$$= \frac{\text{kJ in glass of skim milk}}{\text{kJ / min cycling}}$$

$$= \frac{350}{35}$$

$$= 10$$

It takes 10 minutes to cycle away the energy from a glass of skim milk.

25. $\text{miles per gallon} = \dfrac{\text{number of miles}}{\text{number of gallons}}$

$$= \frac{16{,}935.4 - 16{,}741.3}{10.5}$$

$$= \frac{194.1}{10.5}$$

$$\approx 18.49$$

His car gets about 18.49 miles per gallon.

27. savings = local cost – mail order cost

$$\text{local cost} = \$425 + (0.08)(\$425)$$
$$= \$425 + \$34$$
$$= \$459$$
$$\text{mail cost} = 4(\$62.30 + \$6.20 + \$8)$$
$$= 4(\$76.50)$$
$$= \$306$$
$$\text{savings} = \$459 - \$306$$
$$= \$153$$

Karin saved $153.

29. a. taxes = 15% of income
taxes = 0.15($26,420)
= $3963
Their taxes were $3963.

3/13/2015 Manage Orders

Ship To:

Wayne Bell; Roswell Police Department
39 HILL ST
ROSWELL, GA 30075-4536

Order ID: 106-2218720-6016248

Thank you for buying from RR Camardo Enterprise on Amazon Marketplace.

Shipping Address:
Wayne Bell; Roswell Police Department
39 HILL ST
ROSWELL, GA 30075-4536

Order Date:	Mar 4, 2015
Shipping Service:	Standard
Buyer Name:	Charles Bell
Seller Name:	RR Camardo Enterprise

Quantity	Product Details	Price	Total
1	**Student Solutions Manual for Elementary Algebra Early Graphing for College Students [Paperback] [2007] Angel, Allen R.** **SKU:** 9780130401816	$1.01	Subtotal: $1.01 Shipping: $3.99 Total: $5.00

= 365(gallons per day)
= 365(11.25 gallons)
= 4106.25 gallons

ey spent
wasted)
ey spent

· 4106.25 gallons

o.

37. a. Since people per telephone $= \dfrac{\text{number of people}}{\text{number of telephones}}$, number of telephones $= \dfrac{\text{number of people}}{\text{people per telephone}}$.

$$\begin{aligned}\text{number of telephones in Hong Kong} &= \frac{5.5 \text{ million}}{1.5}\\ &\approx 3.67 \text{ million}\\ \text{number of telephones in U.S.} &= \frac{263.4 \text{ million}}{1.3}\\ &\approx 202.62 \text{ million}\end{aligned}$$

b. The population of the U.S. is greater than the population of Hong Kong.

39. $\text{mean} = \dfrac{\text{total points}}{\text{number of games}}$

$$\begin{aligned}\text{mean} &= \frac{184}{8}\\ &= 23 \text{ points per game}\end{aligned}$$

41. a. $\text{mean} = \dfrac{\text{sum of grades}}{\text{number of exams}}$

$$\begin{aligned}60 &= \frac{50+59+67+80+56+\text{ last}}{6}\\ 360 &= 312 + \text{ last}\\ \text{last} &= 360 - 312\\ &= 48\end{aligned}$$

Lamond needs at least a 48 on the last exam.

b.

$$\begin{aligned}70 &= \frac{312+\text{ last}}{6}\\ 420 &= 312 + \text{ last}\\ \text{last} &= 420 - 312\\ &= 108\end{aligned}$$

Lamond would need 108 points on the last exam, so he cannot get a C.

Exercise Set 1.3

1. a. Variables are letters that represent numbers.

b. Letters often used to represent variables are x, y, and z.

3. $5(x)$, $(5)x$, $(5)(x)$, $5x$, $5 \cdot x$

5. Divide out factors that are common to both the numerator and the denominator.

7. Only b) is correct. In part a) the numerator of one fraction and the denominator of the second fraction are divided by a common factor.

9. c); $\dfrac{4}{5} \cdot \dfrac{1}{4} = \dfrac{1}{5} \cdot \dfrac{1}{1} = \dfrac{1}{5}$. Divide out the common factor, 4. This process can be used only when multiplying fractions and so cannot be used for a) or b). Part d) becomes $\dfrac{4}{5} \cdot \dfrac{4}{1}$ so no common factors can be divided out.

11. Invert the divisor, then multiply.

13. Multiply the denominator by the whole number, then add the numerator. This number is the numerator of the fraction. The denominator is the denominator of the fraction in the mixed number.

15. The greatest common factor of 3 and 9 is 3.
$\frac{3}{9}=\frac{3\div 3}{9\div 3}=\frac{1}{3}$

17. The greatest common factor of 10 and 15 is 5. $\frac{10}{15}=\frac{10\div 5}{15\div 5}=\frac{2}{3}$

19. The greatest common factor of 13 and 13 is 13. $\frac{13}{13}=\frac{13\div 13}{13\div 13}=\frac{1}{1}=1$

21. The greatest common factor of 36 and 76 is 4. $\frac{36}{76}=\frac{36\div 4}{76\div 4}=\frac{9}{19}$

23. The greatest common factor of 40 and 264 is 8. $\frac{40}{264}=\frac{40\div 8}{264\div 8}=\frac{5}{33}$

25. 12 and 25 have no common factors other than 1. Therefore, the fraction is already simplified.

27. $1\frac{7}{8}=\frac{8+7}{8}=\frac{15}{8}$

29. $2\frac{13}{15}=\frac{30+13}{15}=\frac{43}{15}$

31. $3\frac{3}{4}=\frac{12+3}{4}=\frac{15}{4}$

33. $4\frac{13}{19}=\frac{76+13}{19}=\frac{89}{19}$

35. $\frac{4}{3}=1\frac{1}{3}$ because $4\div 3=1$ R 1

37. $\frac{15}{4}=3\frac{3}{4}$ because $15\div 4=3$ R 3

39. $\frac{150}{20}=7\frac{10}{20}=7\frac{1}{2}$ because
$150\div 20=7$ R 10

41. $\frac{32}{7}=4\frac{4}{7}$ because $32\div 7=4$ R 4

43. $\frac{1}{2}\cdot\frac{3}{4}=\frac{1\cdot 3}{2\cdot 4}=\frac{3}{8}$

45. $\frac{5}{12}\cdot\frac{4}{15}=\frac{1}{12}\cdot\frac{4}{3}=\frac{1}{3}\cdot\frac{1}{3}=\frac{1\cdot 1}{3\cdot 3}=\frac{1}{9}$

47. $\frac{3}{4}\div\frac{1}{2}=\frac{3}{4}\cdot\frac{2}{1}=\frac{3}{2}\cdot\frac{1}{1}=\frac{3}{2}$ or $1\frac{1}{2}$

49. $\frac{2}{5}\div\frac{1}{8}=\frac{2}{5}\cdot\frac{8}{1}=\frac{2\cdot 8}{5\cdot 1}=\frac{16}{5}$ or $3\frac{1}{5}$

51. $\frac{5}{12}\div\frac{4}{3}=\frac{5}{12}\cdot\frac{3}{4}=\frac{5}{4}\cdot\frac{1}{4}=\frac{5\cdot 1}{4\cdot 4}=\frac{5}{16}$

53.
$$\begin{aligned}\frac{10}{3}\div\frac{5}{9}&=\frac{10}{3}\cdot\frac{9}{5}\\&=\frac{2}{3}\cdot\frac{9}{1}\\&=\frac{2}{1}\cdot\frac{3}{1}\\&=\frac{2\cdot 3}{1\cdot 1}\\&=\frac{6}{1}\\&=6\end{aligned}$$

55. $\left(2\frac{1}{5}\right)\frac{7}{8}$
$2\frac{1}{5}=\frac{10+1}{5}=\frac{11}{5}$
$\left(2\frac{1}{5}\right)\frac{7}{8}=\left(\frac{11}{5}\right)\frac{7}{8}=\frac{11\cdot 7}{5\cdot 8}=\frac{77}{40}$ or $1\frac{37}{40}$

57. $5\frac{3}{8} \div 1\frac{1}{4}$

$5\frac{3}{8} = \frac{40+3}{8} = \frac{43}{8}$

$1\frac{1}{4} = \frac{4+1}{4} = \frac{5}{4}$

$$\begin{aligned} 5\frac{3}{8} \div 1\frac{1}{4} &= \frac{43}{8} \div \frac{5}{4} \\ &= \frac{43}{8} \cdot \frac{4}{5} \\ &= \frac{43}{2} \cdot \frac{1}{5} \\ &= \frac{43 \cdot 1}{2 \cdot 5} \\ &= \frac{43}{10} \text{ or } 4\frac{3}{10} \end{aligned}$$

59. $\frac{1}{3} + \frac{2}{3} = \frac{1+2}{3} = \frac{3}{3} = 1$

61. $\frac{5}{12} - \frac{1}{12} = \frac{5-1}{12} = \frac{4}{12} = \frac{1}{3}$

63. $\frac{8}{17} + \frac{2}{34}$

$\frac{8}{17} = \frac{8}{17} \cdot \frac{2}{2} = \frac{16}{34}$

$\frac{8}{17} + \frac{2}{34} = \frac{16}{34} + \frac{2}{34} = \frac{16+2}{34} = \frac{18}{34} = \frac{9}{17}$

65. $\frac{4}{5} + \frac{6}{15}$

$\frac{4}{5} = \frac{4}{5} \cdot \frac{3}{3} = \frac{12}{15}$

$\frac{4}{5} + \frac{6}{15} = \frac{12}{15} + \frac{6}{15} = \frac{12+6}{15} = \frac{18}{15} = \frac{6}{5} \text{ or } 1\frac{1}{5}$

67. $\frac{1}{9} - \frac{1}{18}$

$\frac{1}{9} = \frac{1}{9} \cdot \frac{2}{2} = \frac{2}{18}$

$\frac{1}{9} - \frac{1}{18} = \frac{2}{18} - \frac{1}{18} = \frac{2-1}{18} = \frac{1}{18}$

69. $\frac{5}{12} - \frac{1}{8}$

$\frac{5}{12} = \frac{5}{12} \cdot \frac{2}{2} = \frac{10}{24}$

$\frac{1}{8} = \frac{1}{8} \cdot \frac{3}{3} = \frac{3}{24}$

$\frac{5}{12} - \frac{1}{8} = \frac{10}{24} - \frac{3}{24} = \frac{10-3}{24} = \frac{7}{24}$

71. $\frac{7}{12} - \frac{2}{9}$

$\frac{7}{12} = \frac{7}{12} \cdot \frac{3}{3} = \frac{21}{36}$

$\frac{2}{9} = \frac{2}{9} \cdot \frac{4}{4} = \frac{8}{36}$

$\frac{7}{12} - \frac{2}{9} = \frac{21}{36} - \frac{8}{36} = \frac{21-8}{36} = \frac{13}{36}$

73. $\frac{7}{8} - \frac{1}{15}$

$\frac{7}{8} = \frac{7}{8} \cdot \frac{15}{15} = \frac{105}{120}$

$\frac{1}{15} = \frac{1}{15} \cdot \frac{8}{8} = \frac{8}{120}$

$\frac{7}{8} - \frac{1}{15} = \frac{105}{120} - \frac{8}{120} = \frac{105-8}{120} = \frac{97}{120}$

75. $4\frac{1}{3} - 3\frac{2}{9}$

$4\frac{1}{3} = \frac{12+1}{3} = \frac{13}{3} = \frac{13}{3} \cdot \frac{3}{3} = \frac{39}{9}$

$3\frac{2}{9} = \frac{27+2}{9} = \frac{29}{9}$

$4\frac{1}{3} - 3\frac{2}{9} = \frac{39}{9} - \frac{29}{9} = \frac{39-29}{9} = \frac{10}{9} \text{ or } 1\frac{1}{9}$

77. $5\frac{3}{4} - \frac{1}{3}$

$5\frac{3}{4} = \frac{20+3}{4} = \frac{23}{4} = \frac{23}{4} \cdot \frac{3}{3} = \frac{69}{12}$

$\frac{1}{3} = \frac{1}{3} \cdot \frac{4}{4} = \frac{4}{12}$

$5\frac{3}{4} - \frac{1}{3} = \frac{69}{12} - \frac{4}{12} = \frac{65}{12} \text{ or } 5\frac{5}{12}$

79. $60\frac{3}{16}-40\frac{5}{8}$

$60\frac{3}{16}=\frac{960+3}{16}=\frac{963}{16}$

$40\frac{5}{8}=\frac{320+5}{8}=\frac{325}{8}=\frac{325}{8}\cdot\frac{2}{2}=\frac{650}{16}$

$60\frac{3}{16}-40\frac{5}{8}=\frac{963}{16}-\frac{650}{16}=\frac{313}{16}$ or $19\frac{9}{16}$

Jamie has grown $19\frac{9}{16}$ inches.

81. $1-\frac{9}{10}=\frac{10}{10}-\frac{9}{10}=\frac{1}{10}$

In the U.S., $\frac{1}{10}$ of natural disasters do not involve flooding.

83. $1-\frac{3}{10}=\frac{10}{10}-\frac{3}{10}=\frac{7}{10}$

$\frac{7}{10}$ of Earth's surface is covered by water.

85. $4\frac{1}{2}=\frac{8+1}{2}=\frac{9}{2}=\frac{9}{2}\cdot\frac{6}{6}=\frac{54}{12}$

$1\frac{1}{6}=\frac{6+1}{6}=\frac{7}{6}=\frac{7}{6}\cdot\frac{2}{2}=\frac{14}{12}$

$1\frac{3}{4}=\frac{4+3}{4}=\frac{7}{4}=\frac{7}{4}\cdot\frac{3}{3}=\frac{21}{12}$

$4\frac{1}{2}+1\frac{1}{6}+1\frac{3}{4}=\frac{54}{12}+\frac{14}{12}+\frac{21}{12}=\frac{89}{12}=7\frac{5}{12}$

The total weight is $7\frac{5}{12}$ tons.

87. $11\frac{7}{8}=\frac{88+7}{8}=\frac{95}{8}$

$13\frac{3}{4}=\frac{52+3}{4}=\frac{55}{4}=\frac{55}{4}\cdot\frac{2}{2}=\frac{110}{8}$

$13\frac{3}{4}-11\frac{7}{8}=\frac{110}{8}-\frac{95}{8}=\frac{15}{8}$ or $1\frac{7}{8}$

The stock gained $1\frac{7}{8}$ dollars that day.

89. $28\frac{3}{8}=\frac{28\cdot 8+3}{8}=\frac{227}{8}$

$30=\frac{30}{1}\cdot\frac{8}{8}=\frac{240}{8}$

$30-28\frac{3}{8}=\frac{240}{8}-\frac{227}{8}=\frac{13}{8}$ or $1\frac{5}{8}$

The pants will need to be shortened by $1\frac{5}{8}$ inches.

91. $5\frac{1}{2}=\frac{10+1}{2}=\frac{11}{2}$

$5\frac{1}{2}\cdot\frac{1}{4}=\frac{11}{2}\cdot\frac{1}{4}=\frac{11\cdot 1}{2\cdot 4}=\frac{11}{8}$ or $1\frac{3}{8}$

$1\frac{3}{8}$ cups of chopped onions are needed.

93. $15\div\frac{3}{8}=\frac{15}{1}\cdot\frac{8}{3}=\frac{5}{1}\cdot\frac{8}{1}=\frac{5\cdot 8}{1\cdot 1}=\frac{40}{1}=40$

Tierra can wash her hair 40 times.

95. $\frac{1}{4}+\frac{1}{4}+1=\frac{1}{4}+\frac{1}{4}+\frac{4}{4}=\frac{6}{4}=\frac{3}{2}$ or $1\frac{1}{2}$

The total thickness is $1\frac{1}{2}$ inches.

97. $32\frac{1}{2}=\frac{64+1}{2}=\frac{65}{2}$

$2\frac{1}{2}=\frac{4+1}{2}=\frac{5}{2}$

$\frac{65}{2}\div\frac{5}{2}=\frac{65}{2}\cdot\frac{2}{5}=\frac{13}{1}\cdot\frac{1}{1}=\frac{13\cdot 1}{1\cdot 1}=\frac{13}{1}=13$

Rod can fill 13 vials.

99. $4\frac{1}{2}=\frac{9}{2}=\frac{9}{2}\cdot\frac{12}{12}=\frac{108}{24}$

$2\frac{1}{3}=\frac{7}{3}=\frac{7}{3}\cdot\frac{8}{8}=\frac{56}{24}$

$\frac{1}{8}=\frac{1}{8}\cdot\frac{3}{3}=\frac{3}{24}$

$4\frac{1}{2}+2\frac{1}{3}+\frac{1}{8}=\frac{108}{24}+\frac{56}{24}+\frac{3}{24}=\frac{167}{24}$ or $6\frac{23}{24}$

The length of the shaft of the bolt must be $6\frac{23}{24}$ inches.

101. a. $\frac{*}{a}+\frac{?}{a}=\frac{*+?}{a}$

b. $\frac{\odot}{?}-\frac{\square}{?}=\frac{\odot-\square}{?}$

c. $\frac{\triangle}{\square}+\frac{4}{\square}=\frac{\triangle+4}{\square}$

d. $\frac{x}{3}-\frac{2}{3}=\frac{x-2}{3}$

e. $\frac{12}{x}-\frac{4}{x}=\frac{12-4}{x}=\frac{8}{x}$

103. number of pills $=\frac{(\text{mg per day})(\text{days per month})(\text{number of months})}{\text{mg per pill}}$

number of pills $=\frac{(450)(30)(6)}{300}=270$
Dr. Highland should prescribe 270 pills.

105. Answers will vary.

106. $\frac{4+9+17+32+16}{5}=\frac{78}{5}=15.6$
The mean is 15.6.

107. In order, the values are: 4, 9, 16, 17, 32. The median is 16.

108. Variables are letters used to represent numbers.

Exercise Set 1.4

1. A set is collection of elements.

3. Answers will vary. One possible answer is the set of all natural numbers less than 0.

5. The set of natural numbers is also the set of counting numbers and the set of positive integers.

7. a. The natural number 7 is a whole number because it is a member of the set $\{0, 1, 2, 3, \ldots\}$

b. The natural number 7 is a rational number because it can be written as the quotient of two integers, $\frac{7}{1}$.

c. All natural numbers are real numbers.

9. The integers are $\{\ldots, -3, -2, -1, 0, 1, 2, 3, \ldots\}$.

11. The natural numbers are $\{1, 2, 3, 4, \ldots\}$.

13. The whole numbers are $\{0, 1, 2, \ldots\}$.

15. True; the negative integers are $\{-1, -2, -3, \ldots\}$.

17. True; the integers are $\{\ldots, -3, -2, -1, 0, 1, 2, 3, \ldots\}$.

19. False; the integers are $\{\ldots, -3, -2, -1, 0, 1, 2, 3, \ldots\}$.

21. False; $\sqrt{2}$ cannot be expressed as the quotient of two integers.

23. True; $-\frac{1}{5}$ is a quotient of two integers, $\frac{-1}{5}$.

25. True; $-2\frac{1}{3}$ can be expressed as a quotient of two integers, $\frac{-7}{3}$.

27. False; $-\frac{5}{3}$ is rational since it is a quotient of two integers.

29. True; positive numbers are greater than zero.

31. True, either $\varnothing$ or { } is used.

33. True; any negative integer can be represented on a real number line and is therefore real.

35. True; any rational number can be represented on a real number line and is therefore real.

37. True; irrational numbers are real numbers which are not rational.

39. True; the counting numbers are {1, 2, 3, ...}, the whole numbers are {0, 1, 2, ...}.

41. True; the symbol $\mathbb{R}$ represents the set of real numbers.

43. True; this is the definition of a negative number.

45. True; the integers are

$$\left\{\underbrace{\ldots, -2, -1}_{\text{negative integers}}, \underbrace{0}_{\text{zero}}, \underbrace{1, 2, \ldots}_{\text{positive integers}}\right\}.$$

47. **a.** 3 and 77 are positive integers.

b. 0, 3, and 77 are whole numbers

c. 0, –2, 3, and 77 are integers.

d. $-\frac{4}{3}$, 0, -2, 3, $5\frac{1}{2}$, 1.63, and 77 are rational numbers.

e. $\sqrt{8}$ and $-\sqrt{3}$ are irrational numbers.

f. $-\frac{4}{3}$, 0, -2, 3, $5\frac{1}{2}$, $\sqrt{8}$, $-\sqrt{3}$, 1.63, and 77 are real numbers.

For Exercises 49–60, answers will vary. One possible answer is given.

49. $\sqrt{2}$, 1.6, –4.8

51. $\frac{1}{2}$, 6.4, –2.6

53. $\sqrt{2}$, $-\sqrt{2}$, π

55. –1, –7, –24

57. 0, $\sqrt{2}$, –5

59. –7, 1, 5

61. {5, 6, 7, 8, ..., 87}
$87 - 5 + 1 = 82 + 1 = 83$
The set has 83 elements.

63. **a.** $A = \{1, 3, 4, 5, 8\}$

b. $B = \{2, 5, 6, 7, 8\}$

c. A and $B = \{5, 8\}$

d. A or $B = \{1, 2, 3, 4, 5, 6, 7, 8\}$

65. **a.** Set B continues beyond 4.

b. Set A has 4 elements.

c. Set B has an infinite number of elements.

d. Set B is an infinite set.

67. **a.** There are an infinite number of fractions between any 2 numbers.

b. There are an infinite number of fractions between any 2 numbers.

69. $4\frac{2}{3} = \frac{4 \cdot 3 + 2}{3} = \frac{12 + 2}{3} = \frac{14}{3}$

70. $\frac{16}{3} = 5\frac{1}{3}$ because $16 \div 3 = 5$ R 1

71. $\frac{3}{5}+\frac{5}{8}$

$\frac{3}{5}=\frac{3}{5}\cdot\frac{8}{8}=\frac{24}{40}$

$\frac{5}{8}\cdot\frac{5}{5}=\frac{25}{40}$

$\frac{3}{5}+\frac{5}{8}=\frac{24}{40}+\frac{25}{40}=\frac{49}{40}$ or $1\frac{9}{40}$

72. $\left(\frac{5}{9}\right)\left(4\frac{2}{3}\right)$

$4\frac{2}{3}=\frac{12+2}{3}=\frac{14}{3}$

$\left(\frac{5}{9}\right)\left(4\frac{2}{3}\right)=\frac{5}{9}\cdot\frac{14}{3}=\frac{70}{27}$ or $2\frac{16}{27}$

Exercise Set 1.5

1. a. −6 −5 −4 −3 −2 −1 0 1 2 3 4 5 6

b. −6 −5 −4 −3 −2 −1 0 1 2 3 4 5 6

c. −2 is greater than −4 because it is farther to the right on the number line.

d. $-4<-2$

e. $-2>-4$

3. a. 4 is 4 units from 0 on a number line.

b. −4 is 4 units from 0 on a number line.

c. 0 is 0 units from 0 on a number line.

5. Yes; for example, $5>3$ and $3<5$. Also, $-2>-5$ and $-5<-2$.

7. $|7|=7$

9. $|-15|=15$

11. $|0|=0$

13. $-|-3|=-(3)=-3$

15. $-|-15|=-(15)=-15$

17. $5<8$; 5 is to the left of 8 on a number line.

19. $-4<0$; −4 is to the left of 0 on a number line.

21. $\frac{1}{2}>-\frac{2}{3}$; $\frac{1}{2}$ is to the right of $-\frac{2}{3}$ on a number line.

23. $0.9>0.8$; 0.9 is the right of 0.8 on a number line.

25. $-\frac{1}{2}>-1$; $-\frac{1}{2}$ is to the right of −1 on a number line.

27. $3>-3$; 3 is to the right of −3 on a number line.

29. $-2.1<-2$; −2.1 is to the left of −2 on a number line.

31. $\frac{5}{9}>-\frac{5}{9}$; $\frac{5}{9}$ is to the right of $-\frac{5}{9}$ on a number line.

33. $-\frac{3}{2}<\frac{3}{2}$; $-\frac{3}{2}$ is to the left of $\frac{3}{2}$ on a number line.

35. $0.49>0.43$; 0.49 is to the right of 0.43 on a number line.

37. $5>-7$; 5 is to the right of −7 on a number line.

39. $-0.006>-0.007$; −0.006 is to the right of −0.007 on a number line.

41. $\frac{5}{8}>0.6$ because $\frac{5}{8}=0.625$ and 0.625 is to the right of 0.6 on a number line.

43. $-\frac{2}{3}<-\frac{1}{3}$; $-\frac{2}{3}$ is to the left of $-\frac{1}{3}$ on a number line.

45. $-\frac{1}{2}>-\frac{3}{2}$; $-\frac{1}{2}$ is to the right of $-\frac{3}{2}$ on a number line.

47. $2 > |-1|$ since $|-1| = 1$

49. $|-4| > \frac{2}{3}$ since $|-4| = 4$

51. $|0| < |-4|$ since $|0| = 0$ and $|-4| = 4$

53. $4 < \left|-\frac{9}{2}\right|$ since $\left|-\frac{9}{2}\right| = \frac{9}{2}$ or $4\frac{1}{2}$

55. $\left|-\frac{6}{2}\right| > \left|-\frac{2}{6}\right|$ since $\left|-\frac{6}{2}\right| = \frac{6}{2} = 3$ and $\left|-\frac{2}{6}\right| = \frac{2}{6} = \frac{1}{3}$

57. $|-4.6| = \left|-\frac{23}{5}\right|$ since $|-4.6| = 4.6$ and $\left|-\frac{23}{5}\right| = \frac{23}{5} = 4.6$

59. $\frac{2}{3}+\frac{2}{3}+\frac{2}{3}+\frac{2}{3} = 4\cdot\frac{2}{3}$ since $\frac{2}{3}+\frac{2}{3}+\frac{2}{3}+\frac{2}{3} = \frac{2+2+2+2}{3} = \frac{8}{3}$ and $4\cdot\frac{2}{3} = \frac{4}{1}\cdot\frac{2}{3} = \frac{8}{3}$

61. $\frac{1}{2}\cdot\frac{1}{2} < \frac{1}{2}\div\frac{1}{2}$ since $\frac{1}{2}\cdot\frac{1}{2} = \frac{1\cdot1}{2\cdot2} = \frac{1}{4}$ and $\frac{1}{2}\div\frac{1}{2} = \frac{1}{2}\cdot\frac{2}{1} = \frac{1}{1}\cdot\frac{1}{1} = 1$

63. $\frac{5}{8}-\frac{1}{2} < \frac{5}{8}\div\frac{1}{2}$ since $\frac{5}{8}-\frac{1}{2} = \frac{5}{8}-\frac{4}{8} = \frac{1}{8}$ and $\frac{5}{8}\div\frac{1}{2} = \frac{5}{8}\cdot\frac{2}{1} = \frac{10}{8}$

65. 4 and −4 since $|4| = |-4| = 4$

For Exercises 67–74, answers will vary. One possible answer is given.

67. Three numbers greater than 4 and less than 6 are $4\frac{1}{2}$, 5, 5.5.

69. Three numbers less than −2 and greater than −6 are −3, −4, −5.

71. Three numbers greater than −3 and greater than 3 are 4, 5, 6.

73. Three numbers greater than $|-2|$ and less than $|-6|$ are 3, 4, 5.

75. **a.** Between does not include endpoints.

b. Three real numbers between 4 and 6 are 4.1, 5, and $5\frac{1}{2}$.

c. No, 4 is an endpoint.

d. Yes, 5 is greater than 4 and less than 6.

e. True

77. **a.** In January 1995 the interest rate was 9.15%.

b. In January 1996 the interest rate was 7.03%.

c. The interest rate was less than 8% in January, February, March, April, August, October, November, and December of 1996.

d. The interest rate was greater than 8% in May, June, July, and September of 1996.

e. Monthly payment
= ($6.39)(number of $1000s)
Monthly payment = ($6.39)(40)
= $255.60

f. monthly payment
= ($8.70)(number of $1000s)
= ($8.70)(40)
= $348

79. The result of dividing a number by itself is 1. Thus, the result of dividing a number between 0 and 1 by itself is a number, 1, which is greater than the original number.

81. Yes, 0 since $|0| = 0$ and $-|0| = -(0) = 0$

84. $1\frac{2}{3} - \frac{3}{8}$

$1\frac{2}{3} = \frac{3+2}{3} = \frac{5}{3} = \frac{5}{3} \cdot \frac{8}{8} = \frac{40}{24}$

$\frac{3}{8} = \frac{3}{8} \cdot \frac{3}{3} = \frac{9}{24}$

$1\frac{2}{3} - \frac{3}{8} = \frac{40}{24} - \frac{9}{24} = \frac{31}{24}$ or $1\frac{7}{24}$

85. The set of whole numbers is {0, 1, 2, 3, ...}.

86. The set of counting numbers is {1, 2, 3, 4, ...}.

87. **a.** 5 is a natural number.

b. 5 and 0 are whole numbers.

c. 5, –2, and 0 are integers.

d. 5, –2, 0, $\frac{1}{3}$, $-\frac{5}{9}$, and 2.3 are rational numbers.

e. $\sqrt{3}$ is an irrational number.

f. 5, –2, 0, $\frac{1}{3}$, $\sqrt{3}$, $-\frac{5}{9}$, and 2.3 are real numbers.

Exercise Set 1.6

1. The 4 basic operations of arithmetic are addition, subtraction, multiplication, and division.

3. **a.** No; $-\frac{2}{3} + \frac{3}{2}$ does not equal 0.

b. The opposite of $-\frac{2}{3}$ is $\frac{2}{3}$ because $-\frac{2}{3} + \frac{2}{3} = 0$.

5. The sum of a positive number and a negative number can be either positive or negative. The sum of a positive number and a negative number has the same sign as the number that has larger absolute value.

9. **a.** He owed 175, a negative, and then paid a positive amount, 93, toward his debt.

b. $-175 + 93$
The numbers have different signs so find the difference between the absolute values.
$|-175| - |93| = 82$
$|-175|$ is greater so sum is negative.
$-175 + 93 = -82$

c. Answers will vary.

11. The opposite of 5 is –5 since 5 + (–5) = 0.

13. The opposite of –28 is 28 since –28 + 28 = 0.

15. The opposite of 0 is 0 since 0 + 0 = 0.

17. The opposite of $\frac{5}{3}$ is $-\frac{5}{3}$ since $\frac{5}{3} + \left(-\frac{5}{3}\right) = 0$.

19. The opposite of $2\frac{3}{5}$ is $-2\frac{3}{5}$ since $2\frac{3}{5} + \left(-2\frac{3}{5}\right) = 0$.

21. The opposite of 0.47 is –0.47 since $0.47 + (-0.47) = 0$.

23. Numbers have same sign, so add absolute values.
$|3| + |8| = 3 + 8 = 11$
Numbers are positive so sum is positive.
3 + 8 = 11

25. Numbers have different signs so find difference between larger and smaller absolute values.
$|4|-|-3| = 4-3 = 1$. $|4|$ is greater than $|-3|$ so the sum is positive.
$4 + (-3) = 1$

27. Numbers have same sign, so add absolute values.
$|-4|+|-2| = 4+2 = 6$.
Numbers are negative, so sum is negative.
$-4 + (-2) = -6$

29. Numbers have different signs, so find difference between absolute values.
$|6|-|-6| = 6-6 = 0$
$6+(-6) = 0$

31. Numbers have different signs, so find difference between absolute values.
$|-4|-|4| = 4-4 = 0$
$-4 + 4 = 0$

33. Numbers have same sign, so add absolute values. $|-8|+|-2| = 8+2 = 10$. Numbers are negative, so sum is negative.
$-8 + (-2) = -10$

35. Numbers have different signs, so find difference between absolute values.
$|-6|-|6| = 6-6 = 0$
$-6 + 6 = 0$

37. Numbers have same sign, so add absolute values.
$|-4|+|-5| = 4+5 = 9$
Numbers are negative, so sum is negative.
$-4 + (-5) = -9$

39. $0+0=0$

41. $-6+0=-6$

43. Numbers have different signs, so find difference between larger and smaller absolute values. $|18|-|-9| = 18-9 = 9$. $|18|$ is greater than $|-9|$ so sum is positive.
$18+(-9) = 9$

45. Numbers have same sign, so add absolute values.
$|-23|+|-31| = 23+31 = 54$
Numbers are negative, so sum is negative.
$-23+(-31) = -54$

47. Numbers have same sign, so add absolute values. $|-35|+|-9| = 35+9 = 44$. Numbers are negative, so sum is negative.
$-35+(-9) = -44$

49. Numbers have different signs, so find difference between larger and smaller absolute values. $|-30|-|4| = 30-4 = 26$.
$|-30|$ is greater than $|4|$ so sum is negative.
$4+(-30) = -26$

51. Numbers have different signs, so find difference between larger and smaller absolute values. $|40|-|-35| = 40-35 = 5$.
$|40|$ is greater than $|-35|$ so sum is positive.
$-35 + 40 = 5$

53. Numbers have different signs, so find difference between larger and smaller absolute values.
$|-200|-|180| = 200-180 = 20$. $|-200|$ is greater than $|180|$ so sum is negative.
$180+(-200) = -20$

55. Numbers have different signs, so find difference between larger and smaller absolute values. $|-67|-|28| = 67-28 = 39$.
$|-67|$ is greater than $|28|$ so sum is negative.
$-67 + 28 = -39$

57. Numbers have different signs, so find difference between larger and smaller absolute values. $|184|-|-93| = 184-93 = 91$.
$|184|$ is greater than $|-93|$ so sum is positive.
$184+(-93) = 91$

59. Numbers have different signs, so find difference between larger and smaller absolute values.
$|-452|-|312| = 452 - 312 = 140$. $|-452|$ is greater than $|312|$ so sum is negative.
$-452 + 312 = -140$

61. Numbers have same sign, so add absolute values. $|-26| + |-74| = 26 + 79 = 105$.
Numbers are negative so sum is negative.
$-26 + (-79) = -105$

63. a. Positive; $|463|$ is greater than $|-197|$ so sum will be positive.

b. $463 + (-197) = 266$

c. Yes; By part a) we expect a positive sum. The magnitude of the sum is the difference between the larger and smaller absolute values.

65. a. Negative; the sum of 2 negative numbers is always negative.

b. $-84 + (-289) = -373$

c. Yes; the sum of 2 negative numbers should be (and is) a larger negative number.

67. a. Negative; $|-947|$ is greater than $|495|$ so sum will be negative.

b. $-947 + 495 = -452$

c. Yes; by part a) we expect a negative sum. Magnitude of the sum is the difference between the larger and smaller absolute values.

69. a. Negative; the sum of 2 negative numbers is always negative.

b. $-496 + (-804) = -1300$

c. Yes; the sum of 2 negative numbers should be (and is) a larger negative number.

71. a. Negative; $|-285|$ is greater than $|263|$ so sum will be negative.

b. $-285 + 263 = -22$

c. Yes; by part a) we expect a negative sum. The magnitude of the sum is the difference between the larger and smaller absolute values.

73. a. Negative; the sum of 2 negative numbers is always negative.

b. $-1833 + (-2047) = -3880$

c. Yes; The sum of 2 negative numbers should be (and is) a larger negative number.

75. a. Positive; $|3124|$ is greater than $|-2013|$ so sum will be positive.

b. $3124 + (-2013) = 1111$

c. Yes; by part a) we expect a positive sum. Magnitude of sum is difference between larger and smaller absolute values.

77. a. Negative; the sum of 2 negative numbers is always negative.

b. $-1025 + (-1025) = -2050$

c. Yes; the sum of 2 negative numbers should be (and is) a larger negative number.

79. True

81. True; the sum of two positive numbers is always positive.

83. False; the sum has the sign of the number with the larger absolute value.

85. Mr. Peter's balance was –$94. His new balance can be found by adding.
$-94 + (-183) = -277$
Mr. Peter owes the bank $277.

87. Total loss can be represented as –18 + (–3). $|-18|+|-3| = 18+3 = 21$. The total loss in yardage is 21 yards.

89. The depth of the well can be found by adding –27 + (–34) = –61. The well is 61 feet deep.

91. The height of the mountain peak above sea level can be found by adding $33,480+(-19,684) = 13,796$. The mountain peak is 13,796 feet above sea level.

93. a. $11,250+(-18,560) = -7310$
The Frenches had a loss of $7310 the first month.

b. $17,980+(-12,750) = 5230$
The Frenches had a gain of $5230 the second month.

c. $19,420+(-16,980) = 2440$
The Frenches had a gain of $2440 the third month.

95. $(-4)+(-6)+(-12) = (-10)+(-12) = -22$

97. $29+(-46)+37 = (-17)+37 = 20$

99. $(-12)+(-10)+25+(-3)$
$= (-22)+25+(-3)$
$= 3+(-3)$
$= 0$

101. $1+2+3+\cdots+10$
$= (1+10)+(2+9)\cdots+(5+6)$
$= (5)(11)$
$= 55$

103. $1\frac{2}{3} = \frac{3+2}{3} = \frac{5}{3}$
$\left(\frac{3}{5}\right)\left(1\frac{2}{3}\right) = \left(\frac{3}{5}\right)\left(\frac{5}{3}\right) = \frac{1}{1}\cdot\frac{1}{1} = 1$

104. $3 = \frac{3}{1}\cdot\frac{16}{16} = \frac{48}{16}$
$3-\frac{5}{16} = \frac{48}{16}-\frac{5}{16} = \frac{48-5}{16} = \frac{43}{16}$ or $2\frac{11}{16}$

105. $|-3| > 2$ since $|-3| = 3$

106. $8 > |-7|$ since $|-7| = 7$

Exercise Set 1.7

1. $5-8$

3. $\square - *$

5. b. $9-12 = 9+(-12)$

c. $9+(-12) = -3$

7. a. $a-(+b) = a-b$

b. $7-(+9) = 7-9$

c. $7-9 = 7+(-9) = -2$

9. a. $3-(-6)+(-5) = 3+6-5$

b. $3+6-5 = 9-5 = 4$

11. $9-5 = 9+(-5) = 4$

13. $8-9 = 8+(-9) = -1$

15. $-4-2 = -4+(-2) = -6$

17. $-4-(-3) = -4+3 = -1$

19. $-4-4 = -4+(-4) = -8$

21. $0-6 = 0+(-6) = -6$

23. $0-(-6) = 0+6 = 6$

25. $-3-1 = -3+(-1) = -4$

27. $5-3 = 5+(-3) = 2$

29. $6-(-3)=6+3=9$

31. $4-4=4+(-4)=0$

33. $-9-11=-9+(-11)=-20$

35. $(-4)-(-4)=-4+4=0$

37. $-8-(-12)=-8+12=4$

39. $-6-(-2)=-6+2=-4$

41. $-9-2=-9+(-2)=-11$

43. $-24-(-8)=-24+8=-16$

45. $-100-80=-100+(-80)=-180$

47. $-45-37=-45+(-37)=-82$

49. $70-(-70)=70+70=140$

51. $40-62=40+(-62)=-22$

53. $-61-(-9)=-61+9=-52$

55. $-15-3=-15+(-3)=-18$

57. $-8-8=-8+(-8)=-16$

59. $-5-(-3)=-5+3=-2$

61. $9-(-4)=9+4=13$

63. $13-24=13+(-24)=-11$

65. $-4-(-15)=-4+15=11$

67. a. Positive; $296-197=296+(-197)$ $|296|$ is greater than $|-197|$ so the sum will be positive.

b. $296+(-197)=99$

c. Yes; by part a) we expect a positive sum. The size of the sum is the difference between the absolute values of the 2 numbers.

69. a. Negative; $-372-195=-372+(-195)$ The sum of 2 negative numbers is always negative.

b. $-372+(-195)=-567$

c. Yes; the sum of two negative numbers should be (and is) a larger negative number.

71. a. Positive; $843-(-745)=843+745$. The sum of 2 positive numbers is always positive.

b. $843+745=1588$

c. Yes; by part a) we expect a positive answer. The size of the sum is the sum of the absolute values of the numbers.

73. a. Positive; $-408-(-604)=-408+604$. $|604|$ is greater than $|-408|$ so the sum will be positive.

b. $-408+604=196$

c. Yes; by part a) we expect a positive answer. The size of the answer is the difference between the larger and smaller absolute values.

75. a. Negative; $-1024-(-576)=-1024+576$. $|-1024|$ is greater than $|576|$ so the sum will be negative.

b. $-1024+576=-448$

c. Yes; by part a) we expect a negative answer. The size of the answer is the difference between the larger and the smaller absolute values.

77. a. Positive;
$165.7-49.6=165.7+(-49.6)$.
$|165.7|$ is greater than $|-49.6|$ so the sum will be positive.

b. $165.7+(-49.6)=116.1$

c. Yes; by part a) we expect a negative answer. The size of the answer is the difference between the larger and the smaller absolute values.

79. a. Negative; $295-364=295+(-364)$.
Since $|-364|$ is greater than $|295|$ the answer will be negative.

b. $295+(-364)=-69$

c. Yes; by part a) we expect a negative answer. The size of the answer is the difference between the larger and the smaller absolute values.

81. a. Negative;
$-1023-647=-1023+(-647)$.
The sum of two negative numbers is always negative.

b. $-1023+(-647)=-1670$

c. Yes; the sum of two negative numbers should be (and is) a larger negative number.

83. a. Zero; $-7.62-(-7.62)=-7.62+7.62$.
The sum of two opposite numbers is always zero.

b. $-7.62+7.62=0$

c. Yes; by part a) we expect zero.

85. $6+5-(+4)=6+5+(-4)=11+(-4)=7$

87. $-3+(-4)+5=-7+5=-2$

89. $-13-(+5)+3=-13+(-5)+3$
$=-18+3$
$=-15$

91. $-9-(-3)+4=-9+3+4=-6+4=-2$

93. $5-(-9)+(-1)=5+9+(-1)$
$=14+(-1)$
$=13$

95. $25+(-13)-(+5)=25+(-13)+(-5)$
$=12+(-5)$
$=7$

97. $-36-5+9=-36+(-5)+9$
$=-41+9$
$=-32$

99. $-2+7-9=-2+7+(-9)=5+(-9)=-4$

101. $25-19+3=25+(-19)+3=6+3=9$

103. $(-4)+(-3)+5-7=(-4)+(-3)+5+(-7)$
$=-7+5+(-7)$
$=-2+(-7)$
$=-9$

105. $17+(-3)-9-(-7)=17+(-3)+(-9)+7$
$=14+(-9)+7$
$=5+7$
$=12$

107. $-9+(-7)+(-5)-(-3)=-9+(-7)+(-5)+3$
$=-16+(-5)+3$
$=-21+3$
$=-18$

109. a. $300-343=300+(-343)=-43$
They had 43 sweaters on back order.

b. $43+100=143$
They would need to order 143 sweaters.

111. $42-58=42+(-58)=-16$
The bottom of the well is 16 feet below sea level.

113. $44-(-56)=44+56=100$
Thus the temperature dropped 100°F.

115. a. $562-(-273)=562+273=835$
The difference is 835 thousands or 835,000.

b. $473-(-212)=473+212=685$
The difference is 685 thousands or 685,000.

117. $1-2+3-4+5-6+7-8+9-10$
$=(1-2)+(3-4)+(5-6)+(7-8)+(9-10)$
$=(-1)+(-1)+(-1)+(-1)+(-1)$
$=-5$

119. a. 7 units

b. $5-(-2)=7$

121. a. $3+2+2+1+1=9$
The ball travels 9 feet vertically.

b. $-3+2+(-2)+1+(-1)=-3$
The net distance is –3 feet.

122. The integers are $\{\ldots, -2, -1, 0, 1, 2, \ldots\}$.

123. The set of rational numbers together with the set of irrational numbers forms the set of real numbers.

124. $|-3|>-5$ since $|-3|=3$

125. $|-6|<|-7|$ since $|-6|=6$ and $|-7|=7$

Exercise Set 1.8

1. Like signs: product is positive. Unlike signs: product is negative.

3. The product of 0 and any real number is 0.

5. The product (–3)(4)(–5) is positive since there is an even number (2) of negatives

7. The product (–102)(–16)(24)(19) is positive since there is an even number of negatives.

9. The product (–40)(–16)(30)(50)(–13) is negative since there is an odd number (3) of negatives.

11. A fraction of the form $\frac{a}{-b}$ is generally written $-\frac{a}{b}$ or $\frac{-a}{b}$.

13. a. With 3 – 5 you subtract, but with 3(–5) you multiply.

b. $3-5=3+(-5)=-2$
$3(-5)=-15$

15. a. With $x-y$ you subtract, but with $x(-y)$ you multiply.

b. $x-y=5-(-2)=5+2=7$

c. $x(-y)=[-(-2)]=5(2)=10$

d. $-x-y=-5-(-2)=-5+2=-3$

17. Since the numbers have like signs, the product is positive. (–4)(–3) = 12

19. Since the number have unlike signs, the product is negative. 3(–3) = –9

21. Since the numbers have unlike signs, the product is negative. (–2)(4) = –8

23. Zero multiplied by any real number equals zero. 0(–8) = 0

25. Since the numbers have like signs, the product is positive. 6(7) = 42

27. Since the numbers have like signs, the product is positive. (–5)(–8) = 40

29. Since the numbers have like signs, the product is positive. (–5)(–6) = 30

31. Zero multiplied by any real number equals zero.
0(3)(8) = 0(8) = 0

33. Since there is one negative number (an odd number), the product will be negative.
$(21)(-1)(4)=(-21)(4)=-84$

35. Since there are three negative numbers (an odd number), the product will be negative.
$-1(-3)(3)(-8) = 3(3)(-8) = 9(-8) = -72$

37. Since there are two negative numbers (an even number), the product will be positive.
$(-4)(3)(-7)(1) = (-12)(-7)(1) = (84)(1) = 84$

39. Zero multiplied by any real number equals zero.
$(-1)(3)(0)(-7) = (-3)(0)(-7) = 0(-7) = 0$

41. $\left(\frac{-1}{2}\right)\left(\frac{3}{5}\right) = \frac{(-1)(3)}{2\cdot 5} = \frac{-3}{10} = -\frac{3}{10}$

43. $\left(\frac{-8}{9}\right)\left(\frac{-7}{12}\right) = \frac{(-8)(-7)}{9\cdot 12} = \frac{(-2)(-7)}{9\cdot 3} = \frac{14}{27}$

45. $\left(\frac{6}{-3}\right)\left(\frac{4}{-2}\right) = \left(\frac{2}{-1}\right)\left(\frac{2}{-1}\right) = \frac{(2)(2)}{(-1)(-1)} = \frac{4}{1} = 4$

47. $\left(\frac{-3}{8}\right)\left(\frac{5}{6}\right) = \left(\frac{-1}{8}\right)\left(\frac{5}{2}\right)$
$= \frac{(-1)(5)}{(8)(2)}$
$= \frac{-5}{16}$
$= -\frac{5}{16}$

49. Since the numbers have like signs, the quotient is positive. $\frac{6}{2} = 3$

51. Since the numbers have like signs, the quotient is positive. $-16 \div (-4) = \frac{-16}{-4} = 4$

53. Since the numbers have like signs, the quotient is positive. $\frac{-36}{-9} = 4$

55. Since the numbers have unlike signs, the quotient is negative. $\frac{36}{-2} = -18$

57. Since the numbers have like signs, the quotient is positive. $\frac{-12}{-1} = 12$

59. Since the numbers have unlike signs, the quotient is negative. $20/(-4) = \frac{20}{-4} = -5$

61. Since the numbers have unlike signs, the quotient is negative. $\frac{-42}{7} = -6$

63. Since the numbers have unlike signs, the quotient is negative. $\frac{36}{-4} = -9$

65. Since the numbers have like signs, the quotient is positive. $-40 \div (-8) = \frac{-40}{-8} = 5$

67. Zero divided by any nonzero number is zero.
$\frac{0}{4} = 0$

69. Since the numbers have unlike signs, the quotient is negative. $\frac{30}{-10} = -3$

71. Since the numbers have unlike signs, the quotient is negative. $\frac{-180}{30} = -6$

73. $\frac{3}{12} \div \left(\frac{-5}{8}\right) = \frac{3}{12}\cdot\left(\frac{8}{-5}\right)$
$= \frac{1}{1}\cdot\left(\frac{2}{-5}\right)$
$= \frac{1\cdot 2}{1(-5)}$
$= \frac{2}{-5}$
$= -\frac{2}{5}$

75. $\frac{-7}{12} \div (-2) = \frac{-7}{12}\cdot\frac{1}{-2} = \frac{-7(1)}{12(-2)} = \frac{7}{24}$

77. $\frac{-15}{21} \div \left(\frac{-15}{21}\right) = \frac{-15}{21} \cdot \frac{21}{-15}$
$= \frac{-1}{1} \cdot \frac{1}{-1}$
$= \frac{(-1)(1)}{(1)(-1)}$
$= \frac{-1}{-1}$
$= 1$

79. $-12 \div \frac{5}{12} = \frac{-12}{1} \cdot \frac{12}{5}$
$= \frac{(-12)(12)}{(1)(5)}$
$= \frac{-144}{5}$
$= -\frac{144}{5}$

81. Since the numbers have unlike signs, the product is negative. $-4(8) = -32$

83. Since the numbers have unlike signs, the quotient is negative. $\frac{100}{-5} = -20$

85. Since the numbers have unlike signs, the product is negative. $-7(2) = -14$

87. Since the numbers have unlike signs, the quotient is negative. $27 \div (-3) = \frac{27}{-3} = -9$

89. Since the numbers have unlike signs, the quotient is negative. $\frac{-100}{5} = -20$

91. Since the numbers have unlike signs, the quotient is negative. $\frac{60}{-60} = -1$

93. Zero divided by any nonzero number is zero.
$0 \div 6 = \frac{0}{6} = 0$

95. Any nonzero number divided by zero is undefined. $\frac{5}{0}$ is undefined.

97. Zero divided by any nonzero number is zero.
$\frac{0}{1} = 0$

99. Any nonzero number divided by zero is undefined. $\frac{8}{0}$ is undefined.

101. a. Since the numbers have like signs, the product will be positive.

b. $(-212)(-87) = 18{,}444$

c. Yes; as expected the product is positive.

103. a. Since the numbers have unlike signs, the quotient will be negative.

b. $-240/15 = \frac{-240}{15} = -16$

c. Yes; as expected the quotient is negative.

105. a. Since the numbers have unlike signs, the quotient will be negative.

b. $243 \div (-27) = \frac{243}{-27} = -9$

c. Yes; as expected the quotient is negative.

107. a. Since the numbers have unlike signs, the product will be negative.

b. $(171)(-89) = -15{,}219$

c. Yes; as expected the product is negative.

109. a. The quotient will be zero; zero divided by any nonzero number is zero.

b. $\frac{0}{5335} = 0$

c. Yes; as expected the answer is zero.

111. a. Undefined; any nonzero number divided by 0 is undefined.

b. $7.2 \div 0 = \frac{7.2}{0}$ is undefined

c. Yes; as expected the quotient is undefined.

113. a. Since the numbers have like signs, the quotient will be positive.

b. $8 \div 2.5 = \frac{8}{2.5} = 3.2$

c. Yes; as expected the quotient is positive.

115. a. Since there are two negative numbers (an even number), the product will be positive.

b. $(-3.0)(4.2)(-18) = 226.8$

c. Yes; as expected the product is positive.

117. True; the product of two numbers with unlike signs is a negative number.

119. True; the quotient of two numbers with like signs is a positive number.

121. True

123. False; zero divided by any nonzero number is zero.

125. False: one divided by zero is undefined.

127. True; any nonzero number divided by zero is undefined.

129. True: any nonzero number divided by zero is undefined.

131. a. $\frac{1}{3}(450) = \frac{450}{3} = 150$
She paid back \$150.

b. $-450 + 150 = -300$
Her new balance is –\$300.

133. $3\left(-1\frac{1}{2}\right) = \frac{3}{1}\left(-\frac{3}{2}\right) = -\frac{9}{2}$ or $-4\frac{1}{2}$
It has lost $4\frac{1}{2}$ points.

The wind chill was –10°F on Tuesday.

135. a. $(-909{,}050) \div (-290{,}525) \approx 3.13$
The public debt was about 3.13 times greater in 1980 than in 1960.

b. $(-6{,}076{,}600) \div (-909{,}050) \approx 6.68$
The public debt in 2000 is projected to be about 6.68 times that in 1980.

137. a. $220 - 50 = 170$
60% of $170 = 0.6(170) = 102$
75% of $170 = 0.75(170) = 127.5$
Target heart rate is 102 to 128 beats per minute.

b. Answers will vary.

139. $(-2)^3 = (-2)(-2)(-2) = 4(-2) = -8$

141. $1^{100} = 1$

143. The product $(-1)(-2)(-3)(-4)\cdots(-10)$ will be positive because there are an even number (10) of negative numbers.

145. $\frac{5}{7} \div \frac{1}{5} = \frac{5}{7} \cdot \frac{5}{1} = \frac{5 \cdot 5}{7 \cdot 1} = \frac{25}{7}$ or $3\frac{4}{7}$

146. $-20 - (-18) = -20 + 18 = -2$

147. $6 - 3 - 4 - 2 = 3 - 4 - 2 = -1 - 2 = -3$

148.
$$\begin{aligned} 5 - (-2) + 3 - 7 &= 5 + 2 + 3 - 7 \\ &= 7 + 3 - 7 \\ &= 10 - 7 \\ &= 3 \end{aligned}$$

Exercise Set 1.9

1. In the expression a^b, a is the base and b is the exponent.

3. a. Every number has an understood exponent of 1.

b. In $5x^3y^2z$, 5 has exponent of 1, x has an exponent of 3, y has an exponent of 2, and z has an exponent of 1.

5. a. $x+x+x+x+x=5x$

b. $x\cdot x\cdot x\cdot x\cdot x=x^5$

7. The order of operations are parentheses, exponents, multiplication or division, then addition or subtraction.

9. No; $4+5\times 2=4+10=14$, on a scientific calculator.

11. a. $15\div 5-2=3-2=1$

b. $15\div(5-2)=15\div 3=5$

c. The keystrokes in b) are used since the fraction bar is a grouping symbol.

13. b.
$$\begin{aligned}[9-(8\div 2)]^2-6^3&=[9-4]^2-6^3\\&=5^2-6^3\\&=25-216\\&=-191\end{aligned}$$

15. b. When $x=5$:
$$\begin{aligned}-4x^2+3x-6&=-4(5)^2+3(5)-6\\&=-4(25)+3(5)-6\\&=-100+15-6\\&=-85-6\\&=-91\end{aligned}$$

17. $4^2=4\cdot 4=16$

19. $1^5=1\cdot1\cdot1\cdot1\cdot1=1$

21. $-5^2=-(5)(5)=-25$

23. $(-3)^2=(-3)(-3)=9$

25. $(-1)^3=(-1)(-1)(-1)=-1$

27. $(-9)^2=(-9)(-9)=81$

29. $(-6)^2=(-6)(-6)=36$

31. $4^1=4$

33. $(-4)^4=(-4)(-4)(-4)(-4)=256$

35. $-2^4=-(2)(2)(2)(2)=-16$

37. $5^2\cdot 3^2=5\cdot5\cdot3\cdot3=225$

39. $2^3\cdot 3^2=2\cdot2\cdot2\cdot3\cdot3=72$

41. a. Positive; a positive number raised to any power is positive.

b. $7^3=343$

c. Yes; as expected the answer is positive.

43. a. Positive; a positive number raised to any power is positive.

b. $5^4=625$

c. Yes; as expected the answer is positive.

45. a. Negative; a negative number raised to an odd power is negative.

b. $(-3)^5=-243$

c. Yes; as expected the answer is negative.

47. a. Positive; a negative number raised to an even power is positive.

b. $(-6)^4=1296$

c. Yes; as expected the answer is positive.

49. a. Positive; a positive number raised to any power is positive.

b. $(5.3)^4 = 789.0481$

c. Yes; as expected the answer is positive.

51. a. Negative; $\left(\frac{7}{8}\right)^2$ is positive therefore, $-\left(\frac{7}{8}\right)^2$ is negative.

b. $-\left(\frac{7}{8}\right)^2 = -0.765625$

c. Yes; as expected the answer is negative.

53. $2+5\cdot 6=2+30=32$

55. $6-6+8=0+8=8$

57. $1+3\cdot 2^2=1+3\cdot 4=1+12=13$

59. $-3^3+27=-27+27=0$

61. $(4-3)\cdot(5-1)^2=(1)\cdot(4)^2=1\cdot 16=16$

63. $3\cdot 7+4\cdot 2=21+8=29$

65.
$$\begin{aligned}4^2-3\cdot 4-6&=16-3\cdot 4-6\\&=16-12-6\\&=4-6\\&=-2\end{aligned}$$

67.
$$\begin{aligned}(6\div 3)^3+4^2\div 8&=(2)^3+4^2\div 8\\&=8+16\div 8\\&=8+2\\&=10\end{aligned}$$

69.
$$\begin{aligned}[6-(-2-3)]^2&=[6-(-5)]^2\\&=[6+5]^2\\&=(11)^2\\&=121\end{aligned}$$

71.
$$\begin{aligned}\left(3^2-1\right)\div(3+1)^2&=(9-1)\div(3+1)^2\\&=8\div(4)^2\\&=8\div 16\\&=\frac{8}{16}\\&=\frac{1}{2}\end{aligned}$$

73.
$$\begin{aligned}[4+((5-2)^2\div 3)^2]^2&=[4+((3)^2\div 3)^2]^2\\&=[4+(9\div 3)^2]^2\\&=[4+(3)^2]^2\\&=[4+9]^2\\&=(13)^2\\&=169\end{aligned}$$

75.
$$\begin{aligned}&2.5+7.56\div 2.1+(9.2)^2\\&=2.5+7.56\div 2.1+84.64\\&=2.5+3.6+84.64\\&=6.1+84.64\\&=90.74\end{aligned}$$

77.
$$\begin{aligned}&2[1.55+5(3.7)]-3.35\\&=2[1.55+18.5]-3.35\\&=2(20.05)-3.35\\&=40.1-3.35\\&=36.75\end{aligned}$$

79.
$$\begin{aligned}\left(\frac{2}{7}+\frac{3}{8}\right)-\frac{3}{112}&=\left(\frac{16}{56}+\frac{21}{56}\right)-\frac{3}{112}\\&=\frac{37}{56}-\frac{3}{112}\\&=\frac{74}{112}-\frac{3}{112}\\&=\frac{71}{112}\end{aligned}$$

81.
$$\begin{aligned}\frac{3}{4}-4\cdot\frac{5}{40}&=\frac{3}{4}-\frac{4}{1}\cdot\frac{5}{40}\\&=\frac{3}{4}-\frac{4}{8}\\&=\frac{3}{4}-\frac{2}{4}\\&=\frac{1}{4}\end{aligned}$$

83. $\dfrac{5-[3(6\div 3)-2]}{5^2-4^2\div 2}=\dfrac{5-[3(2)-2]}{25-16\div 2}$
$=\dfrac{5-[6-2]}{25-8}$
$=\dfrac{5-4}{25-8}$
$=\dfrac{1}{17}$

85. Substitute 3 for x

a. $x^2=3^2=3\cdot 3=9$

b. $-x^2=-3^2=-(3)(3)=-9$

c. $(-x)^2=(-3)^2=(-3)(-3)=9$

87. Substitute –4 for x

a. $x^2=(-4)^2=(-4)(-4)=16$

b. $-x^2=-(-4)^2$
$=-(-4)(-4)$
$=-(16)$
$=-16$

c. $(-x)^2=4^2=4\cdot 4=16$

89. Substitute 6 for x

a. $x^2=6^2=6\cdot 6=36$

b. $-x^2=-6^2=-(6\cdot 6)=-36$

c. $(-x)^2=(-6)^2=(-6)(-6)=36$

91. Substitute $-\frac{1}{2}$ for x

a. $x^2=\left(-\frac{1}{2}\right)^2=\left(-\frac{1}{2}\right)\left(-\frac{1}{2}\right)=\frac{1}{4}$

b. $-x^2=-\left(-\frac{1}{2}\right)^2=-\left(-\frac{1}{2}\right)\left(-\frac{1}{2}\right)=-\frac{1}{4}$

c. $(-x)^2=\left(\frac{1}{2}\right)^2=\left(\frac{1}{2}\right)\left(\frac{1}{2}\right)=\frac{1}{4}$

93. Substitute –2 for x in the expression.
$x+4=-2+4=2$

95. Substitute 4 for x in the expression.
$5x-1=5(4)-1=20-1=19$

97. Substitute –3 for a in the expression.
$a^2-6=(-3)^2-6=9-6=3$

99. Substitute –1 for each x in the expression.
$-4x^2-2x+1=-4(-1)^2-2(-1)+1$
$=-4(1)-2(-1)+1$
$=-4+2+1$
$=-2+1$
$=-1$

101. Substitute 2 for each p in the expression.
$3p^2-6p-4=3(2)^2-6(2)-4$
$=3(4)-12-4$
$=12-12-4$
$=0-4$
$=-4$

103. Substitute –4 for each x in the expression.
$-x^2-2x+5=-(-4)^2-2(-4)+5$
$=-16+8+5$
$=-8+5$
$=-3$

105. Substitute 5 for each x in the expression.
$4(x+1)^2-6x=4(5+1)^2-6(5)$
$=4(6)^2-30$
$=4(36)-30$
$=144-30$
$=114$

107. Substitute 2 for x and 4 for y in the expression.
$-6x+3y=-6(2)+3(4)=-12+12=0$

109. Substitute –2 for r and –3 for s in the expression.
$r^2-s^2=(-2)^2-(-3)^2=4-9=-5$

111. Substitute 2 for x and –3 for y in the expression.

$$\begin{aligned}&2(x+y)+4x-3y\\&=2[2+(-3)]+4(2)-3(-3)\\&=2(-1)+8-(-9)\\&=-2+8+9\\&=6+9\\&=15\end{aligned}$$

113. Substitute 2 for x and –3 for y in the expression.

$$\begin{aligned}6x^2+3xy-y^2&=6(2)^2+3(2)(-3)-(-3)^2\\&=6(4)+3(2)(-3)-9\\&=24+(-18)-9\\&=6-9\\&=-3\end{aligned}$$

115. $6\cdot 3$ Multiply 6 by 3

$(6\cdot 3)-4$ Subtract 4 from the product

$[(6\cdot 3)-4]-2$ Subtract 2 from the difference

Evaluate:

$[(6\cdot 3)-4]-2=[18-4]-2=14-2=12$

117. $20\div 5$ Divide 20 by 5

$(20\div 5)+12$ Add 12 to the quotient

$[(20\div 5)+12]-8$ Subtract 8 from the sum

$9[[(20\div 5)+12]-8]$ Multiply the difference by 9

Evaluate:

$$\begin{aligned}9[[(20\div 5)+12]-8]&=9[[4+12]-8]\\&=9[16-8]\\&=9(8)\\&=72\end{aligned}$$

119. $\frac{4}{5}+\frac{3}{7}$ Add $\frac{4}{5}$ to $\frac{3}{7}$

$\left(\frac{4}{5}+\frac{3}{7}\right)\cdot\frac{2}{3}$ Multiply the sum by $\frac{2}{3}$

Evaluate:

$$\begin{aligned}\left(\frac{4}{5}+\frac{3}{7}\right)\cdot\frac{2}{3}&=\left(\frac{28}{35}+\frac{15}{35}\right)\cdot\frac{2}{3}\\&=\left(\frac{43}{35}\right)\cdot\left(\frac{2}{3}\right)\\&=\frac{86}{105}\end{aligned}$$

121. $-\left(x^2\right)=-x^2$ is true for all real numbers.

123. When $d=15.99$, $0.07d=0.07(15.99)\approx 1.12$. The sales tax is \$1.12.

125. When $d=15{,}000$,

$$\begin{aligned}d+0.07d&=15{,}000+0.07(15{,}000)\\&=15{,}000+1050\\&=16{,}050\end{aligned}$$

The total cost is \$16,050.

127. **a.** $2\div 5^2=2\div 25=0.08$

b. $(2\div 5)^2=(0.4)^2=0.16$

129. When $R=2$ and $T=70$,

$$\begin{aligned}&0.2R^2+0.003RT+0.0001T^2\\&=0.02(2)^2+0.003(2)(70)+0.0001(70)^2\\&=0.2(4)+0.003(2)(70)+0.0001(4900)\\&=0.8+0.42+0.49=1.71\end{aligned}$$

The growth is 1.71 inches.

131. $12-(4-6)+10=24$

137. **a.** There are 4 houses with 3 occupants.

b.

Occupants	Number of Houses
1	3
2	5
3	4
4	6
5	2

c. $3(1) + 5(2) + 4(3) + 6(4) + 2(5)$
$= 3 + 10 + 12 + 24 + 10$
$= 13 + 12 + 24 + 10$
$= 25 + 24 + 10$
$= 49 + 10$
$= 59$
There are 59 occupants in all.

d. $$\text{Number of houses} = 3+5+4+6+2 = 8+4+6+2 = 12+6+2 = 18+2 = 20$$

$$\text{mean} = \frac{\text{number of occupants}}{\text{number of houses}} = \frac{59}{20} = 2.95$$

There is a mean of 2.95 people per house.

138. $3 = \frac{6}{2} = \frac{1}{2} + \frac{5}{2} = \frac{1}{2} + \frac{20}{8}$
$\text{Cost} = \$2.40 + 20(0.20)$
$= \$2.40 + \$4.00$
$= \$6.40$

139. $(-2)(-4)(6)(-1)(-3) = (8)(6)(-1)(-3)$
$= (48)(-1)(-3)$
$= (-48)(-3)$
$= 144$

140. $$\left(\frac{-5}{7}\right) \div \left(\frac{-3}{14}\right) = \left(\frac{-5}{7}\right) \cdot \left(\frac{14}{-3}\right) = \frac{-5}{1} \cdot \frac{2}{-3} = \frac{(-5)(2)}{(1)(-3)} = \frac{10}{3} \text{ or } 3\frac{1}{3}$$

Exercise Set 1.10

1. The commutative property of addition states that the sum of two numbers is the same regardless of the order in which they are added. One possible example is $3+4=4+3$.

3. The associative property of addition states that the sum of 3 numbers is the same regardless of the way the numbers are grouped. One possible example is $(2+3)+4=2+(3+4)$.

5. The distributive property states that the product of a number with a sum is the same as the sum of the products of the number with each number in the sum. One possible example is $2(3+4)=2(3)+2(4)$.

7. The associative property involves changing parentheses with one operation whereas the distributive property involves distributing a multiplication over an addition.

9. Distributive property

11. Commutative property of multiplication

13. Distributive property

15. Associative property of multiplication

17. Distributive property

19. $3+4=4+3$

21. $-6 \cdot (4 \cdot 2) = (-6 \cdot 4) \cdot 2$

23. $(6)(y) = (y)(6)$

25. $1(x + y) = 1 \cdot x + 1 \cdot y$ or $x + y$

27. $4x + 3y = 3y + 4x$

29. $(3 + x) + y = 3 + (x + y)$

31. $(3x + 4) + 6 = 3x + (4 + 6)$

33. $3(x + y) = (x + y)3$

35. $4(x + y + 12)$
$= 4 \cdot x + 4 \cdot y + 12$ or $4x + 4y + 12$

37. Commutative property of addition

39. Distributive property

41. Commutative property of addition

43. Distributive property

45. Yes; the order does not affect the outcome so the process is commutative.

47. No; the order affects the outcome, so the process is not commutative.

49. No; the order affects the outcome, so the process is not commutative.

51. Yes; the outcome is not affected by whether you do the first two items first or the last two first, so the process is associative.

53. No; the outcome is affected by whether you do the first two items first or the last two first, so the process is not associative.

55. No; the outcome is affected by whether you do the first two items first or the last two first, so the process is not associative.

57. In $(3+4)+x=x+(3+4)$ the $(3+4)$ is treated as one value.

59. This illustrates the commutative property of addition because the change is $3 + 5 = 5 + 3$.

61. No; it illustrates the associative property of addition since the grouping is changed.

63. $2\frac{3}{5}+\frac{2}{3}$

$2\frac{3}{5}=\frac{13}{5}=\frac{3}{3}\cdot\frac{13}{5}=\frac{39}{15}$

$\frac{2}{3}=\frac{2}{3}\cdot\frac{5}{5}=\frac{10}{15}$

$2\frac{3}{5}+\frac{2}{3}=\frac{39}{15}+\frac{10}{15}=\frac{49}{15}$ or $3\frac{4}{15}$

64. $3\frac{5}{8}-2\frac{3}{16}$

$3\frac{5}{8}=\frac{29}{8}=\frac{2}{2}\cdot\frac{29}{8}=\frac{58}{16}$

$2\frac{3}{16}=\frac{35}{16}$

$3\frac{5}{8}-2\frac{3}{16}=\frac{58}{16}-\frac{35}{16}=\frac{23}{16}$ or $1\frac{7}{16}$

65.
$$\begin{aligned}12-24\div 8+4\cdot 3^2 &= 12-24\div 8+4\cdot 9\\ &= 12-3+36\\ &= 9+36\\ &= 45\end{aligned}$$

66. Substitute 2 for x and –3 for y.

$$\begin{aligned}&-4x^2+6xy+3y^2\\ &= -4(2)^2+6(2)(-3)+3(-3)^2\\ &= -4(4)+6(2)(-3)+3(9)\\ &= -16+(-36)+27\\ &= -52+27\\ &= -25\end{aligned}$$

Review Exercises

1.
$$\begin{aligned}&3(15)-8.2-9.2-10.4\\ &= 45-8.2-9.2-10.4\\ &= 36.8-9.2-10.4\\ &= 27.6-10.4\\ &= 17.2\end{aligned}$$

He had 17.2 pounds left.

2. $1.05[1.05(500.00)]=1.05[525]=551.25$

In 2 years the goods will cost $551.25.

3.
$$\begin{aligned}[30+12(25)]-300 &= [30+300]-300\\ &= 330-300\\ &= 30\end{aligned}$$

4. Less than; the increase of 20% of the original price is less than the decrease of 20% of the higher price.

5. a.
$$\begin{aligned}\text{mean} &= \frac{75+79+86+88+64}{5}\\ &= \frac{392}{5}\\ &= 78.4\end{aligned}$$

The mean grade is 78.4.

b. 64, 75, 79, 86, 88

The middle number is 79. The median grade is 79.

6. a. mean

$$=\frac{76+79+84+82+79}{5}=\frac{400}{5}=80$$

The mean temperature is 80°F.

b. 76, 79, 79, 82, 84
The middle number is 79. The median temperature is 79°F.

7. a. The percents were equal around 1991.

b. In 2000, cassettes should equal 10% of shipments and CD's should equal 80%.

c. $80 - 10 = 70$
There is a 70% difference.

d. $\frac{80}{10} = 8$
The percent of CDs is 8 times the percent of cassettes.

8. a. In November 1863, a first class stamp cost 3¢.

b. The cost reached 10¢ in the 1970s.

c. From 1900 to 1910 the price did not increase.

d. $\frac{33}{3} = 11$
The cost of a 33¢ stamp is 11 times the cost of a 3¢ stamp.

9. $\frac{3}{5} \cdot \frac{5}{6} = \frac{1}{1} \cdot \frac{1}{2} = \frac{1 \cdot 1}{1 \cdot 2} = \frac{1}{2}$

10. $\frac{2}{5} \div \frac{10}{9} = \frac{2}{5} \cdot \frac{9}{10} = \frac{1}{5} \cdot \frac{9}{5} = \frac{1 \cdot 9}{5 \cdot 5} = \frac{9}{25}$

11. $\frac{5}{12} \div \frac{3}{5} = \frac{5}{12} \cdot \frac{5}{3} = \frac{5 \cdot 5}{12 \cdot 3} = \frac{25}{36}$

12. $\frac{5}{6} + \frac{1}{3} = \frac{5}{6} + \frac{1}{3} \cdot \frac{2}{2} = \frac{5}{6} + \frac{2}{6} = \frac{7}{6}$ or $1\frac{1}{6}$

13. $\frac{3}{8} - \frac{1}{9} = \frac{3}{8} \cdot \frac{9}{9} - \frac{1}{9} \cdot \frac{8}{8} = \frac{27}{72} - \frac{8}{72} = \frac{19}{72}$

14. $2\frac{1}{3} - 1\frac{1}{5}$
$2\frac{1}{3} = \frac{6+1}{3} = \frac{7}{3} = \frac{7}{3} \cdot \frac{5}{5} = \frac{35}{15}$
$1\frac{1}{5} = \frac{5+1}{5} = \frac{6}{5} = \frac{6}{5} \cdot \frac{3}{3} = \frac{18}{15}$
$2\frac{1}{3} - 1\frac{1}{5} = \frac{35}{15} - \frac{18}{15} = \frac{17}{15}$ or $1\frac{2}{15}$

15. The natural numbers are {1, 2, 3, ...}.

16. The whole numbers are {0, 1, 2, 3, ...}.

17. The integers are {..., –3, –2, –1, 0, 1, 2, ...}.

18. The set of rational numbers is the set of all numbers which can be expressed as the quotient of two integers, denominator not zero.

19. The set of real numbers consists of all numbers that can be represented on a real number line.

20. a. 3 and 426 are positive integers.

b. 3, 0, and 426 are whole numbers.

c. 3, –5, –12, 0, and 426 are integers.

d. 3, –5, –12, 0, $\frac{1}{2}$, –0.62, 426, and $-3\frac{1}{4}$ are rational numbers.

e. $\sqrt{7}$ is an irrational number.

f. 3, –5, –12, 0, $\frac{1}{2}$, –0.62, $\sqrt{7}$, 426, and $-3\frac{1}{4}$ are real numbers.

21. a. 1 is a natural number.

b. 1 is a whole number.

c. –8 and –9 are negative numbers.

d. –8, –9, and 1 are integers.

e. –2.3, –8, –9, $1\frac{1}{2}$, 1, and $-\frac{3}{17}$ are rational numbers.

f. $-2.3, -8, -9,$
$1\frac{1}{2}, \sqrt{2}, -\sqrt{2}, 1,$ and $-\frac{3}{17}$ are real numbers.

22. $-3 > -5$; -3 is to the right of -5 on a number line.

23. $-2.6 > -3.6$; -2.6 is to the right of -3.6 on a number line.

24. $0.50 < 0.509$; 0.50 is to the left 0.509 on a number line.

25. $4.6 > 4.06$; 4.6 is to the right of 4.06 on a number line.

26. $-3.2 < -3.02$; -3.2 is to the left of -3.02 on a number line.

27. $5 > |-3|$ since $|-3|$ equals 3.

28. $-3 < |-7|$ since $|-7|$ equals 7.

29. $|-2.5| = \left|\frac{5}{2}\right|$ since $|-2.5| = \left|\frac{5}{2}\right| = 2.5.$

30. $-4 + (-5) = -9$

31. $-6 + 6 = 0$

32. $0 + (-3) = -3$

33. $-10 + 4 = -6$

34. $-8 - (-2) = -8 + 2 = -6$

35. $-9 - (-4) = -9 + 4 = -5$

36. $4 - (-4) = 4 + 4 = 8$

37. $2 - 12 = 2 + (-12) = -10$

38. $7 - 2 = 5$

39. $2 - 7 = 2 + (-7) = -5$

40. $0 - (-4) = 0 + 4 = 4$

41. $-7 - 5 = -7 + (-5) = -12$

42. $6 - 4 + 3 = 2 + 3 = 5$

43. $-5 + 7 - 6 = 2 - 6 = -4$

44. $-5 - 4 - 3 = -9 - 3 = -12$

45. $-2 + (-3) - 2 = -5 - 2 = -7$

46. $7 - (+4) - (-3) = 7 - 4 + 3 = 3 + 3 = 6$

47. $4 - (-2) + 3 = 4 + 2 + 3 = 6 + 3 = 9$

48. Since the numbers have unlike signs, the product is negative; $-4(7) = -28$

49. Since the numbers have like signs, the product is positive; $(-9)(-3) = 27$

50. Since there are an odd number (3) of negatives the product is negative; $(-4)(-5)(-6) = (20)(-6) = -120$

51. $\left(\frac{3}{5}\right)\left(\frac{-2}{7}\right) = \frac{3(-2)}{5 \cdot 7} = \frac{-6}{35} = -\frac{6}{35}$

52.
$$\left(\frac{10}{11}\right)\left(\frac{3}{-5}\right) = \frac{2}{11} \cdot \frac{3}{-1}$$
$$= \frac{2 \cdot 3}{(11)(-1)}$$
$$= \frac{6}{-11}$$
$$= -\frac{6}{11}$$

53. $\left(\frac{-5}{8}\right)\left(\frac{-3}{7}\right) = \frac{(-5)(-3)}{8 \cdot 7} = \frac{15}{56}$

54. Zero multiplied by any real number is zero.
$0 \cdot \frac{4}{9} = 0$

55. Since there are two negative numbers (an even number), the product is positive.
$4(-2)(-6) = (-8)(-6) = 48$

56. Since there are four negative numbers (an even number), the product is positive.

$$\begin{aligned}(-4)(-6)(-2)(-3) &= (24)(-2)(-3)\\ &= (-48)(-3)\\ &= 144\end{aligned}$$

57. Since the numbers have unlike signs, the quotient is negative; $15 \div (-3) = \frac{15}{-3} = -5$

58. Since the numbers have unlike signs, the quotient is negative; $6 \div (-2) = \frac{6}{-2} = -3$

59. Since the numbers have unlike signs, the quotient is negative; $-20 \div 5 = \frac{-20}{5} = -4$

60. Zero divided by any nonzero number is zero; $0 \div 4 = \frac{0}{4} = 0$

61. Since the numbers have unlike signs, the quotient is negative; $72 \div (-9) = \frac{72}{-9} = -8$

62. $-4 \div \left(\frac{-4}{9}\right) = \frac{-4}{1} \cdot \frac{9}{-4} = \frac{-1}{1} \cdot \frac{9}{-1} = \frac{-9}{-1} = 9$

63. $\frac{28}{-3} \div \left(\frac{9}{-2}\right) = \left(\frac{28}{-3}\right) \cdot \left(\frac{-2}{9}\right) = \frac{-56}{-27} = \frac{56}{27}$

64.
$$\begin{aligned}\frac{14}{3} \div \left(\frac{-6}{5}\right) &= \frac{14}{3} \cdot \left(\frac{5}{-6}\right)\\ &= \frac{7}{3} \cdot \frac{5}{-3}\\ &= \frac{35}{-9}\\ &= -\frac{35}{9}\end{aligned}$$

65.
$$\begin{aligned}\left(\frac{-5}{12}\right) \div \left(\frac{-5}{12}\right) &= \left(\frac{-5}{12}\right) \cdot \left(\frac{12}{-5}\right)\\ &= \frac{-1}{1} \cdot \frac{1}{-1}\\ &= \frac{-1}{-1}\\ &= 1\end{aligned}$$

66. Zero divided by any nonzero number is zero; $0 \div 4 = \frac{0}{4} = 0$

67. Zero divided by any nonzero number is zero; $0 \div (-6) = \frac{0}{-6} = 0$

68. Any real number divided by zero is undefined; $8 \div 0 = \frac{8}{0}$ is undefined.

69. Any real number divided by zero is undefined; $-4 \div 0 = \frac{-4}{0}$ is undefined.

70. Any real number divided by zero is undefined; $\frac{8}{0}$ is undefined

71. Zero divided by any nonzero number is zero; $\frac{0}{-5} = 0$

72. $-4(2-8) = -4(-6) = 24$

73. $2(4-8) = 2(-4) = -8$

74. $(3-6)+4 = -3+4 = 1$

75. $(-4+3)-(2-6) = (-1)-(-4) = -1+4 = 3$

76. $[4+3(-2)]-6 = [4+(-6)]-6 = -2-6 = -8$

77. $(-4-2)(-3) = (-6)(-3) = 18$

78. $[4+(-4)]+(6-8) = (0)+(-2) = -2$

79. $9[3+(-4)]+5=9(-1)+5=-9+5=-4$

80. $-4(-3)+[4\div(-2)]=(12)+(-2)=10$

81. $(-3\cdot 4)\div(-2\cdot 6)=-12\div(-12)=1$

82. $(-3)(-4)+6-3=12+6-3=18-3=15$

83. $[-2(3)+6]-4=[-6+6]-4=0-4=-4$

84. $6^2=(6)(6)=36$

85. $9^3=(9)(9)(9)=729$

86. $3^4=(3)(3)(3)(3)=81$

87. $(-3)^3=(-3)(-3)(-3)=-27$

88. $(-1)^9=(-1)(-1)(-1)(-1)(-1)(-1)(-1)(-1)(-1$
$=-1$

89. $(-2)^5=(-2)(-2)(-2)(-2)(-2)=-32$

90. $\left(\frac{-3}{5}\right)^2=\left(\frac{-3}{5}\right)\left(\frac{-3}{5}\right)=\frac{9}{25}$

91. $\left(\frac{2}{5}\right)^3=\left(\frac{2}{5}\right)\left(\frac{2}{5}\right)\left(\frac{2}{5}\right)=\frac{8}{125}$

92. $xxy=x^2y$

93. $2\cdot 2\cdot 3\cdot 3\cdot 3xyy=2^2\cdot 3^3xy^2$

94. $5\cdot 7\cdot 7\cdot xxy=5\cdot 7^2x^2y$

95. $xyxyz=x^2y^2z$

96. $x^2y=xxy$

97. $xz^3=xzzz$

98. $y^3z=yyyz$

99. $2x^3y^2=2xxxyy$

100. $3+5\cdot 4=3+20=23$

101. $3\cdot 5+4\cdot 2=15+8=23$

102. $(3-7)^2+6=(-4)^2+6=16+6=22$

103. $8-36\div 4\cdot 3=8-9\cdot 3=8-27=-19$

104. $6-3^2\cdot 5=6-9\cdot 5=6-45=-39$

105. $[6-(3\cdot 5)]+5=[6-15]+5=-9+5=-4$

106. $3[9(4^2+3)]\cdot 2=3[9-(16+3)]\cdot 2$
$=3[9-19]\cdot 2$
$=3\cdot(-10)\cdot 2$
$=-30\cdot 2$
$=-60$

107. $(-3^2+4^2)+(3^2\div 3)=(-9+16)+(9\div 3)$
$=(7)+(3)$
$=10$

108. $2^3\div 4+6\cdot 3=8\div 4+6\cdot 3=2+18=20$

109. $(4\div 2)^4+4^2\div 2^2=(2)^4+16\div 4$
$=16+16\div 4$
$=16+4$
$=20$

110. $\left(15-2^2\right)^2-4\cdot 3+10\div 2$
$=(15-4)^2-4\cdot 3+10\div 2$
$=(11)^2-4\cdot 3+10\div 2$
$=121-4\cdot 3+10\div 2$
$=121-12+5$
$=109+5$
$=114$

111. $4^3\div 4^2-5(2-7)\div 5=64\div 16-5(-5)\div 5$
$=4-(-25)\div 5$
$=4-(-5)$
$=4+5$
$=9$

112. Substitute 5 for x;
$4x-6=4(5)-6=20-6=14$

113. Substitute –5 for x;
$6-4x=6-4(-5)=6-(-20)=6+20=26$

114. Substitute 6 for x;
$$\begin{aligned}x^2-5x+3&=(6)^2-5(6)+3\\&=36-30+3\\&=6+3\\&=9\end{aligned}$$

115. Substitute –1 for y;
$$\begin{aligned}5y^2+3y-2&=5(-1)^2+3(-1)-2\\&=5(1)-3-2\\&=5-3-2\\&=2-2\\&=0\end{aligned}$$

116. Substitute 2 for x;
$$\begin{aligned}-x^2+2x-3&=-2^2+2(2)-3\\&=-4+4-3\\&=0-3\\&=-3\end{aligned}$$

117. Substitute –2 for x;
$$\begin{aligned}-x^2+2x-3&=-(-2)^2+2(-2)-3\\&=-4+(-4)-3\\&=-8-3\\&=-11\end{aligned}$$

118. Substitute 1 for x;
$$\begin{aligned}-3x^2-5x+5&=-3(1)^2-5(1)+5\\&=-3(1)-5+5\\&=-3-5+5\\&=-8+5\\&=-3\end{aligned}$$

119. Substitute 3 for x and 4 for y;
$$\begin{aligned}3xy-5x&=3(3)(4)-5(3)\\&=9(4)-15\\&=36-15\\&=21\end{aligned}$$

120. Substitute –3 for x and –2 for y;
$$\begin{aligned}-x^2-8x-12y&=-(-3)^2-8(-3)-12(-2)\\&=-9-(-24)+24\\&=-9+24+24\\&=15+24\\&=39\end{aligned}$$

121. a. $158+(-493)=-335$

b. $|-493|$ is greater than $|158|$ so the sum should be (and is) negative.

122. a. $324-(-29.6)=324+29.6=353.6$

b. The sum of two positive numbers is always positive. As expected, the answer is positive.

123. a. $\frac{-17.28}{6}=-2.88$

b. Since the numbers have unlike signs, the quotient is negative, as expected.

124. a. $(-62)(-1.9)=117.8$

b. Since the numbers have like signs, the product is positive, as expected.

125. a. $(-3)^6=729$

b. A negative number raised to an even power is positive. As expected, the answer is positive.

126. a. $-(4.2)^3=-74.088$

b. Since $(4.2)^3$ is positive, $-(4.2)^3$ should be (and is) negative.

127. Associative property of addition

128. Commutative property of multiplication

129. Distributive property

130. Commutative property of multiplication

131. Commutative property of addition

132. Associative property of addition

Practice Test

1. a. $2(1.30) + 4.75 + 3(1.10)$
$= 2.60 + 4.75 + 3.30$
$= 7.35 + 3.30$
$= 10.65$
The bill is \$10.65 before tax.

b. $0.07(3.30) \approx 0.23$
The tax on the soda is \$0.23.

c. $10.65 + 0.23 = 10.88$
The total bill is \$10.88.

d. $50 - 10.88 = 39.12$
Her change will be \$39.12.

2. a. Levi's share in 1990 was about 31% and in 1997 was about 18%.

b. VF Corp's share was about 24.5% in 1997 and store brand's share was about 24%.

c. $18\% + 24.5\% + 24\% = 66.5\%$

3. a. In 1960, about 58% watched the Kentucky Derby and in 1997, about 19% did.

b. $58 - 19 = 39$
The ratings share was 39% greater.

c. $\frac{58}{19} \approx 3$
The market share in 1960 was about 3 times that in 1997.

4. a. 42 is a natural number.

b. 42 and 0 are whole numbers.

c. –6, 42, 0, –7, and –1 are integers.

d. $-6, 42, -3\frac{1}{2}, 0, 6.52, \frac{5}{9}, -7$, and -1 are rational numbers.

e. $\sqrt{5}$ is an irrational number.

f. $-6, 42, -3\frac{1}{2}, 0, 6.52, \sqrt{5}, \frac{5}{9}, -7$, and -1 are real numbers.

5. $-6 < -3$; -6 is to the left of -3 on a number line.

6. $|-3| > |-2|$ since $|-3| = 3$ and $|-2| = 2$.

7. $-4 + (-8) = -12$

8. $-6 - 5 = -6 + (-5) = -11$

9. $5 - 12 - 7 = -7 - 7 = -14$

10. $(-4 + 6) - 3(-2) = (2) - (-6) = 2 + 6 = 8$

11. $(-4)(-3)(2)(-1) = (12)(2)(-1)$
$= (24)(-1)$
$= -24$

12. $\left(\frac{-2}{9}\right) \div \left(\frac{-7}{8}\right) = \frac{-2}{9} \cdot \frac{8}{-7} = \frac{-16}{-63} = \frac{16}{63}$

13. $\left(-12 \cdot \frac{1}{2}\right) \div 3 = \left(\frac{-12}{1} \cdot \frac{1}{2}\right) \div 3$
$= \left(\frac{-6}{1} \cdot \frac{1}{1}\right) \div 3$
$= -6 \div 3$
$= -2$

14. $3 \cdot 5^2 - 4 \cdot 6^2 = 3 \cdot 25 - 4 \cdot 36$
$= 75 - 144$
$= -69$

15. $-6(-2 - 3) \div 5 \cdot 2 = -6(-5) \div 5 \cdot 2$
$= 30 \div 5 \cdot 2$
$= 6 \cdot 2$
$= 12$

16. $\left(\frac{3}{5}\right)^3 = \left(\frac{3}{5}\right)\left(\frac{3}{5}\right)\left(\frac{3}{5}\right) = \frac{27}{125}$

17. $2 \cdot 2 \cdot 5 \cdot 5 \cdot yyzzz = 2^2 5^2 y^2 z^3$

18. $2^2 3^3 x^4 y^2 = 2 \cdot 2 \cdot 3 \cdot 3 \cdot 3xxxxyy$

19. Substitute –4 for x;

$$\begin{aligned} 2x^2 - 6 &= 2(-4)^2 - 6 \\ &= 2(16) - 6 \\ &= 32 - 6 \\ &= 26 \end{aligned}$$

20. Substitute 3 for x and –2 for y;

$$\begin{aligned} 6x - 3y^2 + 4 &= 6(3) - 3(-2)^2 + 4 \\ &= 6(3) - 3(4) + 4 \\ &= 18 - 12 + 4 \\ &= 6 + 4 \\ &= 10 \end{aligned}$$

21. Substitute –2 for each x;

$$\begin{aligned} -x^2 - 6x + 3 &= -(-2)^2 - 6(-2) + 3 \\ &= -4 - (-12) + 3 \\ &= -4 + 12 + 3 \\ &= 8 + 3 \\ &= 11 \end{aligned}$$

22. Substitute 1 for x and –2 for y;

$$\begin{aligned} -x^2 + xy + y^2 &= -(1)^2 + (1)(-2) + (-2)^2 \\ &= -1 + (-2) + 4 \\ &= -3 + 4 \\ &= 1 \end{aligned}$$

23. Commutative property of addition

24. Distributive property

25. Associative property of addition

Chapter 2

Exercise Set 2.1

1. a. The terms of an expression are the parts that are added.

   b. The terms of $3x-4y-5$ are $3x$, $-4y$, and -5.

   c. The terms of $6xy+3x-y-9$ are $6xy$, $3x$, $-y$, and -9.

3. a. The factors of an expression are the parts that are multiplied.

   b. In $3x$, 3 and x are factors because they are multiplied together.

   c. In $5xy$, 5, x and y are factors because they are multiplied together.

5. a. The numerical part of a term is the numerical coefficient or coefficient of the term.

   b. The coefficient of $4x$ is 4.

   c. Since $x = 1x$, the coefficient of x is 1.

   d. Since $-x = -1x$, the coefficient of $-x$ is -1.

   e. Since $\frac{3x}{5} = \frac{3}{5}x$, the coefficient of $\frac{3x}{5}$ is $\frac{3}{5}$.

   f. Since $\frac{2x+3}{7} = \frac{2x}{7} + \frac{3}{7} = \frac{2}{7}x + \frac{3}{7}$, the coefficient of $\frac{2x+3}{7}$ is $\frac{2}{7}$.

7. a. The parentheses may be removed without having to change the expression within the parentheses.

   b. $+(x-8) = x-8$

9. $5x+3x = 8x$

11. $4x-5x = -x$

13. $y+3+4y = y+4y+3$
    $= 5y+3$

15. $-2x+5x = 3x$

17. $3-8x+5 = -8x+3+5$
    $= -8x+8$

19. $-2x-3x-2-3 = -5x-5$

21. $-x+2-x-2 = -x-x+2-2$
    $= -2x$

23. $x-4x+3 = -3x+3$

25. $5+2x-4x+6 = 2x-4x+5+6$
    $= -2x+11$

27. $x-2-4+2x = x+2x-2-4$
    $= 3x-6$

29. $2-3x-2x+y = -3x-2x+y+2$
    $= -5x+y+2$

31. $-2x+4x-3 = 2x-3$

33. $x+4+\frac{3}{5} = x+\frac{20}{5}+\frac{3}{5}$
    $= x+\frac{23}{5}$

35. $5.1x+6.42-4.3x = 5.1x-4.3x+6.42$
    $= 0.8x+6.42$

37. There are no like terms.
    $\frac{1}{2}x+3y+1$

39. $2x+3y+4x+5y = 2x+4x+3y+5y$
    $= 6x+8y$

41. $-x+2x+y = x+y$

43. $2x-7y-5x+2y=2x-5x-7y+2y$
$=-3x-5y$

45. $6-3x+9-2x=-3x-2x+6+9$
$=-5x+15$

47. $-19.36+40.02x+12.25-18.3x$
$=40.02x-18.3x-19.36+12.25$
$=21.72x-7.11$

49. $\frac{3}{5}x-3-\frac{7}{4}x-2=\frac{3}{5}x-\frac{7}{4}x-3-2$
$=\frac{12}{20}x-\frac{35}{20}x-5$
$=-\frac{23}{20}x-5$

51. $2(x+6)=2x+2(6)$
$=2x+12$

53. $5(x+4)=5x+5(4)$
$=5x+20$

55. $-2(x-4)=-2[x+(-4)]$
$=-2x+(-2)(-4)$
$=-2x+8$

57. $-\frac{1}{2}(2x-4)=-\frac{1}{2}[2x+(-4)]$
$=-\frac{1}{2}(2x)+\left(-\frac{1}{2}\right)(-4)$
$=-x+2$

59. $1(-4+x)=1(-4)+1(x)$
$=-4+x$
$=x-4$

61. $\frac{2}{5}(x-5)=\frac{2}{5}x-\frac{2}{5}(5)$
$=\frac{2}{5}x-2$

63. $-0.3(3x+5)=-0.3(3x)+(-0.3)(5)$
$=-0.9x+(-1.5)$
$=-0.9x-1.5$

65. $\frac{1}{2}(-2x+6)=\frac{1}{2}(-2x)+\frac{1}{2}(6)$
$=-x+3$

67. $0.7(2x+0.5)=0.7(2x)+0.7(0.5)$
$=1.4x+0.35$

69. $-(-x+y)=-1(-x+y)$
$=-1(-x)+(-1)(y)$
$=x+(-y)$
$=x-y$

71. $-(2x+4y-8)$
$=-1[2x+4y+(-8)]$
$=-1(2x)+(-1)(4y)+(-1)(-8)$
$=-2x-4y+8$
$=-2x-4y+8$

73. $1.1(3.1x-5.2y+2.8)$
$=1.1[3.1x+(-5.2y)+2.8]$
$=(1.1)(3.1x)+(1.1)(-5.2y)+(1.1)(2.8)$
$=3.41x+(-5.72y)+3.08$
$=3.41x-5.72y+3.08$

75. $2\left(3x-2y+\frac{1}{4}\right)=2\left[3x+(-2y)+\frac{1}{4}\right]$
$=2(3x)+2(-2y)+2\left(\frac{1}{4}\right)$
$=6x+(-4y)+\frac{1}{2}$
$=6x-4y+\frac{1}{2}$

77. $(x+3y-9)=1[x+3y+(-9)]$
$=1(x)+1(3y)+(1)(-9)$
$=x+3y+(-9)=x+3y-9$

79. $-3(-x+4+2y)$
$=-3(-x)+(-3)(4)+(-3)(2y)$
$=3x+(-12)+(-6y)$
$=3x-12-6y$

81. $4(x-2)-x=4x-8-x$
$=4x-x-8$
$=3x-8$

83. $-2(3-x)+7=-6+2x+7$
$=2x-6+7$
$=2x+1$

85. $6x+2(4x+9)=6x+8x+18$
$=14x+18$

87. $2(x-y)+2x+3=2x-2y+2x+3$
$=2x+2x-2y+3$
$=4x-2y+3$

89. $(2x-y)+2x+5=2x-y+2x+5$
$=2x+2x-y+5$
$=4x-y+5$

91. $8x-(x-3)=8x-x+3$
$=7x+3$

93. $2(x-3)-(x+3)=2x-6-x-3$
$=2x-x-6-3$
$=x-9$

95. $4(x-1)+2(3-x)-4$
$=4x-4(1)+2(3)-2x-4$
$=4x-4+6-2x-4$
$=4x-2x-4+6-4$
$=2x-2$

97. $(x-4)+3x+9=x-4+3x+9$
$=x+3x-4+9$
$=4x+5$

99. $-3(x+1)+5x+6=-3x-3+5x+6$
$=-3x+5x-3+6$
$=2x+3$

101. $4(x+3)-2x-7=4x+4(3)-2x-7$
$=4x+12-2x-7$
$=4x-2x+12-7$
$=2x+5$

103. $0.4+(y+5)+0.6-2=0.4+y+5+0.6-2$
$=y+0.4+5+0.6-2$
$=y+4$

105. $4+(3x-4)-5=4+3x-4-5$
$=3x+4-4-5$
$=3x-5$

107. $4(x+2)-3(x-4)-5$
$=4x+4(2)-3x-3(-4)-5$
$=4x+8-3x+12-5$
$=4x-3x+8+12-5$
$=x+15$

109. $-0.2(6-x)-4(y+0.4)$
$=-0.2(6)-0.2(-x)-4y-4(0.4)$
$=-1.2+0.2x-4y-1.6$
$=0.2x-4y-1.2-1.6$
$=0.2x-4y-2.8$

111. $-6x+7y-(3+x)+(x+3)$
$=-6x+7y-3-x+x+3$
$=-6x-x+x+7y-3+3$
$=-6x+7y$

113. $\frac{1}{2}(x+3)+\frac{1}{3}(3x+6)=\frac{1}{2}x+\frac{3}{2}+\frac{3}{3}x+\frac{6}{3}$
$=\frac{1}{2}x+\frac{3}{2}+x+2$
$=\frac{1}{2}x+x+\frac{3}{2}+2$
$=\frac{3}{2}x+\frac{7}{2}$

115. $\square+\ominus+\ominus+\square+\ominus$
$=\square+\square+\ominus+\ominus+\ominus$
$=2\square+3\ominus$

117. $x+y+\triangle+\triangle+x+y+y$
$=x+x+y+y+y+\triangle+\triangle$
$=2x+3y+2\triangle$

119. $1\cdot 12,\ 2\cdot 6,\ 3\cdot 4$
positive factors: 1, 2, 3, 4, 6, 12

121. $3\triangle + 5\square - \triangle - 3\square = 3\triangle - \triangle + 5\square - 3\square$
$= 2\triangle + 2\square$

123. $4x + 5y + 6(3x - 5y) - 4x + 3$
$= 4x + 5y + 18x - 30y - 4x + 3$
$= 4x + 18x - 4x + 5y - 30y + 3$
$= 18x - 25y + 3$

125. $x^2 + 2y - y^2 + 3x + 5x^2 + 6y^2 + 5y$
$= x^2 + 5x^2 - y^2 + 6y^2 + 3x + 2y + 5y$
$= 6x^2 + 5y^2 + 3x + 7y$

127. $|-7| = 7$

128. $-|-16| = -(16) = -16$

129. Answers will vary. The answer should include that the order is parentheses, exponents, multiplication and division from left to right, and addition and subtraction from left to right.

130. Substitute –1 for each x in the expression.
$-x^2 + 5x - 6 = -(-1)^2 + 5(-1) - 6$
$= -1 + (-5) - 6$
$= -6 - 6$
$= -12$

Exercise Set 2.2

1. An equation is a statement that shows two algebraic expressions are equal.

3. A solution to an equation may be checked by substituting the value in the equation and determining if it results in a true statement.

5. Equivalent equations are two or more equations with the same solution.

7. Add 4 to both sides of the equation to get the variable by itself.

9. One example is $x + 2 = 1$.

11. Subtraction is defined in terms of addition.

13. Substitute 4 for x, $x = 4$.
$3x - 3 = 9$
$3(4) - 3 = 9$
$12 - 3 = 9$
$9 = 9$ True
Since we obtain true statement, 4 is a solution.

15. Substitute –3 for x, $x = -3$.
$2x - 5 = 5(x + 2)$
$2(-3) - 5 = 5[(-3) + 2]$
$-6 - 5 = 5(-1)$
$-11 = -5$ False
Since we obtain a false statement, –3 is not a solution.

17. Substitute 0 for x, $x = 0$.
$3x - 5 = 2(x + 3) - 11$
$3(0) - 5 = 2(0 + 3) - 11$
$0 - 5 = 2(3) - 11$
$-5 = 6 - 11$
$-5 = -5$ True
Since we obtain a true statement, 0 is a solution.

19. Substitute 3.4 for x, $x = 3.4$.
$3(x + 2) - 3(x - 1) = 9$
$3(3.4 + 2) - 3(3.4 - 1) = 9$
$3(5.4) - 3(2.4) = 9$
$16.2 - 7.2 = 9$
$9 = 9$ True
Since we obtain a true statement, 3.4 is a solution.

21. Substitute $\frac{1}{2}$ for x, $x = \frac{1}{2}$.
$4x - 4 = 2x - 2$
$4\left(\frac{1}{2}\right) - 4 = 2\left(\frac{1}{2}\right) - 2$
$2 - 4 = 1 - 2$
$-2 = -1$ False
Since we obtain a false statement, $\frac{1}{2}$ is not a solution.

23. Substitute $\frac{11}{2}$ for x, $x = \frac{11}{2}$.

$$3(x+2) = 5(x-1)$$
$$3\left(\frac{11}{2}+2\right) = 5\left(\frac{11}{2}-1\right)$$
$$3\left(\frac{15}{2}\right) = 5\left(\frac{9}{2}\right)$$
$$\frac{45}{2} = \frac{45}{2} \text{ True}$$

Since we obtain a true statement, $\frac{11}{2}$ is a solution.

25. a. $x+5=8$

b. $x+5+(-5) = 8+(-5)$
Add –5 to both sides.
$x+0=3$
$x=3$

27. a. $12 = x+3$

b. $12+(-3) = x+3+(-3)$
Add –3 to both sides.
$9 = x+0$
$9 = x$

29. a. $10 = x+7$

b.
$10+(-7) = x+7+(-7)$
Add –7 to both sides.
$3 = x+0$
$3 = x$

31. a. $x+6 = 4+11$

b. $x+6 = 15$
$x+6+(-6) = 15+(-6)$
Add –6 to both sides.
$x = 9$

33. $x+2=6$
$x+2-2 = 6-2$
$x+0=4$
$x=4$
Check: $x+2=6$
$4+2=6$
$6=6$ True

35. $x+1=-6$
$x+1-1 = -6-1$
$x+0=-7$
$x=-7$
Check: $x+1=-6$
$-7+1=-6$
$-6=-6$ True

37. $x+4=-5$
$x+4-4 = -5-4$
$x+0=-9$
$x=-9$
Check: $x+4=-5$
$-9+4=-5$
$-5=-5$ True

39. $x+9=52$
$x+9-9 = 52-9$
$x+0=43$
$x=43$
Check: $x+9=52$
$43+9=52$
$52=52$ True

41. $-8+x=14$
$-8+8+x = 14+8$
$0+x=22$
$x=22$
Check: $-8+x=14$
$-8+22=14$
$14=14$ True

43. $27 = x+16$
$27-16 = x+16-16$
$11 = x+0$
$11 = x$
Check: $27 = x+16$
$27 = 11+16$
$27 = 27$ True

45. $-18 = -14 + x$
$-18 + 14 = -14 + 14 + x$
$-4 = 0 + x$
$-4 = x$
Check: $-18 = -14 + x$
$-18 = -14 + (-4)$
$-18 = -18$ True

47. $9 + x = 4$
$9 - 9 + x = 4 - 9$
$0 + x = -5$
$x = -5$
Check: $9 + x = 4$
$9 + (-5) = 4$
$4 = 4$ True

49. $4 + x = -9$
$4 - 4 + x = -9 - 4$
$0 + x = -13$
$x = -13$
Check: $4 + x = -9$
$4 + (-13) = -9$
$-9 = -9$ True

51. $5 + x = -18$
$5 - 5 + x = -18 - 5$
$0 + x = -23$
$x = -23$
Check: $5 + x = -18$
$5 + (-23) = -18$
$-18 = -18$ True

53. $6 = 4 + x$
$6 - 4 = 4 - 4 + x$
$2 = 0 + x$
$2 = x$
Check: $6 = 4 + x$
$6 = 4 + 2$
$6 = 6$ True

55. $-4 = x - 3$
$-4 + 3 = x - 3 + 3$
$-1 = x + 0$
$-1 = x$
Check: $-4 = x - 3$
$-4 = -1 - 3$
$-4 = -4$ True

57. $12 = 16 + x$
$12 - 16 = 16 - 16 + x$
$-4 = 0 + x$
$-4 = x$
Check: $12 = 16 + x$
$12 = 16 + (-4)$
$12 = 12$ True

59. $15 + x = -5$
$15 - 15 + x = -5 - 15$
$0 + x = -20$
$x = -20$
Check: $15 + x = -5$
$15 + (-20) = -5$
$-5 = -5$ True

61. $-10 = -10 + x$
$-10 + 10 = -10 + 10 + x$
$0 = 0 + x$
$0 = x$
Check: $-10 = -10 + x$
$-10 = -10 + 0$
$-10 = -10$ True

63. $5 = x - 12$
$5 + 12 = x - 12 + 12$
$17 = x + 0$
$17 = x$
Check: $5 = x - 12$
$5 = 17 - 12$
$5 = 5$ True

65. $-50 = x - 24$
$-50 + 24 = x - 24 + 24$
$-26 = x + 0$
$-26 = x$
Check: $-50 = x - 24$
$-50 = -26 - 24$
$-50 = -50$ True

67. $16 + x = -20$
$16 - 16 + x = -20 - 16$
$0 + x = -36$
$x = -36$
Check: $16 + x = -20$
$16 + (-36) = -20$
$-20 = -20$ True

69. $40.2 + x = -5.9$
$40.2 - 40.2 + x = -5.9 - 40.2$
$0 + x = -46.1$
$x = -46.1$
Check: $40.2 + x = -5.9$
$40.2 + (-46.1) = -5.9$
$-5.9 = -5.9$ True

71. $-37 + x = 9.5$
$-37 + 37 + x = 9.5 + 37$
$0 + x = 46.5$
$x = 46.5$
Check: $-37 + x = 9.5$
$-37 + 46.5 = 9.5$
$9.5 = 9.5$ True

73. $x - 8.77 = -17$
$x - 8.77 + 8.77 = -17 + 8.77$
$x + 0 = -8.23$
$x = -8.23$
Check: $x - 8.77 = -17$
$-8.23 - 8.77 = -17$
$-17 = -17$ True

75. $9.32 = x + 3.75$
$9.32 - 3.75 = x + 3.75 - 3.75$
$5.57 = x + 0$
$5.57 = x$
Check: $9.32 = x + 3.75$
$9.32 = 5.57 + 3.75$
$9.32 = 9.32$ True

77. No; there are no real numbers that can make $x + 1 = x + 2$.

79. **a.** $2x = 8$

b. $\frac{2x}{2} = \frac{8}{2}$
$x = 4$

81. **a.** $20 = 4x$

b. $\frac{20}{4} = \frac{4x}{4}$
$5 = x$ or $x = 5$

83. $x - \triangle = \square$
$x - \triangle + \triangle = \square + \square$
$x + 0 = \square + \triangle$
$x = \square + \triangle$

85. $☺ = \square + \triangle$
$☺ - \triangle = \square + \triangle - \triangle$
$☺ - \triangle = \square + 0$
$☺ - \triangle = \square$

88. Substitute 4 for each x in the expression.
$3x + 4(x - 3) + 2 = 3(4) + 4(4 - 3) + 2$
$= 12 + 4(1) + 2$
$= 12 + 4 + 2$
$= 16 + 2$
$= 18$

89. Substitute –3 for x in the expression.
$6x - 2(2x + 1) = 6(-3) - 2[2(-3) + 1]$
$= -18 - 2(-6 + 1)$
$= -18 - 2(-5)$
$= -18 + 10$
$= -8$

90. $4x + 3(x - 2) - 5x - 7 = 4x + 3x - 6 - 5x - 7$
$= 4x + 3x - 5x - 6 - 7$
$= 2x - 13$

91. $-(x - 3) + 7(2x - 5) - 3x$
$= -x + 3 + 14x - 35 - 3x$
$= -x + 14x - 3x + 3 - 35$
$= 10x - 32$

Exercise Set 2.3

1. Answers will vary. Answer should include that both sides of an equation can be multiplied by the same nonzero number without changing the solution to the equation.

3. **a.** $-x = a$
$-1x = a$
$(-1)(-1x) = (-1)a$
$1x = -a$
$x = -a$

b.
$$\begin{aligned} -x &= 5 \\ -1x &= 5 \\ (-1)(-1x) &= (-1)5 \\ 1x &= -5 \\ x &= -5 \end{aligned}$$

c.
$$\begin{aligned} -x &= -5 \\ -1x &= -5 \\ (-1)(-1x) &= (-1)(-5) \\ 1x &= 5 \\ x &= 5 \end{aligned}$$

5. Divide by –2 to isolate the variable.

7. Multiply both sides by 3 because $3 \cdot \frac{x}{3} = x$.

9. a. $2x = 10$

b.
$$\begin{aligned} \frac{2x}{2} &= \frac{10}{2} \\ x &= 5 \end{aligned}$$

11. a. $6 = 3x$

b.
$$\begin{aligned} \frac{6}{3} &= \frac{3x}{3} \\ 2 &= x \end{aligned}$$

13. a. $2x = 5$

b.
$$\begin{aligned} \frac{2x}{2} &= \frac{5}{2} \\ x &= \frac{5}{2} \end{aligned}$$

15. a. $4 = 3x$

b.
$$\begin{aligned} \frac{4}{3} &= \frac{3x}{3} \\ \frac{4}{3} &= x \end{aligned}$$

17.
$$\begin{aligned} 2x &= 6 \\ \frac{2x}{2} &= \frac{6}{2} \\ x &= 3 \end{aligned}$$
Check: $2x = 6$
$$\begin{aligned} 2(3) &= 6 \\ 6 &= 6 \text{ True} \end{aligned}$$

19.
$$\begin{aligned} \frac{x}{2} &= 4 \\ 2\left(\frac{x}{2}\right) &= 2(4) \\ x &= 8 \end{aligned}$$
Check: $\frac{x}{2} = 4$
$$\begin{aligned} \frac{8}{2} &= 4 \\ 4 &= 4 \text{ True} \end{aligned}$$

21.
$$\begin{aligned} -4x &= 12 \\ \frac{-4x}{-4} &= \frac{12}{-4} \\ x &= -3 \end{aligned}$$
Check: $-4x = 12$
$$\begin{aligned} -4(-3) &= 12 \\ 12 &= 12 \text{ True} \end{aligned}$$

23.
$$\begin{aligned} \frac{x}{4} &= -2 \\ 4\left(\frac{x}{4}\right) &= 4(-2) \\ x &= 4(-2) \\ x &= -8 \end{aligned}$$
Check: $\frac{x}{4} = -2$
$$\begin{aligned} \frac{-8}{4} &= -2 \\ -2 &= -2 \text{ True} \end{aligned}$$

25. $\frac{x}{5} = 1$

$5\left(\frac{x}{5}\right) = 5(1)$

$x = 5$

Check: $\frac{x}{5} = 1$

$\frac{5}{5} = 1$

$1 = 1$ True

27. $-32x = -96$

$-32x = -96$

$\frac{-32x}{-32} = \frac{-96}{-32}$

$x = 3$

Check: $-32x = -96$

$-32(3) = -96$

$-96 = -96$ True

29. $-6 = 4z$

$\frac{-6}{4} = \frac{4z}{4}$

$-\frac{3}{2} = z$

Check: $-6 = 4z$

$-6 = 4\left(-\frac{3}{2}\right)$

$-6 = -6$ True

31. $-x = -11$

$-1x = -11$

$(-1)(-1x) = (-1)(-11)$

$1x = 11$

$x = 11$

Check: $-x = -11$

$-11 = -11$ True

33. $10 = -y$

$10 = -1y$

$(-1)(10) = (-1)(-1y)$

$-10 = 1y$

$-10 = y$

Check: $10 = -y$

$10 = -(-10)$

$10 = 10$ True

35. $-\frac{x}{7} = -7$

$\frac{x}{-7} = -7$

$(-7)\left(\frac{x}{-7}\right) = (-7)(-7)$

$x = 49$

Check: $-\frac{x}{7} = -7$

$-\frac{49}{7} = -7$

$-7 = -7$ True

37. $4 = -12x$

$\frac{4}{-12} = \frac{-12x}{-12}$

$-\frac{1}{3} = x$

Check: $4 = -12x$

$4 = -12\left(-\frac{1}{3}\right)$

$4 = 4$ True

39. $-\frac{x}{3} = -2$

$\frac{x}{-3} = -2$

$(-3)\left(\frac{x}{-3}\right) = (-3)(-2)$

$x = 6$

Check: $-\frac{x}{3} = -2$

$-\frac{6}{3} = -2$

$-2 = -2$ True

41. $13x = 10$

$\frac{13x}{13} = \frac{10}{13}$

$x = \frac{10}{13}$

Check: $13x = 10$

$13\left(\frac{10}{13}\right) = 10$

$10 = 10$ True

43. $-4.2x = -8.4$

$$\frac{-4.2x}{-4.2} = \frac{-8.4}{-4.2}$$

$$x = 2$$

Check: $-4.2x = -8.4$

$$-4.2(2) = -8.4$$

$$-8.4 = -8.4 \text{ True}$$

45. $7x = -7$

$$\frac{7x}{7} = \frac{-7}{7}$$

$$x = -1$$

Check: $7x = -7$

$$7(-1) = -7$$

$$-7 = -7 \text{ True}$$

47. $5x = -\frac{3}{8}$

$$\frac{1}{5} \cdot 5x = \left(\frac{1}{5}\right) \cdot \left(-\frac{3}{8}\right)$$

$$x = \frac{(1) \cdot (-3)}{(5) \cdot (8)}$$

$$x = -\frac{3}{40}$$

Check: $5x = -\frac{3}{8}$

$$5\left(-\frac{3}{40}\right) = -\frac{3}{8}$$

$$-\frac{3}{8} = -\frac{3}{8} \text{ True}$$

49. $15 = -\frac{x}{4}$

$$15 = \frac{x}{-4}$$

$$(-4)(15) = (-4) \cdot \left(\frac{x}{-4}\right)$$

$$-60 = x$$

Check: $15 = -\frac{x}{4}$

$$15 = -\frac{(-60)}{4}$$

$$15 = 15 \text{ True}$$

51. $-\frac{x}{2} = -75$

$$\frac{x}{-2} = -75$$

$$(-2)\left(\frac{x}{-2}\right) = (-2)(-75)$$

$$x = 150$$

Check: $-\frac{x}{2} = -75$

$$-\frac{150}{2} = -75$$

$$-75 = -75 \text{ True}$$

53. $\frac{x}{5} = -7$

$$5\left(\frac{x}{5}\right) = 5(-7)$$

$$x = -35$$

Check: $\frac{x}{5} = -7$

$$\frac{-35}{5} = -7$$

$$-7 = -7 \text{ True}$$

55. $5 = \frac{x}{4}$

$$4 \cdot 5 = 4\left(\frac{x}{4}\right)$$

$$20 = x$$

Check: $5 = \frac{x}{4}$

$$5 = \frac{20}{4}$$

$$5 = 5 \text{ True}$$

57. $\frac{3}{5}d = -30$

$$\frac{5}{3} \cdot \frac{3}{5}d = \frac{5}{3}(-30)$$

$$d = -50$$

Check: $\frac{3}{5}d = -30$

$$\frac{3}{5}(-50) = -30$$

$$-30 = -30 \text{ True}$$

59.
$$\frac{y}{-2} = -6$$
$$(-2)\left(\frac{y}{-2}\right) = (-2)(-6)$$
$$y = 12$$
Check: $\frac{4}{-2} = -6$
$$\frac{12}{-2} = -6$$
$$-6 = -6 \text{ True}$$

61.
$$\frac{-7}{8}w = 0$$
$$\frac{8}{-7}\left(\frac{-7}{8}w\right) = \frac{8}{-7}\cdot 0$$
$$w = 0$$
Check: $\frac{-7}{8}w = 0$
$$\frac{-7}{8}(0) = 0$$
$$0 = 0 \text{ True}$$

63.
$$\frac{1}{5}x = 4.5$$
$$5\left(\frac{1}{5}x\right) = 5(4.5)$$
$$x = 22.5$$
Check: $\frac{1}{5}x = 4.5$
$$\frac{1}{5}(22.5) = 4.5$$
$$4.5 = 4.5 \text{ True}$$

65.
$$-4 = -\frac{2}{3}z$$
$$\left(-\frac{3}{2}\right)(-4) = \left(-\frac{3}{2}\right)\left(-\frac{2}{3}\right)z$$
$$6 = z$$
Check: $-4 = -\frac{2}{3}z$
$$-4 = -\frac{2}{3}\cdot 6$$
$$-4 = -4 \text{ True}$$

67.
$$-1.4x = 28.28$$
$$\frac{-1.4x}{-1.4} = \frac{28.28}{-1.4}$$
$$x = -20.2$$
Check: $-1.4x = 28.28$
$$-1.4(-20.2) = 28.28$$
$$28.28 = 28.28 \text{ True}$$

69.
$$6x = \frac{5}{10}$$
$$\frac{1}{6}\cdot 6x = \frac{1}{6}\cdot\frac{5}{10}$$
$$x = \frac{1}{12}$$
Check: $6x = \frac{5}{10}$
$$6\left(\frac{1}{12}\right) = \frac{5}{10}$$
$$\frac{6}{12} = \frac{5}{10}$$
$$\frac{1}{2} = \frac{1}{2} \text{ True}$$

71.
$$\frac{2}{3}x = 6$$
$$\frac{3}{2}\cdot\frac{2}{3}x = \frac{3}{2}\cdot 6$$
$$x = 9$$
Check: $\frac{2}{3}x = 6$
$$\frac{2}{3}(9) = 6$$
$$6 = 6 \text{ True}$$

73. **a.** In $5 + x = 10$, 5 is added to the variable, whereas in $5x = 10$, 5 is multiplied by the variable.

b.
$$5x = 10$$
$$5 + x - 5 = 10 - 5$$
$$x = 5$$

c.
$$5x = 10$$
$$\frac{5x}{5} = \frac{10}{5}$$
$$x = 2$$

75. a. To get kisses by themselves, add –6 to both sides. Two kisses balance with 8 so one kiss is 4.

b. $2x+6=14$

c.
$$\begin{aligned} 2x+6&=14\\ 2x+6-6&=14-6\\ 2x&=8\\ \frac{2x}{2}&=\frac{8}{2}\\ x&=4 \end{aligned}$$

77. a. To get kisses by themselves, add –4 to both sides. Two kisses balance with 2 so each kiss is 1.

b. $6=2x+4$

c.
$$\begin{aligned} 6&=2x+4\\ 6-4&=2x+4-4\\ 2&=2x\\ \frac{2}{2}&=\frac{2x}{2}\\ 1&=x \text{ or } x=1 \end{aligned}$$

79. Multiplying by $\frac{3}{2}$ is easier because the equation involves fractions.
$$\begin{aligned} \frac{2}{3}x&=4\\ \left(\frac{3}{2}\right)\left(\frac{2}{3}\right)x&=\left(\frac{3}{2}\right)\left(\frac{4}{1}\right)\\ x&=\frac{12}{2}\\ x&=6 \end{aligned}$$

81. Multiplying by $\frac{7}{3}$ is easier because the equation involves fractions.
$$\begin{aligned} \frac{3}{7}x&=\frac{4}{5}\\ \left(\frac{7}{3}\right)\frac{3}{7}x&=\left(\frac{7}{3}\right)\frac{4}{5}\\ x&=\frac{28}{15} \end{aligned}$$

83. a. ⊡

b. Divide both sides of the equation by △.

c.
☺ = △⊡

☺/△ = △⊡/△

☺/△ = ⊡

85.
$$\begin{aligned} -8-(-4)&=-8+4\\ &=-4 \end{aligned}$$

86.
$$\begin{aligned} 6-(-3)-5-4&=6+3-5-4\\ &=9-5-4\\ &=4-4\\ &=0 \end{aligned}$$

87.
$$\begin{aligned} &-(x+3)-5(2x-7)+6\\ &=-x-3-10x+35+6\\ &=-x-10x-3+35+6\\ &=-11x+38 \end{aligned}$$

88.
$$\begin{aligned} -48&=x+9\\ -48-9&=x+9-9\\ -57&=x+0\\ -57&=x \end{aligned}$$

Exercise Set 2.4

1. No; the variable x is on both sides of the equal sign.

3. $x=\frac{5}{8}$ because $1x=x$.

5. $x=-\frac{1}{2}$ because $-x=-1x$.

7. $x=\frac{3}{5}$ because $-x=-1x$.

9. You solve an equation. An equation that contains a variable is true for certain values of that variable. We solve an equation to find those values.

11. a. Answers will vary.

b. Answers will vary.

13. a. Use the distributive property.
Subtract 8 from both sides of the equation.
Divide both sides of the equation by 6.

b.
$$\begin{aligned} 2(3x+4) &= -4 \\ 6x+8 &= -4 \\ 6x+8-8 &= -4-8 \\ 6x &= -12 \\ \frac{6x}{6} &= -\frac{12}{6} \\ x &= -2 \end{aligned}$$

15. a. $2x + 4 = 16$

b.
$$\begin{aligned} 2x+4-4 &= 16-4 \\ 2x &= 12 \\ \frac{2x}{2} &= \frac{12}{2} \\ x &= 6 \end{aligned}$$

17. a. $30 = 2x + 12$

b.
$$\begin{aligned} 30-12 &= 2x+12-12 \\ 18 &= 2x \\ \frac{18}{2} &= \frac{2x}{2} \\ 9 &= x \end{aligned}$$

19. a. $3x + 10 = 4$

b.
$$\begin{aligned} 3x+10-10 &= 4-10 \\ 3x &= -6 \\ \frac{3x}{3} &= \frac{-6}{3} \\ x &= -2 \end{aligned}$$

21. a. $5 + 3x = 12$

b.
$$\begin{aligned} 5-5+3x &= 12-5 \\ 3x &= 7 \\ \frac{3x}{3} &= \frac{7}{3} \\ x &= \frac{7}{3} \end{aligned}$$

23.
$$\begin{aligned} 2x+4 &= 10 \\ 2x+4-4 &= 10-4 \\ 2x &= 6 \\ \frac{2x}{2} &= \frac{6}{2} \\ x &= 3 \end{aligned}$$

25.
$$\begin{aligned} -2x-5 &= 7 \\ -2x-5+5 &= 7+5 \\ -2x &= 12 \\ \frac{-2x}{-2} &= \frac{12}{-2} \\ x &= -6 \end{aligned}$$

27.
$$\begin{aligned} 5x-6 &= 19 \\ 5x-6+6 &= 19+6 \\ 5x &= 25 \\ \frac{5x}{5} &= \frac{25}{5} \\ x &= 5 \end{aligned}$$

29.
$$\begin{aligned} 5x-2 &= 10 \\ 5x-2+2 &= 10+2 \\ 5x &= 12 \\ \frac{5x}{5} &= \frac{12}{5} \\ x &= \frac{12}{5} \end{aligned}$$

31.
$$\begin{aligned} -x+4 &= 15 \\ -x+4-4 &= 15-4 \\ -x &= 11 \\ (-1)(-x) &= (-1)(11) \\ x &= -11 \end{aligned}$$

33.
$$\begin{aligned} 12-x &= 9 \\ 12-12-x &= 9-12 \\ -x &= -3 \\ (-1)(-x) &= (-1)(-3) \\ x &= 3 \end{aligned}$$

35.
$$\begin{aligned} 8+3x &= 19 \\ 8-8+3x &= 19-8 \\ 3x &= 11 \\ \frac{3x}{3} &= \frac{11}{3} \\ x &= \frac{11}{3} \end{aligned}$$

37.
$$\begin{aligned} 16x+5&=-14 \\ 16x+5-5&=-14-5 \\ 16x&=-19 \\ \frac{16x}{16}&=\frac{-19}{16} \\ x&=-\frac{19}{16} \end{aligned}$$

39.
$$\begin{aligned} -4.2&=3x+25.8 \\ -4.2-25.8&=3x+25.8-25.8 \\ -30&=3x \\ \frac{-30}{3}&=\frac{3x}{3} \\ -10&=x \end{aligned}$$

41.
$$\begin{aligned} 6x-29&=7 \\ 6x-29+29&=7+29 \\ 6x&=36 \\ \frac{6x}{6}&=\frac{36}{6} \\ x&=6 \end{aligned}$$

43.
$$\begin{aligned} 56&=-6x+2 \\ 56-2&=-6x+2-2 \\ 54&=-6x \\ \frac{54}{-6}&=\frac{-6x}{-6} \\ -9&=x \end{aligned}$$

45.
$$\begin{aligned} -2x-7&=-13 \\ -2x-7+7&=-13+7 \\ -2x&=-6 \\ \frac{-2x}{-2}&=\frac{-6}{-2} \\ x&=3 \end{aligned}$$

47.
$$\begin{aligned} 2.3x-9.34&=6.3 \\ 2.3x-9.34+9.34&=6.3+9.34 \\ 2.3x&=15.64 \\ \frac{2.3x}{2.3}&=\frac{15.64}{2.3} \\ x&=6.8 \end{aligned}$$

49.
$$\begin{aligned} x+0.07x&=16.05 \\ 1.07x&=16.05 \\ \frac{1.07x}{1.07}&=\frac{16.05}{1.07} \\ x&=15 \end{aligned}$$

51.
$$\begin{aligned} 28.8&=x+1.40x \\ 28.8&=2.40x \\ \frac{28.8}{2.40}&=\frac{2.40x}{2.40} \\ 12&=x \end{aligned}$$

53.
$$\begin{aligned} 3(x+2)&=6 \\ 3x+6&=6 \\ 3x+6-6&=6-6 \\ 3x&=0 \\ \frac{3x}{3}&=\frac{0}{3} \\ x&=0 \end{aligned}$$

55.
$$\begin{aligned} 4(3-x)&=12 \\ 12-4x&=12 \\ 12-12-4x&=12-12 \\ -4x&=0 \\ \frac{-4x}{-4}&=\frac{0}{-4} \\ x&=0 \end{aligned}$$

57.
$$\begin{aligned} -4&=-(x+5) \\ -4&=-x-5 \\ -4+5&=-x-5+5 \\ 1&=-x \\ (-1)(1)&=(-1)(-x) \\ -1&=x \end{aligned}$$

59.
$$\begin{aligned} 12&=4(x-3) \\ 12&=4x-12 \\ 12+12&=4x-12+12 \\ 24&=4x \\ \frac{24}{4}&=\frac{4x}{4} \\ 6&=x \end{aligned}$$

61.
$$\begin{aligned} 22&=-(3x-4) \\ 22&=-3x+4 \\ 22-4&=-3x+4-4 \\ 18&=-3x \\ \frac{18}{-3}&=\frac{-3x}{-3} \\ -6&=x \end{aligned}$$

63. $2x+3(x+2)=11$
$$2x+3x+6=11$$
$$5x+6=11$$
$$5x+6-6=11-6$$
$$5x=5$$
$$\frac{5x}{5}=\frac{5}{5}$$
$$x=1$$

65. $x-3(2x+3)=11$
$$x-6x-9=11$$
$$-5x-9=11$$
$$-5x-9+9=11+9$$
$$-5x=20$$
$$\frac{-5x}{-5}=\frac{20}{-5}$$
$$x=-4$$

67. $5x+3x-4x-7=9$
$$4x-7=9$$
$$4x-7+7=9+7$$
$$4x=16$$
$$\frac{4x}{4}=\frac{16}{4}$$
$$x=4$$

69.
$$0.7(x-3)=1.4$$
$$0.7x-2.1=1.4$$
$$0.7x-2.1+2.1=1.4+2.1$$
$$0.7x=3.5$$
$$\frac{0.7x}{0.7}=\frac{3.5}{0.7}$$
$$x=5$$

71.
$$1.4(5x-4)=-1.4$$
$$7x-5.6=-1.4$$
$$7x-5.6+5.6=-1.4+5.6$$
$$7x=4.2$$
$$\frac{7x}{7}=\frac{4.2}{7}$$
$$x=0.6$$

73. $3-2(x+3)+2=1$
$$3-2x-6+2=1$$
$$-2x-1=1$$
$$-2x-1+1=1+1$$
$$-2x=2$$
$$\frac{-2x}{-2}=\frac{2}{-2}$$
$$x=-1$$

75. $1+(x+3)+6x=6$
$$1+x+3+6x=6$$
$$7x+4=6$$
$$7x+4-4=6-4$$
$$7x=2$$
$$\frac{7x}{7}=\frac{2}{7}$$
$$x=\frac{2}{7}$$

77.
$$4.85-6.4x+1.11=22.6$$
$$-6.4x+5.96=22.6$$
$$-6.4x+5.96-5.96=22.6-5.96$$
$$-6.4x=16.64$$
$$\frac{-6.4x}{-6.4}=\frac{16.64}{-6.4}$$
$$x=-2.6$$

79. a. By subtracting first, you will not have to work with fractions.

b.
$$3x+2=11$$
$$3x+2-2=11-2$$
$$3x=9$$
$$\frac{3x}{3}=\frac{9}{3}$$
$$x=3$$

81. a. $2x=x+3$

b.
$$2x=x+3$$
$$2x-x=x-x+3$$
$$x=3$$

83. a. $2x+3=4x+2$

b. $$\begin{aligned} 2x-2x+3 &= 4x-2x+2 \\ 3 &= 2x+2 \\ 3-2 &= 2x+2-2 \\ 1 &= 2x \\ \frac{1}{2} &= \frac{2x}{2} \\ \frac{1}{2} &= x \text{ or } x=\frac{1}{2} \end{aligned}$$

85. $$\begin{aligned} 3(x-2)-(x+5)-2(3-2x) &= 18 \\ 3x-6-x-5-6+4x &= 18 \\ 3x-x+4x-6-5-6 &= 18 \\ 6x-17 &= 18 \\ 6x-17+17 &= 18+17 \\ 6x &= 35 \\ \frac{6x}{6} &= \frac{35}{6} \\ x &= \frac{35}{6} \end{aligned}$$

87. $$\begin{aligned} 4[3-2(x+4)]-(x+3) &= 13 \\ 4(3-2x-8)-x-3 &= 13 \\ 4(-2x-5)-x-3 &= 13 \\ -8x-20-x-3 &= 13 \\ -9x-23 &= 13 \\ -9x-23+23 &= 13+23 \\ -9x &= 36 \\ \frac{-9x}{-9} &= \frac{36}{-9} \\ x &= -4 \end{aligned}$$

91. $\frac{5}{8}+\frac{3}{5}$

$\frac{5}{8}=\frac{5}{8}\cdot\frac{5}{5}=\frac{25}{40}$ and $\frac{3}{5}=\frac{3}{5}\cdot\frac{8}{8}=\frac{24}{40}$

$\frac{5}{8}+\frac{3}{5}=\frac{25}{40}+\frac{24}{40}=\frac{49}{40}$ or $1\frac{9}{40}$

92. $$\begin{aligned} \left[5(2-6)+3(8\div 4)^2\right]^2 &= \left[5(-4)+3(2)^2\right]^2 \\ &= [-20+3(4)]^2 \\ &= [-20+12]^2 \\ &= [-8]^2 \\ &= 64 \end{aligned}$$

93. To solve an equation, we need to isolate the variable on one side of the equation.

94. To solve the equation, we divide both sides of the equation by –4.

Exercise Set 2.5

1. Answers will vary.

3. a. An identity is an equation that is true for all real numbers.

b. The solution is all real numbers.

5. The equation is an identity because both sides of the equation are identical.

7. An equation has no solution if it simplifies to a false statement.

9. a. Use the distributive property.
Subtract $4x$ from both sides of the equation.
Add 30 to both sides of the equation.
Divide both sides of the equation by 2.

b. $$\begin{aligned} 4(x+3) &= 6(x-5) \\ 4x+12 &= 6x-30 \\ 12 &= 2x-30 \\ 42 &= 2x \\ 21 &= x \text{ or } x=21 \end{aligned}$$

11. a. $2x=x+6$

b. $$\begin{aligned} 2x-x &= x-x+6 \\ x &= 6 \end{aligned}$$

13. a. $5+2x=x+19$

b. $$\begin{aligned} 5+2x-x &= x-x+19 \\ 5+x &= 19 \\ 5-5+x &= 19-5 \\ x &= 14 \end{aligned}$$

15. a. $5+x=2x+5$

b. $$\begin{aligned} 5+x-x &= 2x-x+5 \\ 5 &= x+5 \\ 5-5 &= x+5-5 \\ 0 &= x \end{aligned}$$

17. a. $2x+8=x+4$

b. $$\begin{aligned}2x-x+8&=x-x+4\\x+8&=4\\x+8-8&=4-8\\x&=-4\end{aligned}$$

19. $$\begin{aligned}8x&=6x+30\\8x-6x&=6x-6x+30\\2x&=30\\\frac{2x}{2}&=\frac{30}{2}\\x&=15\end{aligned}$$

21. $$\begin{aligned}-4x+10&=6x\\-4x+4x+10&=6x+4x\\10&=10x\\\frac{10}{10}&=\frac{10x}{10}\\1&=x\end{aligned}$$

23. $$\begin{aligned}5x+3&=6\\5x+3-3&=6-3\\5x&=3\\\frac{5x}{5}&=\frac{3}{5}\\x&=\frac{3}{5}\end{aligned}$$

25. $$\begin{aligned}15-5x&=3x-2x\\15-5x&=x\\15-5x+5x&=x+5x\\15&=6x\\\frac{15}{6}&=\frac{6x}{6}\\\frac{5}{2}&=x\end{aligned}$$

27. $$\begin{aligned}2x-4&=3x-6\\2x-2x-4&=3x-2x-6\\-4&=x-6\\-4+6&=x-6+6\\2&=x\end{aligned}$$

29. $$\begin{aligned}3-2y&=9-8y\\3-2y+8y&=9-8y+8y\\3+6y&=9\\3-3+6y&=9-3\\6y&=6\\\frac{6y}{6}&=\frac{6}{6}\\y&=1\end{aligned}$$

31. $$\begin{aligned}9-0.5x&=4.5x+8.50\\9-0.5x+0.5x&=4.5x+0.5x+8.50\\9&=5x+8.50\\9-8.50&=5x+8.50-8.50\\0.5&=5x\\\frac{0.5}{5}&=\frac{5x}{5}\\0.1&=x\end{aligned}$$

33. $$\begin{aligned}5x+3&=2(x+6)\\5x+3&=2x+12\\5x-2x+3&=2x-2x+12\\3x+3&=12\\3x+3-3&=12-3\\3x&=9\\\frac{3x}{3}&=\frac{9}{3}\\x&=3\end{aligned}$$

35. $$\begin{aligned}x-25&=12x+9+3x\\x-25&=15x+9\\x-x-25&=15x-x+9\\-25&=14x+9\\-25-9&=14x+9-9\\-34&=14x\\\frac{-34}{14}&=\frac{14x}{14}\\-\frac{17}{7}&=x\end{aligned}$$

37. $$\begin{aligned}2(x-2)&=4x-6-2x\\2x-4&=2x-6\\2x-2x-4&=2x-2x-6\\-4&=-6 \text{ False}\end{aligned}$$

Since a false statement is obtain, there is no solution.

39.
$$\begin{aligned}-(w+2)&=-6w+32\\ -w-2&=-6w+32\\ -w+w-2&=-6w+w+32\\ -2&=-5w+32\\ -2-32&=-5w+32-32\\ -34&=-5w\\ \frac{-34}{-5}&=\frac{-5w}{-5}\\ \frac{34}{5}&=w\end{aligned}$$

41.
$$\begin{aligned}4-(2x-5)&=3x+13\\ 4-2x+5&=3x+13\\ -2x+9&=3x+13\\ -2x+2x+9&=3x+2x+13\\ 9&=5x+13\\ 9-13&=5x+13-13\\ -4&=5x\\ \frac{-4}{5}&=\frac{5x}{5}\\ -\frac{4}{5}&=x\end{aligned}$$

43.
$$\begin{aligned}0.1(x+10)&=0.3x-4\\ 0.1x+1&=0.3x-4\\ 0.1x-0.1x+1&=0.3x-0.1x-4\\ 1+4&=0.2x-4+4\\ 5&=0.2x\\ \frac{5}{0.2}&=\frac{0.2x}{0.2}\\ 25&=x\end{aligned}$$

45.
$$\begin{aligned}2(x+4)&=4x+3-2x+5\\ 2x+8&=4x+3-2x+5\\ 2x+8&=2x+8\end{aligned}$$
Since the left side of the equation is identical to the right side, the equation is true for all values of x. Thus the solution is all real numbers.

47.
$$\begin{aligned}9(-y-3)&=6y-15+3y+6\\ -9y-27&=9y-9\\ -9y+9y-27y&=9y+9y-9\\ -27&=18y-9\\ -27+9&=18y-9+9\\ -18&=18y\\ \frac{-18}{18}&=\frac{18y}{18}\\ -1&=y\end{aligned}$$

49.
$$\begin{aligned}-(3-p)&=-(2p+3)\\ -3+p&=-2p-3\\ -3+p+2p&=-2p+2p-3\\ -3+3p&=-3\\ -3+3+3p&=-3+3\\ 3p&=0\\ \frac{3p}{3}&=\frac{0}{3}\\ p&=0\end{aligned}$$

51.
$$\begin{aligned}-(x+4)+5&=4x+1-5x\\ -x-4+5&=4x+1-5x\\ -x+1&=-x+1\end{aligned}$$
Since the left side of the equation is identical to the right side, the equation is true for all values of x. Thus the solution is all real numbers.

53.
$$\begin{aligned}35(2x-1)&=7(x+4)+3x\\ 70x-35&=7x+28+3x\\ 70x-35&=10x+28\\ 70x-10x-35&=10x-10x+28\\ 60x-35&=28\\ 60x-35+35&=28+35\\ 60x&=63\\ \frac{60x}{60}&=\frac{63}{60}\\ x&=\frac{21}{20}\end{aligned}$$

55.
$$\begin{aligned} 0.4(x+0.7) &= 0.6(x-4.2) \\ 0.4x+0.28 &= 0.6x-2.52 \\ 0.4x-0.4x+0.28 &= 0.6x-0.4x-2.52 \\ 0.28 &= 0.2x-2.52 \\ 0.28+2.52 &= 0.2x-2.52+2.52 \\ 2.8 &= 0.2x \\ \frac{2.8}{0.2} &= \frac{0.2x}{0.2} \\ 14 &= x \end{aligned}$$

57.
$$\begin{aligned} -(x-5)+2 &= 3(4-x)+5x \\ -x+5+2 &= 12-3x+5x \\ -x+7 &= 12+2x \\ -x+x+7 &= 12+2x+x \\ 7 &= 12+3x \\ 7-12 &= 12-12+3x \\ -5 &= 3x \\ \frac{-5}{3} &= \frac{3x}{3} \\ -\frac{5}{3} &= x \end{aligned}$$

59.
$$\begin{aligned} 2(x-6)+(x+1) &= x-11 \\ 2x-12+x+1 &= x-11 \\ 3x-11 &= x-11 \\ 3x-x-11 &= x-x-11 \\ 2x-11 &= -11 \\ 2x-11+11 &= -11+11 \\ 2x &= 0 \\ \frac{2x}{2} &= \frac{0}{2} \\ x &= 0 \end{aligned}$$

61.
$$\begin{aligned} 5+2x &= 6(x+1)-5(x-3) \\ 5+2x &= 6x+6-5x+15 \\ 5+2x &= x+21 \\ 5+2x-x &= x-x+21 \\ 5+x &= 21 \\ 5-5+x &= 21-5 \\ x &= 16 \end{aligned}$$

63.
$$\begin{aligned} 5-(x-5) &= 2(x+3)-6(x+1) \\ 5-x+5 &= 2x+6-6x-6 \\ -x+10 &= -4x \\ -x+x+10 &= -4x+x \\ 10 &= -3x \\ \frac{10}{-3} &= \frac{-3x}{-3} \\ -\frac{10}{3} &= x \end{aligned}$$

65. **a.** One example is $x+x+1=x+2$.

b. It has a single solution.

c. Answers will vary. For equation given in part **a**):
$$\begin{aligned} x+x+1 &= x+2 \\ 2x+1 &= x+2 \\ 2x-x+1 &= x-x+2 \\ x+1 &= 2 \\ x+1-1 &= 2-1 \\ x &= 1 \end{aligned}$$

67. **a.** One example is $x+x+1=2x+1$.

b. Both sides simplify to the same expression.

c. The solution is all real numbers.

69. **a.** One example is $x+x+1=2x+2$.

b. It simplifies to a false statement.

c. The solution is that there is no solution.

71.
$$\begin{aligned} 5*-1 &= 4*+5 \\ 5*-4*-1 &= 4*-4*+5 \\ *-1 &= 5 \\ *-1+1 &= 5+1 \\ * &= 6 \end{aligned}$$

73. 3☺ − 5 = 2☺ − 5 + ☺
3☺ − 5 = 3☺ − 5
The left side of the equation is identical to the right side. The solution is all real numbers.

75.
$$\begin{aligned} 4-[5-3(x+2)] &= x-3 \\ 4-(5-3x-6) &= x-3 \\ 4-5+3x+6 &= x-3 \\ 3x+5 &= x-3 \\ 3x-x+5 &= x-x-3 \\ 2x+5 &= -3 \\ 2x+5-5 &= -3-5 \\ 2x &= -8 \\ \frac{2x}{2} &= \frac{-8}{2} \\ x &= -4 \end{aligned}$$

79. $\left(\frac{2}{3}\right)^5 \approx 0.131687243$

80. Factors are expressions that are multiplied together; terms are expressions that are added together.

81.
$$\begin{aligned} 2(x-3)+4x-(4-x) &= 2x-6+4x-4+x \\ &= 7x-10 \end{aligned}$$

82.
$$\begin{aligned} 2(x-3)+4x-(4-x) &= 0 \\ 2x-6+4x-4+x &= 0 \\ 7x-10 &= 0 \\ 7x-10+10 &= 0+10 \\ 7x &= 10 \\ \frac{7x}{7} &= \frac{10}{7} \\ x &= \frac{10}{7} \end{aligned}$$

83.
$$\begin{aligned} (x+4)-(4x-3) &= 16 \\ x+4-4x+3 &= 16 \\ -3x+7 &= 16 \\ -3x+7-7 &= 16-7 \\ -3x &= 9 \\ \frac{-3x}{-3} &= \frac{9}{-3} \\ x &= -3 \end{aligned}$$

Exercise Set 2.6

1. A ratio is a quotient of two quantities.

3. The ratio of c to d can be written as c to d, c:d, and $\frac{c}{d}$.

5. To set up and solve a proportion, we need a given ratio and one of the two parts of a second ratio.q

7. Yes; The terms in each ratio are in the same order.

9. No; The terms in each ratio are not in the same order.

11. No, similar figures have the same shape but not necessarily the same size.

13. 6:9 = 2:3

15. 3:2

17. Total grades $= 6+4+9+3+2 = 24$
Ratio of total grades to D's = 24:3 = 8:1

19. 7:4

21. 16:24 = 2:3

23. 3 hours $= 3\times 60 = 180$ minutes
Ratio is $\frac{180}{30} = \frac{6}{1}$ or 6:1.

25. 4 pounds is $4\times 16 = 64$ ounces
Ratio is $\frac{26}{64} = \frac{13}{32}$ or 13:32.

27. Gear ratio
$$= \frac{\text{number of teeth on driving gear}}{\text{number of teeth on driven gear}}$$
$$= \frac{40}{5} = \frac{8}{1}$$
Gear ratio is 8:1.

29. a. 430:320 = 43:32

b. Since $43 \div 32 \approx 1.34$, $43{:}32 \approx 1.34{:}1$.

31. **a.** 1,001,000:798,000 = 143:114

b. Since $143 \div 114 \approx 1.25$, $143:114 \approx 1.25:1$.

33. **a.** 40:9

b. 33:53

35. **a.** 63:13

b. 57:13

37.
$$\frac{4}{x} = \frac{5}{20}$$
$$4 \cdot 20 = x \cdot 5$$
$$80 = 5x$$
$$\frac{80}{5} = x$$
$$16 = x$$

39.
$$\frac{5}{3} = \frac{75}{x}$$
$$5 \cdot x = 3 \cdot 75$$
$$5x = 225$$
$$x = \frac{225}{5} = 45$$

41.
$$\frac{90}{x} = \frac{-9}{10}$$
$$90 \cdot 10 = x(-9)$$
$$900 = -9x$$
$$\frac{900}{-9} = x$$
$$-100 = x$$

43.
$$\frac{15}{45} = \frac{x}{-6}$$
$$45 \cdot x = -6 \cdot 15$$
$$45x = -90$$
$$x = \frac{-90}{45} = -2$$

45.
$$\frac{3}{z} = \frac{-1.5}{27}$$
$$-1.5 \cdot z = 3 \cdot 27$$
$$-1.5z = 81$$
$$z = \frac{81}{-1.5} = -54$$

47.
$$\frac{15}{20} = \frac{x}{8}$$
$$15 \cdot 8 = 20 \cdot x$$
$$120 = 20x$$
$$\frac{120}{20} = x$$
$$6 = x$$

49.
$$\frac{3}{12} = \frac{8}{x}$$
$$3x = (8)(12)$$
$$3x = 96$$
$$x = \frac{96}{3} = 32$$
Thus the side is 32 inches in length.

51.
$$\frac{2}{1.8} = \frac{0.8}{x}$$
$$2x = (1.8)(0.8)$$
$$2x = 1.44$$
$$x = \frac{1.44}{2} = 0.72$$
Thus the side is 0.72 feet in length.

53.
$$\frac{16}{12} = \frac{26}{x}$$
$$16x = (12)(26)$$
$$16x = 312$$
$$x = \frac{312}{16} = 19.5$$
Thus the side is 19.5 inches in length.

55. Let x = number of loads one bottle can do.

$$\frac{4 \text{ fl ounces}}{1 \text{ load}} = \frac{100 \text{ fl ounces}}{x \text{ loads}}$$
$$\frac{4}{1} = \frac{100}{x}$$
$$4x = 100$$
$$x = \frac{100}{4} = 25$$

One bottle can do 25 loads.

57. Let x = number of miles that can be driven with a full tank.

$$\frac{23 \text{ miles}}{1 \text{ gallon}} = \frac{x}{15.7 \text{ gallons}}$$
$$\frac{23}{1} = \frac{x}{15.7}$$
$$x = 23 \cdot 15.7$$
$$x = 361.1$$

It can travel 361.1 miles on a full tank.

59. Let x = width of corridor on blueprint in feet.

$$\frac{1 \text{ foot}}{120 \text{ feet}} = \frac{x \text{ feet}}{90 \text{ feet}}$$
$$\frac{1}{120} = \frac{x}{90}$$
$$120x = 90$$
$$x = \frac{90}{120} = \frac{3}{4}$$

To convert to inches, multiply by 12 or $\frac{12}{1}$.

$$\frac{3}{4} \cdot \frac{12}{1} = \frac{3}{1} \cdot \frac{3}{1} = \frac{9}{1} = 9$$

The width on the blueprint is $\frac{3}{4}$ feet or 9 inches.

61. Let x = number of teaspoons needed for sprayer.

$$\frac{3 \text{ teaspoons}}{1 \text{ gallon water}} = \frac{x \text{ teaspoons}}{8 \text{ gallons water}}$$
$$\frac{3}{1} = \frac{x}{8}$$
$$3 \cdot 8 = 1 \cdot x$$
$$24 = x$$

Thus 24 teaspoons are needed for the sprayer.

63. Let x = length of monument.

$$\frac{0.875 \text{ inch}}{64 \text{ inches}} = \frac{3.117 \text{ inches}}{x \text{ inches}}$$
$$\frac{0.875}{64} = \frac{3.117}{x}$$
$$0.875x = 64 \cdot 3.117$$
$$0.875x = 199.488$$
$$\frac{0.875x}{0.875} = \frac{199.488}{0.875}$$
$$x \approx 228$$

The monument is about 228 inches tall which is about 19 feet.

65. Let x = number of trees saved.

$$\frac{17 \text{ trees}}{1 \text{ ton recycled paper}} = \frac{x \text{ trees}}{20 \text{ tons recycled paper}}$$
$$\frac{17}{1} = \frac{x}{20}$$
$$17 \cdot 20 = 1 \cdot x$$
$$340 = x$$

Thus 340 trees have been saved.

67. Let x = length of the model in feet.

$$\frac{1}{3} = \frac{x \text{ feet}}{26.24 \text{ feet}}$$
$$\frac{1}{3} = \frac{x}{26.24}$$
$$3x = 26.24$$
$$x = \frac{26.24}{3} \approx 8.75$$

The model was about 8.75 feet long.

69. Let x = number of milliliters to be given.

$$\frac{1 \text{ milliliter}}{400 \text{ micrograms}} = \frac{x \text{ milliliter}}{220 \text{ micrograms}}$$
$$\frac{1}{400} = \frac{x}{220}$$
$$1 \cdot 220 = 400 \cdot x$$
$$220 = 400x$$
$$\frac{220}{400} = x$$
$$0.55 = x$$

Thus 0.55 milliliter should be given.

71. Let x = number of seconds to go from 0 to 800.

$$\frac{30 \text{ seconds}}{250 \text{ counts on VCR}} = \frac{x \text{ seconds}}{800 \text{ counts on VCR}}$$
$$\frac{30}{250} = \frac{x}{800}$$
$$30 \cdot 800 = x \cdot 250$$
$$24,000 = 250x$$
$$\frac{24,000}{250} = x$$
$$96 = x$$

Thus you should fast forward for 96 sec or 1 min 36 sec.

73. Let x = number of children born with Prader-Willi Syndrome.

$$\frac{15,000 \text{ births}}{1 \text{ baby with syndrome}} = \frac{4,179,200 \text{ births}}{x \text{ babies with syndrome}}$$
$$\frac{15,000}{1} = \frac{4,179,200}{x}$$
$$15,000x = 4,179,200$$
$$x = \frac{4,179,200}{15,000} = 278.6$$

Thus, about 279 children were born with Prader-Willi Syndrome.

75. $\frac{12 \text{ inches}}{1 \text{ foot}} = \frac{42 \text{ inches}}{x \text{ feet}}$

$$\frac{12}{1} = \frac{42}{x}$$
$$12x = 42$$
$$x = \frac{42}{12} = 3.5$$

Thus 42 inches equals 3.5 feet.

77. $\frac{9 \text{ square feet}}{1 \text{ square yard}} = \frac{26.1 \text{ square feet}}{x \text{ square yards}}$

$$\frac{9}{1} = \frac{26.1}{x}$$
$$9x = 26.1$$
$$x = \frac{26.1}{9} = 2.9$$

Thus 26.1 square feet equals 2.9 square yards.

79. $\frac{2.54 \text{ cm}}{1 \text{ inch}} = \frac{26.67 \text{ cm}}{x \text{ inches}}$

$$\frac{2.54}{1} = \frac{26.67}{x}$$
$$2.54x = 26.67$$
$$x = \frac{26.67}{2.54} = 10.5$$

Thus the length of the book is 10.5 inches.

81. a. $\frac{1 \text{ kilogram}}{\$3.00} = \frac{3.5 \text{ kilograms}}{x \text{ dollars}}$

$$\frac{1}{3} = \frac{3.5}{x}$$
$$x = 3 \cdot 3.5$$
$$x = 10.5$$

Thus, the cost of 3.5 kilograms of Mung Dhau is \$10.50.

b. $\frac{2.2 \text{ pounds}}{\$3.00} = \frac{1 \text{ pound}}{x \text{ dollars}}$

$$\frac{2.2}{3} = \frac{1}{x}$$
$$2.2x = 3$$
$$x = \frac{3}{2.2}$$
$$x \approx 1.36$$

Thus, the cost of 1 pound of Mung Dhau is \$1.36.

83. $\frac{480 \text{ grains}}{408 \text{ dollars}} = \frac{1 \text{ grain}}{x \text{ dollars}}$

$$\frac{480}{408} = \frac{1}{x}$$
$$480x = 408$$
$$x = \frac{408}{480} = 0.85$$

Thus the cost per grain is \$0.85.

85. $\frac{3.75 \text{ standard deviations}}{15 \text{ points}} = \frac{1 \text{ standard deviation}}{x \text{ points}}$

$$\frac{3.75}{15} = \frac{1}{x}$$
$$3.75x = 15$$
$$x = \frac{15}{3.75} = 4$$

Thus 1 standard deviation equals 4 points.

87. The ratio of Mrs. Ruff's low density to high density cholesterol is $\frac{127}{60}$. If we divide 127 by 60 we obtain approximately 2.12. Thus Mrs. Ruff's ratio is approximately equivalent to 2.12:1. Therefore her ratio is less than the desired 4:1 ratio.

91. Let x = number of miles remaining on the life of each tire.
Inches remaining on the life of each tire:

$$0.31 - 0.06 = 0.25$$
$$\frac{0.03 \text{ inches}}{5000 \text{ miles}} = \frac{0.25 \text{ miles}}{x \text{ miles}}$$
$$\frac{0.03}{5000} = \frac{0.25}{x}$$
$$0.03x = 5000 \cdot 0.25$$
$$0.03x = 1250$$
$$x = \frac{1250}{0.03}$$
$$x \approx 41{,}667$$

The tires will last about 41,667 more miles.

93. Dodge Viper: P335/35R17
35% of 335 = 0.35(335) = 117.25
The height of a Dodge Viper tire is 117.25 mm.
Cadillac Seville: P235/60R16
60% of 235 = 0.6(235) = 141
The height of a Cadillac Seville tire is 141 mm.

98. Commutative property of addition

99. Associative property of multiplication

100. Distributive property

101.

$$-(2x+6) = 2(3x-6)$$
$$-2x-6 = 6x-12$$
$$-2x+2x-6 = 6x+2x-12$$
$$-6 = 8x-12$$
$$-6+12 = 8x-12+12$$
$$6 = 8x$$
$$\frac{6}{8} = \frac{8x}{8}$$
$$\frac{3}{4} = x$$

Exercise Set 2.7

1. >: is greater than;
   ≥: is greater than or equal to;
   <: is less than;
   ≤: is less than or equal to

3. a. $3 > 3$ is false because 3 is not greater than 3.

   b. $3 \geq 3$ is true because 3 is greater than or equal to 3.

5. The direction of the inequality symbol is changed when multiplying or dividing by a negative number.

7. Since 3 is always less than 5, the solution is all real numbers.

9. Since 5 is never less than 2, the answer is no solution.

11. $$\begin{aligned} x+2 &> 6 \\ x+2-2 &> 6-2 \\ x &> 4 \end{aligned}$$
    4

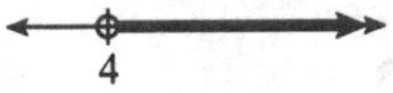

13. $$\begin{aligned} x+7 &\geq 4 \\ x+7-7 &\geq 4-7 \\ x &\geq -3 \end{aligned}$$
    −3

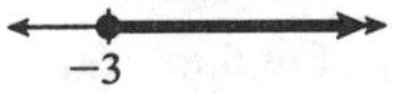

15. $$\begin{aligned} -x+3 &< 8 \\ -x+3-3 &< 8-3 \\ -x &< 5 \\ (-1)(-x) &> (-1)(5) \\ x &> -5 \end{aligned}$$
    −5

17. $$\begin{aligned} 6 &> x-4 \\ 6+4 &> x-4+4 \\ 10 &> x \\ x &< 10 \end{aligned}$$
    10

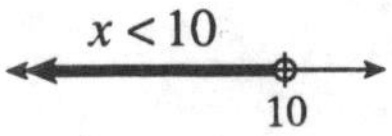

19. $$\begin{aligned} 9 &\leq 2-x \\ 9-2 &\leq 2-x-2 \\ 7 &\leq -x \\ (-1)(7) &\geq (-1)(-x) \\ -7 &\geq x \\ x &\leq -7 \end{aligned}$$
    −7

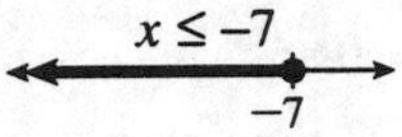

21. $$\begin{aligned} -2x &< 3 \\ \frac{-2x}{-2} &> \frac{3}{-2} \\ x &> -\frac{3}{2} \end{aligned}$$
    $-\frac{3}{2}$

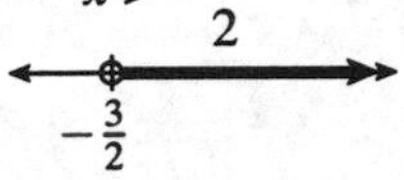

23. $$\begin{aligned} 2x+3 &\leq 5 \\ 2x+3-3 &\leq 5-3 \\ 2x &\leq 2 \\ \frac{2x}{2} &\leq \frac{2}{2} \\ x &\leq 1 \end{aligned}$$
    1

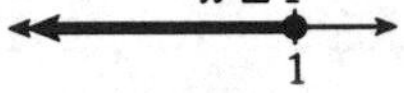

25. $$\begin{aligned} 12x-12 &< -12 \\ 12x-12+12 &< -12+12 \\ 12x &< 0 \\ \frac{12x}{12} &< \frac{0}{12} \\ x &< 0 \end{aligned}$$
    0

27. $$\begin{aligned} 4-6x &> -5 \\ 4-4-6x &> -5-4 \\ -6x &> -9 \\ \frac{-6x}{-6} &< \frac{-9}{-6} \\ x &< \frac{3}{2} \end{aligned}$$
    $\frac{3}{2}$

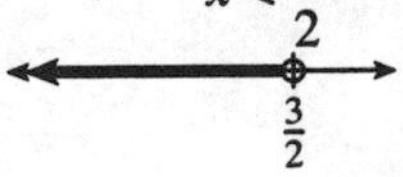

29.
$$15 > -9x + 50$$
$$9x + 15 > -9x + 9x + 50$$
$$9x + 15 > 50$$
$$9x + 15 - 15 > 50 - 15$$
$$9x > 35$$
$$\frac{9x}{9} > \frac{35}{9}$$
$$x > \frac{35}{9}$$

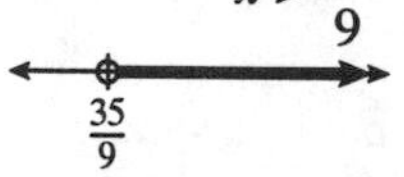

31.
$$4 < 3x + 12$$
$$4 - 12 < 3x + 12 - 12$$
$$-8 < 3x$$
$$\frac{-8}{3} < \frac{3x}{3}$$
$$\frac{-8}{3} < x$$
$$x > -\frac{8}{3}$$

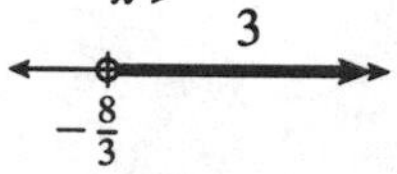

33.
$$6x + 2 \le 3x - 9$$
$$6x + 2 - 2 \le 3x - 9 - 2$$
$$6x \le 3x - 11$$
$$6x - 3x \le 3x - 3x - 11$$
$$3x \le -11$$
$$\frac{3x}{3} \le \frac{-11}{3}$$
$$x \le -\frac{11}{3}$$

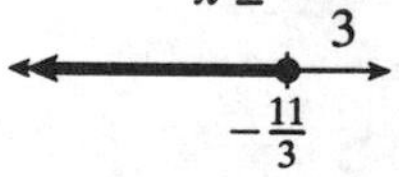

35.
$$x - 4 \le 3x + 8$$
$$x - 4 - 8 \le 3x + 8 - 8$$
$$x - 12 \le 3x$$
$$x - x - 12 \le 3x - x$$
$$-12 \le 2x$$
$$\frac{-12}{2} \le \frac{2x}{2}$$
$$-6 \le x$$
$$x \ge -6$$

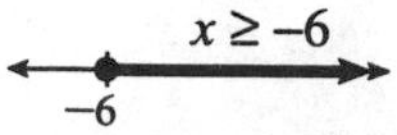

37.
$$-x + 4 < -3x + 6$$
$$-x + 4 - 4 \le -3x + 6 - 4$$
$$-x < -3x + 2$$
$$-x + 3x < -3x + 3x + 2$$
$$2x < 2$$
$$\frac{2x}{2} < \frac{2}{2}$$
$$x < 1$$

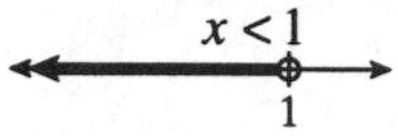

39.
$$-3(2x - 4) > 2(6x - 12)$$
$$-6x + 12 > 12x - 24$$
$$-6x + 12 + 24 > 12x - 24 + 24$$
$$-6x + 36 > 12x$$
$$-6x + 6x + 36 > 12x + 6x$$
$$36 > 18x$$
$$\frac{36}{18} > \frac{18x}{18}$$
$$2 > x$$
$$x < 2$$

41.
$$x + 3 < x + 4$$
$$x - x + 3 < x - x + 4$$
$$3 < 4$$

Since 3 is always less than 4, the solution is all real numbers.

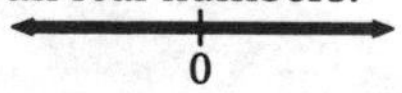

43.
$$6(3 - x) < 2x + 12$$
$$18 - 6x < 2x + 12$$
$$18 - 12 - 6x < 2x + 12 - 12$$
$$6 - 6x < 2x$$
$$6 - 6x + 6x < 2x + 6x$$
$$6 < 8x$$
$$\frac{6}{8} < \frac{8x}{8}$$
$$\frac{3}{4} < x$$
$$x > \frac{3}{4}$$

45. $$\begin{aligned} -21(2-x)+3x &> 4x+4 \\ -42+21x+3x &> 4x+4 \\ -42+24x &> 4x+4 \\ -42+42+24x &> 4x+4+42 \\ 24x &> 4x+46 \\ 24x-4x &> 4x-4x+46 \\ 20x &> 46 \\ \frac{20x}{20} &> \frac{46}{20} \\ x &> \frac{23}{10} \end{aligned}$$

$\frac{23}{10}$

47. $$\begin{aligned} 4x-4 &< 4(x-5) \\ 4x-4 &< 4x-20 \\ 4x-4x-4 &< 4x-4x-20 \\ -4 &< -20 \end{aligned}$$

Since –4 is never less than –20, there is no solution.

0

49. $$\begin{aligned} 5(2x+3) &\geq 6+(x+2)-2x \\ 10x+15 &\geq 6+x+2-2x \\ 10x+15 &\geq 8-x \\ 10x+15-15 &\geq 8-15-x \\ 10x &\geq -7-x \\ 10x+x &\geq -7-x+x \\ 11x &\geq -7 \\ \frac{11x}{11} &\geq \frac{-7}{11} \\ x &\geq -\frac{7}{11} \end{aligned}$$

$-\frac{7}{11}$

51. **a.** The average high temperature was greater than 65°F in May, June, July, August, and September.

b. The average high temperature was less than or equal to 59°F in January, February, March, April, November, and December.

c. The average low temperature was less than 29°F in January, February, and December.

d. The average low temperature was greater than or equal to 58°F in June, July, and August.

53. $\neq$

55. We cannot divide both sides of an inequality by *y* because we do not know that *y* is positive. If *y* is negative, we must reverse the sign of the inequality.

57. $$\begin{aligned} 6x-6 &> -4(x+3)+5(x+6)-x \\ 6x-6 &> -4x-12+5x+30-x \\ 6x-6 &> 18 \\ 6x-6+6 &> 18+6 \\ 6x &> 24 \\ \frac{6x}{6} &> \frac{24}{6} \\ x &> 4 \end{aligned}$$

58. Substitute 3 for *x*.

$$\begin{aligned} -x^2 &= -(3)^2 \\ &= -(3)(3) \\ &= -9 \end{aligned}$$

59. Substitute –5 for *x*.

$$\begin{aligned} -x^2 &= -(-5)^2 \\ &= -(-5)(-5) \\ &= -25 \end{aligned}$$

60.
$$\begin{aligned} 4-3(2x-4) &= 5-(x+3) \\ 4-6x+12 &= 5-x-3 \\ -6x+16 &= 2-x \\ -6x+6x+16 &= 2-x+6x \\ 16 &= 2+5x \\ 16-2 &= 2-2+5x \\ 14 &= 5x \\ \frac{14}{5} &= \frac{5x}{5} \\ \frac{14}{5} &= x \\ x &= \frac{14}{5} \text{ or } 2\frac{4}{5} \end{aligned}$$

61. Let x = number of kilowatt-hours of electricity used.

$$\frac{\$0.174}{1 \text{ kilowatt - hour}} = \frac{\$87}{x \text{ kilowatt - hours}}$$
$$\frac{0.174}{1} = \frac{87}{x}$$
$$0.174x = 87$$
$$x = \frac{87}{0.174} = 500$$

Thus the Cisneros used 500 kilowatt-hours of electricity in July.

Review Exercises

1. $3(x+4) = 3x+3(4)$
$= 3x+12$

2. $3(x-2) = 3[x+(-2)]$
$= 3x+3(-2)$
$= 3x+(-6)$
$= 3x-6$

3. $-2(x+4) = -2x+(-2)(4)$
$= -2x+(-8)$
$= -2x-8$

4. $-(x+2) = -1(x+2)$
$= (-1)(x)+(-1)(2)$
$= -x+(-2)$
$= -x-2$

5. $-(x-2) = -1(x-2)$
$= -1[x+(-2)]$
$= (-1)(x)+(-1)(-2)$
$= -x+2$

6. $-4(4-x) = -4[4+(-x)]$
$= (-4)(4)+(-4)(-x)$
$= -16+4x$

7. $3(6-2x) = 3[6+(-2x)]$
$= 3(6)+3(-2x)$
$= 18+(-6x)$
$= 18-6x$

8. $6(4x-5) = 6(4x)-6(5)$
$= 24x-30$

9. $-5(5x-5) = -5[5x+(-5)]$
$= -5(5x)+(-5)(-5)$
$= -25x+25$

10. $4(-x+3) = 4(-x)+4(3)$
$= -4x+12$

11. $-2(3x-2) = -2[3x+(-2)]$
$= -2(3x)+(-2)(-2)$
$= -6x+4$

12. $-(3+2y) = -1(3+2y)$
$= (-1)(3)+(-1)(2y)$
$= -3+(-2y)$
$= -3-2y$

13. $-(x+2y-z) = -1[x+2y+(-z)]$
$= -1(x)+(-1)(2y)+(-1)(-z)$
$= -x+(-2y)+z$
$= -x-2y+z$

14. $-2(2x-3y+7)$
$= -2[2x+(-3y)+7]$
$= -2(2x)+(-2)(-3y)+(-2)(7)$
$= -4x+6y+(-14)$
$= -4x+6y-14$

15. $5x+3x=8x$

16. $2y+3y+2=5y+2$

17. $5-3y+3=-3y+5+3$
$=-3y+8$

18. $1+3x+2x=1+5x$
$=5x+1$

19. $-2x-x+3y=-3x+3y$

20. $2x+3y+4x+5y=2x+4x+3y+5y$
$=6x+8y$

21. There are no like terms.
$9x+3y+2$ cannot be further simplified.

22. $6x-2x+3y+6=4x+3y+6$

23. $x+8x-9x+3=9x-9x+3$
$=3$

24. $-4x-8x+3=-12x+3$

25. $-2(x+3)+6=-2x-6+6$
$=-2x$

26. $2x+3(x+4)-5=2x+3x+12-5$
$=5x+7$

27. $4(3-2x)-2x=12-8x-2x$
$=-8x-2x+12$
$=-10x+12$

28. $6-(-x+6)-x=6+x-6-x$
$=x-x+6-6$
$=0$

29. $2(2x+5)-10-4=4x+10-10-4$
$=4x-4$

30. $-6(4-3x)-18+4x=-24+18x-18+4x$
$=18x+4x-24-18$
$=22x-42$

31. $6-3(x+y)+6x=6-3x-3y+6x$
$=-3x+6x-3y+6$
$=3x-3y+6$

32. $3x-6y+2(4y+8)=3x-6y+8y+16$
$=3x+2y+16$

33. $3-(x-y)+(x-y)=3-x+y+x-y$
$=-x+x+y-y+3$
$=3$

34. $(x+y)-(2x+3y)+4=x+y-2x-3y+4$
$=x-2x+y-3y+4$
$=-x-2y+4$

35. $4x=4$
$\frac{4x}{4}=\frac{4}{4}$
$x=1$

36. $x+6=-7$
$x+6-6=-7-6$
$x=-13$

37. $x-4=7$
$x-4+4=7+4$
$x=11$

38. $\frac{x}{3}=-9$
$3\left(\frac{x}{3}\right)=3(-9)$
$x=-27$

39. $2x+4=8$
$2x+4-4=8-4$
$2x=4$
$\frac{2x}{2}=\frac{4}{2}$
$x=2$

40.
$$\begin{aligned} 14 &= 3+2x \\ 14-3 &= 3-3+2x \\ 11 &= 2x \\ \frac{11}{2} &= \frac{2x}{2} \\ \frac{11}{2} &= x \end{aligned}$$

41.
$$\begin{aligned} 8x-3 &= -19 \\ 8x-3+3 &= -19+3 \\ 8x &= -16 \\ \frac{8x}{8} &= \frac{-16}{8} \\ x &= -2 \end{aligned}$$

42.
$$\begin{aligned} 6-x &= 9 \\ 6-x+x &= 9+x \\ 6 &= 9+x \\ 6-9 &= 9-9+x \\ -3 &= x \end{aligned}$$

43.
$$\begin{aligned} -x &= -12 \\ -1x &= -12 \\ (-1)(-1x) &= (-1)(-12) \\ 1x &= 12 \\ x &= 12 \end{aligned}$$

44.
$$\begin{aligned} 3(x-2) &= 6 \\ 3x-6 &= 6 \\ 3x-6+6 &= 6+6 \\ 3x &= 12 \\ \frac{3x}{3} &= \frac{12}{3} \\ x &= 4 \end{aligned}$$

45.
$$\begin{aligned} -3(2x-8) &= -12 \\ -6x+24 &= -12 \\ -6x+24-24 &= -12-24 \\ -6x &= -36 \\ \frac{-6x}{-6} &= \frac{-36}{-6} \\ x &= 6 \end{aligned}$$

46.
$$\begin{aligned} 4(6+2x) &= 0 \\ 24+8x &= 0 \\ 24-24+8x &= 0-24 \\ 8x &= -24 \\ \frac{8x}{8} &= \frac{-24}{8} \\ x &= -3 \end{aligned}$$

47.
$$\begin{aligned} 3x+2x+6 &= -15 \\ 5x+6 &= -15 \\ 5x+6-6 &= -15-6 \\ 5x &= -21 \\ \frac{5x}{5} &= \frac{-21}{5} \\ x &= -\frac{21}{5} \end{aligned}$$

48.
$$\begin{aligned} 4 &= -2(x+3) \\ 4 &= -2x-6 \\ 4+6 &= -2x-6+6 \\ 10 &= -2x \\ \frac{10}{-2} &= \frac{-2x}{-2} \\ -5 &= x \end{aligned}$$

49.
$$\begin{aligned} 27 &= 46+2x-x \\ 27 &= 46+x \\ 27-46 &= 46-46+x \\ -19 &= x \end{aligned}$$

50.
$$\begin{aligned} 4x+6-7x+9 &= 18 \\ -3x+15 &= 18 \\ -3x+15-15 &= 18-15 \\ -3x &= 3 \\ \frac{-3x}{-3} &= \frac{3}{-3} \\ x &= -1 \end{aligned}$$

51.
$$\begin{aligned} 4+3(x+2) &= 10 \\ 4+3x+6 &= 10 \\ 3x+10 &= 10 \\ 3x+10-10 &= 10-10 \\ 3x &= 0 \\ \frac{3x}{3} &= \frac{0}{3} \\ x &= 0 \end{aligned}$$

52.
$$\begin{aligned} -3+3x &= -2(x+1) \\ -3+3x &= -2x-2 \\ -3+3x+3 &= -2x-2+3 \\ 3x &= -2x+1 \\ 3x+2x &= -2x+1+2x \\ 5x &= 1 \\ \frac{5x}{5} &= \frac{1}{5} \\ x &= \frac{1}{5} \end{aligned}$$

53.
$$\begin{aligned} 9x-6 &= -3x+30 \\ 9x+3x-6 &= -3x+3x+30 \\ 12x-6 &= 30 \\ 12x-6+6 &= 30+6 \\ 12x &= 36 \\ \frac{12x}{12} &= \frac{36}{12} \\ x &= 3 \end{aligned}$$

54.
$$\begin{aligned} -(x+2) &= 2(3x-6) \\ -x-2 &= 6x-12 \\ -x-2+12 &= 6x-12+12 \\ -x+10 &= 6x \\ -x+x+10 &= 6x+x \\ 10 &= 7x \\ \frac{10}{7} &= \frac{7x}{7} \\ \frac{10}{7} &= x \end{aligned}$$

55.
$$\begin{aligned} 2x+6 &= 3x+9-3 \\ 2x+6 &= 3x+6 \\ 2x-2x+6 &= 3x-2x+6 \\ 6 &= x+6 \\ 6-6 &= x+6-6 \\ 0 &= x \end{aligned}$$

56.
$$\begin{aligned} -5x+3 &= 2x+10 \\ -5x+3-10 &= 2x+10-10 \\ -5x-7 &= 2x \\ -5x+5x-7 &= 2x+5x \\ -7 &= 7x \\ \frac{-7}{7} &= \frac{7x}{7} \\ -1 &= x \end{aligned}$$

57.
$$\begin{aligned} 3x-12x &= 24-9x \\ -9x &= 24-9x \\ -9x+9x &= 24-9x+9x \\ 0 &= 24 \text{ False} \end{aligned}$$

Since a false statement is obtained, there is no solution.

58.
$$\begin{aligned} 2(x+4) &= -3(x+5) \\ 2x+8 &= -3x-15 \\ 2x+3x+8 &= -3x+3x-15 \\ 5x+8 &= -15 \\ 5x+8-8 &= -15-8 \\ 5x &= -23 \\ \frac{5x}{5} &= \frac{-23}{5} \\ x &= -\frac{23}{5} \end{aligned}$$

59.
$$\begin{aligned} 4(2x-3)+4 &= 8x-8 \\ 8x-12+4 &= 8x-8 \\ 8x-8 &= 8x-8 \end{aligned}$$

Since the equation is true for all values of x, the solution is all real numbers.

60.
$$\begin{aligned} 6x+11 &= -(6x+5) \\ 6x+11 &= -6x-5 \\ 6x+11-11 &= -6x-5-11 \\ 6x &= -6x-16 \\ 6x+6x &= -6x+6x-16 \\ 12 &= -16 \\ \frac{12x}{12} &= \frac{-16}{12} \\ x &= -\frac{4}{3} \end{aligned}$$

61.
$$\begin{aligned} 2(x+7) &= 6x+9-4x \\ 2x+14 &= 6x+9-4x \\ 2x+14 &= 2x+9 \\ 2x-2x+14 &= 2x-2x+9 \\ 14 &= 9 \text{ False} \end{aligned}$$

Since a false statement is obtained, there is no solution.

62. $-5(3-4x)=-6+20x-9$
$-15+20x=-6+20x-9$
$-15+20x=-15+20x$
The statement is true for all values of x, thus the solution is all real numbers.

63. $4(x-3)-(x+5)=0$
$4x-12-x-5=0$
$3x-17=0$
$3x-17+17=0+17$
$3x=17$
$\frac{3x}{3}=\frac{17}{3}$
$x=\frac{17}{3}$

64. $-2(4-x)=6(x+2)+3x$
$-8+2x=6x+12+3x$
$-8+2x=9x+12$
$-8-12+2x=9x+12-12$
$-20+2x=9x$
$-20+2x-2x=9x-2x$
$-20=7x$
$\frac{-20}{7}=\frac{7x}{7}$
$-\frac{20}{7}=x$

65. 12: 20 = 3: 5

66. 80 ounces $=\frac{80}{16}=5$ pounds
The ratio of 80 ounces to 12 pounds is thus 5:12.

67. 32 ounces $=\frac{32}{16}=2$ pounds
The ratio of 32 ounces to 2 pounds is $\frac{2}{2}=\frac{1}{1}$.
The ratio is 1:1.

68. $\frac{x}{4}=\frac{8}{16}$
$16\cdot x=8\cdot 4$
$16x=32$
$x=\frac{32}{16}=2$

69. $\frac{5}{20}=\frac{x}{80}$
$20\cdot x=80\cdot 5$
$20x=400$
$x=\frac{400}{20}=20$

70. $\frac{3}{x}=\frac{15}{45}$
$3\cdot 45=15\cdot x$
$135=15x$
$\frac{135}{15}=x$
$9=x$

71. $\frac{20}{45}=\frac{15}{x}$
$20\cdot x=15\cdot 45$
$20x=675$
$x=\frac{675}{20}=\frac{135}{4}$

72. $\frac{6}{5}=\frac{-12}{x}$
$6\cdot x=-12\cdot 5$
$6x=-60$
$x=\frac{-60}{6}=-10$

73. $\frac{x}{9}=\frac{8}{-3}$
$-3\cdot x=9\cdot 8$
$-3x=72$
$x=\frac{72}{-3}=-24$

74. $\frac{-4}{9}=\frac{-16}{x}$
$-4\cdot x=-16\cdot 9$
$-4x=-144$
$x=\frac{-144}{-4}=36$

75. $\frac{x}{-15}=\frac{30}{-5}$
$-5\cdot x=-15\cdot 30$
$-5x=-450$
$x=\frac{-450}{-5}=90$

76. $\frac{6}{8}=\frac{30}{x}$
$6\cdot x=8\cdot 30$
$6x=240$
$x=\frac{240}{6}=40$
The length of the side is thus 40 in.

77. $\frac{7}{3.5}=\frac{2}{x}$
$7\cdot x=2\cdot 3.5$
$7x=7$
$x=\frac{7}{7}=1$
The length of the side is thus 1 ft.

78. $3x+4\geq 10$
$3x+4-4\geq 10-4$
$3x\geq 6$
$\frac{3x}{3}\geq\frac{6}{3}$
$x\geq 2$

2

79. $8-6x>4x-12$
$8-6x+6x>4x+6x-12$
$8>10x-12$
$8+12>10x-12+12$
$20>10x$
$\frac{20}{10}>\frac{10x}{10}$
$2>x$
$x<2$

2

80. $6-3x\leq 2x+18$
$6-18-3x\leq 2x+18-18$
$-12-3x\leq 2x$
$-12-3x+3x\leq 2x+3x$
$-12\leq 5x$
$\frac{-12}{5}<\frac{5x}{5}$
$\frac{-12}{5}\leq x$
$x\geq -\frac{12}{5}$

$-\frac{12}{5}$

81. $2(x+4)\leq 2x-5$
$2x+8\leq 2x-5$
$2x-2x+8\leq 2x-2x-5$
$8\leq -5$
Since 8 is never less than or equal to –5, there is no solution.

0

82. $2(x+3)>6x-4x+4$
$2x+6>6x-4x+4$
$2x+6>2x+4$
$2x-2x+6>2x-2x+4$
$6>4$
Since 6 is always greater than 4, the answer is all real numbers.

0

83. $x+6>9x+30$
$x+6-30>9x+30-30$
$x-24>9x$
$x-x-24>9x-x$
$-24>8x$
$\frac{-24}{8}>\frac{8x}{8}$
$-3>x$
$x<-3$

–3

84.
$$x-2 \le -4x+7$$
$$x-2+2 \le -4x+7+2$$
$$x \le -4x+9$$
$$x+4x \le -4x+4x+9$$
$$5x \le 9$$
$$\frac{5x}{5} \le \frac{9}{5}$$
$$x \le \frac{9}{5}$$

$\frac{9}{5}$

85.
$$-(x+2) < -2(-2x+5)$$
$$-x-2 < 4x-10$$
$$-x-2+10 < 4x-10+10$$
$$-x+8 < 4x$$
$$-x+x+8 < 4x+x$$
$$8 < 5x$$
$$\frac{8}{5} < \frac{5x}{5}$$
$$\frac{8}{5} < x$$
$$x > \frac{8}{5}$$

$\frac{8}{5}$

86.
$$-2(x-4) \le 3x+6-5x$$
$$-2x+8 \le 3x+6-5x$$
$$-2x+8 \le -2x+6$$
$$-2x+2x+8 \le -2x+2x+6$$
$$8 \le 6$$
Since 8 is never less than or equal to 6, there is no solution.

0

87.
$$2(2x+4) > 4(x+2)-6$$
$$4x+8 > 4x+8-6$$
$$4x+8 > 4x+2$$
$$4x-4x+8 > 4x-4x+2$$
$$8 > 2$$
Since 8 is always greater than 2, the solution is all real numbers.

0

88. Let x = number of calories in a 6-ounce piece of cake.
$$\frac{4 \text{ ounces}}{160 \text{ calories}} = \frac{6 \text{ ounces}}{x \text{ calories}}$$
$$\frac{4}{160} = \frac{6}{x}$$
$$4 \cdot x = 160 \cdot 6$$
$$4x = 960$$
$$x = \frac{960}{4} = 240$$
Thus, a 6-ounce piece of cake has 240 calories.

89. Let x = number of pages that can be copied in 22 minutes.
$$\frac{1 \text{ minutes}}{20 \text{ pages}} = \frac{22 \text{ minutes}}{x \text{ pages}}$$
$$\frac{1}{20} = \frac{22}{x}$$
$$1 \cdot x = 22 \cdot 20$$
$$x = 440$$
440 pages can be copied in 22 minutes.

90. Let x = number of inches representing 380 miles.
$$\frac{60 \text{ miles}}{1 \text{ inch}} = \frac{380 \text{ miles}}{x \text{ inches}}$$
$$\frac{60}{1} = \frac{380}{x}$$
$$60 \cdot x = 380 \cdot 1$$
$$60x = 380$$
$$x = \frac{380}{60} = 6\frac{1}{3}$$
$6\frac{1}{3}$ inches on the map represent 380 miles.

91. Let x = size of actual car in feet
$$\frac{1 \text{ inch}}{0.9 \text{ feet}} = \frac{10.5 \text{ inches}}{x \text{ feet}}$$
$$\frac{1}{0.9} = \frac{10.5}{x}$$
$$1 \cdot x = 0.9 \cdot 10.5$$
$$x = 9.45$$
The size of the actual car is 9.45 ft.

92. Let x = the value of 1 peso in terms of U.S. dollars.

$$\frac{\$1 \text{ U.S.}}{8.5410 \text{ pesos}} = \frac{x \text{ dollars}}{1 \text{ peso}}$$
$$\frac{1}{8.5410} = \frac{x}{1}$$
$$8.5410 \cdot x = 1 \cdot 1$$
$$8.5410x = 1$$
$$x = \frac{1}{8.5410} \approx 0.1171$$

1 peso equals about \$0.1171.

93. Let x = number of degrees in 1 radian.

$$\frac{3 \text{ radians}}{171.9 \text{ degrees}} = \frac{1 \text{ radian}}{x \text{ degrees}}$$
$$\frac{3}{171.9} = \frac{1}{x}$$
$$3 \cdot x = 1 \cdot 171.9$$
$$3x = 171.9$$
$$x = \frac{171.9}{3} = 57.3$$

There are 57.3 degrees in 1 radian.

94. Let x = number of bottles the machine can fill and cap in 2 minutes.

$$2 \text{ minutes} = 120 \text{ seconds}$$
$$\frac{50 \text{ seconds}}{80 \text{ bottles}} = \frac{120 \text{ seconds}}{x \text{ bottles}}$$
$$\frac{50}{80} = \frac{120}{x}$$
$$50 \cdot x = 80 \cdot 120$$
$$50x = 9600$$
$$x = \frac{9600}{50} = 192$$

The machine can fill and cap 192 bottles in 2 minutes.

Practice Test

1. $-6(4-2x) = -6[4+(-2x)]$
$= -6(4)+(-6)(-2x)$
$= -24+12x$ or $12x-24$

2. $-(x+3y-4) = -[x+3y+(-4)]$
$= -1[x+3y+(-4)]$
$= (-1)(x)+(-1)(3y)+(-1)(-4)$
$= -x+(-3y)+4$
$= -x-3y+4$

3. $5x-8x+4 = -3x+4$

4. $4+2x-3x+6 = 2x-3x+4+6$
$= -x+10$

5. $-y-x-4x-6 = -x-4x-y-6$
$= -5x-y-6$

6. $x-4y+6x-y+3 = x+6x-4y-y+3$
$= 7x-5y+3$

7. $2x+3+2(3x-2) = 2x+3+6x-4$
$= 2x+6x+3-4$
$= 8x-1$

8.
$$2x+4 = 12$$
$$2x+4-4 = 12-4$$
$$2x = 8$$
$$\frac{2x}{2} = \frac{8}{2}$$
$$x = 4$$

9.
$$-x-3x+4 = 12$$
$$-4x+4 = 12$$
$$-4x+4-4 = 12-4$$
$$-4x = 8$$
$$\frac{-4x}{-4} = \frac{8}{-4}$$
$$x = -2$$

10.
$$2x-2 = 4x+4$$
$$2x-2x-2 = 4x-2x+4$$
$$-2 = 2x+4$$
$$-2-4 = 2x+4-4$$
$$-6 = 2x$$
$$\frac{-6}{2} = \frac{2x}{2}$$
$$-3 = x$$

11.
$$\begin{aligned}3(x-2) &= -(5-4x)\\ 3x-6 &= -5+4x\\ 3x-6+5 &= -5+5+4x\\ 3x-1 &= 4x\\ 3x-3x-1 &= 4x-3x\\ -1 &= x\end{aligned}$$

12.
$$\begin{aligned}2x-3(-2x+4) &= -13+x\\ 2x+6x-12 &= -13+x\\ 8x-12 &= -13+x\\ 8x-12+12 &= -13+12+x\\ 8x &= -1+x\\ 8x-x &= -1+x-x\\ 7x &= -1\\ \frac{7x}{7} &= \frac{-1}{7}\\ x &= -\frac{1}{7}\end{aligned}$$

13.
$$\begin{aligned}3x-4-x &= 2(x+5)\\ 3x-4-x &= 2x+10\\ 2x-4 &= 2x+10\\ 2x-2x-4 &= 2x-2x+10\\ -4 &= 10 \quad \text{False}\end{aligned}$$

Since a false statement is obtained, there is no solution.

14.
$$\begin{aligned}-3(2x+3) &= -2(3x+1)-7\\ -6x-9 &= -6x-2-7\\ -6x-9 &= -6x-9\end{aligned}$$

Since the equation is true for all values of x, the solution is all real numbers.

15.
$$\begin{aligned}\frac{9}{x} &= \frac{3}{-15}\\ 9(-15) &= 3x\\ -135 &= 3x\\ \frac{-135}{3} &= x\\ -45 &= x\end{aligned}$$

16.
$$\begin{aligned}4(x-3)-2 &= 2x-14\\ 4x-12-2 &= 2x-14\\ 4x-14 &= 2x-14\\ 4x-2x-14 &= 2x-2x-14\\ 2x-14 &= -14\\ 2x-14+14 &= -14+14\\ 2x &= 0\\ \frac{2x}{2} &= \frac{0}{2}\\ x &= 0\end{aligned}$$

17. **a.** An equation that has exactly one solution is a conditional equation.

b. An equation that has no solution is a contradiction.

c. An equation that has all real numbers as its solution is an identity.

18.
$$\begin{aligned}2x-4 &< 4x+10\\ 2x-4-10 &< 4x+10-10\\ 2x-14 &< 4x\\ 2x-2x-14 &< 4x-2x\\ -14 &< 2x\\ \frac{-14}{2} &< \frac{2x}{2}\\ -7 &< x\\ x &> -7\end{aligned}$$

−7

19.
$$\begin{aligned}3(x+4) &\ge 5x-12\\ 3x+12 &\ge 5x-12\\ 3x+12+12 &\ge 5x-12+12\\ 3x+24 &\ge 5x\\ 3x-3x+24 &\ge 5x-3x\\ 24 &\ge 2x\\ \frac{24}{2} &\ge \frac{2x}{2}\\ 12 &\ge x\\ x &\le 12\end{aligned}$$

12

20. $4(x+3)+2x<6x-3$
$4x+12+2x<6x-3$
$6x+12<6x-3$
$6x-6x+12<6x-6x-3$
$12<-3$
Since 12 is never less than –3, the answer is no solution.

0

21. $-(x-2)-3x=4(1-x)-2$
$-x+2-3x=4-4x-2$
$-4x+2=-4x+2$ True
This equation is true for all real numbers.

0

22. $\frac{3}{4}=\frac{8}{x}$
$3x=4\cdot 8$
$3x=32$
$x=\frac{32}{3}$

The length of side x is $\frac{32}{3}$ feet or $10\frac{2}{3}$ feet.

23. Let x = number of gallons needed.
$$\frac{3\text{ acres}}{6\text{ gallons}}=\frac{75\text{ acres}}{x\text{ gallons}}$$
$\frac{3}{6}=\frac{75}{x}$
$3x=6\cdot 75$
$3x=450$
$x=\frac{450}{3}=150$
150 gallons are needed to treat 75 acres.

24. Let x = number of gallons he needs to sell.
$$\frac{\$0.40}{1\text{ gallon}}=\frac{\$20,000}{x\text{ gallons}}$$
$\frac{0.40}{1}=\frac{20,000}{x}$
$0.40x=20,000$
$x=\frac{20,000}{0.40}=50,000$
He needs to sell 50,000 gallons.

25. Let x = number of minutes it will take.
$$\frac{25\text{ miles}}{35\text{ minutes}}=\frac{125\text{ miles}}{x\text{ minutes}}$$
$\frac{25}{35}=\frac{125}{x}$
$25x=35\cdot 125$
$25x=4375$
$x=\frac{4375}{25}=175$
It would take 175 minutes or 2 hours 55 minutes.

Cumulative Review Test

1. $\frac{16}{20}\cdot\frac{4}{5}=\frac{16}{5}\cdot\frac{1}{5}$
$=\frac{16\cdot 1}{5\cdot 5}$
$=\frac{16}{25}$

2. $\frac{8}{24}\div\frac{2}{3}=\frac{8}{24}\cdot\frac{3}{2}$
$=\frac{4}{8}\cdot\frac{1}{1}$
$=\frac{4\cdot 1}{8\cdot 1}$
$=\frac{4}{8}$
$=\frac{1}{2}$

3. $|-2|>1$ since $|-2|=2$ and $2>1$.

4. $-7-(-4)+5-8=-7+4+5-8$
$=-3+5-8$
$=2-8$
$=-6$

5. $-7-(-6)=-7+6$
$=-1$

6. $20-6\div 3\cdot 2=20-2\cdot 2$
$=20-4$
$=16$

7. $$\begin{aligned} 3\left[6-\left(4-3^2\right)\right]-30 &= 3[6-(4-9)]-30 \\ &= 3[6-(-5)]-30 \\ &= 3[6+5]-30 \\ &= 3(11)-30 \\ &= 33-30 \\ &= 3 \end{aligned}$$

8. Substitute –2 for each x.
$$\begin{aligned} -3x^2-4x+5 &= -3(-2)^2-4(-2)+5 \\ &= -3(4)-(-8)+5 \\ &= -12+8+5 \\ &= -4+5 \\ &= 1 \end{aligned}$$

9. Associative property of addition

10. $$\begin{aligned} 8x+2y+4x-y &= 8x+4x+2y-y \\ &= 12x+y \end{aligned}$$

11. $$\begin{aligned} 3x-2x+16+2x &= 3x-2x+2x+16 \\ &= 3x+16 \end{aligned}$$

12. $$\begin{aligned} 4x-2 &= 10 \\ 4x-2+2 &= 10+2 \\ 4x &= 12 \\ \frac{4x}{4} &= \frac{12}{4} \\ x &= 3 \end{aligned}$$

13. $$\begin{aligned} \frac{1}{4}x &= -10 \\ 4\left(\frac{1}{4}x\right) &= 4(-10) \\ x &= -40 \end{aligned}$$

14. $$\begin{aligned} -6x-5x+6 &= 28 \\ -11x+6 &= 28 \\ -11x+6-6 &= 28-6 \\ -11x &= 22 \\ \frac{-11x}{-11} &= \frac{22}{-11} \\ x &= -2 \end{aligned}$$

15. $$\begin{aligned} 4(x-2) &= 5(x-1)+3x+2 \\ 4x-8 &= 5x-5+3x+2 \\ 4x-8 &= 8x-3 \\ 4x-4x-8 &= 8x-4x-3 \\ -8 &= 4x-3 \\ -8+3 &= 4x-3+3 \\ -5 &= 4x \\ \frac{-5}{4} &= \frac{4x}{4} \\ -\frac{5}{4} &= x \end{aligned}$$

16. $$\begin{aligned} \frac{15}{30} &= \frac{3}{x} \\ 15x &= 3\cdot 30 \\ 15x &= 90 \\ x &= \frac{90}{15} = 6 \end{aligned}$$

17. $$\begin{aligned} x-3 &> 7 \\ x-3+3 &> 7+3 \\ x &> 10 \end{aligned}$$

10

18. $$\begin{aligned} 2x-7 &\le 3x+5 \\ 2x-7-5 &\le 3x+5-5 \\ 2x-12 &\le 3x \\ 2x-2x-12 &\le 3x-2x \\ -12 &\le x \\ x &\ge -12 \end{aligned}$$

–12

19. Let x = number of pounds of fertilizer needed.
$$\begin{aligned} \frac{5000 \text{ square feet}}{36 \text{ pounds}} &= \frac{22{,}000 \text{ square feet}}{x \text{ pounds}} \\ \frac{5000}{36} &= \frac{22{,}000}{x} \\ 5000x &= (36)(22{,}000) \\ 5000x &= 792{,}000 \\ x &= \frac{792{,}000}{5000} = 158.4 \end{aligned}$$
158.4 pounds are needed to fertilize a 22,000-square-foot lawn.

20. Let x = amount he earns after 8 hours.

$$\frac{2 \text{ hours}}{\$10.50} = \frac{8 \text{ hours}}{x \text{ dollars}}$$

$$\frac{2}{10.5} = \frac{8}{x}$$

$$2x = (10.5)(8)$$

$$2x = 84$$

$$x = \frac{84}{2} = 42$$

He earns \$42 after 8 hours.

Chapter 3

Exercise Set 3.1

1. A formula is an equation used to express a relationship mathematically.

3. The simple interest formula is:
$i = prt$ where i is interest, p is principle, r is the interest rate, and t is time.

5. The diameter of a circle is 2 times its radius.

7. When you multiply a unit by the same unit, you get a square unit.

9. Substitute 4 for s.
$P = 4s$
$P = 4(4) = 16$

11. Substitute 12 for l and 8 for w.
$A = lw$
$A = 12(8) = 96$

13. Substitute 12 for i.
$c = 2.54i$
$c = 2.54(12) = 30.48$

15. Substitute 4 for r.
$A = \pi r^2$
$A = \pi(4)^2$
$A = 16\pi \approx 50.27$

17. Substitute 100 for x, 80 for m, and 10 for s.

$$z = \frac{x - m}{s}$$
$$z = \frac{100 - 80}{10}$$
$$z = \frac{20}{10} = 2$$

19. Substitute 60 for V and 12 for B.

$$V = \frac{1}{3}Bh$$
$$60 = \frac{1}{3}(12)h$$
$$60 = 4h$$
$$\frac{60}{4} = \frac{4h}{4}$$
$$15 = h$$

21. Substitute 36 for A and 16 for m.

$$A = \frac{m+n}{2}$$
$$36 = \frac{16+n}{2}$$
$$2(36) = 2\left(\frac{16+n}{2}\right)$$
$$72 = 16 + n$$
$$72 - 16 = n$$
$$56 = n$$

23. Substitute 1050 for A, 1 for t, and 1000 for P.

$$A = P(1 + rt)$$
$$1050 = 1000[1 + r(1)]$$
$$1050 = 1000(1 + r)$$
$$\frac{1050}{1000} = \frac{1000(1+r)}{1000}$$
$$1.05 = 1 + r$$
$$1.05 - 1 = 1 - 1 + r$$
$$0.05 = r$$

25. Substitute 6 for r.

$$V = \frac{4}{3}\pi r^3$$
$$V = \frac{4}{3}\pi(6)^3$$
$$V = \frac{4}{3}\pi(216) \approx 904.78$$

27. Substitute 24 for B and 61 for h.

$$B = \frac{703w}{h^2}$$
$$24 = \frac{703w}{(61)^2}$$
$$24(61)^2 = \frac{703w}{61^2}(61)^2$$
$$89,304 = 703w$$
$$\frac{89,304}{703} = \frac{703w}{703}$$
$$127.03 \approx w$$

29. Substitute 160 for C and 0.12 for r.

$$S = C + rC$$
$$S = 160 + (0.12)(160)$$
$$S = 160 + 19.20 = 179.20$$

31. $3x + y = 5$

a. $3x + y - 3x = 5 - 3x$
$$y = 5 - 3x$$

b. Substitute 2 for x.
$$y = 5 - 3(2) = 5 - 6 = -1$$

33. $4x = 6y - 8$

a.
$$4x + 8 = 6y - 8 + 8$$
$$4x + 8 = 6y$$
$$2(2x + 4) = 6y$$
$$\frac{2(2x+4)}{6} = \frac{6y}{6}$$
$$\frac{2x+4}{3} = y$$

b. Substitute 10 for x.
$$y = \frac{2(10)+4}{3}$$
$$= \frac{20+4}{3}$$
$$= \frac{24}{3}$$
$$= 8$$

35. $2y = 6 - 3x$

a.
$$2y = -3x + 6$$
$$\frac{2y}{2} = \frac{-3x+6}{2}$$
$$y = \frac{-3x+6}{2}$$

b. Substitute 2 for x.
$$y = \frac{-3(2)+6}{2} = \frac{-6+6}{2} = \frac{0}{2} = 0$$

37. $-3x + 5y = -10$

a.
$$-3x + 5y + 3x = -10 + 3x$$
$$5y = -10 + 3x$$
$$\frac{5y}{5} = \frac{-10+3x}{5}$$
$$y = \frac{-10+3x}{5}$$

b. Substitute 4 for x.
$$y = \frac{-10+3(4)}{5}$$
$$= \frac{-10+12}{5}$$
$$= \frac{2}{5}$$

39. $-3x = 18 - 6y$

a.
$$-3x - 18 = 18 - 18 - 6y$$
$$-3x - 18 = -6y$$
$$\frac{-3x-18}{-6} = \frac{-6y}{-6}$$
$$\frac{-3x-18}{-6} = y$$
$$\frac{-3(x+6)}{(-1)(6)} = y$$
$$\frac{x+6}{2} = y$$

b. Substitute 0 for x.
$$y = \frac{(0)+6}{2} = \frac{6}{2} = 3$$

41. $-8 = -x - 2y$

a.
$$x - 8 = x - x - 2y$$
$$x - 8 = -2y$$
$$\frac{x-8}{-2} = \frac{-2y}{-2}$$
$$\frac{-x+8}{2} = y$$

b. Substitute –4 for x.
$$y = \frac{-(-4)+8}{2} = \frac{4+8}{2} = \frac{12}{2} = 6$$

43.
$$P = 4s$$
$$\frac{P}{4} = \frac{4s}{4}$$
$$\frac{P}{4} = s$$

45.
$$d = rt$$
$$\frac{d}{t} = \frac{rt}{t}$$
$$\frac{d}{t} = r$$

47.
$$C = \pi d$$
$$\frac{C}{\pi} = \frac{\pi d}{\pi}$$
$$\frac{C}{\pi} = d$$

49.
$$A = \frac{1}{2}bh$$
$$2A = 2\left(\frac{1}{2}bh\right)$$
$$2A = bh$$
$$\frac{2A}{b} = \frac{bh}{b}$$
$$\frac{2A}{b} = h$$

51.
$$P = 2l + 2w$$
$$P - 2l = 2l - 2l + 2w$$
$$P - 2l = 2w$$
$$\frac{P-2l}{2} = \frac{2w}{2}$$
$$\frac{P-2l}{2} = w$$

53.
$$6n + 3 = m$$
$$6n + 3 - 3 = m - 3$$
$$6n = m - 3$$
$$\frac{6n}{6} = \frac{m-3}{6}$$
$$n = \frac{m-3}{6}$$

55.
$$y = mx + b$$
$$y - mx = mx - mx + b$$
$$y - mx = b$$

57.
$$A = P + Prt$$
$$A - P = P - P + Prt$$
$$A - P = Prt$$
$$\frac{A-P}{Pt} = \frac{Prt}{Pt}$$
$$\frac{A-P}{Pt} = r$$

59.
$$A = \frac{m+2d}{3}$$
$$3A = 3\left(\frac{m+2d}{3}\right)$$
$$3A = m + 2d$$
$$3A - m = m - m + 2d$$
$$3A - m = 2d$$
$$\frac{3A-m}{2} = \frac{2d}{2}$$
$$\frac{3A-m}{2} = d$$

61.
$$d = a + b + c$$
$$d - a = a - a + b + c$$
$$d - a = b + c$$
$$d - a - c = b + c - c$$
$$d - a - c = b$$

63.
$$ax + by = c$$
$$ax - ax + by = -ax + c$$
$$by = -ax + c$$
$$\frac{by}{b} = \frac{-ax + c}{b}$$
$$y = \frac{-ax + c}{b}$$

65.
$$V = \pi r^2 h$$
$$\frac{V}{\pi r^2} = \frac{\pi r^2 h}{\pi r^2}$$
$$\frac{V}{\pi r^2} = h$$

67. Substitute 10 for n.
$$d = \frac{1}{2}n^2 - \frac{3}{2}n$$
$$d = \frac{1}{2}(10)^2 - \frac{3}{2}(10)$$
$$= \frac{1}{2}(100) - 15$$
$$= 50 - 15$$
$$= 35$$

69. Substitute 50 for F.
$$C = \frac{5}{9}(F - 32)$$
$$C = \frac{5}{9}(50 - 32)$$
$$= \frac{5}{9}(18)$$
$$= 10$$
The equivalent temperature is 10°C.

71. Substitute 35 for C.
$$F = \frac{9}{5}C + 32$$
$$F = \frac{9}{5}(35) + 32$$
$$= 63 + 32$$
$$= 95$$
The equivalent temperature is 95°F.

73.
$$P = \frac{KT}{V}$$
$$P = \frac{(1)(10)}{1} = \frac{10}{1} = 10$$

75.
$$P = \frac{KT}{V}$$
$$80 = \frac{K(100)}{5}$$
$$80 = 20K$$
$$\frac{80}{20} = \frac{20K}{20}$$
$$4 = K$$

77.
$$A = s^2$$
$$A = (2s)^2 = 4s^2$$
The area is 4 times as large as the original area.

79. Substitute 5 for n.
$$S = n^2 + n$$
$$S = (5)^2 + 5 = 25 + 5 = 30$$

81.
$$i = prt$$
$$i = (6000)(0.08)(3) = 1440$$
He will pay $1440 interest.

83.
$$i = prt$$
$$1050 = p(0.07)(3)$$
$$1050 = 0.21p$$
$$\frac{1050}{0.21} = \frac{0.21p}{0.21}$$
$$5000 = p$$
She placed $5000 in the savings account.

85. $P = a + b + c$
$P = 5 + 12 + 8 = 25$
The perimeter of the table top is 25 feet.

87. $A = \frac{1}{2}bh$
$A = \frac{1}{2}(36)(31) = 558$
The area is 558 square inches.

89. $A = \pi r^2$
$A = \pi(26)^2 = 676\pi \approx 2123.72$
Each mat has an area of about 2123.72 square centimeters.

91. $A = \frac{1}{2}h(b+d)$
$A = \frac{1}{2}(2)(4+3)$
$= \frac{1}{2}(2)(7)$
$= (1)7$
$= 7$
The area of the sign is 7 square feet.

93. $A = \frac{1}{2}bh$
$A = \frac{1}{2}(6)(6) = 3(6) = 18$
The area is 18 square feet.

95. $A = \frac{1}{2}h(b+d)$
$A = \frac{1}{2}(100)(80+200)$
$= \frac{1}{2}(100)(280)$
$= (50)(280)$
$= 14,000$
The seating area is 14,000 square feet.

97. The radius is half the diameter, so $r = \frac{3}{2} = 1.5$ inches.
$V = \frac{1}{3}\pi r^2 h$
$V = \frac{1}{3}\pi(1.5)^2(5)$
$= \frac{1}{3}\pi(2.25)5$
$= 3.75\pi$
≈ 11.78
The volume of the cone is about 11.78 cubic inches.

99. **a.** $B = \frac{703w}{h^2}$

b. 5 feet 3 inches $= 5(12) + 3$
$= 60 + 3$
$= 63$ inches
$B = \frac{703(135)}{(63)^2} = \frac{94,905}{3969} \approx 23.91$

101. **a.** $C = 2\pi r$
$\pi = \frac{C}{2r}$ or $\pi = \frac{C}{d}$

b. π or about 3.14.

c. About 3.14.

103. **a.** $V = lwh$
$V = (3x)(x)(6x-1)$
$= 3x^2(6x-1)$
$= 18x^3 - 3x^2$

b. $V = 18x^3 - 3x^2$
$V = 18(7)^3 - 3(7)^2$
$= 6174 - 147$
$= 6027$
Volume is 6027 cm^3.

c. $S = 2lw + 2lh + 2wh$
$S = 2(3x)(x) + 2(3x)(6x-1) + 2(x)(6x-1)$
$= 6x^2 + 36x^2 - 6x + 12x^2 - 2x$
$= 54x^2 - 8x$

d. $S = 54x^2 - 8x$
$S = 54(7)^2 - 8(7)$
$= 2646 - 56$
$= 2590$
Surface area is 2590 cm^2.

106. $\left[4\left(12 \div 2^2 - 3\right)^2\right]^2 = \left[4(12 \div 4 - 3)^2\right]^2$
$= \left[4(3-3)^2\right]^2$
$= \left[4(0)^2\right]^2$
$= [0]^2$
$= 0$

107. $\frac{6}{4} = \frac{3}{2}$ so the ratio of Arabians to Morgans is 3:2.

108. Let x = number of minutes to siphon 13,500 gallons
$\frac{25 \text{ gallons}}{3 \text{ minutes}} = \frac{13,500 \text{ gallons}}{x \text{ minutes}}$
$\frac{25}{3} = \frac{13,500}{x}$
$25 \cdot x = 3(13,500)$
$25x = 40,500$
$x = \frac{40,500}{25} = 1620$
It will take 1620 minutes or 27 hours to empty the pool.

109. $2(x-4) \geq 3x + 9$
$2x - 8 \geq 3x + 9$
$2x - 17 \geq 3x$
$-17 \geq x$
$x \leq -17$

Exercise Set 3.2

1. Added to, more than, increased by, and sum indicate the operation of addition.

3. Multiplied by, product of, twice, and three times indicate the operation of multiplication.

5. The cost is increased by 25% of the cost, so the expression needs $0.25c$.

7. $n + 5$

9. $4x$

11. $\frac{x}{2}$

13. $s + 1.2$

15. $0.16P$

17. $10P - 9$

19. $\frac{8}{9}m + 16,000$

21. $45 + 0.40x$

23. $25x$

25. $16x + y$

27. $n + 0.04n$

29. $t - 0.18t$

31. $220 + 80x$

33. $275x + 25y$

35. Six less than a number

37. One more than four times a number

39. Seven less than five times a number

41. Four times a number, decreased by two

43. Three times a number subtracted from two

45. Twice the difference between a number and one

47. a. Let c = Chuck's age.

b. Then Dana's age is $c + 3$.

49. a. Let l = Lois' age.

b. Then her son's age is $\frac{1}{3}l$.

51. a. The first consecutive even integer is x.

b. The next consecutive even integer is $x + 2$.

53. a. Let c = the cost of a Camaro.

b. Then a Firebird costs $1.1c$.

55. a. Let t = the percent of the profits Tabitha receives.

b. Then the percent Shari receives is $100 - t$.

57. a. Let l = the average monthly rental rate in Los Angeles.

b. Then the rate in San Jose is $2l - 177$.

59. a. Let x be the number of coupons, in billions, redeemed in 1997.

b. Then the number of coupons redeemed in 1999 is $2x - 4.4$.

61. a. Let n = the number of subscribers in 1985.

b. Then the number of subscribers in 1996 was $125n + 1,500,000$.

63. a. Let s = Dianne's 1998 sales.

b. 1999 sales = $s + 0.60s$

65. a. Let s = Kevin's salary last year.

b. Kevin's salary this year = $s + 0.15s$

67. a. Let c = the cost of the car.

b. Cost of car with tax = $c + 0.07c$

69. a. Let p = the original pollution level.

b. Current pollution level = $p - 0.50p$

71. a. Let x = first number, then $5x$ = second number.

b. First number + second number = 18
$x + 5x = 18$

73. a. Let x = smaller integer, then $x + 1$ = larger consecutive integer.

b. Smaller + larger = 47
$x + (x + 1) = 47$

75. a. Let x = the number.

b. Twice the number decreased by 8 is 12.
$2x - 8 = 12$

77. a. Let x = the number.

b. One-fifth of the sum of the number and 10 is 150.
$\frac{1}{5}(x + 10) = 150$

79. a. Let s = the distance traveled by the Southern Pacific train.

b. $s + (2s - 4) = 890$

81. a. Let c = the cost of the car.

b. $c + 0.07c = 26,200$

83. **a.** Let c = the cost of the meal.

b. $c+0.15c=32.50$

85. **a.** Let x = the average player's salary.

b. $7.69x-x=87,000$

87. **a.** Let s = average teacher's salary in South Dakota.

b. $s+(3s-28,784)=76,600$

89. Three more than a number is six.

91. Three times a number, decreased by one, is four more than twice the number.

93. Four times the difference between a number and one is six.

95. Six more than five times a number is the difference between six times the number and one.

97. The sum of a number and the number increased by four is eight.

99. The sum of twice a number and the number increased by three is five.

101. Answers will vary.

103. **a.** 1 minute = 60 seconds
1 hour = 60 minutes = 3600 seconds
1 day = 24 hours
= 1440 minutes
= 86,400 seconds
$86,400d+3600h+60m+s$

b. $86,400d+3600h+60m+s$
$=86,400(4)+3600(6)+60(15)+25$
$=368,125$ seconds

109. $\frac{\frac{1}{2}}{1}=\frac{x}{6.7}$

$x=\frac{1}{2}(6.7)$

$x=3.35$

3.35 teaspoons of thyme should be used.

110. $\frac{3 \text{ cups cat food}}{1 \text{ cup water}}=\frac{\frac{1}{2} \text{ cup cat food}}{x \text{ cups water}}$

$\frac{3}{1}=\frac{\frac{1}{2}}{x}$

$3x=\frac{1}{2}$

$\frac{1}{3}(3x)=\left(\frac{1}{3}\right)\left(\frac{1}{2}\right)$

$x=\frac{1}{6}$

Cindy should add $\frac{1}{6}$ cup water to $\frac{1}{2}$ cup dry cat food.

111. Substitute 40 for P and 5 for w.

$P=2l+2w$
$40=2l+2(5)$
$40=2l+10$
$30=2l$
$15=l$

112.

$3x-2y=6$

$3x-3x-2y=-3x+6$

$-2y=-3x+6$

$\frac{-2y}{-2}=\frac{-3x+6}{-2}$

$y=\frac{3x-6}{2}$

$y=\frac{3}{2}x-3$

Substitute 6 for x.

$y=\frac{3(6)-6}{2}=\frac{18-6}{2}=\frac{12}{2}=6$

Exercise Set 3.3

1. Answers will vary.

3. Use the equation
Volume · percent = amount.

5. **a.** Let x = smaller integer, then
$x + 1$ = next consecutive integer.
Smaller number + larger number = 85.
$x + (x + 1) = 85$

b. $2x+1=85$
$2x=84$
$x=42$
Smaller number = 42
Larger number = $x + 1 = 42 + 1 = 43$

7. a. Let x = one number.
Then $2x + 3$ = second number.
First number + second number = 27
$x+(2x+3)=27$

b. $3x+3=27$
$3x=24$
$x=8$
First number = 8
Second number = $2x + 3 = 2(8) + 3 = 19$

9. a. Let x = smallest integer. The next two consecutive integers are $x + 1$ and $x + 2$.
First number + second number + third number = 39
$x+(x+1)+(x+2)=39$

b. $3x+3=39$
$3x=36$
$x=12$
First number = 12
Second number $= x+1=12+1=13$
Third number $= x+2=12+2=14$

11. Let x = smaller integer, then larger integer = $2x - 8$.
Larger integer – smaller integer = 17
$(2x-8)-x=17$
$2x-8-x=17$
$x-8=17$
$x=25$
Smaller number = 25
Larger number $= 2x-8=2(25)-8=42$

13. a. Let x = noise level of a vacuum cleaner. Then $x + 40$ = noise level of a rock concert.
vacuum's noise level + concert's noise level
= 200 db
$x+(x+40)=200$

b. $2x+40=200$
$2x=160$
$x=80$
Noise level of vacuum cleaner = 80 decibels;
Noise level of rock concert $= x + 40$
$= 80 + 40$
$= 120$
decibels

15. a. Let x = amount of caffeine in a cup of decaffeinated coffee, then
$26x$ = amount of caffeine in a cup of regular coffee.
Caffeine in decaffeinated coffee + caffeine in regular coffee = 121.5
$x+26x=121.5$

b. $27x=121.5$
$x=4.5$
There are 4.5 milligrams of caffeine in decaffeinated coffee and
$26x = 26(4.5) = 117$ milligrams of caffeine in regular coffee.

17. a. Let x = cost of mailing a letter in Germany, and $2x-0.44$ = cost of mailing a letter in Japan.
Cost in Germany + cost in Japan = \$1.27
$x+(2x-0.44)=1.27$

b. $3x-0.44=1.27$
$3x=1.71$
$x=0.57$
The cost of mailing a letter in Germany is \$0.57 or 57¢. The cost of mailing a letter in Japan is
$2x-0.44=2(0.57)-0.44=\$0.70$ or 70¢.

19. a. Let x = percent of pure gold by weight.
Total weight × percent gold
= amount of gold
$20x=15$

b. $x = \frac{15}{20} = 0.75$
18 karat gold is 75% pure gold by weight.

21. a. Let x = number of years it will take to donate \$10,050.
Amount already donated + (amount each year) · (number of years) = \$10,050
$6000 + 450 \cdot x = 10{,}050$

b.
$$6000 + 450x = 10{,}050$$
$$450x = 4050$$
$$x = 9$$
It will take her 9 years.

23. a. Let x = number of successful students in the first section. Then $2x - 60$ = number of successful students in the second section.
successful students in first section + successful students in second section = 153
$x + (2x - 60) = 153$

b.
$$3x - 60 = 153$$
$$3x = 213$$
$$x = 71$$
There were 71 successful students in the first section and $2x - 60 = 2(71) - 60 = 82$ in the second section.

25. a. Let x = total amount collected at door; 3000 + 3% of admission fees = total amount received.
$3000 + 0.03x = 3750$

b.
$$0.03x = 750$$
$$x = \frac{7.50}{0.03} = 25{,}000$$
The total amount collected at the door was \$25,000.

27. a. Let x = original price of hat.
Then discount = $0.20x$
Original price – discount = 25.99
$x - 0.20x = 25.99$

b.
$$0.08x = 25.99$$
$$x = \frac{25.99}{0.80} \approx 32.49$$
The regular price of the hat is \$32.49.

29. a. Let x = Dana's recent salary, then his salary increase = $0.04x$
Present salary + increase in salary = future salary
$x + 0.04x = 36{,}400$

b.
$$1.04x = 36{,}400$$
$$x = \frac{36{,}400}{1.04} = 35{,}000$$
His present salary is \$35,000.

31. a. Let x = cost of car before tax. Then $0.07x$ = tax on the car
Cost of car before tax + tax on car = 22,800
$x + 0.07x = 22{,}800$

b.
$$1.07x = 22{,}800$$
$$x = \frac{22{,}800}{1.07} \approx 21{,}308.41$$
The car costs \$21,308.41 before tax.

33. a. Let x = the budget amount. Then $2x + 5$ = the projected gross earnings.
(Projected gross earnings) – (budgeted amount) = \$65 million
$(2x + 5) - x = 65$

b.
$$2x + 5 - x = 65$$
$$x + 5 = 65$$
$$x = 60$$
The budget was \$60 million and the projected gross earnings were $2x + 5 = 2(60) + 5 = 120 + 5 = \$125$ million.

35. a. Let x = the time it takes the Boxster and $2x - 8.6$ = the time it takes the Corvette.
Boxster's time – Corvette's time = 1.5 seconds

$x-(2x-8.6)=1.5$

b. $x-2x+8.6=1.5$

$-x+8.6=1.5$

$-x=-7.1$

$x=7.1$

It takes the Boxster 7.1 seconds and it takes the Corvette
$2x-8.6=2(7.1)-8.6=5.6$ seconds.

37. a. Let x = number of years until the salaries are the same.
yearly salary = base salary + (yearly increase) · (number of years)
yearly salary at Data Tech. = yearly salary at Nuteck

$40,000+2400x=49,600+800x$

b. $40,000+2400x=49,600+800x$

$40,000+1600x=49,600$

$1600x=9600$

$x=6$

It will take 6 years for the two salaries to be the same.

39. a. Let x = number of hours assigned to each of the younger workers. Then $2x$ = number of hours assigned to third worker and $3x$ = number of hours assigned to fourth worker. Total number of hours assigned to the hour workers = 91

$x+x+2x+3x=91$

b. $7x=91$

$x=13$

Each of the younger workers is assigned 13 hours. Third worker is assigned $2x=2(13)=26$ hours. Fourth worker is assigned $3x=3(13)=39$ hours.

41. a. Let x = amount of water that would be used without the special shower head, then amount of water saved = $0.60x$
(amount of water without special shower head)
– (amount of water saved) = (amount of water used with special shower head)

$x-0.60x=24$

b. $0.40x=24$

$x=\frac{24}{0.40}=60$

Without the special shower head, 60 gallons of water would be used.

43. a. Let x = tax rate. Then tax on bill = $22x$
Bill before tax + tax on bill = 23.76

$22+22x=23.76$

b. $22+22x=23.76$

$22x=1.76$

$x=\frac{1.76}{22}=0.08$

The local sales tax rate is 8%.

45. a. Let x = regular membership fee, then amount of reduction = $0.10x$.
regular fee – reduction – 20 = new fee on a Monday

$x-0.10x-20=250$

b. $x-0.10x-20=250$

$x-0.10x=270$

$0.90x=270$

$x=\frac{270}{0.90}=300$

Regular fee is $300.

47. a. Let x = the amount of toxic chemicals released at the Asarco Plant.
(Amount of chemicals released at Magnesium Corp. of America)
= (Amount of chemicals released at Asarco) + (Increase in chemicals released)

$64.3=x+0.63x$

b. $64.3 = 1.63x$
$39.45 \approx x$
There were approximately 39.45 million pounds of toxic chemicals released at the Asarco plant.

49. a. Let x = the number of pieces of personal mail.
Then $x + 1$ = the number of bills and statements and $5x + 2$ = the number of advertisements.
(Personal mail) + (bills and statements) + (advertisements) = total mail
$x + (x + 1) + (5x + 2) = 24$

b. $7x + 3 = 24$
$7x = 21$
$x = 3$
There were 3 pieces of personal mail, $3 + 1 = 4$ bills and statements and $5x + 2 = 5(3) + 2 = 17$ advertisements.

51. a. Let x = the number of minutes of calling to make both plans equal in cost.
(cost of One Rate Plan) = $0.15x$
(cost of One Rate Plus Plan) $= 0.10x + 4.95$
cost of One Rate Plan = cost of One Rate Plus Plan
$0.15x = 0.10x + 4.95$
$0.05x = 4.95$
$x = 99$
99 minutes of calling will make the plans equal in cost.

b. One Rate Plan = 0.15(200) = 30
One Rate Plus Plan $= 0.10(200) + 4.95$
$= 24.95$
The One Rate Plus Plan results in a lower cost. It saves 30 – 24.95 = \$5.05 each month.

53. a. With the 9.50% mortgage, the monthly payments are 50(9.33) = \$466.50.

b. With the 8.00% mortgage, the monthly payments are 50(8.37) = \$418.50.

c. In addition Quincy must pay 0.04(50,000) = \$2000 because of the 4 points.

d. Let x = number of months when total payments from both mortgages are equal. Monthly payments for 9.5% mortgage = monthly payments for 8.00% mortgage + points.
$466.50x = 418.50x + 2000$
$48x = 2000$
$x = \frac{2000}{48} \approx 41.7$
The total payments are the same after about 42 months or 3.5 years.

e. After 3.5 years the Citibank mortgage is cheaper because of the lower monthly payments. If he plans to live in the house for 20 years the Citibank mortgage is less expensive.

55. a. With the 9.00% mortgage, the monthly payments are 100(9.00) = \$900.
With the 8.875% mortgage, the monthly payments are 100(8.92) = \$892.
Let x = number of months when total payments from the two mortgages are equal. Total payments from Key Mortgage = total payments form Countrywide plus additional fees
$900x = 892x + 150$
$8x = 150$
$x = \frac{150}{8} = 18.75$
The total costs from the two mortgages are equal after 18.75 months or about 1.56 years.

b. After about 1.56 years the Countrywide mortgage is cheaper because of the lower monthly payments. If Jean-Pierre plans to live in the house for 10 years, the Countrywide Mortgage is less expensive.

57. a. If they refinance, the monthly payments will be $50(8.29) = \$414.50$. Their monthly savings will be $\$740 - \$414.50 = \$325.50$.
Let x = number of months when money saved = closing cost.
Money saved = closing cost
$325.50x = 3000$

$$x = \frac{3000}{325.50} \approx 9.2$$

It will take about 9.2 months for the money saved to equal the closing costs.

b. Their monthly payments will be $\$740 - \$414.50 = \$325.50$ lower

59. a. $\dfrac{74+88+76+x}{4} = 80$

b.
$$\frac{74+88+76+x}{4} = 80$$
$$74+88+76+x = 320$$
$$238+x = 320$$
$$x = 82$$

Paul must receive an 82 on his fourth exam.

61. a. Yearly cost with 10% discount $= 600 - 0.10(600) = \$540$. Let x = the number of years it will take for the costs to be equal.
Total cost with driver's ed = Total cost without driver's ed.
$$45 + 540x = 600x$$
$$45 = 60x$$
$$0.75 = x$$

It will take 0.75 years, or 9 months for the costs to be equal.

b. $25 - 18 = 7$ years. The cost with driver's ed is $45 + 540(7) = \$3825$. The cost without driver's ed is $600 \cdot (7) = \$4200$. He will save $4200 - 3825 = \$375$.

63.
$$\frac{1}{4}+\frac{3}{4}\div\frac{1}{2}-\frac{1}{3} = \frac{3}{12}+\frac{9}{12}\div\frac{6}{12}-\frac{4}{12}$$
$$= \frac{3}{12}+\frac{9}{12}\cdot\frac{12}{6}-\frac{4}{12}$$
$$= \frac{3}{12}+\frac{9}{6}-\frac{4}{12}$$
$$= \frac{3}{12}+\frac{18}{12}-\frac{4}{12}$$
$$= \frac{21}{12}-\frac{4}{12}$$
$$= \frac{17}{12}$$

64. Associative property of addition

65. Commutative property of multiplication

66. Distributive property

67. Let x = number of pounds of coleslaw needed.
$$\frac{5 \text{ people}}{\frac{1}{2} \text{ pound coleslaw}} = \frac{560 \text{ people}}{x \text{ pounds coleslaw}}$$
$$\frac{5}{\frac{1}{2}} = \frac{560}{x}$$
$$5x = \left(\frac{1}{2}\right)(560)$$
$$5x = 280$$
$$x = \frac{280}{5} = 56$$

He will need 56 pounds of coleslaw.

68.
$$M = \frac{a+b}{2}$$
$$2M = 2\left(\frac{a+b}{2}\right)$$
$$2M = a+b$$
$$2M - a = a - a + b$$
$$2M - a = b$$

Exercise Set 3.4

1. $A = (2l)\cdot\left(\dfrac{w}{2}\right) = lw$

The area remains the same.

3. $V = 2l \cdot 2w \cdot 2h = 8(lwh)$
The volume is eight times as great.

5. An isosceles triangle is a triangle with 2 equal sides.

7. The sum of the measures of the angles in a triangle is 180°.

9. Let x = length of each side of the triangle, then $P = x + x + x = 3x$.
Perimeter = 28.5
$$x = \frac{28.5}{3} = 9.5$$
The length of each side is 9.5 inches.

11. Let x = measure of angle A, then $3x - 8$ = measure of angle B.
Sum of the 2 angles = 180
$$x + (3x - 8) = 180$$
$$4x - 8 = 180$$
$$4x = 188$$
$$x = \frac{188}{4} = 47$$
Measure of angle A = 47°
Measure of angle B = 3(47) – 8
= 141 – 8
= 133°

13. Let x = measure of smallest angle. Then second angle = x + 10 and third angle = $2x$ – 30.
Sum of the 3 angles = 180
$$x + (x + 10) + (2x - 30) = 180$$
$$4x - 20 = 180$$
$$4x = 200$$
$$x = \frac{200}{4} = 50$$
The first angle is 50°.
The second angle is 50 + 10 = 60°.
The third angle is 2(50) – 30 = 70°.

15. Let x = length of each of the two equal sides, then $x - 2$ = length of third side,
Sum of the three sides = perimeter
$$x + x + (x - 2) = 10$$
$$3x - 2 = 10$$
$$3x = 12$$
$$x = 4$$
The length of the two equal sides is 4 meters.
The length of the third side is 4 – 2 = 2 meters.

17. Let x = width of tennis court.
Then $2x + 6$ = length of tennis court.
$$P = 2l + 2w$$
$$228 = 2(2x + 6) + 2x$$
$$228 = 4x + 12 + 2x$$
$$228 = 6x + 12$$
$$216 = 6x$$
$$36 = x$$
The width is 36 feet and the length is $2(36) + 6 = 78$ feet.

19. Let x = measure of each smaller angle. Then $3x - 20$ = measure of each larger angle.
(measure of the two smaller angles) + (measure of the two larger angles) = 360°
$$x + x + (3x - 20) + (3x - 20) = 360$$
$$8x - 40 = 360$$
$$8x = 400$$
$$x = 50$$
Each smaller angle is 50°. Each larger angle is $3(50) - 20 = 130°$.

21. Let x = measure of the smallest angle. Then x + 10 = measure of the second angle, $2x$ + 14 = measure of third angle, and x + 21 = measure of fourth angle.
Sum of the four angles = 360°
$$x + (x + 10) + (2x + 14) + (x + 21) = 360$$
$$5x + 45 = 360$$
$$5x = 315$$
$$x = 63$$
Thus the angles are 63°, 63 + 10 = 73°
2(63) + 14 = 140° and 63 + 21 = 84°.

23. Let x = width of bookcase shelf. Then $x + 3$ = height of bookcase.
4 shelves + 2 sides = total lumber available.

$$\begin{aligned} 4x+2(x+3) &= 30 \\ 4x+2x+6 &= 30 \\ 6x+6 &= 30 \\ 6x &= 24 \\ x &= 4 \end{aligned}$$

The width of each shelf is 4 feet and the height is 4 + 3 = 7 feet.

25. Let x = length of a shelf. Then $2x$ = height of bookcase.
4 shelves + 2 sides = total lumber available

$$\begin{aligned} 4x+2(2x) &= 20 \\ 4x+4x &= 20 \\ 8x &= 20 \\ x &= \frac{20}{8} = 2.5 \end{aligned}$$

The width of the bookcase is 2.5 feet. The height of the bookcase is $2(2.5) = 5$ feet.

27. Let x = width of fenced in area. Then $x + 4$ = length of fenced in area
Five "widths" + one "length" = total fencing

$$\begin{aligned} 5x+(x+4) &= 64 \\ 6x+4 &= 64 \\ 6x &= 60 \\ x &= 10 \end{aligned}$$

Width is 10 feet and length is 10 + 4 = 14 feet.

29. a. Area of the outer square is S^2.
Area of the inner square is s^2.
Area of shaded region = Area of outer square – Area of inner square.
$A = S^2 - s^2$

b. $A = 9^2 - 6^2 = 81 - 36 = 45$ square inches.

32. $-|-6| < |-4|$ since $-|-6| = -6$ and $|-4| = 4$

33. $|-3| > -|3|$ since $|-3| = 3$ and $-|3| = -3$

34.
$$\begin{aligned} -6-(-2)+(-4) &= -6+2+(-4) \\ &= -4+(-4) \\ &= -8 \end{aligned}$$

35.
$$\begin{aligned} &-6y+x-3(x-2)+2y \\ &= -6y+x-3x+6+2y \\ &= x-3x-6y+2y+6 \\ &= -2x-4y+6 \end{aligned}$$

36.
$$\begin{aligned} 2x+3y &= 9 \\ 2x-2x+3y &= -2x+9 \\ 3y &= -2x+9 \\ \frac{3y}{3} &= \frac{-2x+9}{3} \\ y &= \frac{-2x+9}{3} \text{ or } y = -\frac{2}{3}x+3 \end{aligned}$$

Substitute 3 for x.

$$y = \frac{-2x+9}{3} = \frac{-6+9}{3} = \frac{3}{3} = 1$$

Exercise Set 3.5

1. Rate $= \frac{\text{distance}}{\text{time}} = \frac{150}{3} = 50$. Therefore, his average speed was 50 mph.

3. Thickness = rate · time = (0.2)(12) = 2.4. The door is 2.4 cm thick.

5. Rate $= \frac{\text{distance}}{\text{time}} = \frac{238,000}{87} \approx 2736$.
Therefore, the average speed was about 2736 mph.

7. Rate $= \frac{\text{volume}}{\text{time}} = \frac{1500}{6} = 250$.
Therefore, the flow rate should be 250 cm^3/hr.

9. Time $= \frac{\text{distance}}{\text{rate}} = \frac{5280}{82} \approx 64.4$
It will take about 64.4 days.

11. Rate $= \frac{\text{distance}}{\text{time}} = \frac{500}{2.635} \approx 189.75$. His average speed was approximately 189.75 mph.

13. Let t be the time it takes for Willie and Shanna to be 16.8 miles apart.

Person	Rate	Time	Distance
Willie	3	t	$3t$
Shanna	4	t	$4t$

$$3t + 4t = 16.8$$
$$7t = 16.8$$
$$t = 2.4$$

It will take 2.4 hours.

15. Let r be the second rate of the machine.

Machine	Rate	Time	Distance
First	60	7.2	432
Second	r	6.8	$6.8r$

$$432 + 6.8r = 908$$
$$6.8r = 476$$
$$r = 70$$

The second speed the machine was set at was 70 miles per hour.

17. Let t be the time they have been jogging.

Jogger	Rate	Time	Distance
Reba	5	t	$5t$
Kantrell	7	t	$7t$

$$5t + 4 = 7t$$
$$4 = 7t - 5t$$
$$4 = 2t$$
$$2 = t$$

They have been jogging for 2 hours.

19. a. Let r be the speed of the cutter coming from the east (westbound). Then $r + 5$ is the speed of the cutter coming from the west (eastbound).

Cutter	Rate	Time	Distance
Eastbound	$r + 5$	3	$3(r + 5)$
Westbound	r	3	$3r$

$$3(r+5) + 3r = 225$$
$$3r + 15 + 3r = 225$$
$$6r = 210$$
$$r = 35$$

The speed of the westbound cutter is 35 mph and the speed of the eastbound cutter is 40 mph.

b. $35 \text{ mph} \cdot \dfrac{1 \text{ knot}}{1.15 \text{ mph}} \approx 30.43 \text{ knots}$,

$40 \text{ mph} \cdot \dfrac{1 \text{ knot}}{1.15 \text{ mph}} \approx 34.78 \text{ knots}$

The speed of the westbound cutter is about 30.43 knots and the speed of the eastbound cutter is about 34.78 knots.

21. a. Distance = rate · time
= (2.35)(1.03)
≈ 2.4 miles

b. Distance = rate · time
= (20.82)(5.38)
≈ 112.0 miles

c. Distance = rate · time
= (8.45)(3.1)
≈ 26.2 miles

d. 2.4 + 112.0 + 26.2 = 140.6 miles

e. 1.03 + 5.38 + 3.1 = 9.51 hours

23. Let r be the rate of *Apollo*. Then $r + 4$ is the rate of *Pythagoras*.

Boat	Rate	Time	Distance
Apollo	r	0.7	$0.7r$
Pythagoras	$r + 4$	0.7	$0.7(r + 4)$

$$0.7r + 0.7(r + 4) = 9.8$$
$$0.7r + 0.7r + 2.8 = 9.8$$
$$1.4r = 7$$
$$r = 5$$

The speed of *Apollo* is 5 mph, and the speed of *Pythagoras* is 9 mph.

25. a. Let t be the time it takes for the Coast Guard to catch the bank robber.

Boat	Rate	Time	Distance
Robber	25	$t+\frac{1}{2}$	$25\left(t+\frac{1}{2}\right)$
Coast Guard	35	t	$35t$

$$25\left(t+\frac{1}{2}\right)=35t$$
$$25t+\frac{25}{2}=35t$$
$$\frac{25}{2}=10t$$
$$1.25=t$$

It will take 1.25 hours for the Coast Guard to catch the bank robber.

b. Distance = rate · time
= (35)(1.25)
= 43.75 miles.

27. a. Let t be the time it takes for Ray's pass to reach Pete.

Player	Rate	Time	Distance
Pete	25	$t+2$	$25(t+2)$
Ray	50	t	$50t$

$$25(t+2)=50t$$
$$25t+50=50t$$
$$50=25t$$
$$2=t$$

It will take 2 seconds for Ray's pass to reach Pete.

b. Distance = rate · time = 50 · 2 = 100 feet

29. Let r = speed of moving walkway (in ft/min)

	Rate	Time	Distance
Derek on foot	100 ft/min	2.75 min	275 ft
Derek on walkway	$100 + r$	1.25 min	$(100 + r)(1.25)$

a. 275 ft

b. distance going (on foot) = distance returning (walking on moving walkway)

$$275=(100+r)(1.25)$$
$$275=125+1.25r$$
$$150=1.25r$$
$$120=r$$

The moving walkway moves at 120 ft/min.

31. Let r be the planned speed of the plane. Then $r + 30$ is the new speed of the plane.

Speed	Rate	Time	Distance
Planned	r	4	$4r$
New	$r+30$	$4-0.2$	$3.8(r+30)$

$$4r=3.8(r+30)$$
$$4r=3.8r+114$$
$$0.2r=114$$
$$r=570$$

The plane's planned speed is 570 miles per hour. The plane's new speed is 600 miles per hour.

33. Let r be the rate of clearing the bridge. Then $1.2 + r$ is the rate of clearing the road.

	Rate	Time	Distance
Road	$1.2 + r$	20	$20(1.2 + r)$
Bridge	r	60	$60r$

$$20(1.2+r)+60r=124$$
$$24+20r+60r=124$$
$$80r=100$$
$$r=1.25$$

The crew will clear the bridge at a rate of 1.25 feet/day, and they will clear the road at a rate of 2.45 feet/day.

35. Let x be the amount invested at 5%. Then $9400 - x$ is the amount invested at 7%.

Principal	Rate	Time	Interest
x	5%	1	$0.05x$
$9400-x$	7%	1	$0.07(9400-x)$

$$0.05x + 0.07(9400 - x) = 610$$
$$0.05x + 658 - 0.07x = 610$$
$$-0.02x = -48$$
$$x = 2400$$

He invested \$2400 at 5% and \$7000 at 7%.

37. Let x be the amount invested at 6%. Then $6000 - x$ is the amount invested at 4%.

Principal	Rate	Time	Interest
x	6%	1	$0.06x$
$6000 - x$	4%	1	$0.04(6000 - x)$

$$0.06x = 0.04(6000 - x)$$
$$0.06x = 240 - 0.04x$$
$$0.10x = 240$$
$$x = 2400$$

She invested \$2400 at 6% and \$3600 at 4%.

39. Let t be the time, in months, during which Patricia paid \$15.10 per month. Then $12 - t$ is the time during which she paid \$16.40 per month.

Rate	Time	Amount
15.10	t	$15.10t$
16.40	$12 - t$	$16.40(12 - t)$

$$15.10t + 16.40(12 - t) = 183.80$$
$$15.10t + 196.80 - 16.40t = 183.80$$
$$-1.30t = -13$$
$$t = 10$$

She paid \$15.10 for the first 10 months of the year, and paid \$16.40 for the remainder of the year. The rate increase took effect in November.

41. Let x be the number of hours worked at Home Depot (\$6.50 per hour). Then $18 - x$ is the number of hours worked at the veterinary clinic (\$7.00 per hour).

Rate	Hours	Total
\$6.50	x	$6.5x$
\$7.00	$18 - x$	$7(18 - x)$

$$6.5x + 7(18 - x) = 122$$
$$6.5x + 126 - 7x = 122$$
$$-0.5x = -4$$
$$x = 8$$

Mihàly worked 8 hours at Home Depot and 10 hours at the clinic.

43. Let x be the cost per pound of the mixture.

Type	Cost	Pounds	Total
Bavarian	\$6.50	2	\$13.00
Columbian	\$5.90	5	\$29.50
Mixture	x	7	$7x$

$$13 + 29.50 = 7x$$
$$42.50 = 7x$$
$$6.07 \approx x$$

The mixture should cost \$6.07 per pound.

45. Let x be the percentage of alcohol in the mixture.

Percentage	Liters	Amount of Alcohol
12%	5	0.6
9%	2	0.18
x%	7	$\left(\frac{x}{100}\right) \cdot 7$

$$\left(\frac{x}{100}\right) \cdot 7 = 0.6 + 0.18$$
$$0.07x = 0.78$$
$$x \approx 11.1$$

The alcohol content of the mixture is about 11.1%.

47. Let x be the amount of 12 % sulfuric acid solution.

Solution	Strength	Liters	Amount
20%	0.20	1	0.20
12%	0.12	x	$0.12x$
Mixture	0.15	$x + 1$	$0.15(x + 1)$

$$0.20 + 0.12x = 0.15(x+1)$$
$$0.20 + 0.12x = 0.15x + 0.15$$
$$-0.03x = -0.05$$
$$x = \frac{5}{3}$$

$1\frac{2}{3}$ liters of 12% sulfuric acid should be used.

49. Let x be the amount of the swimming pool shock treatment to be added to a quart of water.

Product	Percentage	Ounces	Amount
Clorox	5.25%	8	$(0.0525)\cdot 8$
Shock Treatment	10.5%	x	$0.105x$

$$(0.0525)\cdot 8 = 0.105x$$
$$0.42 = 0.105x$$
$$4 = x$$

Add 4 ounces of the shock treatment to a quart of water.

51. Let x be the amount of Clorox to be used. Then $4 - x$ is the amount of the shock treatment to be used.

Product	Percentage	Gallons	Amount
Clorox	5.25%	x	$0.0525x$
Shock Treatment	10.5%	$4-x$	$0.105(4-x)$
Fungus Remover	8%	4	$(0.08)\cdot 4$

$$0.0525x + 0.105(4-x) = (0.08)\cdot 4$$
$$0.0525x + 0.42 - 0.105x = 0.32$$
$$-0.0525x = -0.10$$
$$x \approx 1.9$$

She should mix 1.9 gallons of Clorox and 2.1 gallons of the shock treatment.

53. Let x be the percentage of milkfat in whole milk.

Type	Percentage	Gallons	Amount
Whole	$x\%$	4	$\left(\frac{x}{100}\right)\cdot 4$
Low fat	1%	5	$(0.01)\cdot 5$
Reduced fat	2%	9	$(0.02)\cdot 9$

$$\left(\frac{x}{100}\right)\cdot 4 + (0.01)\cdot 5 = (0.02)\cdot 9$$
$$\frac{4x}{100} + 0.05 = 0.18$$
$$\frac{4x}{100} = 0.13$$
$$4x = 13$$
$$x = 3.25$$

The milkfat content of whole milk is 3.25%.

55. Let x be the percentage of pure juice before mixing.

Solution	Percentage	Amount of Solution	Amount of Juice
Concentrate	x	12	$12\left(\frac{x}{100}\right)$
Mixture	10%	48	4.8

$$12\left(\frac{x}{100}\right) = 4.8$$
$$0.12x = 4.8$$
$$x = 40$$

40% of the concentrate is pure juice.

57. a. Let x be the number of shares of General Electric. Then $5x$ is the number of shares of PepsiCo.

Stock	Price	Shares	Total
GE	\$74	x	$74x$
PepsiCo	\$35	$5x$	$35\cdot 5x$

$$\begin{aligned} 74x + 35 \cdot 5x &= 8000 \\ 74x + 175x &= 8000 \\ 249x &= 8000 \\ x &\approx 32.1 \end{aligned}$$

Since only whole shares can be purchased, he will purchase 32 shares of GE and 160 shares of PepsiCo.

b. Mr. Gilbert spent
$32 \cdot 74 + 160 \cdot 35 = \$7968$.
He has \$8000 – \$7968 = \$32 left over.

59. The time it takes for the transport to make the trip is:
$\text{Time} = \dfrac{\text{Distance}}{\text{Rate}} = \dfrac{1720}{370} \approx 4.65$ hours
The time it takes for the Hornets to make the trip is:
$\text{Time} = \dfrac{\text{Distance}}{\text{Rate}} = \dfrac{1720}{900} \approx 1.91$ hours
It takes the transport $4.65 - 1.91 = 2.74$ hours longer to make the trip. Since it needs to arrive 3 hours before the Hornets, it should leave about 2.74 + 3 = 5.74 hours before them.

63. a.
$$\begin{aligned} 2\frac{3}{4} \div 1\frac{5}{8} &= \frac{11}{4} \div \frac{13}{8} \\ &= \frac{11}{4} \cdot \frac{8}{13} \\ &= \frac{22}{13} \text{ or } 1\frac{9}{13} \end{aligned}$$

b.
$$\begin{aligned} 2\frac{3}{4} + 1\frac{5}{8} &= \frac{11}{4} + \frac{13}{8} \\ &= \frac{22}{8} + \frac{13}{8} \\ &= \frac{35}{8} \text{ or } 4\frac{3}{8} \end{aligned}$$

64.
$$\begin{aligned} 6(x-3) &= 4x - 18 + 2x \\ 6x - 18 &= 6x - 18 \end{aligned}$$
All real numbers are solutions.

65.
$$\begin{aligned} \frac{6}{x} &= \frac{72}{9} \\ 6 \cdot 9 &= 72x \\ 54 &= 72x \\ x &= \frac{54}{72} = \frac{3}{4} \text{ or } 0.75 \end{aligned}$$

66.
$$\begin{aligned} 3x - 4 &\le -4x + 3(x-1) \\ 3x - 4 &\le -4x + 3x - 3 \\ 3x - 4 &\le -x - 3 \\ 4x &\le 1 \\ x &\le \frac{1}{4} \end{aligned}$$

Review Exercises

1. Substitute 8 for d.
$$\begin{aligned} C &= \pi d \\ C &= \pi(8) \\ &\approx 25.13 \end{aligned}$$

2. Substitute 4 for l and 5 for w.
$$\begin{aligned} P &= 2l + 2w \\ P &= 2(4) + 2(5) = 8 + 10 = 18 \end{aligned}$$

3. Substitute 8 for b and 12 for h.
$$\begin{aligned} A &= \frac{1}{2}bh \\ A &= \frac{1}{2}(8)(12) = 48 \end{aligned}$$

4. Substitute 200 for K and 4 for v.
$$\begin{aligned} K &= \frac{1}{2}mv^2 \\ 200 &= \frac{1}{2}m(4)^2 \\ 200 &= \frac{1}{2}m(16) \\ 200 &= 8m \\ \frac{200}{8} &= \frac{8m}{8} \\ 25 &= m \end{aligned}$$

5. Substitute 15 for y, 3 for m, and –2 for x.

$$y = mx + b$$
$$15 = (3)(-2) + b$$
$$15 = -6 + b$$
$$15 + 6 = -6 + 6 + b$$
$$21 = b$$

6. Substitute 4716.98 for P and 0.06 for i.

$$P = \frac{f}{1+i}$$
$$4716.98 = \frac{f}{1+0.06}$$
$$4716.98 = \frac{f}{1.06}$$
$$1.06 \times 4716.98 = \frac{f}{1.06} \times 1.06$$
$$5000 \approx f$$

7. a.

$$2x = 2y + 4$$
$$2x - 4 = 2y + 4 - 4$$
$$2x - 4 = 2y$$
$$\frac{2x-4}{2} = \frac{2y}{2}$$
$$\frac{2x}{2} - \frac{4}{2} = y$$
$$x - 2 = y$$

b. Substitute 10 for x.

$$10 - 2 = y$$
$$8 = y$$

8. a.

$$6x + 3y = -9$$
$$6x - 6x + 3y = -9 - 6x$$
$$3y = -9 - 6x$$
$$\frac{3y}{3} = \frac{-9-6x}{3}$$
$$y = \frac{-9}{3} - \frac{6x}{3}$$
$$y = -3 - 2x$$

b. Substitute 12 for x.

$$y = -3 - 2(12) = -3 - 24 = -27$$

9. a.

$$4x - 3y = 15$$
$$4x - 4x - 3y = 15 - 4x$$
$$-3y = 15 - 4x$$
$$\frac{-3y}{-3} = \frac{15-4x}{-3}$$
$$y = \frac{15-4x}{-3}$$
$$y = \frac{4}{3}x - 5$$

b. Substitute 3 for x.

$$y = \frac{4}{3}(3) - 5 = 4 - 5 = -1$$

10. a.

$$2x = 3y + 12$$
$$2x - 12 = 3y + 12 - 12$$
$$2x - 12 = 3y$$
$$\frac{2x-12}{3} = \frac{3y}{3}$$
$$\frac{2x-12}{3} = y$$
$$y = \frac{2}{3}x - 4$$

b. Substitute –6 for x.

$$y = \frac{2}{3}(-6) - 4$$
$$= -4 - 4$$
$$= -8$$

11.

$$F = ma$$
$$\frac{F}{a} = \frac{ma}{a}$$
$$\frac{F}{a} = m$$

12. $A = \frac{1}{2}bh$

$2A = 2\left(\frac{1}{2}bh\right)$

$2A = bh$

$\frac{2A}{b} = \frac{bh}{b}$

$\frac{2A}{b} = h$

13. $i = prt$

$\frac{i}{pr} = \frac{prt}{pr}$

$\frac{i}{pr} = t$

14. $P = 2l + 2w$

$P - 2l = 2l - 2l + 2w$

$P - 2l = 2w$

$\frac{P - 2l}{2} = \frac{2w}{2}$

$\frac{P - 2l}{2} = w$

15. $V = \pi r^2 h$

$\frac{V}{\pi r^2} = \frac{\pi r^2 h}{\pi r^2}$

$\frac{V}{\pi r^2} = h$

16. $A = \frac{B + C}{2}$

$2A = 2\left(\frac{B + C}{2}\right)$

$2A = B + C$

$2A - C = B + C - C$

$2A - C = B$

17. Substitute 600 for p, 0.09 for r, and 2 for t.
$i = prt$
$i = (600)0.09(2) = 108$
Tom will pay $108 interest.

18. $P = 2l + 2w$
$16 = 2l + 2(2)$
$16 = 2l + 4$
$12 = 2l$
$6 = l$
The length of the rectangle is 6 inches.

19. The sum of a number and the number increased by 5 is 9.

20. The sum of a number and twice the number decreased by 1 is 10.

21. Let x = the smaller number. Then
$x + 8$ = the larger number.
Smaller number + larger number = 74
$x + (x + 8) = 74$
$2x + 8 = 74$
$2x = 66$
$x = 33$
The smaller number is 33 and the larger number is 33 + 8 = 41.

22. Let x = smaller integer. Then
$x + 1$ = next consecutive integer.
Smaller number + larger number = 237
$x + (x + 1) = 237$
$2x + 1 = 237$
$2x = 236$
$x = 118$
The smaller number is 118 and the larger number is 118 + 1 = 119.

23. Let x = the smaller integer. Then
$5x + 3$ = the larger integer
Larger number – smaller number = 31
$(5x + 3) - x = 31$
$4x + 3 = 31$
$4x = 28$
$x = 7$
The smaller number is 7 and the larger number is 5(7) + 3 = 38.

24. Let x = cost of car before tax. Then $0.07x$ = amount of tax.
Cost of car before tax + tax on car = cost of car after tax

$$x + 0.07x = 19,260$$
$$1.07x = 19,260$$
$$x = \frac{19,260}{1.07} = 18,000$$

The cost of the car before tax is $18,000.

25. Let x = weekly dollar sales that would make total salaries from both companies the same.
The commission at present company $= 0.03x$ and commission at new company $= 0.08x$
Salary + commission for present company = salary + commission for new company

$$500 + 0.03x = 400 + 0.08x$$
$$100 + 0.03x = 0.08x$$
$$100 = 0.05x$$
$$\frac{100}{0.05} = x$$
$$2000 = x$$

Ron's weekly sales would have to be $2000 for the total salaries from both companies to be the same.

26. Let x = original price of camcorder. Then $0.20x$ = reduction during first week.
Original price – first reduction – second reduction = price during second week

$$x - 0.20x - 25 = 495$$
$$0.8x - 25 = 495$$
$$0.8x = 520$$
$$x = \frac{520}{0.8} = 650$$

The original price of the camcorder was $650.

27. a. The monthly payments with the 8.875% mortgage are 60(7.96) = $477.60. With the 8.625% mortgage, the monthly payments are 60(7.78) = $466.80. In addition the Johnsons must pay 0.03(60,000) = $1800 because of the three points. Let x = number of months when total payments from the two mortgages are the same.
Monthly payments for Comerica = monthly payments + points for Mellon

$$477.60x = 466.80x + 1800$$
$$10.8x = 1800$$
$$x = \frac{1800}{10.8} \approx 166.7$$

Total payments from the two mortgages are the same after about 166.7 months or 13.9 years.

b. After about 13.9 years the Mellon mortgage is cheaper because of the lower monthly payments. If the Johnsons plan to keep the house for 20 years, they should apply to the Mellon Bank.

28. a. If she refinances, her monthly payments will be 70(8.68) = $607.60. Her monthly savings will be $750 – $607.60 = $142.40.
The cost of the point will be 0.01(70,000) = $700
Let x = number of months when savings equals cost of the point and the closing cost.

$$142.40x = 700 + 3200$$
$$142.40x = 3900$$
$$x = \frac{3900}{142.40} \approx 27.4$$

The money she saves will equal the cost of the point and the closing cost after about 27.4 months or 2.3 years.

b. After about 2.3 years the money she saves on the monthly payments exceeds the cost of the point and the closing cost. If she plans to live in the house for 10 more years it does pay her to refinance.

29. Let x = measure of the smallest angle. Then
$x + 10$ = measure of second angle and
$2x - 10$ = measure of third angle.
Sum of the three angles = 180°

$$x+(x+10)+(2x-10)=180$$
$$4x=180$$
$$x=45$$

The angles are 45°, 45 + 10 = 55°, and
2(45) − 10 = 80°.

30. Let x = measure of the smallest angle. Then
$x + 10$ = measure of second angle,
$5x$ = measure of third angle,
$4x + 20$ = measure of the fourth angle.
Sum of the four angles = 360°

$$x+(x+10)+5x+(4x+20)=360$$
$$11x+30=360$$
$$11x=330$$
$$x=\frac{330}{11}=30$$

The angles are 30°, 30 + 10 = 40°.
5(30) = 150°, and 4(30) + 20 = 140°.

31. Let w = width of garden. Then
$w + 4$ = length of garden.

$$P=2l+2w$$
$$70=2(w+4)+2w$$
$$70=4w+8$$
$$62=4w$$
$$15.5=w$$

The width is 15.5 feet and the length is
15.5 + 4 = 19.5 feet.

32. Let x = the width of the room. Then
$x + 30$ = the length of the room. The amount of string used is the perimeter of the room plus the wall separating the two rooms.

$$P=2l+2w+w$$
$$P=2l+3w$$
$$310=2(x+30)+3x$$
$$310=2x+60+3x$$
$$310=5x+60$$
$$250=5x$$
$$50=x$$

The width of the room is 50 feet and the length is 50 + 30 = 80 feet.

33. Let r be the flow rate of the water.
Amount = rate · time

$$105=r\cdot(3.5)$$
$$\frac{105}{3.5}=\frac{r\cdot(3.5)}{3.5}$$
$$30=r$$

The flow rate of the water is 30 gallons per hour.

34. $\text{Speed} = \frac{\text{distance}}{\text{time}} = \frac{26}{4} = 6.5 \text{ mph}$

35. Let t be the time it takes for the joggers to be 4 kilometers apart.

Jogger	Rate	Time	Distance
Harold	8	t	$8t$
Susan	6	t	$6t$

$$8t-6t=4$$
$$2t=4$$
$$t=2$$

So, it takes the joggers 2 hours to be 4 kilometers apart.

36. Let t be the amount of time it takes for the trains to be 440 miles apart.

Train	Rate	Time	Distance
First	50	t	$50t$
Second	60	t	$60t$

$$50t+60t=440$$
$$110t=440$$
$$t=4$$

After 4 hours, the two trains will be 440 miles apart.

37. Let x be the amount invested at 8%. Then
$12{,}000 - x$ is the amount invested at $7\frac{1}{4}\%$.

Principal	Rate	Time	Interest
x	8%	1	$0.08x$
$12{,}000 - x$	$7\frac{1}{4}\%$	1	$0.0725(12{,}000 - x)$

$$\begin{aligned} 0.08x + 0.0725(12,000 - x) &= 900 \\ 0.08x + 870 - 0.0725x &= 900 \\ 0.0075x &= 30 \\ x &= 4000 \end{aligned}$$

Kelly should invest $4000 at 8% and $8000 at $7\frac{1}{4}\%$.

38. Let x be the number of liters of the 10% solution. Then $2 - x$ is the number of liters of the 5% acid solution.

Solution	Strength	Liters	Amount
10%	0.10	x	$0.10x$
5%	0.05	$2 - x$	$0.05(2 - x)$
Mixture	0.08	2	0.16

$$\begin{aligned} 0.10x + 0.05(2 - x) &= 0.16 \\ 0.10x + 0.10 - 0.05x &= 0.16 \\ 0.05x &= 0.06 \\ x &= 1.2 \end{aligned}$$

The chemist should mix 1.2 liters of 10% solution with 0.8 liters of 5% solution.

39. Let x = smaller odd integer. Then $x + 2$ = next consecutive odd integer.
Smaller number + larger number = 208

$$\begin{aligned} x + (x + 2) &= 208 \\ 2x + 2 &= 208 \\ 2x &= 206 \\ x &= 103 \end{aligned}$$

The smaller number is 103 and the larger number is 103 + 2 = 105.

40. Let x = cost of television before tax. Then amount of tax = $0.06x$. Cost of television before tax + tax on television = cost of television after tax.

$$\begin{aligned} x + 0.06x &= 477 \\ 1.06x &= 477 \\ x &= \frac{477}{1.06} = 450 \end{aligned}$$

The cost of the television before tax is $450.

41. Let x = his dollar sales. Then $0.05x$ = amount of commission.
Salary + commission = 900

$$\begin{aligned} 300 + 0.05x &= 900 \\ 0.05x &= 600 \\ x &= \frac{600}{0.05} = 12,000 \end{aligned}$$

His sales last week were $12,000.

42. Let x = measure of the smallest angle. Then $x + 8$ = measure of second angle and $2x + 4$ = measure of third angle.
Sum of the three angles = 180°

$$\begin{aligned} x + (x + 8) + (2x + 4) &= 180 \\ 4x + 12 &= 180 \\ 4x &= 168 \\ x &= 42 \end{aligned}$$

The angles are 42°, 42 + 8 = 50°, and 2(42) + 4 = 88°.

43. Let t = number of years. Then $25t$ = increase in employees over t years
Present number of employees + increase in employees = future number of employees

$$\begin{aligned} 427 + 25t &= 627 \\ 25t &= 200 \\ t &= \frac{200}{25} = 8 \end{aligned}$$

It will take 8 years before they reach 627 employees.

44. Let x = measure of each smaller angle. Then $x + 40$ = measure of each larger angle
(measure of the two smaller angles) +(measure of the two larger angles) = 360°

$$\begin{aligned} x + x + (x + 40) + (x + 40) &= 360 \\ 4x + 80 &= 360 \\ 4x &= 280 \\ x &= 70 \end{aligned}$$

Each of the smaller angles is 70° and each of the two larger angles is 70 + 40 = 110°.

45. a. Let x = number of copies that would result in both centers charging the same. Then charge for copies at Copy King = $0.04x$ and charge for copies at King Kopie = $0.03x$
Monthly fee + charge for copies at Copy King = monthly fee + charge for

copies at King Kopie.

$$20 + 0.04x = 25 + 0.03x$$
$$0.04x = 5 + 0.03x$$
$$0.01x = 5$$
$$x = \frac{5}{0.01} = 500$$

500 copies would result in both centers charging the same.

b. Charge for 1000 copies at Copy King $= 20 + (0.04)(1000) = 20 + 40 = \$60$.
Charge for copies at King Kopie $= 25 + (0.03)(1000) = 25 + 30 = \$55$
King Kopie would cost less for 1000 copies by $60 - 55 = \$5$.

46. a. Let t = time the sisters meet after Chris starts swimming.

Person	Rate	Time	Distance
Chris	60	t	$60t$
Kathy	50	$t+2$	$50(t+2)$

$$60t = 50(t+2)$$
$$60t = 50t + 100$$
$$10t = 100$$
$$t = 10$$

The sisters will meet 10 mintutes after Chris starts swimming.

b. rate · time = distance
$60 \cdot 10 = 600$
The sisters will be 600 feet from the boat when they meet.

47. Let x be the amount of \$3.50 per pound hamburger. Then $80 - x$ is the amount of \$4.10 per pound hamburger.

Hamburger	Price	Amount	Total
\$3.50	3.50	x	$3.50x$
\$4.10	4.10	$80 - x$	$4.10(80 - x)$
Mixture	3.65	80	292

$$3.50x + 4.10(80 - x) = 292$$
$$3.50x + 328 - 4.10x = 292$$
$$-0.60x = -36$$
$$x = 60$$

The butcher mixed 60 lbs of \$3.50 per pound hamburger with 20 lbs of \$4.10 per pound hamburger.

48. Let x = the rate the older brother travels. Then
$x + 5$ = the rate the younger brother travels.

Brother	Rate	Time	Distance
Younger	$x+5$	2	$2(x+5)$
Older	x	2	$2x$

Younger brother's distance + older brother's distance = 230 miles.

$$2(x+5) + 2x = 230$$
$$2x + 10 + 2x = 230$$
$$4x + 10 = 230$$
$$4x = 220$$
$$x = 55$$

The older brother travels at 55 miles per hour and the younger brother travels at $55 + 5 = 60$ miles per hour.

49. Let x = the number of liters of 30% solution.

Percent	Liters	Amount
30%	x	$0.30x$
12%	2	$(0.12)(2)$
15%	$x+2$	$0.15(x+2)$

$$0.30x + (0.12)(2) = 0.15(x+2)$$
$$0.30x + 0.24 = 0.15x + 0.30$$
$$0.15x + 0.24 = 0.30$$
$$0.15x = 0.06$$
$$x = 0.4$$

0.4 liters of the 30% acid solution need to be added.

Practice Test

1. Let r = interest rate in decimal form.

$$i = prt$$
$$3240 = (12,000)(r)(3)$$
$$3240 = 36,000r$$
$$\frac{3240}{36,000} = r$$
$$0.09 = r$$

The interest rate is 9%.

2. $P = 2l + 2w$

$$P = 2(6) + 2(3) = 12 + 6 = 18$$

The perimeter is 18 feet.

3. $A = P + Prt$

$$A = 100 + (100)(0.15)(3)$$
$$= 100 + 45$$
$$= 145$$

4.

$$A = \frac{m+n}{2}$$
$$79 = \frac{73+n}{2}$$
$$2 \times 79 = \left(\frac{73+n}{2}\right) \times 2$$
$$158 = 73 + n$$
$$85 = n$$

5.

$$C = 2\pi r$$
$$50 = 2\pi r$$
$$\frac{50}{2\pi} = \frac{2\pi r}{2\pi}$$
$$7.96 \approx r$$

6. a.

$$4x = 3y + 9$$
$$4x - 9 = 3y + 9 - 9$$
$$4x - 9 = 3y$$
$$\frac{4x-9}{3} = \frac{3y}{3}$$
$$\frac{4x-9}{3} = y$$
$$y = \frac{4}{3}x - 3$$

b. Substitute 12 for x.

$$y = \frac{4}{3}(12) - 3$$
$$= 4(4) - 3$$
$$= 16 - 3$$
$$= 13$$

7. $P = IR$

$$\frac{P}{I} = \frac{IR}{I}$$
$$\frac{P}{I} = R$$

8.

$$A = \frac{a+b}{3}$$
$$3A = 3\left(\frac{a+b}{3}\right)$$
$$3A = a + b$$
$$3A - b = a + b - b$$
$$3A - b = a$$

9.

$$D = R(c+a)$$
$$\frac{D}{R} = \frac{R(c+a)}{R}$$
$$\frac{D}{R} = c + a$$
$$\frac{D}{R} - a = c + a - a$$
$$\frac{D}{R} - a = c$$

$$c = \frac{D}{R} - a \text{ or } \frac{D - Ra}{R}$$

10. The area of the skating rink is the area of a rectangle plus the area of two half circles, or one full circle. The radius of the circle is $\frac{30}{2} = 15$ feet.

Area of rectangle $= l \cdot w = 80 \cdot 30 = 2400 \text{ ft}^2$

Area of circle $= \pi r^2$
$= \pi(15)^2$
$= 225\pi$
$\approx 706.86 \text{ ft}^2$

Total area $= 2400 + 706.86 = 3106.86 \text{ ft}^2$

11. a. Let x = the amount of money Matthew receives.

b. Then $300 - x$ = the amount of money Karen receives.

12. a. Let z = the projected gross income from *Mask of Zorro*.

b. Then $2z - 45$ = the projected gross income from *Deep Impact*.

13. The sum of a number and the number increased by 3 is 7.

14. a. Let x = smaller integer.

b. Then $2x - 10$ = larger integer.
Smaller number + larger number = 158
$x + (2x - 10) = 158$

c. $3x - 10 = 158$
$3x = 168$
$x = 56$
The smaller number is 56 and the larger number is $2(56) - 10 = 102$

15. a. Let x = smallest integer.

b. Then $x + 1$ and $x + 2$ are the next two consecutive integers. Sum of the three integers = 42
$x + (x+1) + (x+2) = 42$

c. $3x + 3 = 42$
$3x = 39$
$x = 13$
The integers are 13, 13 + 1 = 14, and 13 + 2 = 15.

16. a. Let c = the cost of the furniture before tax.

b. Tax amount = (cost) · (tax rate) = $0.06c$
Total cost = cost before tax + tax amount
$2650 = c + 0.06c$

c. $2640 = 1.06c$
$\frac{2650}{1.06} = \frac{1.06c}{1.06}$
$2500 = c$
The cost of the furniture before tax was \$2500.

17. a. Let x = price of most expensive meal he can order.

b. Then $0.15x$ = tip and $0.07x$ = tax
Price of meal + tip + tax = 40
$x + 0.15x + 0.07x = 40$

c. $1.22x = 40$
$x = \frac{40}{1.22} \approx 32.79$
The price of the most expensive meal he can order is \$32.79.

18. a. Let x = the amount of profit Kathleen and Corrina each receive.

b. Then $2x$ = the amount of profit Kristen receives. Kathleen's profit + Corrina's profit + Kristin's profit = Total profit
$x + x + 2x = 120,000$

c. $4x = 120,000$
$x = 30,000$
Kathleen and Corrinna each receive \$30,000, and Kristin receives 2(30,000) = \$60,000.

19. a. Let x = the number of times the plow is needed for the costs to be equal.

b. Antoinette Payne's charge = $80 + 5x$
Ray Mesing's charge = $50 + 10x$
The charges are equal when:
$80 + 5x = 50 + 10x$

c. $80 + 5x = 50 + 10x$
$30 + 5x = 10x$
$30 = 5x$
$6 = x$
The snow would need to be plowed 6 times for the costs to be the same.

20. a. Let x = number of months for the total cost of both mortgages to be equal.

b. Then $451.20x$ = monthly payments for x months at Bank One,
and $430.20x$ = monthly payments for x months at NationsBank.
Points from Bank One = 0.01(60,000) = 600,
and points from NationsBank = 0.04(60,000) = 2400
$451.20x + 600 = 430.20x + 2400$

c. $451.20x + 600 = 430.20x + 2400$
$21x + 600 = 2400$
$21x = 1800$
$x \approx 85.7$
It would take about 85.7 months or 7.1 years for the costs to be the same.

21. a. Let x = length of smallest side.

b. Then $x + 15$ = length of second side and $2x$ = length of third side.
Sum of the three sides = perimeter
$x + (x + 15) + 2x = 75$

c. $4x + 15 = 75$
$4x = 60$
$x = 15$
The three sides are 15 inches, 15 + 15 = 30 inches, and 2(15) = 30 inches.

22. a. Let x = measure of each smaller angle.

b. Then $2x + 30$ = measure of each larger angle
(measure of the two smaller angles) + (measure of the two larger angles) = 360°
$x + x + (2x + 30) + (2x + 30) = 360$

c. $6x + 60 = 360$
$6x = 300$
$x = 50$
Each of the smaller angles is 50° and each of the larger angles is 2(50) + 30 = 130°.

23. a. Let x = width of the driveway.

b. Then $4x + 12$ = length.
Perimeter $= 2l + 2w$
$144 = 2(4x + 12) + 2x$

c. $144 = 8x + 24 + 2x$
$144 = 10x + 24$
$120 = 10x$
$12 = x$
The width is 12 feet and the length is $4(12) + 12 = 48 + 12 = 60$ feet

24. a. Let x = the rate Carrie digs.

b. Then $x + 0.2$ is the rate Don digs.

Name	Rate	Time	Distance
Carrie	x	84	$84x$
Don	$x + 0.2$	84	$84(x + 0.2)$

Distance Carrie digs + distance Don digs = total length of trench
$84x + 84(x + 0.2) = 67.2$

c. $84x + 84x + 16.8 = 67.2$
$168x = 50.4$
$x = 0.3$
Carrie digs at 0.3 feet per minute and Don digs at 0.3 + 2 = 0.5 feet per minute.

25. a. Let x = amount of 20% salt solution to be added.

b.

Percent	Liters	Amount
20%	x	$0.20x$
40%	60	(0.40)(60)
35%	$x + 60$	$0.35(x + 60)$

$0.20x + (0.40)(60) = 0.35(x + 60)$

c. $0.20x + 24 = 0.35x + 21$
$-0.15x + 24 = 21$
$-0.15x = -3$
$x = 20$
20 liters of 20% solution must be added.

Cumulative Review Test

1. 40% of 88.4 million tons = (0.40)(88.4) = 35.36 million tons

2. a. $192,585 - 63,323 = 129,262$

b. $192,585 + 63,323 = 255,908$

c. $63,323n = 192,585$
$\frac{63,323n}{63,323} = \frac{192,585}{63,323}$
$n \approx 3.04$ times greater

3. a. $\frac{5+6+8+12+5}{5} = 7.2$
The mean level was 7.2 parts per million.

b. Carbon dioxide levels in order:
5, 5, 6, 8, 12
The median is 6 parts per million.

4. $\frac{5}{12} \div \frac{3}{4} = \frac{5}{12} \cdot \frac{4}{3} = \frac{20}{36} = \frac{5}{9}$

5. $\frac{2}{3} - \frac{3}{8} = \frac{2 \cdot 8}{3 \cdot 8} - \frac{3 \cdot 3}{8 \cdot 3} = \frac{16}{24} - \frac{9}{24} = \frac{7}{24}$
$\frac{2}{3}$ inch is $\frac{7}{24}$ inch greater than $\frac{3}{8}$ inch.

6. a. $\{1, 2, 3, 4, \ldots\}$

b. $\{0, 1, 2, 3, \ldots\}$

c. A rational number is a quotient of two integers, denominator not 0.

7. a. $|-3| = 3$

b. $|-5| = 5$ and $|-3| = 3$. Since $5 > 3$, $|-5| > |-3|$.

8. $12 - 6 \div 2 \cdot 2 = 12 - 3 \cdot 2 = 12 - 6 = 6$

9. $4(2x - 3) - 2(3x + 5) - 6$
$= 8x - 12 - 6x - 10 - 6$
$= 2x - 28$

10. $3x - 6 = x + 12$
$3x - 6 - x = x - x + 12$
$2x - 6 = 12$
$2x - 6 + 6 = 12 + 6$
$2x = 18$
$\frac{2x}{2} = \frac{18}{2}$
$x = 9$

11. $6r = 2(r + 3) - (r + 5)$
$6r = 2r + 6 - r - 5$
$6r = r + 1$
$6r - r = r - r + 1$
$5r = 1$
$\frac{5r}{5} = \frac{1}{5}$
$r = \frac{1}{5}$

12. $2(x + 5) = 3(2x - 4) - 4x$
$2x + 10 = 6x - 12 - 4x$
$2x + 10 = 2x - 12$
$2x - 2x + 10 = 2x - 2x - 12$
$10 = -12$
The equation has no solution.

13. $\frac{50\text{ miles}}{2\text{ gallons}} = \frac{225\text{ miles}}{x\text{ gallons}}$

$\frac{50}{2} = \frac{225}{x}$

$50x = 450$

$x = 9$

It will need 9 gallons.

14. $3x - 4 \le -1$

$3x - 4 + 4 \le -1 + 4$

$3x \le 3$

$\frac{3x}{3} \le \frac{3}{3}$

$x \le 1$

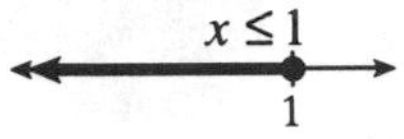

15. Substitute 6 for r.

$A = \pi r^2$

$A = \pi(6)^2 = 36\pi \approx 113.10$

16. **a.** $3x + 6y = 12$

$3x - 3x + 6y = 12 - 3x$

$6y = 12 - 3x$

$\frac{6y}{6} = \frac{12 - 3x}{6}$

$y = \frac{12}{6} - \frac{3x}{6}$

$y = 2 - \frac{1}{2}x$

$y = -\frac{1}{2}x + 2$

b. Substitute –4 for x.

$y = -\frac{1}{2}(-4) + 2 = 2 + 2 = 4$

17. $P = 2l + 2w$

$P - 2w = 2l + 2w - 2w$

$P - 2w = 2l$

$\frac{P - 2w}{2} = \frac{2l}{2}$

$\frac{P - 2w}{2} = l$

18. Let x = smaller number. Then $2x + 11$ = larger number.

smaller number + larger number = 29

$x + (2x + 11) = 29$

$3x + 11 = 29$

$3x = 18$

$x = 6$

The smaller number is 6 and the larger number is $2(6) + 11 = 23$.

19. Let x = number of minutes for the two plans to have the same cost.

Cost for Plan A = $19.95 + 0.35x$

Cost for Plan B = $29.95 + 0.10x$

The costs will be equal when the cost for Plan A = cost for Plan B.

$19.95 + 0.35x = 29.95 + 0.10x$

$0.35x = 10 + 0.10x$

$0.25x = 10$

$x = 40$

Lori would need to talk 40 minutes in a month for the plans to have the same cost.

20. Let x = smallest angle. Then $x + 5$ = second angle, $x + 50$ = third angle, and $4x + 25$ = fourth angle.

Sum of the angle measures = 360°

$x + (x + 5) + (x + 50) + (4x + 25) = 360$

$7x + 80 = 360$

$7x = 280$

$x = 40$

The angle measures are 40°, 40 + 5 = 45°, 40 + 50 = 90°, and 4(40) + 25 = 185°.

Chapter 4

Exercise Set 4.1

1. The x-coordinate is always listed first.

3. **a.** The horizontal axis is the x-axis.

   b. The vertical axis is the y-axis.

5. Axis is singular, while axes is plural.

7. The graph of a linear equation is an illustration of the set of points whose coordinates satisfy the equation.

9. The graph of a linear equation looks like a straight line.

11. $ax + by = c$

13. 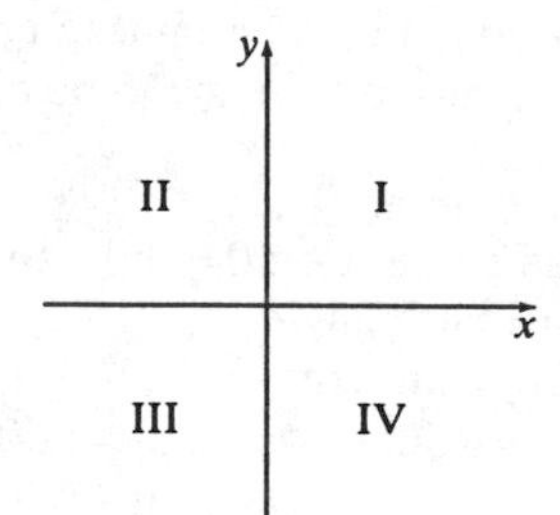

15. I

17. IV

19. II

21. III

23. III

25. II

27. $A(3, 1)$; $B(-3, 0)$; $C(1, -3)$; $D(-2, -3)$; $E(0, 3)$; $F\left(\frac{3}{2}, -1\right)$

29. 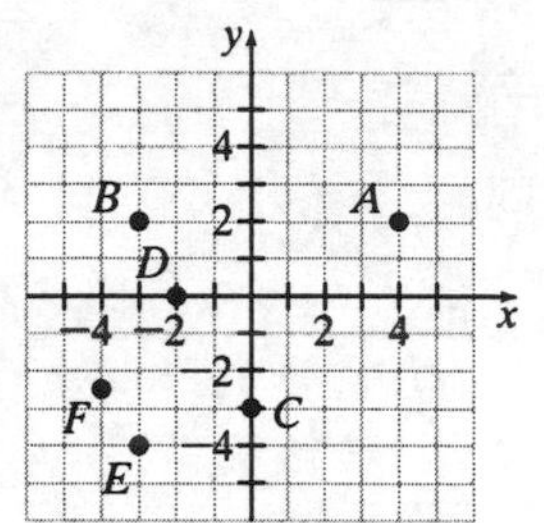

31. 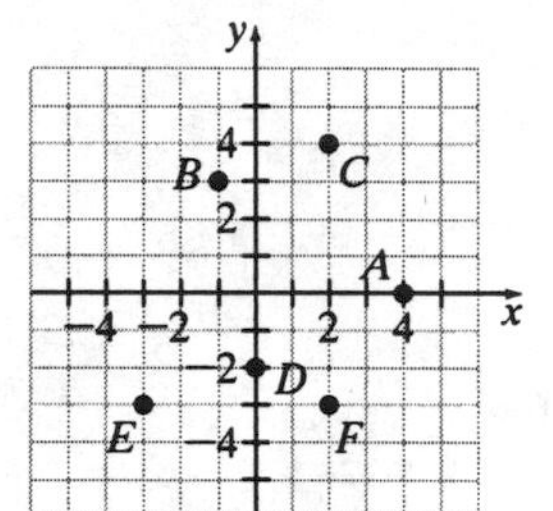

33. 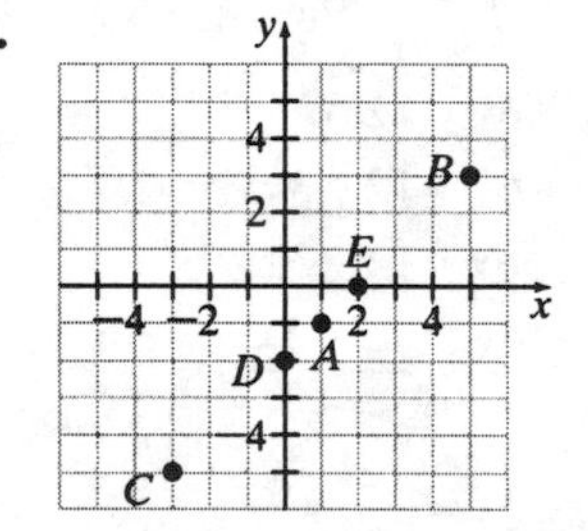

The points are collinear.

35. 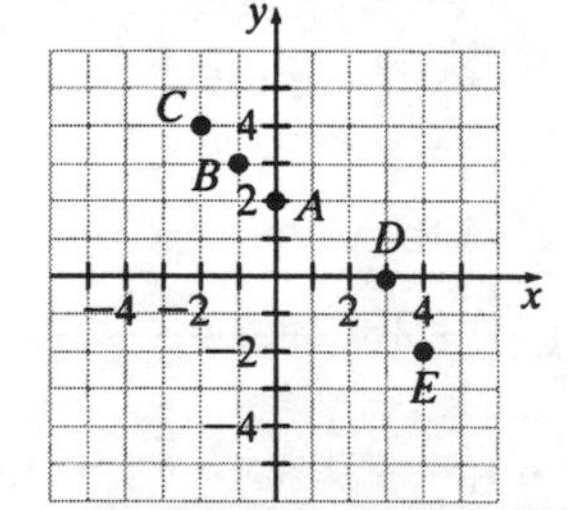

The points are not collinear since $(3, 0)$ is not on the line.

37. a. $y = x + 1$ $\quad$ $y = x + 1$

$1 = 0 + 1$ $\quad$ $0 = -1 + 1$

$1 = 1$ True $\quad$ $0 = 0$ True

$y = x + 1$ $\quad$ $y = x + 1$

$3 = 2 + 1$ $\quad$ $1 = 1 + 1$

$3 = 3$ True $\quad$ $1 = 2$ False

Point D does not satisfy the equation.

b.

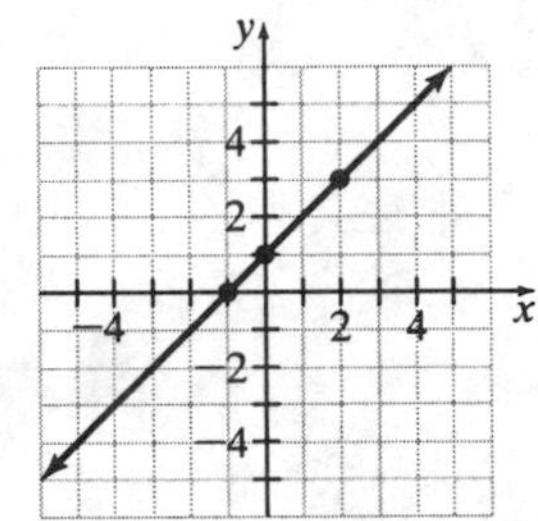

39. a. $3x - 2y = 6$ $\quad$ $3x - 2y = 6$

$3(4) - 2(0) = 6$ $\quad$ $3(2) - 2(0) = 6$

$12 = 6$ False $\quad$ $6 = 6$ True

$3x - 2y = 6$ $\quad$ $3x - 2y = 6$

$3\left(\frac{2}{3}\right) - 2(-2) = 6$ $\quad$ $3\left(\frac{4}{3}\right) - 2(-1) = 6$

$2 + 4 = 6$ $\quad$ $4 + 2 = 6$

$6 = 6$ True $\quad$ $6 = 6$ True

Point a) does not satisfy the equation.

b.

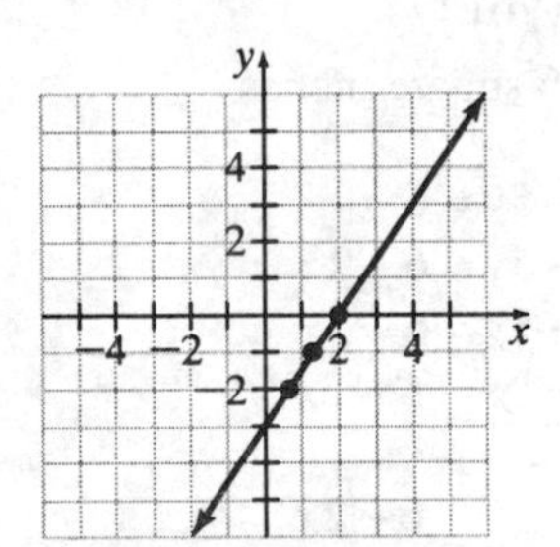

41. a. $\frac{1}{2}x + 4y = 4$ $\quad$ $\frac{1}{2}x + 4y = 4$

$\frac{1}{2}(2) + 4(-1) = 4$ $\quad$ $\frac{1}{2}(2) + 4\left(\frac{3}{4}\right) = 4$

$1 - 4 = 4$ $\quad$ $1 + 3 = 4$

$-3 = 4$ False $\quad$ $4 = 4$ True

$$\begin{aligned}\frac{1}{2}x+4y&=4\\ \frac{1}{2}(0)+4(1)&=4\\ 0+4&=4\\ 4&=4 \text{ True}\end{aligned} \qquad \begin{aligned}\frac{1}{2}x+4y&=4\\ \frac{1}{2}(-4)+4\left(\frac{3}{2}\right)&=4\\ -2+6&=4\\ 4&=4 \text{ True}\end{aligned}$$

Point a) does not satisfy the equation.

b.

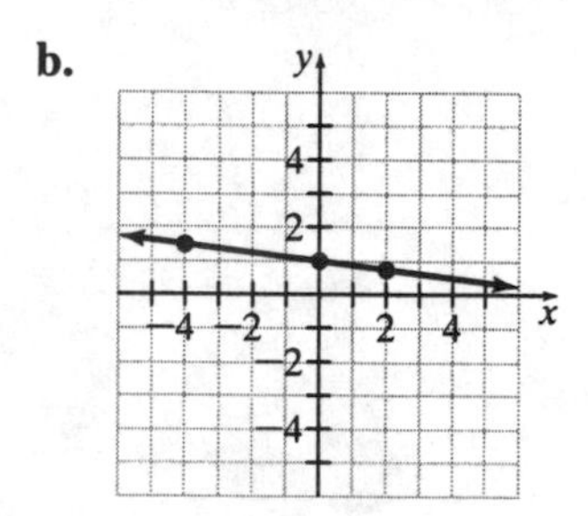

43. $y=3x-2$
$y=3(2)-2$
$y=6-2$
$y=4$

45. $y=3x-2$
$y=3(0)-2$
$y=0-2$
$y=-2$

47. $2x+3y=12$
$2(3)+3y=12$
$6+3y=12$
$3y=6$
$y=2$

49. $2x+3y=12$
$2\left(\frac{1}{2}\right)+3y=12$
$1+3y=12$
$3y=11$
$y=\frac{11}{3}$

51. The value of y is 0 when a straight line crosses the x-axis, because any point on the x-axis is neither above or below the origin.

58. A linear equation in one variable is an equation of the form $ax+b=c$.

59. A conditional linear equation is a linear equation that has only one solution.

60. An identity is an equation that is true for all real numbers.

61. $C=2\pi r$
$C\approx 2(3.14)(3)$
$C\approx 18.84$ inches
$A=\pi r^2$
$A\approx(3.14)(3)^2$
$A\approx(3.14)(9)$
$A\approx 28.26$ square inches

62. $2x-5y=6$
$2x-2x-5y=6-2x$
$-5y=6-2x$
$\frac{-5y}{-5}=\frac{6-2x}{-5}$
$y=\frac{6-2x}{-5}$
$y=\frac{2x-6}{5}$
$y=\frac{2}{5}x-\frac{6}{5}$

Exercise Set 4.2

1. To find the x-intercept, substitute 0 for y and find the corresponding value of x. To find the y-intercept, substitute 0 for x and find the corresponding value of y.

3. The graph of $y = b$ is a horizontal line.

5. You may not be able to read exact answers from a graph.

7. Yes. The equation goes through the origin because the point (0, 0) satisfies the equation.

9. $$\begin{aligned} 2x+y&=6 \\ 2(3)+y&=6 \\ 6+y&=6 \\ y&=0 \end{aligned}$$

11. $$\begin{aligned} 2x+y&=6 \\ 2x-5&=6 \\ 2x&=11 \\ x&=\frac{11}{2} \end{aligned}$$

13. $$\begin{aligned} 2x+y&=6 \\ 2x+0&=6 \\ 2x&=6 \\ x&=3 \end{aligned}$$

15. $$\begin{aligned} 3x-2y&=8 \\ 3\cdot 4-2y&=8 \\ 12-2y&=8 \\ -2y&=-4 \\ y&=2 \end{aligned}$$

17. $$\begin{aligned} 3x-2y&=8 \\ 3x-2(0)&=8 \\ 3x&=8 \\ x&=\frac{8}{3} \end{aligned}$$

19. $$\begin{aligned} 3x-2y&=8 \\ 3(-3)-2y&=8 \\ -9-2y&=8 \\ -2y&=17 \\ y&=-\frac{17}{2} \end{aligned}$$

21. An equation of the form $x = c$ is a vertical line with x-intercept at $(c, 0)$.

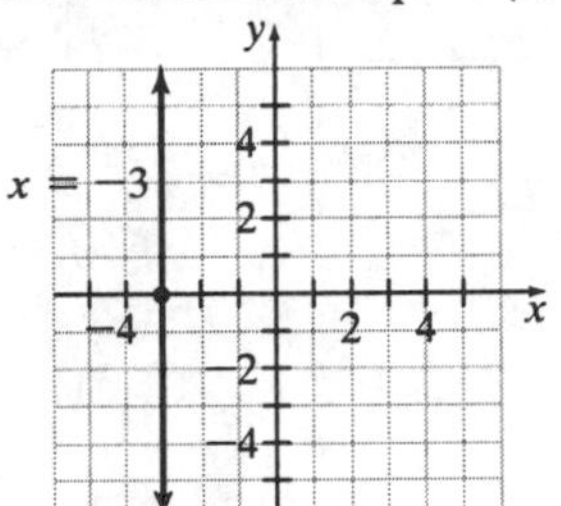

23. An equation of the form $y = c$ is a horizontal line with y-intercept at $(0, c)$.

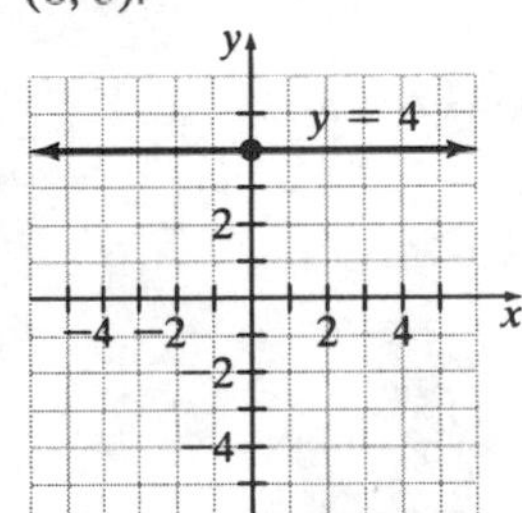

25. Let $x = 0$, $y = 3(0) - 1 = -1$, $(0, -1)$
Let $x = 1$, $y = 3(1) - 1 = 2$, $(1, 2)$
Let $x = 2$, $y = 3(2) - 1 = 5$, $(2, 5)$

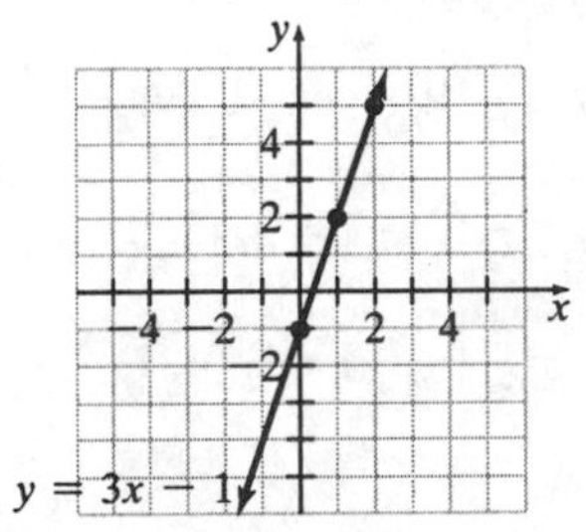

27. Let $x = -1$, $y = 6(-1) + 2 = -4$, $(-1, -4)$
Let $x = 0$, $y = 6(0) + 2 = 2$, $(0, 2)$
Let $x = 1$, $y = 6(1) + 2 = 8$, $(1, 8)$, $(1, 8)$

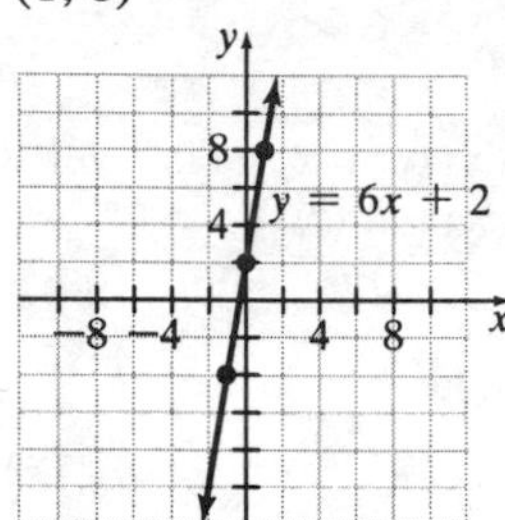

29. Let $x = 0$, $y = -\frac{1}{2}(0) + 3 = 3$, $(0, 3)$

Let $x = 2$, $y = -\frac{1}{2}(2) + 3 = 2$, $(2, 2)$

Let $x = 4$, $y = -\frac{1}{2}(4) + 3 = 1$, $(4, 1)$

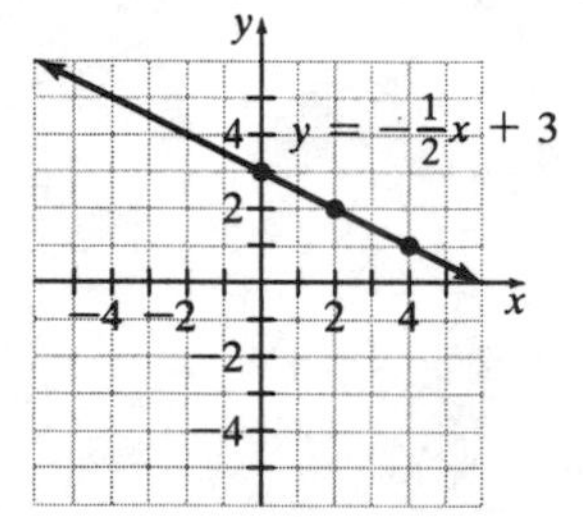

31. $2x - 4y = 4$
$-4y = -2x + 4$
$y = \frac{1}{2}x - 1$

Let $x = 0$, $y = \frac{1}{2}(0) - 1 = -1$, $(0, -1)$

Let $x = 2$, $y = \frac{1}{2}(2) - 1 = 0$, $(2, 0)$

Let $x = 4$, $y = \frac{1}{2}(4) - 1 = 1$, $(4, 1)$

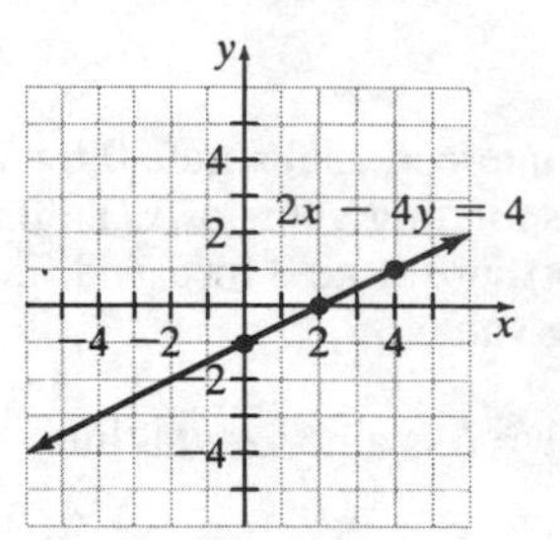

33. $5x - 2y = 8$
$-2y = -5x + 8$
$y = \frac{5}{2}x - 4$

Let $x = 0$, $y = \frac{5}{2}(0) - 4 = -4$, $(0, -4)$

Let $x = 2$, $y = \frac{5}{2}(2) - 4 = 1$, $(2, 1)$

Let $x = 4$, $y = \frac{5}{2}(4) - 4 = 6$, $(4, 6)$

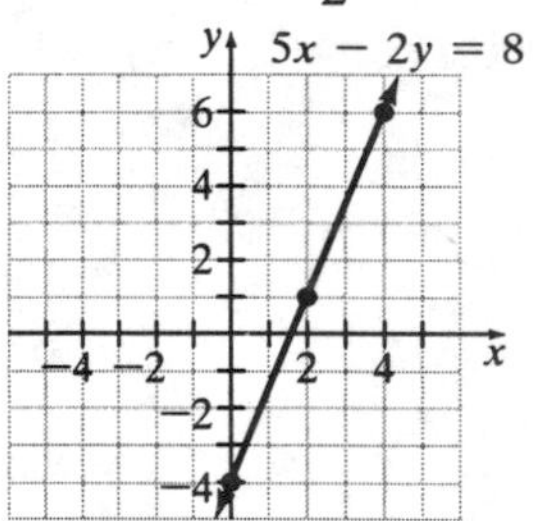

35. $6x + 5y = 30$
$5y = -6x + 30$
$y = -\frac{6}{5}x + 6$

Let $x = 0$, $y = -\frac{6}{5}(0) + 6 = 6$, $(0, 6)$

Let $x = 5$, $y = -\frac{6}{5}(5) + 6 = 0$, $(5, 0)$

Let $x = 10$, $y = -\frac{6}{5}(10) + 6 = -6$, $(10, -6)$

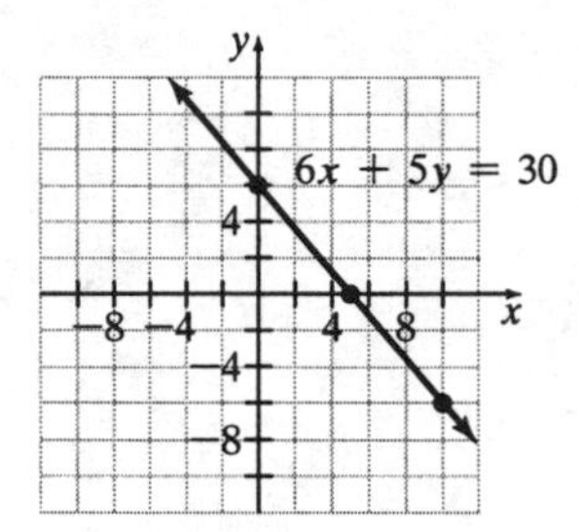

37. $-4x+5y=0$

$5y=4x$

$y=\frac{4}{5}x$

Let $x=-5$, $y=\frac{4}{5}(-5)=-4$, $(-5,-4)$

Let $x=0$, $y=\frac{4}{5}(0)=0$, $(0, 0)$

Let $x=5$, $y=\frac{4}{5}(5)=4$, $(5, 4)$

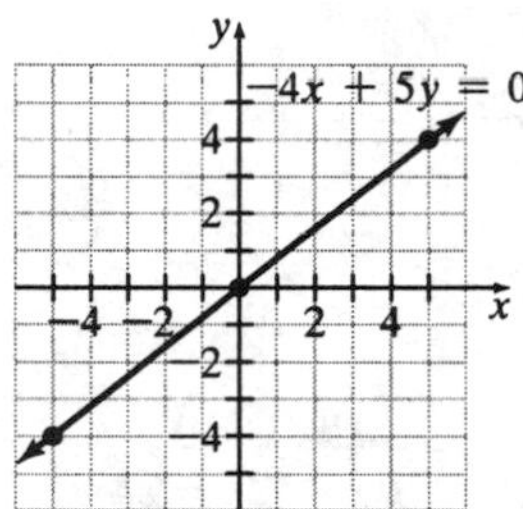

39. Let $x=0$, $y=-20(0)+60=60$, $(0, 60)$
Let $x=2$, $y=-20(2)+60=20$, $(2, 20)$
Let $x=4$, $y=-20(4)+60=-20$,
$(4, -20)$

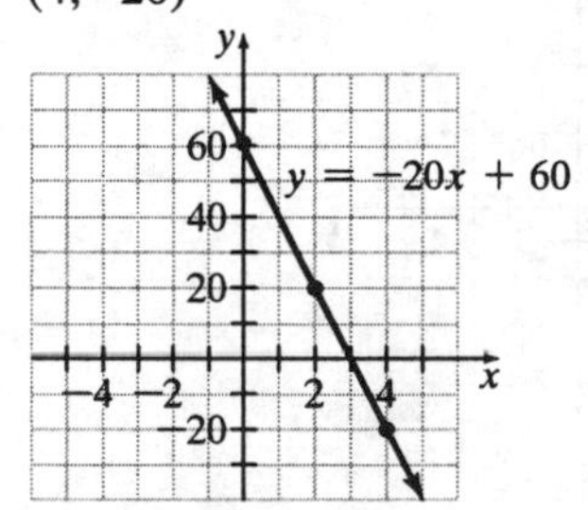

41. Let $x=-3$, $y=\frac{2}{3}(-3)=-2$, $(-3,-2)$

Let $x=0$, $y=\frac{2}{3}(0)=0$, $(0, 0)$

Let $x=3$, $y=\frac{2}{3}(3)=2$, $(3, 2)$

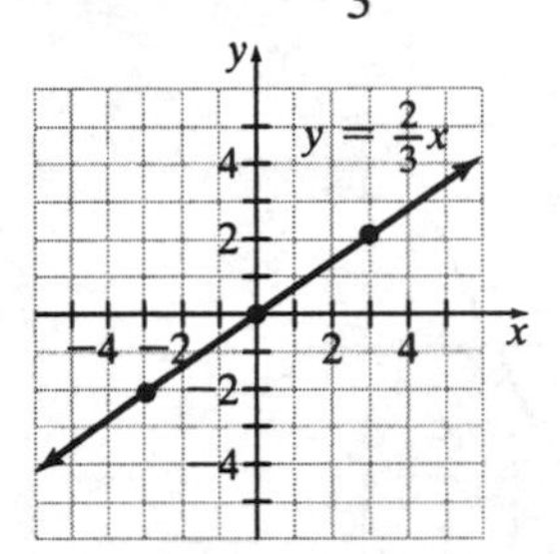

43. Let $x=0$, $y=\frac{1}{2}(0)+4=4$, $(0, 4)$

Let $x=2$, $y=\frac{1}{2}(2)+4=5$, $(2, 5)$

Let $x=4$, $y=\frac{1}{2}(4)+4=6$, $(4, 6)$

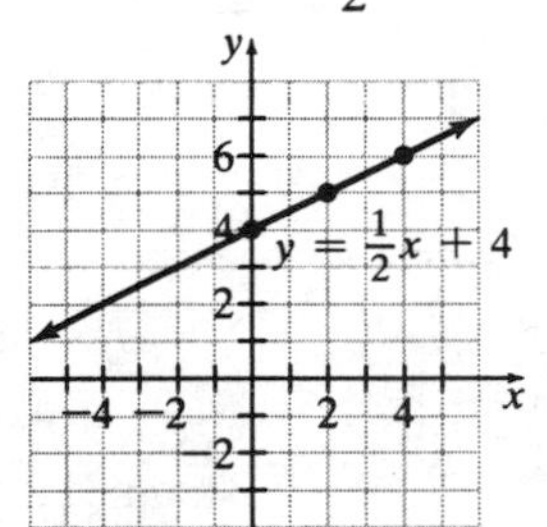

45.

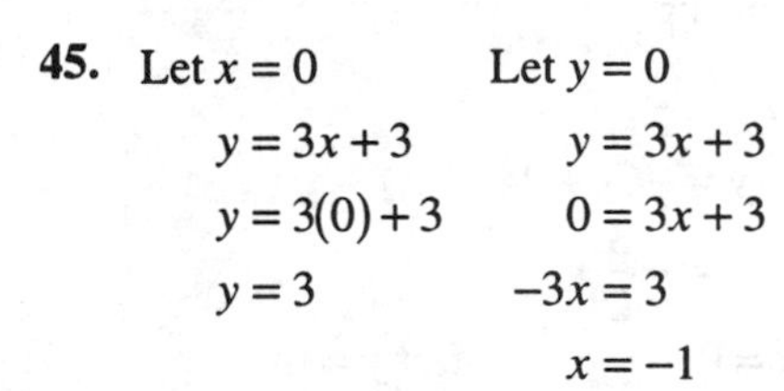

Let $x=0$	Let $y=0$
$y=3x+3$	$y=3x+3$
$y=3(0)+3$	$0=3x+3$
$y=3$	$-3x=3$
	$x=-1$

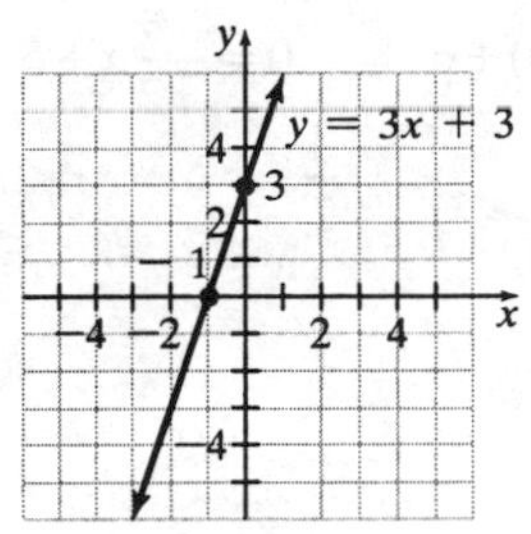

47. Let $x = 0$

$y = 2x - 3$

$y = 2(0) - 3$

$y = -3$

Let $y = 0$

$y = 2x - 3$

$0 = 2x - 3$

$-2x = -3$

$x = \frac{3}{2}$

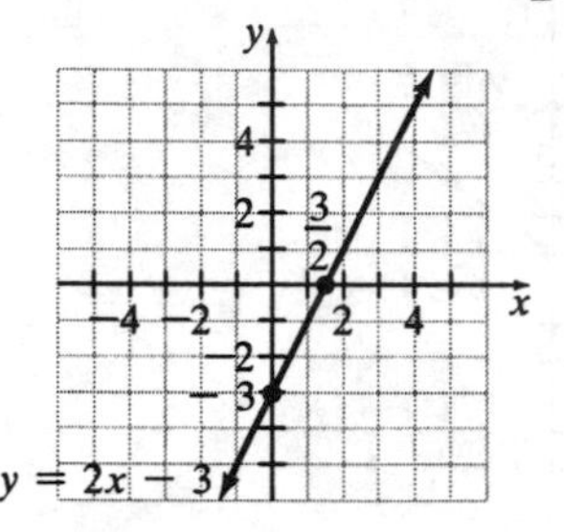

49. Let $x = 0$

$y = -6x + 5$

$y = -6(0) + 5$

$y = 5$

Let $y = 0$

$y = -6x + 5$

$0 = -6x + 5$

$6x = 5$

$x = \frac{5}{6}$

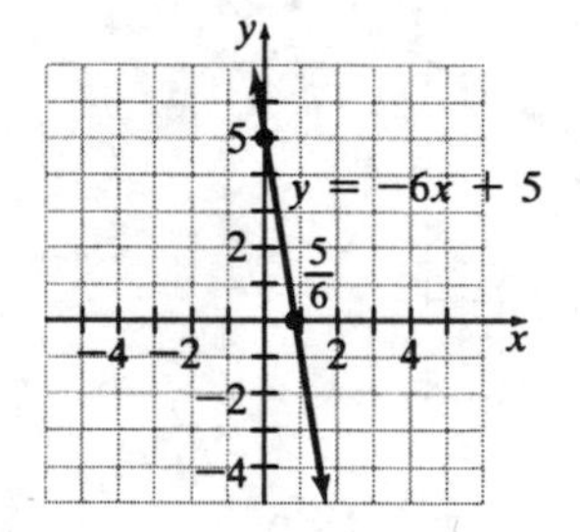

51. $4y + 6x = 24$

$4y = -6x + 24$

$y = -\frac{3}{2}x + 6$

Let $x = 0$

$y = -\frac{3}{2}(0) + 6$

$y = 6$

Let $y = 0$

$0 = -\frac{3}{2}x + 6$

$\frac{3}{2}x = 6$

$x = 4$

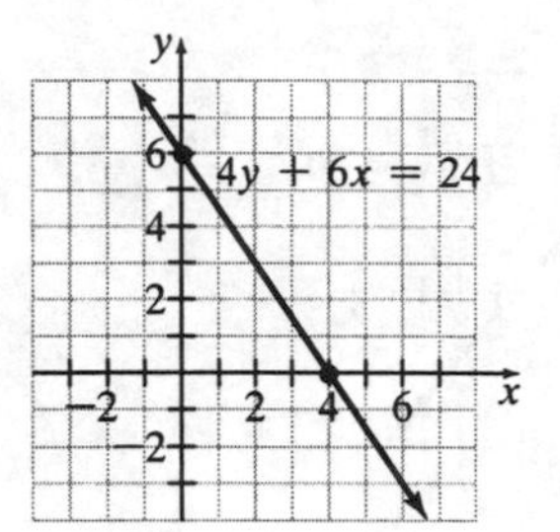

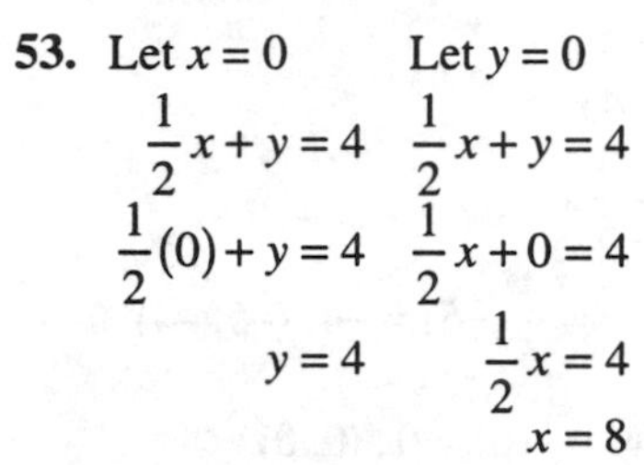

53. Let $x = 0$

$\frac{1}{2}x + y = 4$

$\frac{1}{2}(0) + y = 4$

$y = 4$

Let $y = 0$

$\frac{1}{2}x + y = 4$

$\frac{1}{2}x + 0 = 4$

$\frac{1}{2}x = 4$

$x = 8$

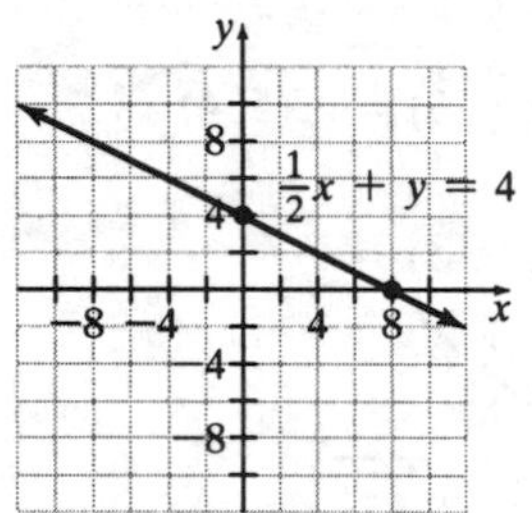

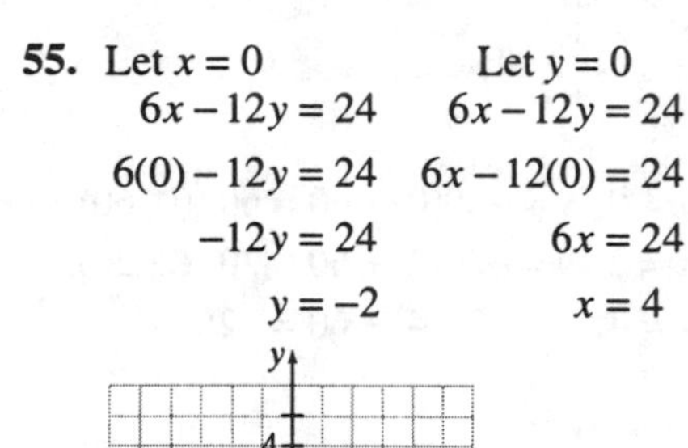

55. Let $x = 0$

$6x - 12y = 24$

$6(0) - 12y = 24$

$-12y = 24$

$y = -2$

Let $y = 0$

$6x - 12y = 24$

$6x - 12(0) = 24$

$6x = 24$

$x = 4$

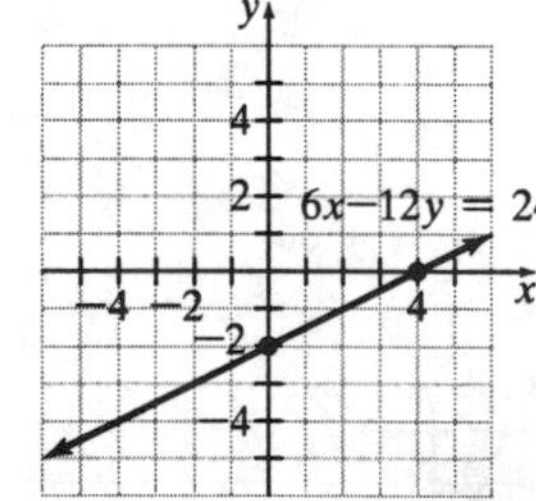

57. Let $x = 0$

$$8y = 6x - 12$$
$$8y = 6(0) - 12$$
$$8y = -12$$
$$y = -\frac{3}{2}$$

Let $y = 0$

$$8y = 6x - 12$$
$$8(0) = 6x - 12$$
$$0 = 6x - 12$$
$$-6x = -12$$
$$x = 2$$

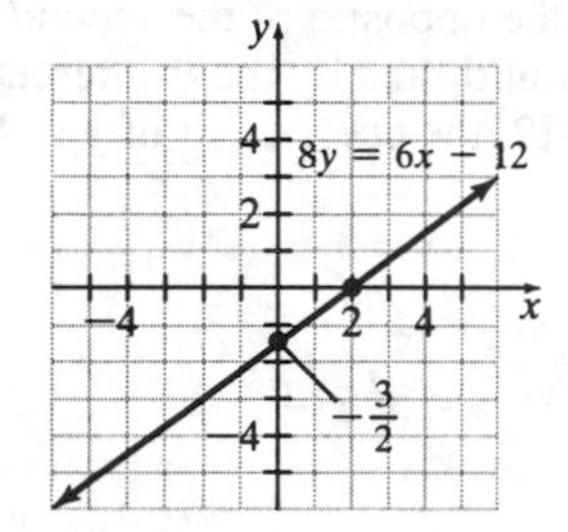

59. Let $x = 0$

$$30y + 10x = 45$$
$$30y + 10(0) = 45$$
$$30y = 45$$
$$y = \frac{3}{2}$$

Let $y = 0$

$$30y + 10x = 45$$
$$30(0) + 10x = 45$$
$$10x = 45$$
$$x = \frac{9}{2}$$

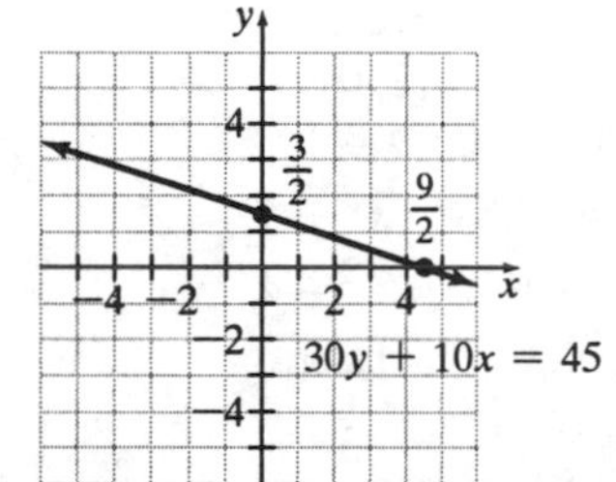

61. Let $x = 0$

$$\frac{1}{3}x + \frac{1}{4}y = 12$$
$$\frac{1}{3}(0) + \frac{1}{4}y = 12$$
$$\frac{1}{4}y = 12$$
$$y = 48$$

Let $y = 0$

$$\frac{1}{3}x + \frac{1}{4}y = 12$$
$$\frac{1}{3}x + \frac{1}{4}(0) = 12$$
$$\frac{1}{3}x = 12$$
$$x = 36$$

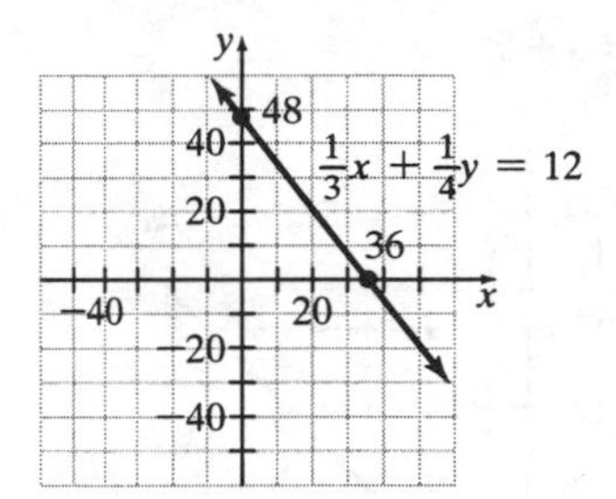

63. Let $x = 0$

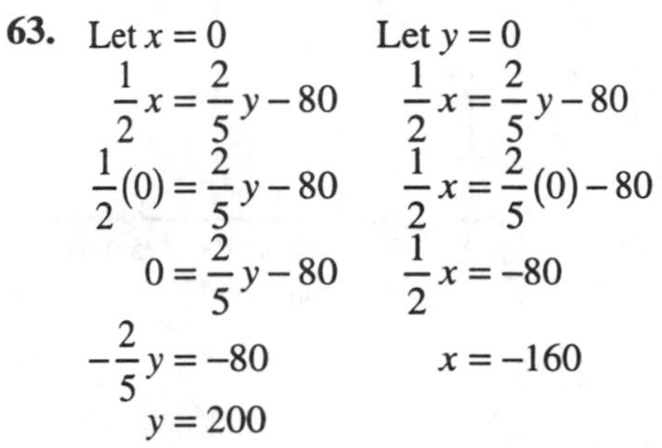

$$\frac{1}{2}x = \frac{2}{5}y - 80$$
$$\frac{1}{2}(0) = \frac{2}{5}y - 80$$
$$0 = \frac{2}{5}y - 80$$
$$-\frac{2}{5}y = -80$$
$$y = 200$$

Let $y = 0$

$$\frac{1}{2}x = \frac{2}{5}y - 80$$
$$\frac{1}{2}x = \frac{2}{5}(0) - 80$$
$$\frac{1}{2}x = -80$$
$$x = -160$$

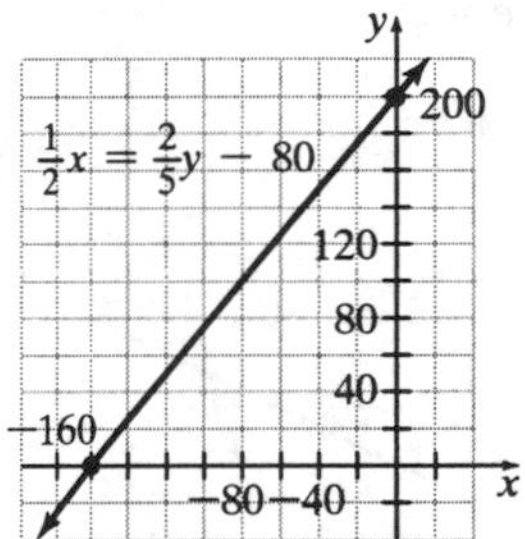

65. $x = -3$

67. $y = 6$

69.
$$ax + 4y = 8$$
$$a(2) + 4(0) = 8$$
$$2a + 0 = 8$$
$$2a = 8$$
$$a = 4$$

71.
$$3x + by = 10$$
$$3(0) + b(5) = 10$$
$$0 + 5b = 10$$
$$5b = 10$$
$$b = 2$$

73. Yes. For each 15 minutes of time, the number of calories burned increases by 200 calories.

75. **a.** $C = s + 50$

b.

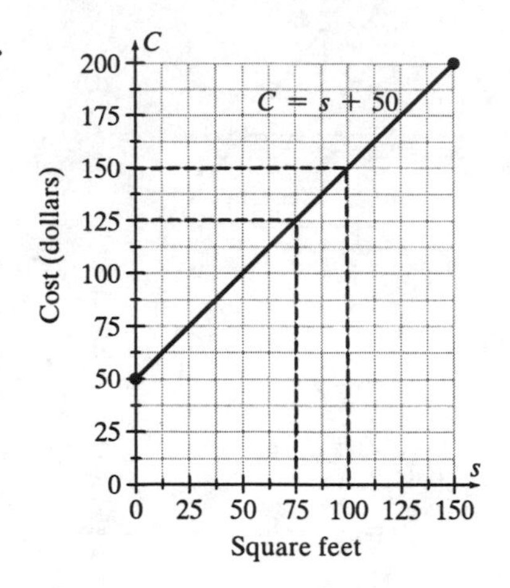

c. \$150

d. 75 square feet

77. **a.** $C = m + 40$

b.

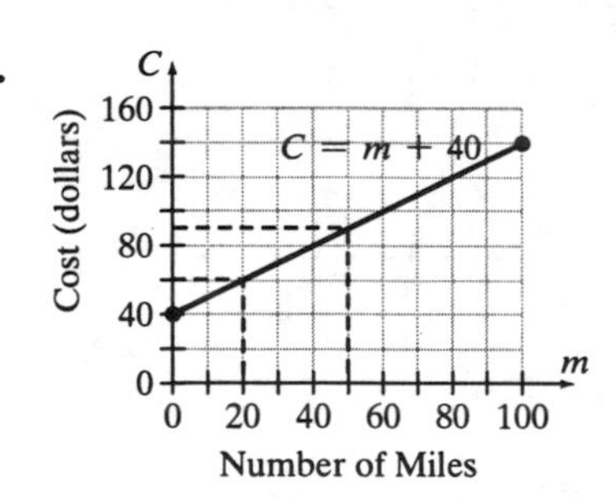

c. \$90

d. 20 miles

79. **a.**

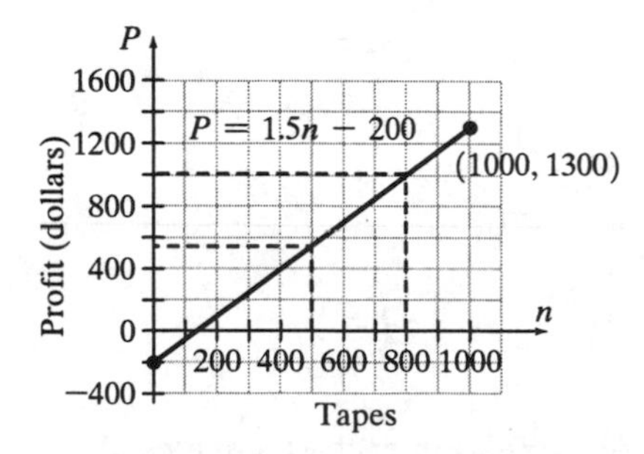

b. \$550

c. 800 tapes

81. Since each shaded area multiplied by the corresponding intercept must equal 20, the coefficients are 5 and 4 respectively.

83. Since the first shaded area multiplied by the x-intercept must equal –12, the coefficient of x is 6. Since the opposite of the second shaded area multiplied by the y-intercept must equal –12, the coefficient of y is 4.

85. Let $x = -3$ Let $x = 3$

$y = x^2 - 4$ $y = x^2 - 4$

$y = (-3)^2 - 4$ $y = 3^2 - 4$

$= 9 - 4$ $= 9 - 4$

$= 5$ $= 5$

$(-3, 5)$ $(3, 5)$

Let $x = -2$ Let $x = 2$

$y = x^2 - 4$ $y = x^2 - 4$

$y = (-2)^2 - 4$ $y = 2^2 - 4$

$= 4 - 4$ $= 4 - 4$

$= 0$ $= 0$

$(-2, 0)$ $(2, 0)$

Let $x = -1$ Let $x = 1$

$y = x^2 - 4$ $y = x^2 - 4$

$y = (-1)^2 - 4$ $= x^2 - 4$

$= 1 - 4$ $= 1 - 4$

$= -3$ $= -3$

$(-1, -3)$ $(1, -3)$

Let $x = 0$

$y = x^2 - 4$

$y = 0^2 - 4$

$= 0 - 4$

$= -4$

$(0, -4)$

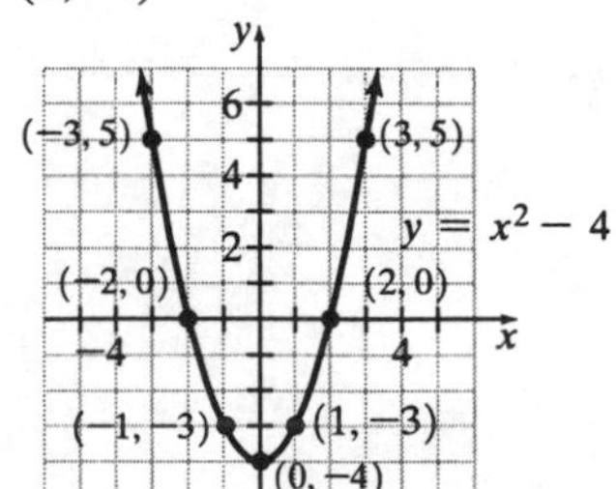

88. $2[6-(4-5)]\div 2-5^2$
$=2[6-(-1)]\div 2-25$
$=2[7]\div 2-25$
$=14\div 2-25$
$=7-25$
$=-18$

89. $\dfrac{-3^2\cdot 4\div 2}{\sqrt{9}-2^2}=\dfrac{-9\cdot 4\div 2}{3-4}$
$=\dfrac{-36\div 2}{3-4}$
$=\dfrac{-18}{-1}$
$=18$

90. $\dfrac{8 \text{ ounces}}{3 \text{ gallons}}=\dfrac{x \text{ ounces}}{2.5 \text{ gallons}}$
$\dfrac{8}{3}=\dfrac{x}{2.5}$
$20=3x$
$6.67\approx x$
You should use 6.67 ounces of cleaner.

91. Let x = smaller integer. Then $3x + 1$ = larger integer.
$x+(3x+1)=37$
$4x+1=37$
$4x=36$
$x=9$
The smaller integer is 9.
The larger is $3(9)+1=27+1=28$.

Exercise Set 4.3

1. The slope of a line is the ratio of the vertical change to the horizontal change between any two points on the line.

3. A line with a positive slope rises from left to right.

5. Lines that rise from the left to right have a positive slope. Lines that fall from left to right have a negative slope.

7. No, since we cannot divide by 0, the slope is undefined.

9. $m=\dfrac{6-1}{5-4}$
$=\dfrac{5}{1}$
$=5$

11. $m=\dfrac{-2-0}{5-9}$
$=\dfrac{-2}{-4}$
$=\dfrac{1}{2}$

13. $m=\dfrac{\frac{1}{2}-\frac{1}{2}}{-3-3}$
$=\dfrac{0}{-6}$
$=0$

15. $m=\dfrac{6-6}{-2-(-4)}$
$=\dfrac{0}{2}$
$=0$

17. $m=\dfrac{-2-4}{3-3}$
$=\dfrac{-6}{0}$ undefined

19. $m=\dfrac{3-0}{-2-6}$
$=\dfrac{3}{-8}$
$=-\dfrac{3}{8}$

21. $m = \frac{1-\frac{3}{2}}{-\frac{3}{4}-0}$
$= \frac{-\frac{1}{2}}{-\frac{3}{4}}$
$= \frac{-1}{2} \cdot \frac{4}{-3}$
$= \frac{-4}{-6}$
$= \frac{2}{3}$

23. $m = \frac{6}{3}$
$= 2$

25. $m = \frac{3}{-2}$
$= -\frac{3}{2}$

27. $m = \frac{6}{-4}$
$= -\frac{3}{2}$

29. $m = \frac{7}{4}$

31. $m = \frac{0}{3}$
$= 0$

33. $m = \frac{-2}{6}$
$= -\frac{1}{3}$

35. Horizontal line, slope is 0.

37. The first graph appears to pass through the points (–1, 0) and (0, 6). It's slope is $m = \frac{6-0}{0-(-1)} = \frac{6}{1} = 6$. The second graph appears to pass through the points (–4, 0) and (0, 6). It's slope is $m = \frac{6-0}{0-(-4)} = \frac{6}{4} = \frac{6}{4} = \frac{3}{2}$. The first graph has the greater slope.

39. a. $m = \frac{410-280}{65-60} = \frac{130}{5} = 26$

b. $m = \frac{618-545}{80-75} = \frac{73}{5} = 14.6$

41. Answers will vary. Sample:

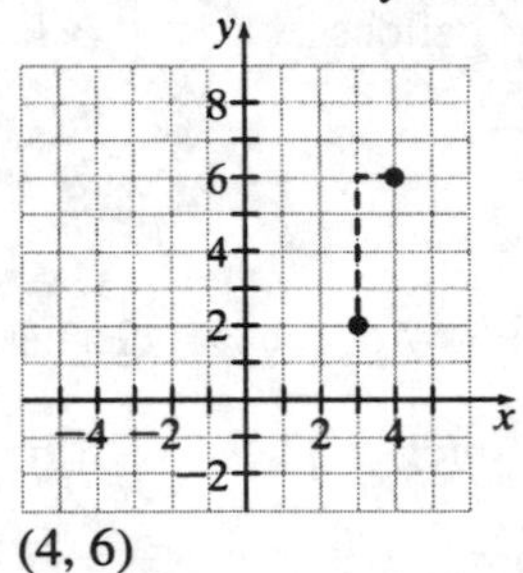

(4, 6)

43. Answers will vary. Sample:

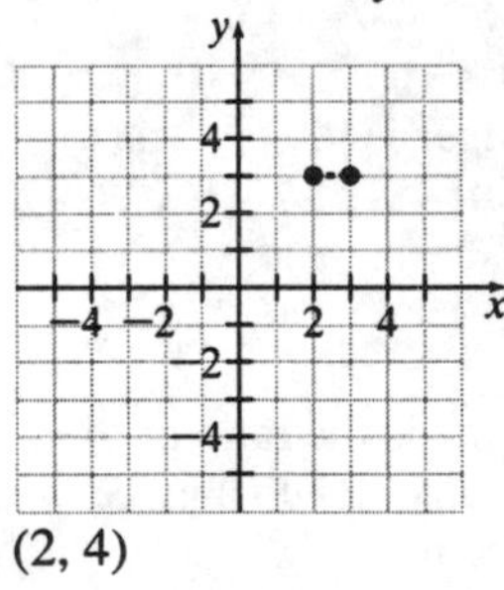

(2, 4)

45. $m = \dfrac{-\frac{7}{2} - \left(-\frac{3}{8}\right)}{-\frac{4}{9} - \frac{1}{2}}$

$= \dfrac{-\frac{28}{8} + \frac{3}{8}}{-\frac{8}{18} - \frac{9}{18}}$

$= \dfrac{-\frac{25}{8}}{-\frac{17}{18}}$

$= \left(-\dfrac{25}{8}\right)\left(-\dfrac{18}{17}\right)$

$= \dfrac{(-25)(-9)}{(4)(17)}$

$= \dfrac{225}{68}$

47. a.

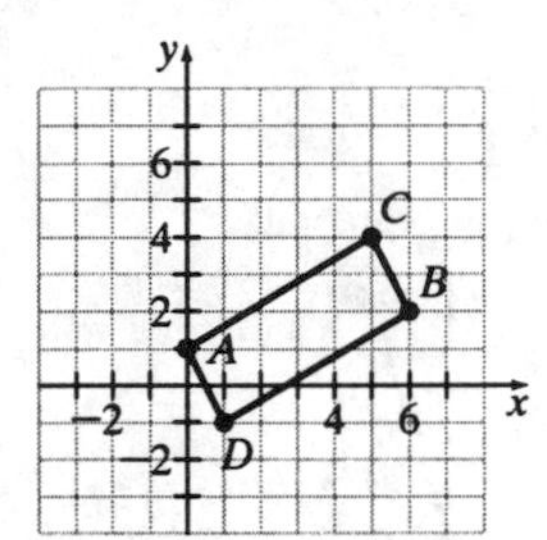

b. $AC;\ m = \dfrac{4-1}{5-0} = \dfrac{3}{5}$

$CB;\ m = \dfrac{4-2}{5-6} = \dfrac{2}{-1} = -2$

$DB;\ m = \dfrac{2-(-1)}{6-1} = \dfrac{3}{5}$

$AD;\ m = \dfrac{-1-1}{1-0} = \dfrac{-2}{1} = -2$

c. Yes; opposite sides are parallel.

51. $4x^2 + 3x + \dfrac{x}{2} = 4(0)^2 + 3(0) + \dfrac{0}{2}$

$= 0 + 0 + 0$

$= 0$

52. a. $-x = -\dfrac{3}{2}$

$(-1)(-x) = (-1)\left(-\dfrac{3}{2}\right)$

$x = \dfrac{3}{2}$

b. $5x = 0$

$\dfrac{5x}{5} = \dfrac{0}{5}$

$x = 0$

53. $2x - 3(x-2) = x + 2$

$2x - 3x + 6 = x + 2$

$-x + 6 = x + 2$

$-2x = -4$

$x = 2$

54. $5x - 3y = 15$

$x = 0$	$y = 0$
$5(0) - 3y = 15$	$5x - 3(0) = 15$
$-3y = 15$	$5x = 15$
$y = -5$	$x = 5$
$(0, -5)$	$(5, 0)$

Exercise Set 4.4

1. $y = mx + b$

3. $y = 4x - 2$

5. Compare their slopes: If slopes are the same and their y-intercepts are different, the lines are parallel.

7. $y - y_1 = m(x - x_1)$

9. $m = 3$; y-intercept: $(0, -7)$

11. $m = \dfrac{4}{3}$; y-intercept: $(0, -5)$

13. $m = 1$; y-intercept: $(0, -1)$

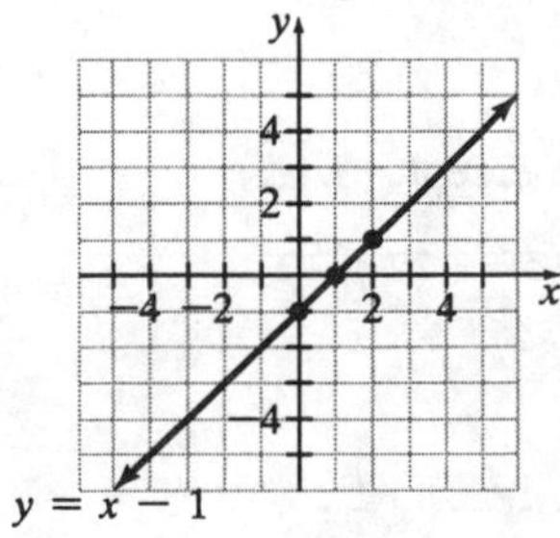

15. $m = 3$; y-interept: (0, 2)

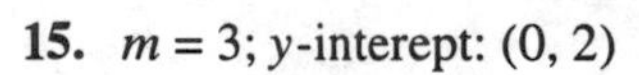

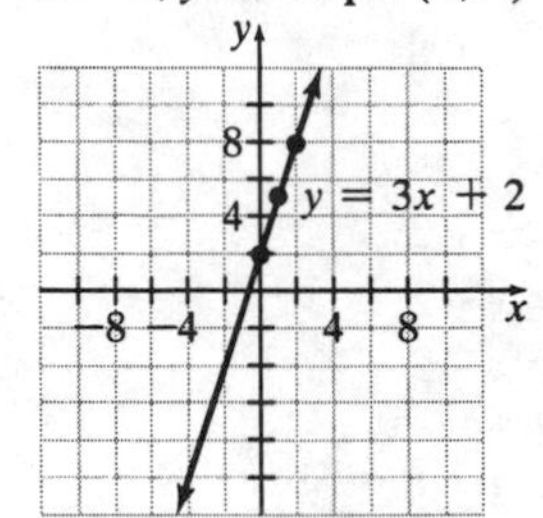

17. $m = -4$; y-intercept: (0, 0)

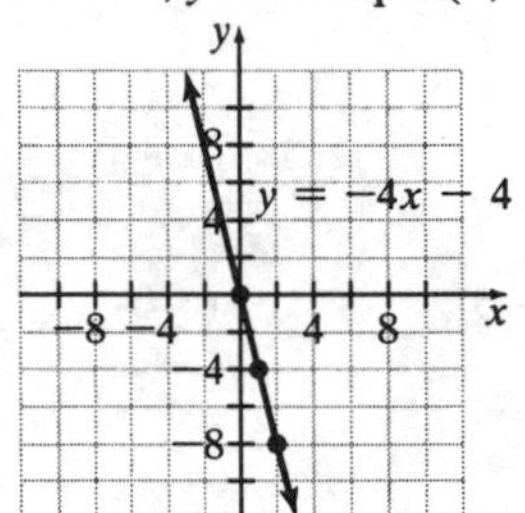

19. $-2x + y = -3$

$y = 2x - 3$

$m = 2$; y-intercept: (0, –3)

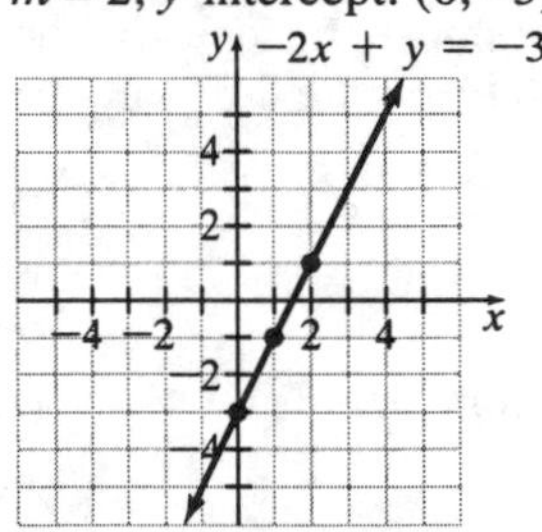

21. $5x - 2y = 10$

$-2y = -5x + 10$

$y = \frac{5}{2}x - 5$

$m = \frac{5}{2}$; y-intercept: (0, –5)

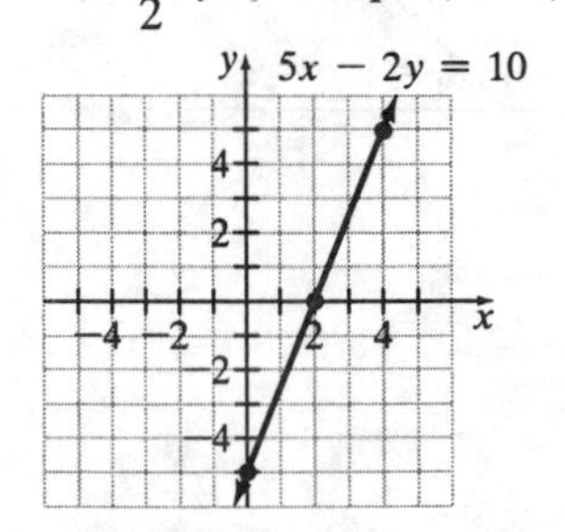

23. $5x + 10y = 15$

$10y = -5x + 15$

$y = -\frac{1}{2}x + \frac{3}{2}$

$m = -\frac{1}{2}$; y-intercept: $\left(0, \frac{3}{2}\right)$

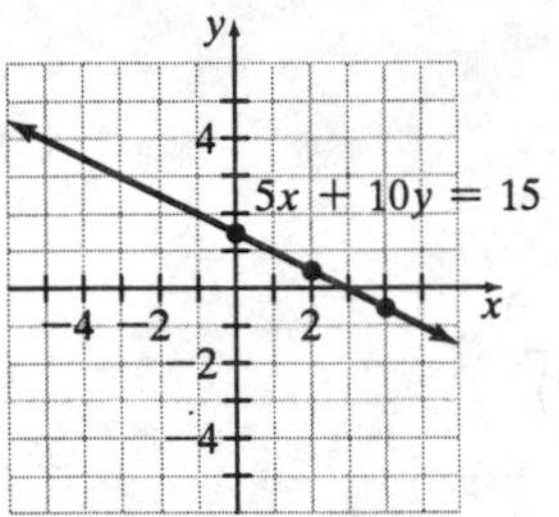

25. $-6x = -2y + 8$

$2y = 6x + 8$

$y = 3x + 4$

$m = 3$; y-intercept: (0, 4)

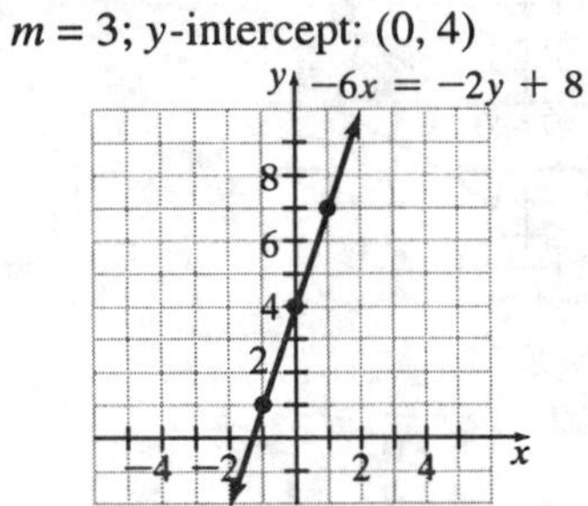

27. $3x = 2y - 4$

$-2y = -3x - 4$

$y = \frac{3}{2}x + 2$

$m = \frac{3}{2}$; y-intercept: (0, 2)

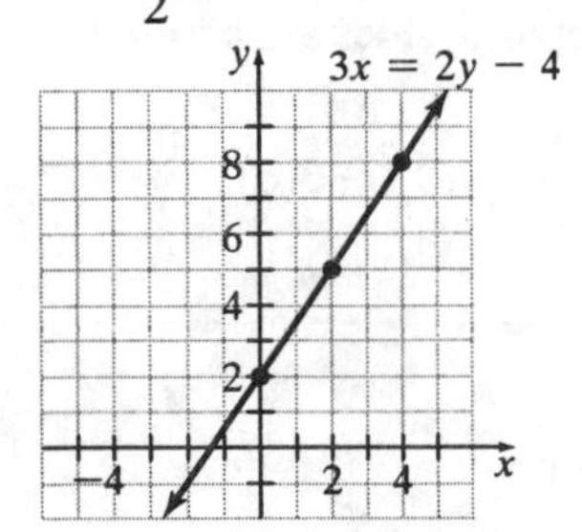

29. $m = \frac{4}{4} = 1,\ b = 2$

$y = x + 2$

31. $m = \frac{-2}{6} = -\frac{1}{3},\ b = 2$

$y = -\frac{1}{3}x + 2$

33. $m = \frac{10}{30} = \frac{1}{3},\ b = 5$

$y = \frac{1}{3}x + 5$

35. $m = \frac{2}{1} = 2,\ b = -1$

$y = 2x - 1$

37. Since the slopes of the lines are the same and *y*-intercepts are different, the lines are parallel.

39. $4x + 2y = 9$ $\quad$ $8x = 4y + 4$

$2y = -4x + 9$ $\quad$ $-4y = -8x + 4$

$y = -2x + \frac{9}{2}$ $\quad$ $y = 2x - 1$

Since the lines do not have the same slope, they are not parallel.

41. $3x + 5y = 9$ $\quad$ $6x = -10y + 9$

$5y = -3x + 9$ $\quad$ $10y = -6x + 9$

$y = -\frac{3}{5}x + \frac{9}{5}$ $\quad$ $y = \frac{-6}{10}x + \frac{9}{10}$

$y = -\frac{3}{5}x + \frac{9}{10}$

No

Since the slopes of the lines are the same and *y*-intercepts are different, the lines are parallel.

43. $y = \frac{1}{2}x - 6$ $\quad$ $3y = 6x + 9$

$y = 2x + 3$

No

Since the lines do not have the same slope, they are not parallel.

45. $y - 2 = 3(x - 0)$

$y = 3x + 2$

47. $y - 5 = -2[x - (-4)]$

$y - 5 = -2(x + 4)$

$y - 5 = -2x - 8$

$y = -2x - 3$

49. $y - (-5) = \frac{1}{2}[x - (-1)]$

$y + 5 = \frac{1}{2}(x + 1)$

$y + 5 = \frac{1}{2}x + \frac{1}{2}$

$y = \frac{1}{2}x - \frac{9}{2}$

51. $y = \frac{2}{5}x + 6$

53. $m = \frac{4 - (-2)}{-2 - (-4)} = \frac{6}{2} = 3$

$y - (-2) = 3[x - (-4)]$

$y + 2 = 3(x + 4)$

$y + 2 = 3x + 12$

$y = 3x + 10$

55. $m = \frac{-9 - 9}{6 - (-6)} = \frac{-18}{12} = -\frac{3}{2}$

$y - 9 = -\frac{3}{2}(x - (-6))$

$y - 9 = -\frac{3}{2}(x + 6)$

$y - 9 = -\frac{3}{2}x - 9$

$y = -\frac{3}{2}x$

57. $m = \frac{-2-3}{0-10} = \frac{-5}{-10} = \frac{1}{2}$

$y - 3 = \frac{1}{2}(x - 10)$

$y - 3 = \frac{1}{2}x - 5$

$y = \frac{1}{2}x - 2$

59. $y = 5.2x - 1.6$

61. a. $y = 5x + 60$

b. $y = 5(30) + 60 = 150 + 60 = \$210$

63. a. Use the slope-intercept form.

b. Use the point-slope form.

c. Use the point-slope form but first find the slope.

65. a. No. The equations will look different because different points are used.

b. $y - (-4) = 2[x - (-5)]$

$y + 4 = 2(x + 5)$

c. $y - 10 = 2(x - 2)$

d. $y + 4 = 2(x + 5)$

$y + 4 = 2x + 10$

$y = 2x + 6$

e. $y - 10 = 2(x - 2)$

$y - 10 = 2x - 4$

$y = 2x + 6$

f. Yes

67. First, find the slope of the line $2x + y = 6$.

$2x + y = 6$

$y = -2x + 6$

$m = 2$

Use $m = 2$ and $b = 4$ in the slope-intercept equation.

$y = 2x + 4$

69. $3x - 4y = 6$

$4y = 3x - 6$

$y = \frac{3}{4}x - \frac{3}{2}$

$m = \frac{3}{4}$

$y - (-1) = \frac{3}{4}[x - (-4)]$

$y + 1 = \frac{3}{4}(x + 4)$

$y + 1 = \frac{3}{4}x + 3$

$y = \frac{3}{4}x + 2$

74. $|-4| < |-6|$ because $4 < 6$.

75. False. For example $(-2)(-3) = 6$.

76. True

77. False. For example, $-1 - (-3) = 2$.

78. False. For example, $\frac{-6}{-2} = 3$.

79. $5^3 = 5 \cdot 5 \cdot 5 = 125$

Exercise Set 4.5

1. Points on the line satisfy the = part of the inequality.

3. The shadings are on opposite sides of the line.

5.

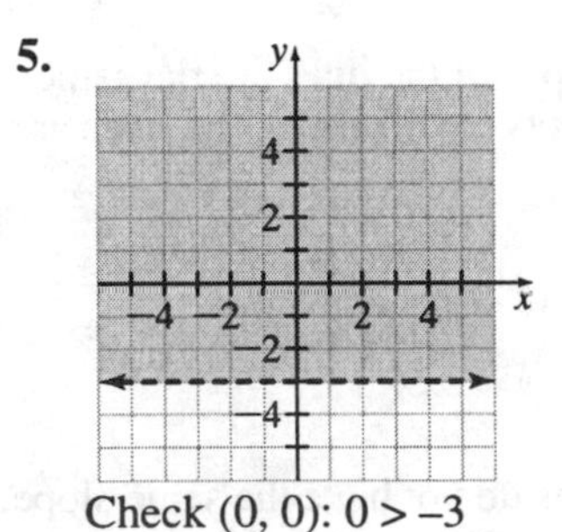

Check (0, 0): $0 > -3$　　True

7.

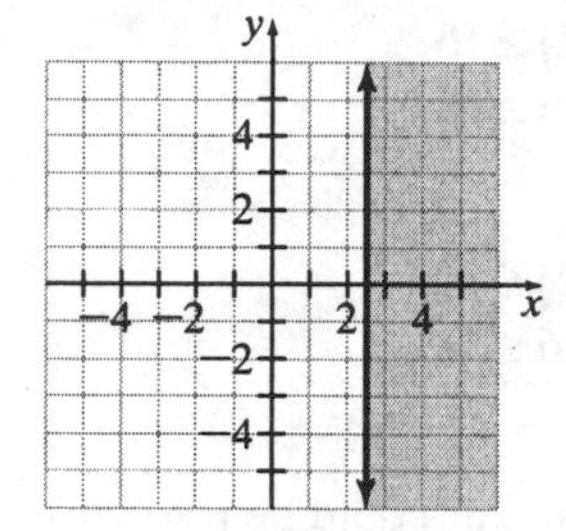

Check (0, 0): $x \geq \frac{5}{2}$

$0 \geq \frac{5}{2}$ False

9.

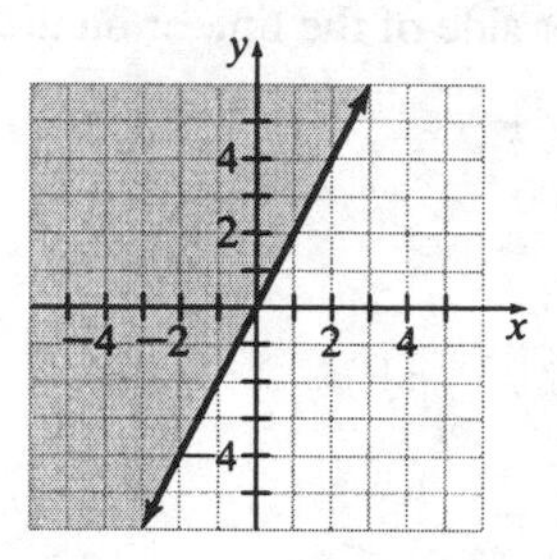

Check (1, 0): $y \geq 2x$

$0 \geq 2 \cdot 1$

$0 \geq 2$ False

11.

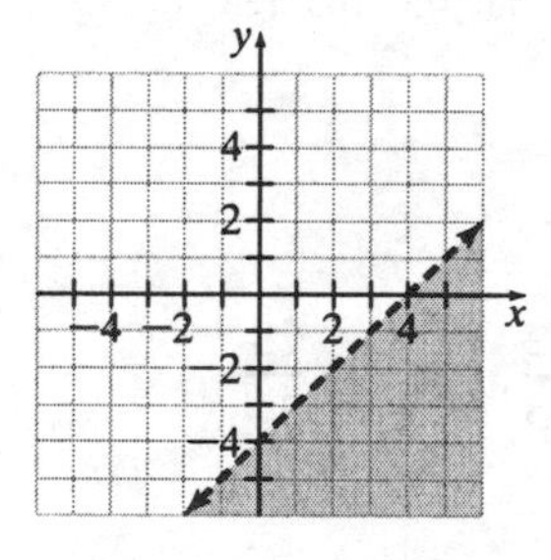

Check (0, 0): $y < x - 4$

$0 < 0 - 4$

$0 < -4$ False

13.

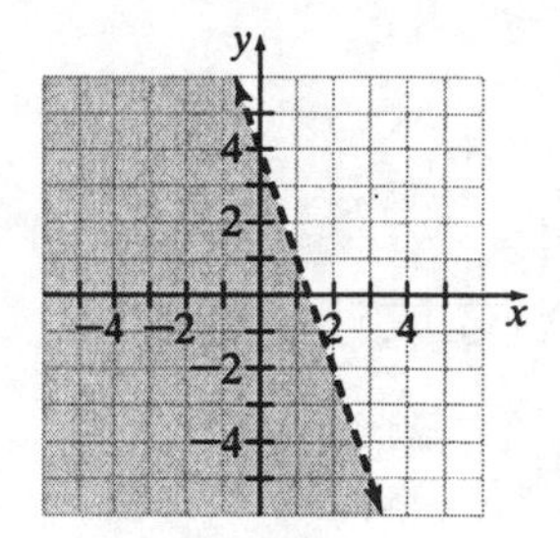

Check (0, 0): $y < -3x + 4$

$0 < -3 \cdot 0 + 4$

$0 < 4$ True

15.

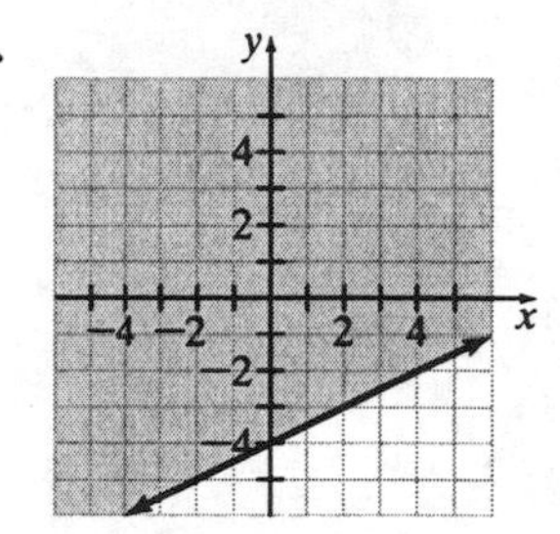

Check (0, 0): $y \geq \frac{1}{2}x - 4$

$0 \geq \frac{1}{2} \cdot 0 - 4$

$0 \geq -4$ True

17.

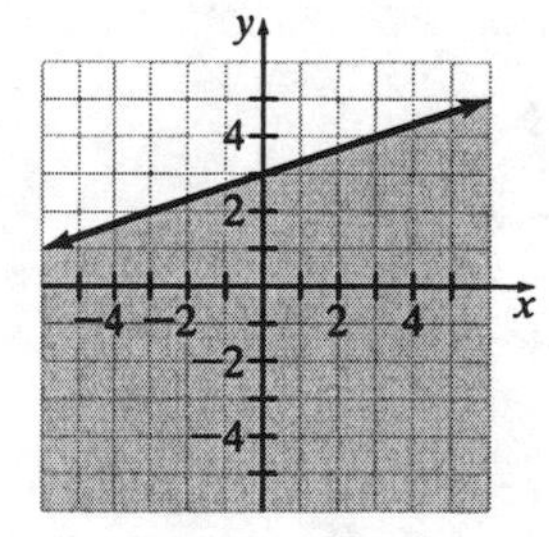

Check (0, 0): $y \leq \frac{1}{3}x + 3$

$0 \leq \frac{1}{3} \cdot 0 + 3$

$0 \leq 3$ True

19.

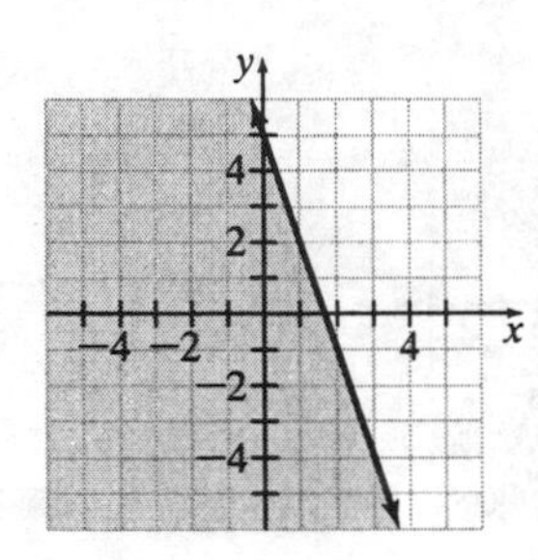

Check (0, 0): $3x + y \le 5$
$3 \cdot 0 + 0 \le 5$
$0 \le 5$ True

21.

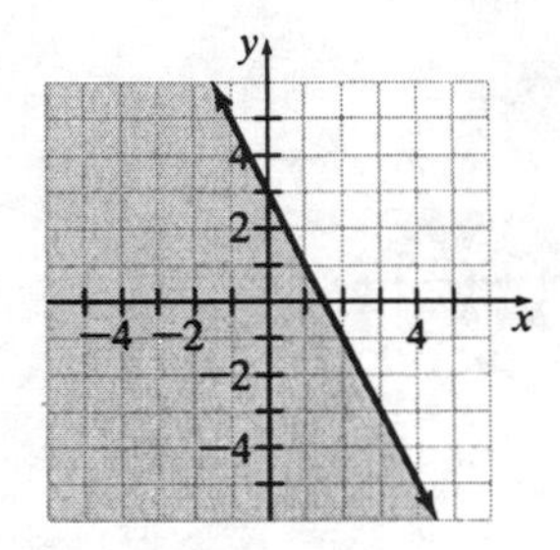

Check (0, 0): $2x + y \le 3$
$2 \cdot 0 + 0 \le 3$
$0 \le 3$ True

23.

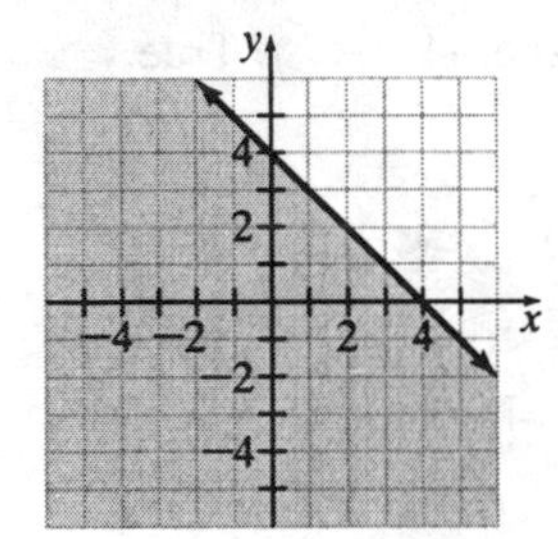

Check (0, 0): $y - 4 \le -x$
$0 - 4 \le -0$
$-4 \le 0$ True

25. **a.** $2(4) + 4(2) < 16$
$8 + 8 < 16$
$16 < 16$
No

b. $2(4) + 4(2) > 16$
$16 > 16$
No

c. $2(4) + 4(2) \ge 16$
$16 \ge 16$
Yes

d. $2(4) + 4(2) \le 16$
$16 \le 16$
Yes

27. No, the ordered pair could be a solution to $ax + by = c$.

29. No. If an ordered pair satisfies $ax + by > c$, it means the ordered pair lies on one side of the line $ax + by = c$. The ordered pair cannot lie on the other side of the line or on the line itself.

31. **a.** $x + y \ge 100$

b.

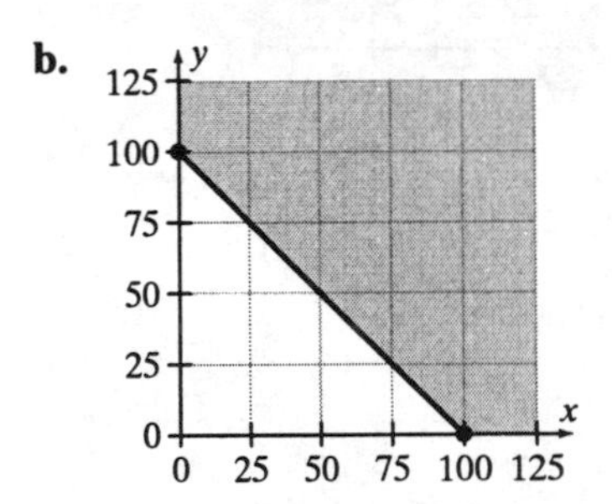

35. **a.** 2

b. 2, 0

c. $2, -5, 0, \frac{2}{5}, -6.3, -\frac{23}{34}$

d. $\sqrt{7}, \sqrt{3}$

e. $2, -5, 0, \sqrt{7}, \frac{2}{5}, -6.3, \sqrt{3}, -\frac{23}{34}$

36. **a.** 0

b. Undefined

37. When evaluating a mathematical expression, the order of operations is parentheses, exponents, multiplication or division (left to right), and addition or subtraction (left to right).

38. $$\begin{aligned} 2(x+3)+2x &= x+4 \\ 2x+6+2x &= x+4 \\ 4x+6 &= x+4 \\ 3x+6 &= 4 \\ 3x &= -2 \\ x &= -\frac{2}{3} \end{aligned}$$

Exercise Set 4.6

1. A relation is any set of ordered pairs.

3. A function is a set of ordered pairs in which each first component corresponds to exactly one second component.

5. a. The domain is the set of first components in the set of ordered pairs.

b. The range is the set of second components in the set of ordered pairs.

7. No, each x must have a unique y for it to be a function.

9. Function
Domain {1, 2, 3, 4, 5}
Range {1, 2, 3, 4, 5}

11. Relation
Domain {1, 2, 3, 6, 7}
Range {–2, 0, 2, 4, 5}

13. Relation
Domain {0, 1, 3, 5}
Range {–4, –1, 0, 1, 2}

15. Relation
Domain {0, 1, 3}
Range {–3, 0, 2, 5}

17. Function
Domain {0, 1, 2, 3, 4}
Range {3}

19. a. {(1, 4), (2, 5), (3, 5), (4, 7)}

b. The relation is a function; every element of the domain corresponds to exactly one element of the range.

21. a. {(–5, 4), (0, 7), (6, 9), (6, 3)}
The relation is not a function; 6, a first component, is paired with more than 1 value.

23. Function

25. Function

27. Since a vertical line drawn at $x = 2$ will intersect the graph in 2 points, the relation is not a function.

29. Function

31. Since a vertical line drawn at $x > -1$ will intersect the graph at more than one point, the relation is not a function.

33. Function

35. a. $f(3) = 4 \cdot 3 + 2 = 14$

b. $f(-1) = 4(-1) + 2 = -2$

37. a. $f(3) = 3^2 - 3 = 6$

b. $f(-2) = (-2)^2 - 3 = 1$

39. a. $f(0) = 2 \cdot 0^2 - 0 + 5 = 5$

b. $f(2) = 2 \cdot 2^2 - 2 + 5 = 11$

41. a. $f(2) = \frac{2+4}{2} = \frac{6}{2} = 3$

b. $f(6) = \frac{6+4}{2} = \frac{10}{2} = 5$

43. Let $x = 0$, $y = f(0) = 0 + 1 = 1$, (0, 1)
Let $x = 1$, $y = f(1) = 1 + 1 = 2$, (1, 2)
Let $x = 2$, $y = f(2) = 2 + 1 = 3$, (2, 3)

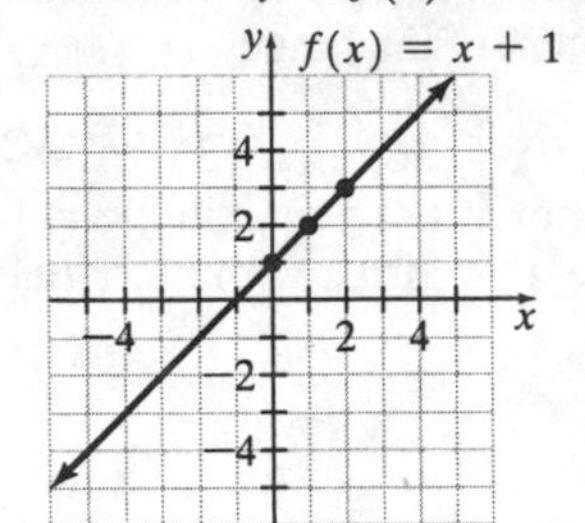

45. Let $x = -2$, $y = f(-2) = 2(-2) - 1 = -5$, (–2, –5)
Let $x = 0$, $y = f(0) = 2 \cdot 0 - 1 = -1$, (0, –1)
Let $x = 2$, $y = f(2) = 2 \cdot 2 - 1 = 3$, (2, 3)

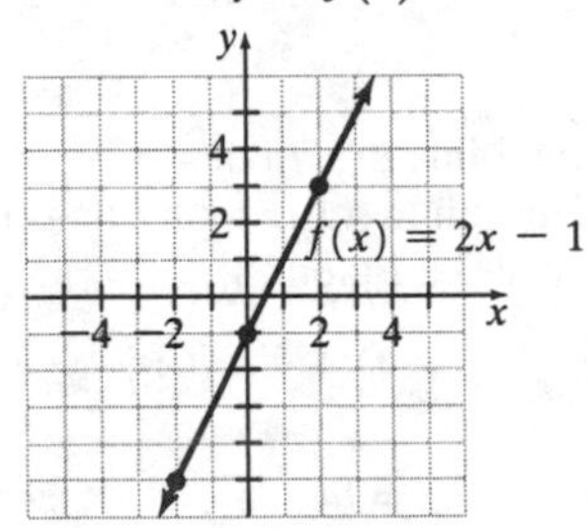

47. Let $x = 0$, $y = f(0) = -2 \cdot 0 + 4 = 4$, (0, 4)
Let $x = 1$, $y = f(1) = -2 \cdot 1 + 4 = 2$, (1, 2)
Let $x = 2$, $y = f(2) = -2 \cdot 2 + 4 = 0$, (2, 0)

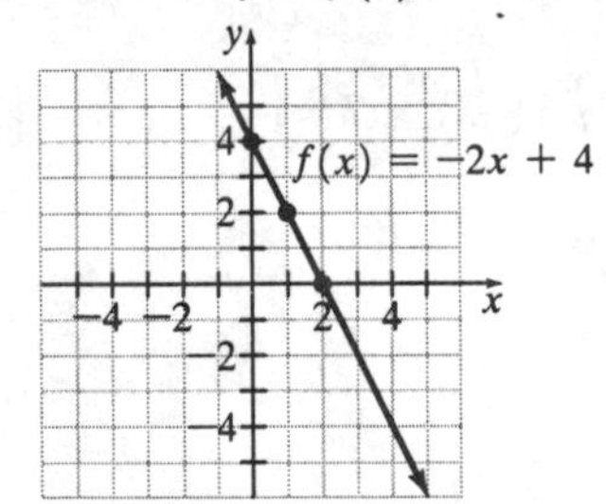

49. Let $x = -2$, $y = f(-2) = -3(-2) - 3 = 3$, (–2, 3)
Let $x = -1$, $y = f(-1) = -3(-1) - 3 = 0$, (–1, 0)
Let $x = 0$, $y = f(0) = -3 \cdot 0 - 3 = -3$, (0, –3)

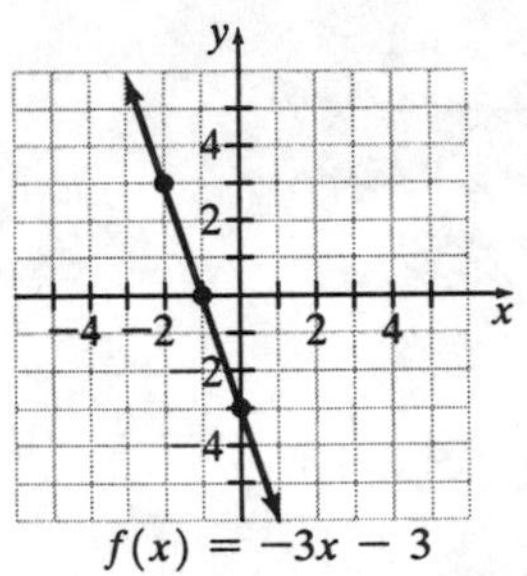

51. No, each x cannot have a unique y.

53. Yes, each year has only one value for the average salary of baseball players.

55. a.

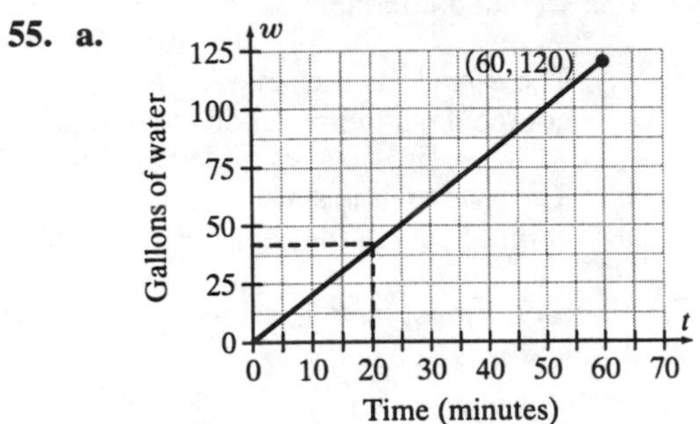

b. 40 gallons

57. a.

(6, 38,000)
Cost (dollars)
Number of Miles

b. $14,000

59. **a.**

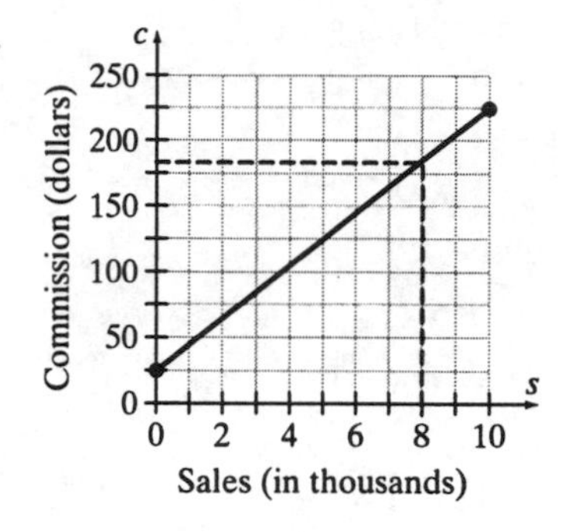

b. $185

61. **a.**

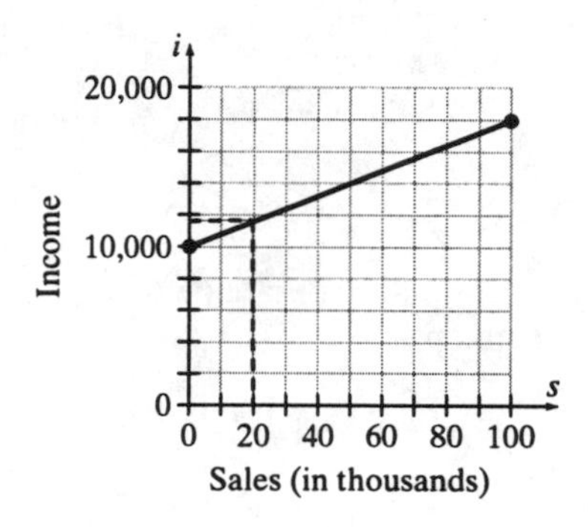

b. $11,600

63. Yes.

65. No. At $x = 1$, there are two y values: $y = 1$ and $y = 2$.

67. **a.**
$$\begin{aligned} f\left(\frac{1}{2}\right) &= \frac{1}{2}\left(\frac{1}{2}\right)^2 - 3\left(\frac{1}{2}\right) + 5 \\ &= \frac{1}{8} - \frac{3}{2} + 5 \\ &= \frac{1}{8} - \frac{12}{8} + \frac{40}{8} \\ &= \frac{29}{8} \end{aligned}$$

b.
$$\begin{aligned} f\left(\frac{2}{3}\right) &= \frac{1}{2}\left(\frac{2}{3}\right)^2 - 3\left(\frac{2}{3}\right) + 5 \\ &= \frac{4}{18} - 2 + 5 \\ &= \frac{2}{9} + 3 \\ &= \frac{2}{9} + \frac{27}{9} \\ &= \frac{29}{9} \end{aligned}$$

c.
$$\begin{aligned} f(0.2) &= \frac{1}{2}(0.2)^2 - 3(0.2) + 5 \\ &= 0.02 - 0.6 + 5 \\ &= 4.42 \end{aligned}$$

71. $\frac{5}{9} - \frac{3}{7} = \frac{35}{63} - \frac{27}{63} = \frac{8}{63}$

72. **a.** Commutative property of multiplication

b. Associative property of addition

c. Distributive property

73.
$$\begin{aligned} 2x - 3(x + 2) &= 8 \\ 2x - 3x - 6 &= 8 \\ -x - 6 &= 8 \\ -x &= 14 \\ x &= -14 \end{aligned}$$

74. Let x = the number of additional miles.
Then
$$\begin{aligned} \text{Cost} &= \$2.00 + \$1.50x \\ 20.00 &= 2.00 + 1.50x \\ 20 &= 2 + 1.5x \\ 18 &= 1.5x \\ 12 &= x \end{aligned}$$
Andrew can travel 12 additional miles for a total of 13 miles.

Review Exercises

1.

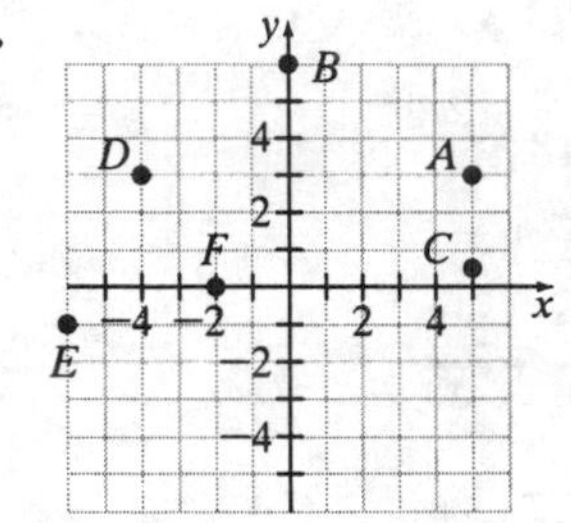

2.

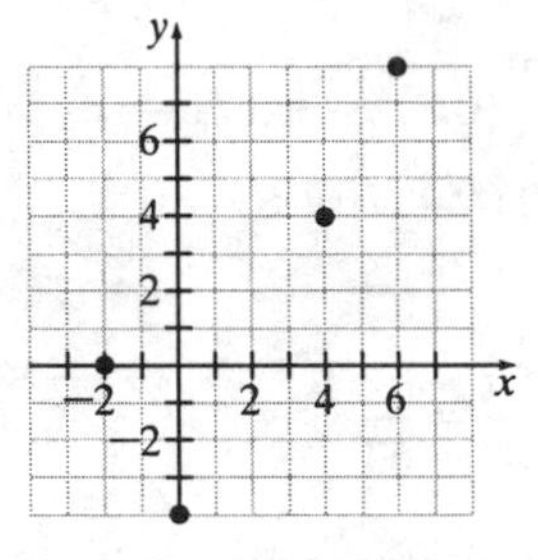

The points are not collinear.

3. a.
$$\begin{aligned} 2x+3y&=9 \\ 2\cdot 4+3\cdot 3&=9 \\ 8+9&=9 \\ 17&=9 \quad \text{False} \end{aligned}$$

b.
$$\begin{aligned} 2x+3y&=9 \\ 2\cdot 0+3\cdot 3&=9 \\ 9&=9 \quad \text{True} \end{aligned}$$

c.
$$\begin{aligned} 2x+3y&=9 \\ 2(-1)+3\cdot 4&=9 \\ -2+12&=9 \\ 10&=9 \quad \text{False} \end{aligned}$$

d.
$$\begin{aligned} 2x+3y&=9 \\ 2\cdot 2+3\left(\frac{5}{3}\right)&=9 \\ 4+5&=9 \\ 9&=9 \quad \text{True} \end{aligned}$$

4. a.
$$\begin{aligned} 3x-2y&=8 \\ 3\cdot 4-2y&=8 \\ 12-2y&=8 \\ -2y&=-4 \\ y&=2 \end{aligned}$$

b.
$$\begin{aligned} 3x-2y&=8 \\ 3\cdot 0-2y&=8 \\ -2y&=8 \\ y&=-4 \end{aligned}$$

c.
$$\begin{aligned} 3x-2y&=8 \\ 3x-2\cdot 4&=8 \\ 3x-8&=8 \\ 3x&=16 \\ x&=\frac{16}{3} \end{aligned}$$

d.
$$\begin{aligned} 3x-2y&=8 \\ 3x-2\cdot 0&=8 \\ 3x&=8 \\ x&=\frac{8}{3} \end{aligned}$$

5. $y = -3$ is a horizontal line with y-intercept = (0, −3).

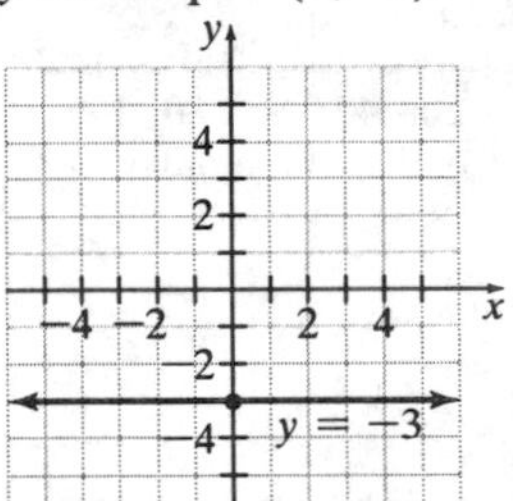

6. $x = 2$ is a vertical line with x-intercept = (2, 0).

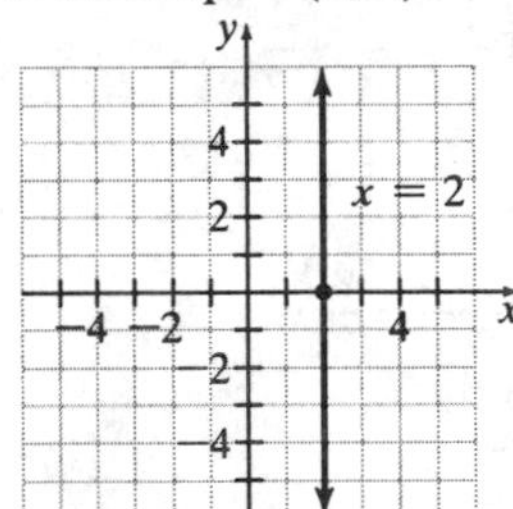

7. Let $x = -1,\ y = 3(-1) = -3,\ (-1, -3)$
Let $x = 0,\ y = 3 \cdot 0 = 0,\ (0, 0)$
Let $x = 1,\ y = 3 \cdot 1 = 3,\ (1, 3)$

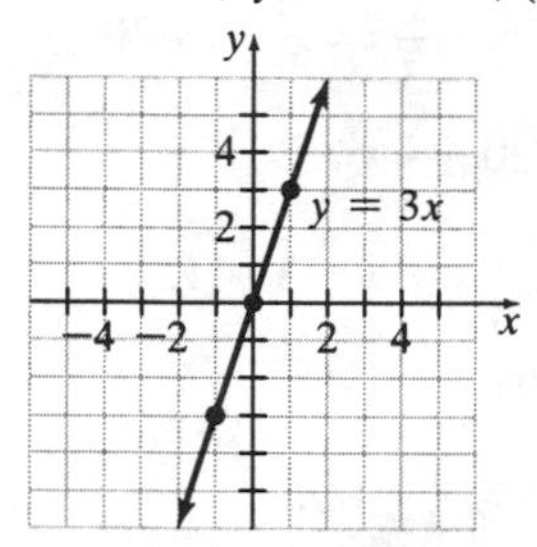

8. Let $x = 0,\ y = 2 \cdot 0 - 1 = -1,\ (0, -1)$
Let $x = 1,\ y = 2 \cdot 1 - 1 = 1,\ (1, 1)$
Let $x = 2,\ y = 2 \cdot 2 - 1 = 3,\ (2, 3)$

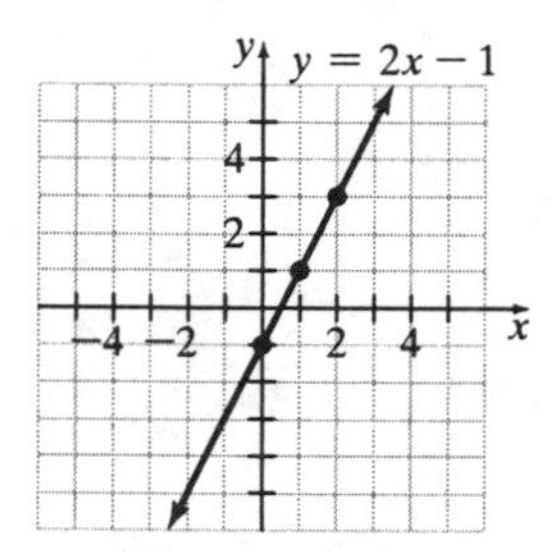

9. Let $x = 0,\ y = -3 \cdot 0 + 4 = 4,\ (0, 4)$
Let $x = 1,\ y = -3 \cdot 1 + 4 = 1,\ (1, 1)$
Let $x = 2,\ y = -3 \cdot 2 + 4 = -2,\ (2, -2)$

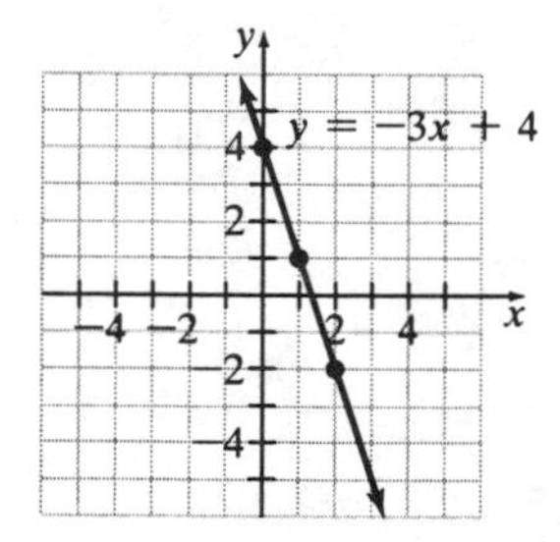

10. Let $x = 0,\ y = -\frac{1}{2}(0) + 4 = 4,\ (0, 4)$

Let $x = 2,\ y = -\frac{1}{2}(2) + 4 = 3,\ (2, 3)$

Let $x = 4,\ y = -\frac{1}{2}(4) + 4 = 2,\ (4, 2)$

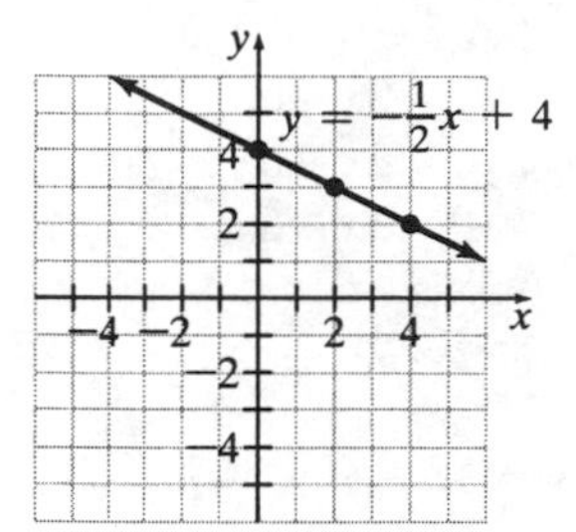

11. Let $x = 0$ Let $y = 0$

$$2x + 3y = 6 \qquad 2x + 3y = 6$$
$$2 \cdot 0 + 3y = 6 \qquad 2x + 3 \cdot 0 = 6$$
$$3y = 6 \qquad 2x = 6$$
$$y = 2 \qquad x = 3$$

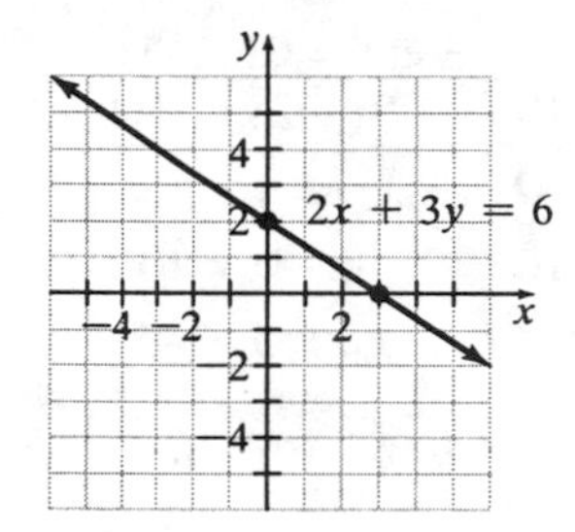

12. Let $x = 0$ Let $y = 0$

$$3x - 2y = 12 \qquad 3x - 2y = 12$$
$$3 \cdot 0 - 2y = 12 \qquad 3x - 2 \cdot 0 = 12$$
$$-2y = 12 \qquad 3x = 12$$
$$y = -6 \qquad x = 4$$

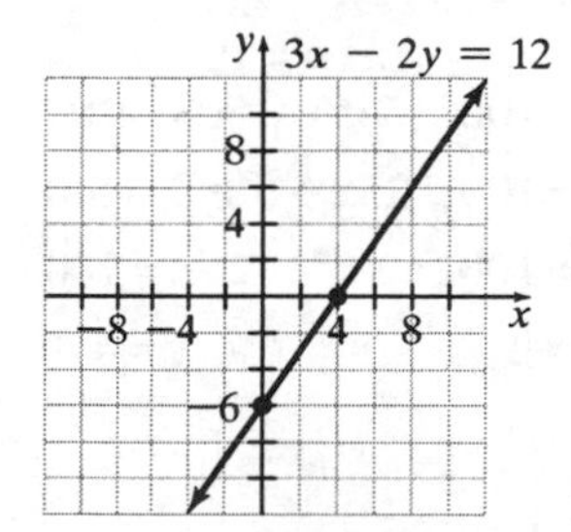

13. Let $x = 0$ Let $y = 0$

$$\begin{aligned} 2y &= 3x - 6 & 2y &= 3x - 6 \\ 2y &= 3 \cdot 0 - 6 & 2 \cdot 0 &= 3x - 6 \\ 2y &= -6 & 0 &= 3x - 6 \\ y &= -3 & -3x &= -6 \\ & & x &= 2 \end{aligned}$$

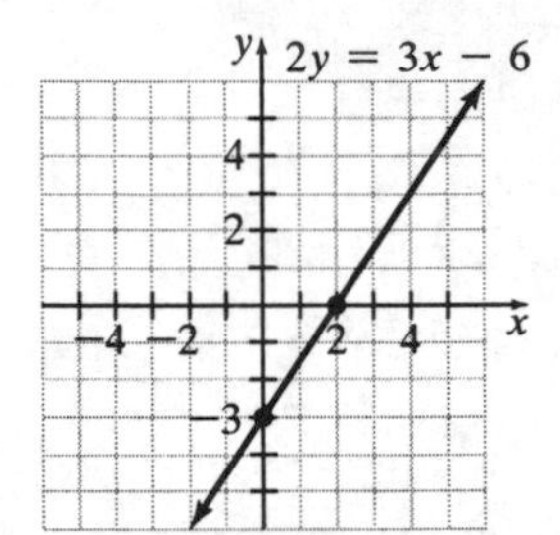

14. Let $x = 0$ Let $y = 0$

$$\begin{aligned} -5x - 2y &= 10 & -5x - 2y &= 10 \\ -5 \cdot 0 - 2y &= 10 & -5x - 2 \cdot 0 &= 10 \\ -2y &= 10 & -5x &= 10 \\ y &= -5 & x &= -2 \end{aligned}$$

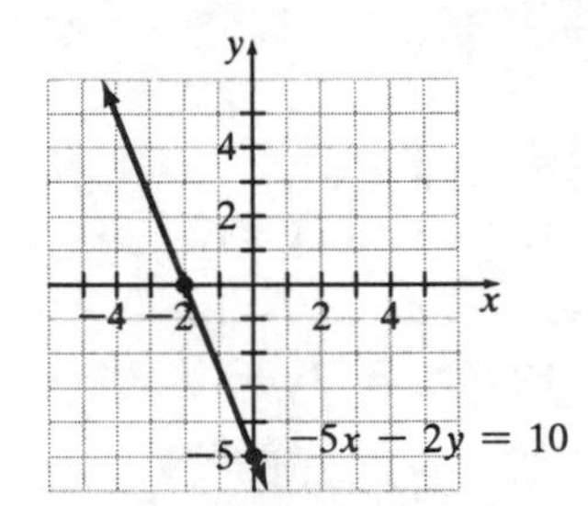

15. Let $x = 0$ Let $y = 0$

$$\begin{aligned} 25x + 50y &= 100 & 25x + 50y &= 100 \\ 25 \cdot 0 + 50y &= 100 & 25x + 50 \cdot 0 &= 100 \\ 50y &= 100 & 25x &= 100 \\ y &= 2 & x &= 4 \end{aligned}$$

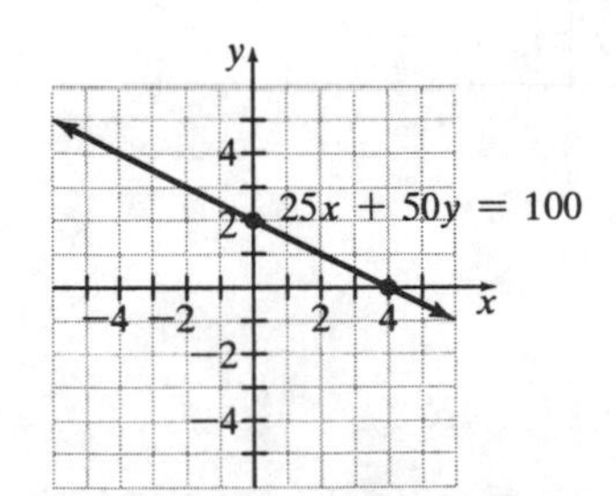

16. Let $x = 0$ Let $y = 0$

$$\begin{aligned} \frac{2}{3}x &= \frac{1}{4}y + 20 & \frac{2}{3}x &= \frac{1}{4}y + 20 \\ \frac{2}{3} \cdot 0 &= \frac{1}{4}y + 20 & \frac{2}{3}x &= \frac{1}{4} \cdot 0 + 20 \\ 0 &= \frac{1}{4}y + 20 & \frac{2}{3}x &= 20 \\ -\frac{1}{4}y &= 20 & x &= 30 \\ y &= -80 & & \end{aligned}$$

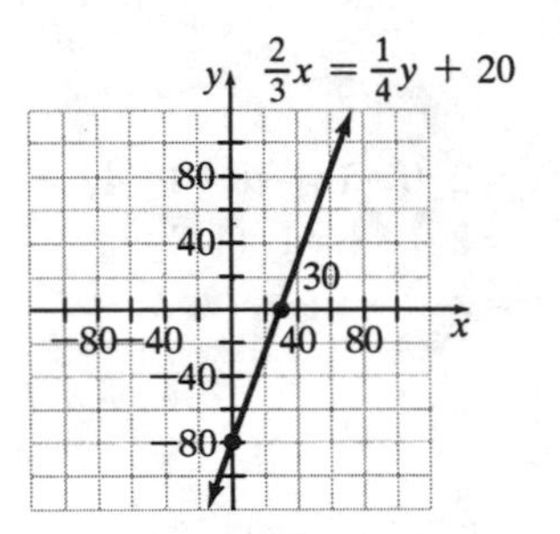

17. $m = \frac{5 - (-7)}{-2 - 8} = \frac{12}{-10} = -\frac{6}{5}$

18. $m = \frac{-3 - (-2)}{8 - (-4)} = \frac{-1}{12} = -\frac{1}{12}$

19. $m = \frac{3 - (-1)}{-4 - (-2)} = \frac{4}{-2} = -2$

20. The slope of a horizontal line is 0.

21. The slope of a vertical line is undefined.

22. The slope of a straight line is the ratio of the vertical change to the horizontal change between any two points on the line.

23. $m=\frac{-5}{7}$
$=-\frac{5}{7}$

24. $m=\frac{2}{8}$
$=\frac{1}{4}$

25. a. $m=\frac{28,000-7,000}{89-85}$
$=\frac{21,000}{4}$
$=5250$

b. $m=\frac{44,000-23,000}{95-96}$
$=\frac{21,000}{-1}$
$=-21,000$

26. $9x+7y=15$
$7y=-9x+15$
$y=-\frac{9}{7}x+\frac{15}{7}$
$m=-\frac{9}{7},\ b=\frac{15}{7}$

The slope is $-\frac{9}{7}$; the y-intercept is $\left(0,\ \frac{15}{17}\right)$.

27. $2x+5=0$
$2x=-5$
$x=-\frac{5}{2}$

This is a vertical line, so the slope is undefined and there is no y-intercept.

28. $3y+9=0$
$3y=-9$
$y=-3$

This is a horizontal line, so the slope is 0 and the y-intercept is (0, –3).

29. $m=\frac{2}{1}=2,\ b=2$
$y=2x+2$

30. $m=\frac{-2}{4}=-\frac{1}{2},\ b=2$
$y=-\frac{1}{2}x+2$

31. $y=2x-6 \quad 6y=12x+6$
$y=2x+1$

Since the slopes are the same and the y-intercepts are different, the lines are parallel.

32. $2x-3y=9 \qquad 3x-2y=6$
$-3y=-2x+9 \qquad -2y=-3x+6$
$y=\frac{2}{3}x-3 \qquad y=\frac{3}{2}x-3$

Since the slopes are different, the lines are not parallel.

33. $y-4=2(x-3)$
$y-4=2x-6$
$y=2x-2$

34. $y-2=-\frac{2}{3}(x-3)$
$y-2=-\frac{2}{3}x+2$
$y=-\frac{2}{3}x+4$

35. $y-2=0(x-4)$
$y-2=0$
$y=2$

36. Lines with undefined slopes are vertical and have the form $x=c$ where c is the value of x for any point on the line.
$x=4$

37. $m = \dfrac{-4-3}{0-(-2)} = \dfrac{-7}{2} = -\dfrac{7}{2}$

$y-3 = -\dfrac{7}{2}[x-(-2)]$

$y-3 = -\dfrac{7}{2}(x+2)$

$y-3 = -\dfrac{7}{2}x - 7$

$y = -\dfrac{7}{2}x - 4$

38. $m = \dfrac{3-(-2)}{-4-(-4)} = \dfrac{5}{0} =$ undefined

Lines with undefined slopes are vertical and have the form $x = c$ where c is the value of x for any point on the line.

$x = -4$

39.

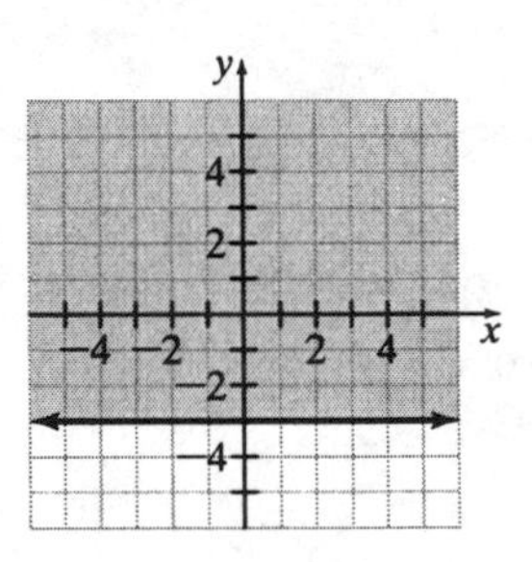

Check (0, 0): $y \ge -3$

$0 \ge -3$ True

40.

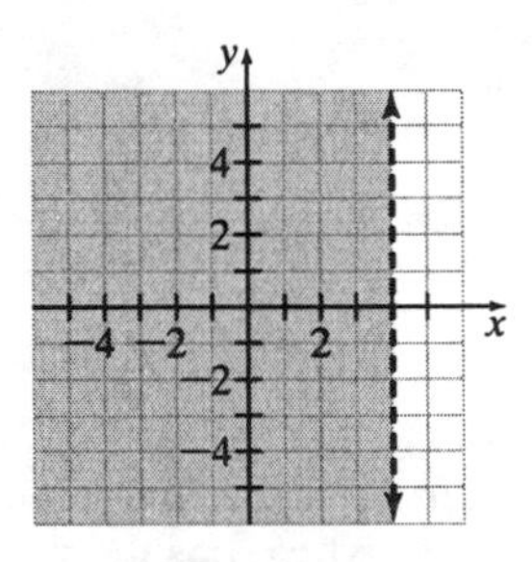

Check (0, 0): $x < 4$

$0 < 4$ True

41.

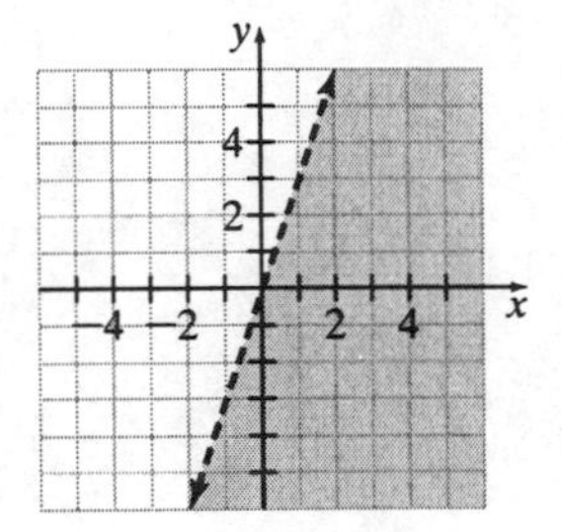

Check (1, –1): $y < 3x$

$-1 < 3 \cdot 1$

$-1 < 3$ True

42.

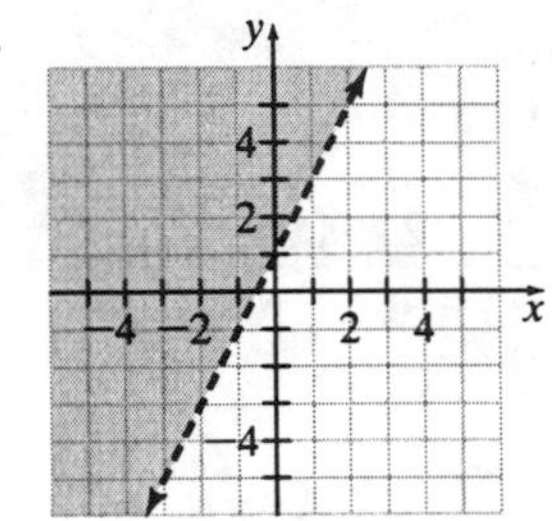

Check (0, 0): $y > 2x + 1$

$0 > 2 \cdot 0 + 1$

$0 > 1$ False

43.

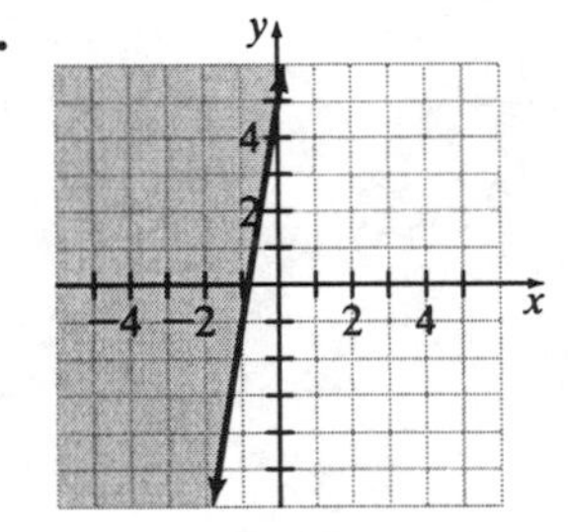

Check (0, 0): $-6x + y \ge 5$

$-6 \cdot 0 + 0 \ge 5$

$0 \ge 5$ False

44.

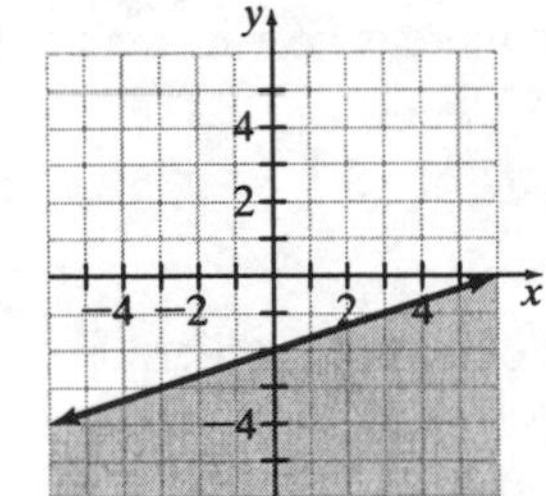

45. Function
Domain {0, 1, 2,4, 6}
Range {–3, –1, 2, 4, 5}

46. Not a function
Domain {3, 4, 6, 7}
Range {0, 1, 2, 5}

47. Not a function
Domain {3, 4, 5, 6}
Range {–3, 1, 2}

48. Function
Domain {–2, 3, 4, 5, 9}
Range {–2}

49. a. {(1, 3), (4, 5), (7, 2), (9, 2)}

b. The relation is a function; every element of the domain corresponds to exactly one element of the range.

50. a. {(4, 1), (6, 3), (6, 5), (8, 7)}

b. The relation is not a function; 6 is paired with more than 1 value.

51. Function

52. Since a vertical line at $x = 0$ will intersect the graph more than once, the graph is not a function.

53. Function

54. Function

55. a. $f(2) = 5 \cdot 2 - 3 = 7$

b. $f(-5) = 5(-5) - 3 = -28$

56. a. $f(-4) = -4(-4) - 5 = 11$

b. $f(8) = -4 \cdot 8 - 5 = -37$

57. a. $f(3) = \frac{1}{3}(3) - 5 = -4$

b. $f(-9) = \frac{1}{3}(-9) - 5 = -8$

58. a. $f(3) = 2 \cdot 3^2 - 4 \cdot 3 + 6 = 12$

b. $f(-5) = 2(-5)^2 - 4(-5) + 6 = 76$

59. Yes, it is a function since it passes the vertical line test.

60. Yes, it is a function since it passes the vertical line test.

61. Let $x = 0$, $y = f(0) = 3 \cdot 0 - 5 = -5$, (0, –5)
Let $x = 1$, $y = f(1) = 3 \cdot 1 - 5 = -2$, (1, –2)
Let $x = 2$, $y = f(2) = 3 \cdot 2 - 5 = 1$, (2, 1)

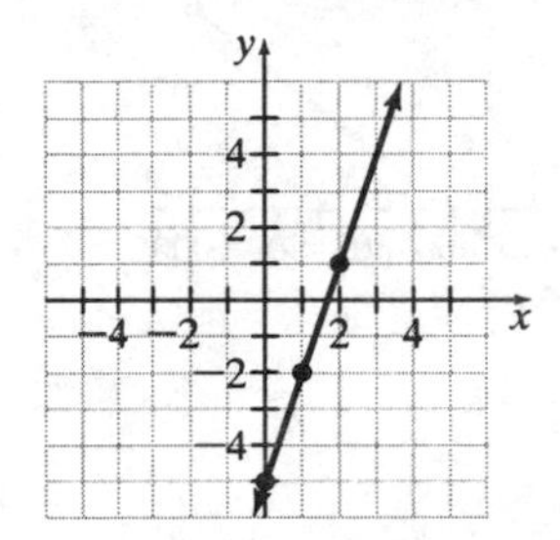

62. Let $x = 0$, $y = f(0) = -2 \cdot 0 + 3 = 3$, (0, 3)
Let $x = 1$, $y = f(1) = -2 \cdot 1 + 3 = 1$, (1, 1)
Let $x = 2$, $y = f(2) = -2 \cdot 2 + 3 = -1$, (2, –1)

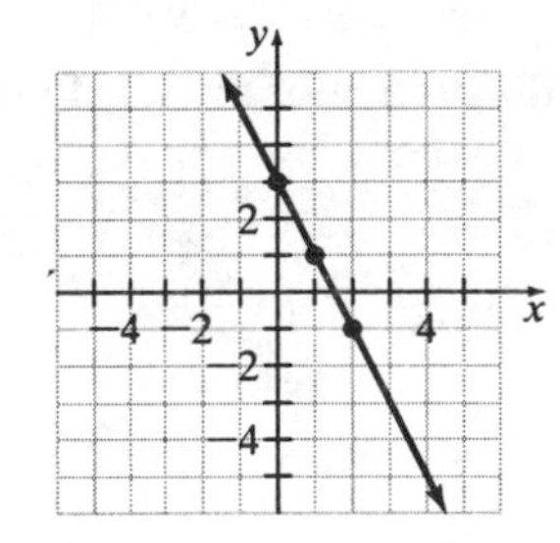

63. a.

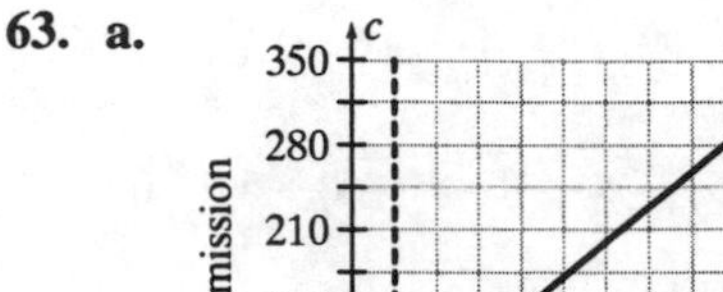
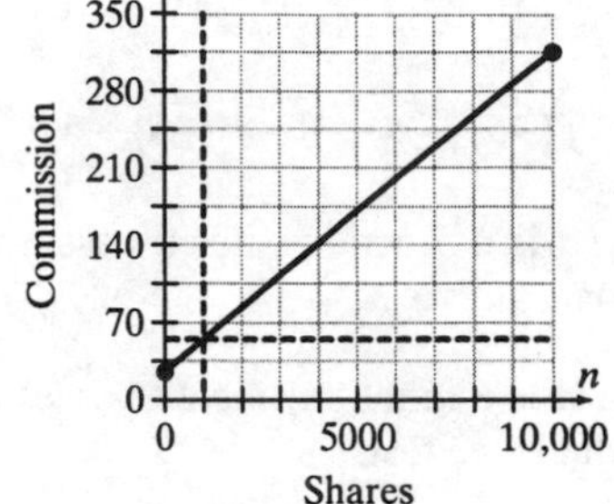

b. \$55

64. a.

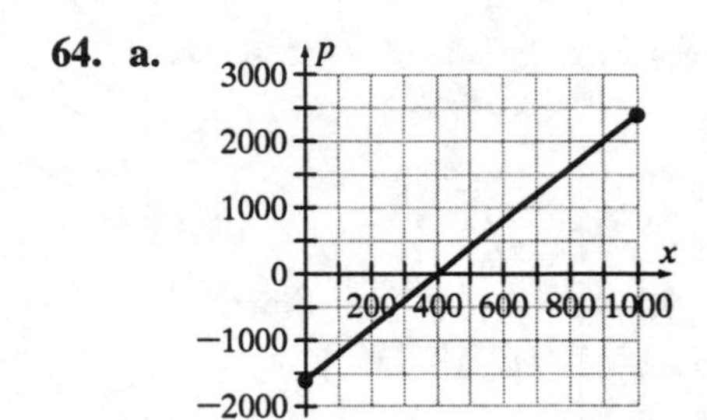

b. \$0

Check (0, 0): $3y + 6 \le x$

$$3 \cdot 0 + 6 \le 0$$

$$6 \le 0 \quad \text{False}$$

Practice Test

1. A graph is an illustration of the set of points that satisfy an equation.

2. a. III

b. IV

3. a. $ax + by = c$

b. $y = mx + b$

c. $y - y_1 = m(x - x_1)$

4. a.

$$3y = 5x - 9$$
$$3 \cdot 2 = 5 \cdot 3 - 9$$
$$6 = 15 - 9$$
$$6 = 6 \quad \text{True}$$

b.

$$3y = 5x - 9$$
$$3 \cdot 0 = 5 \cdot \frac{9}{5} - 9$$
$$0 = 9 - 9$$
$$0 = 0 \quad \text{True}$$

c.

$$3y = 5x - 9$$
$$3(-6) = 5(-2) - 9$$
$$-18 = -10 - 9$$
$$-18 = -19 \quad \text{False}$$

d.

$$3y = 5x - 9$$
$$3 \cdot 3 = 5 \cdot 0 - 9$$
$$9 = -9 \quad \text{False}$$

(3, 2) and $\left(\frac{9}{5}, 0\right)$ satisfy the equation.

5. $m = \dfrac{3-(-5)}{-4-2} = \dfrac{8}{-6} = -\dfrac{4}{3}$

6.

$$4x - 9y = 15$$
$$-9y = -4x + 15$$
$$y = \frac{4}{9}x - \frac{5}{3}$$

$m = \dfrac{4}{9}$, $b = -\dfrac{5}{3}$

The slope is $\dfrac{4}{9}$; the y-intercept is $\left(0, -\dfrac{5}{3}\right)$.

7. $m = \dfrac{-1}{1} = -1$, $b = -1$

$y = -x - 1$

8. $x = -5$ is a vertical line with x-intercept = (−5, 0).

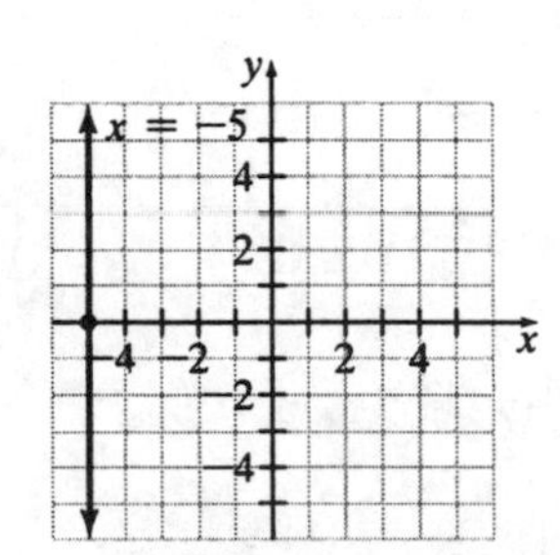

9. $y = 2$ is a horizontal line with y-intercept = (0, 2).

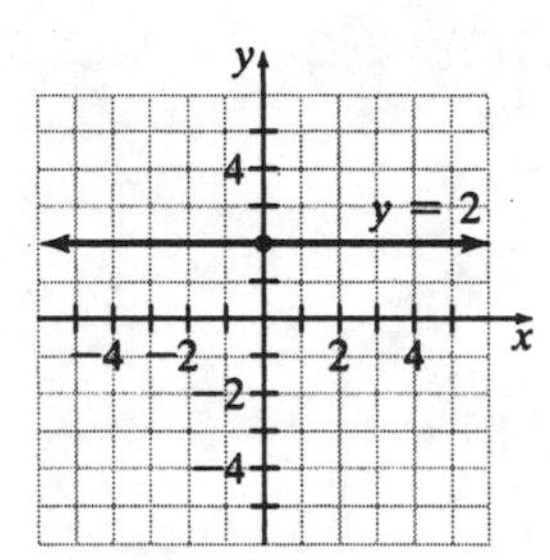

10. Let $x = 0$, $y = 3\cdot 0 - 2 = -2$, (0, −2)
Let $x = 1$, $y = 3\cdot 1 - 2 = 1$, (1, 1)
Let $x = 2$, $y = 3\cdot 2 - 2 = 4$, (2, 4)

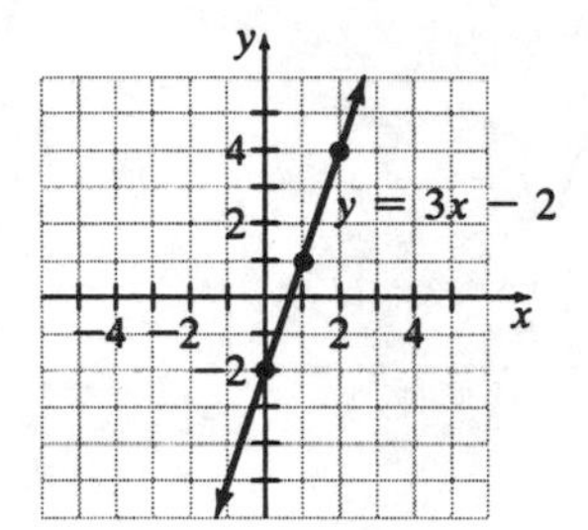

11. a. $2x - 4y = 8$
$-4y = -2x + 8$
$y = \frac{1}{2}x - 2$

b. Let $x = 0$, $y = \frac{1}{2}(0) - 2 = -2$, (0, −2)

Let $x = 2$, $y = \frac{1}{2}(2) - 2 = -1$, (2, −1)

Let $x = 4$, $y = \frac{1}{2}(4) - 2 = 0$, (4, 0)

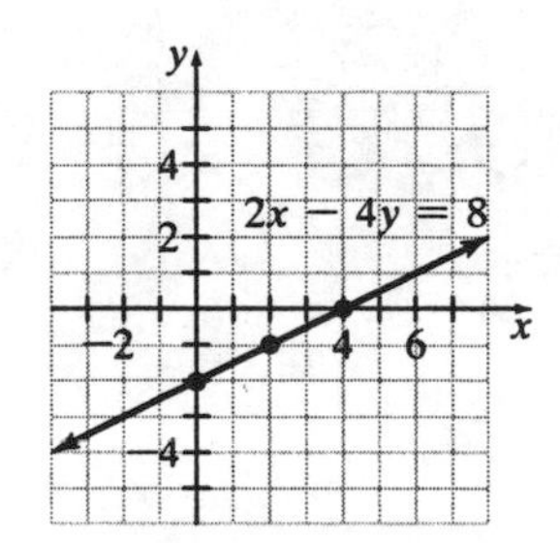

12. $3x + 5y = 15$

Let $x = 0$	Let $y = 0$
$3(0) + 5y = 15$	$3x + 5(0) = 15$
$5y = 15$	$3x = 15$
$y = 3$	$x = 5$

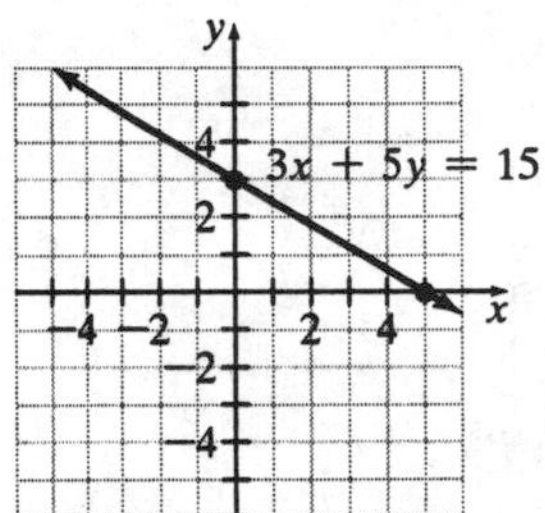

13. $y - 3 = 3(x - 1)$
$y - 3 = 3x - 3$
$y = 3x$

14. $m = \frac{2-(-1)}{-4-3} = \frac{3}{-7} = -\frac{3}{7}$
$y - (-1) = -\frac{3}{7}(x - 3)$
$y + 1 = -\frac{3}{7}x + \frac{9}{7}$
$y = -\frac{3}{7}x + \frac{2}{7}$

15. $2y = 3x - 6$ $\quad y - \frac{3}{2}x = -5$
$y = \frac{3}{2}x - 3$ $\quad y = \frac{3}{2}x - 5$
The lines are parallel since they have the same slope but different y-intercepts.

16. slope = 3, y intercept is (0, −4).

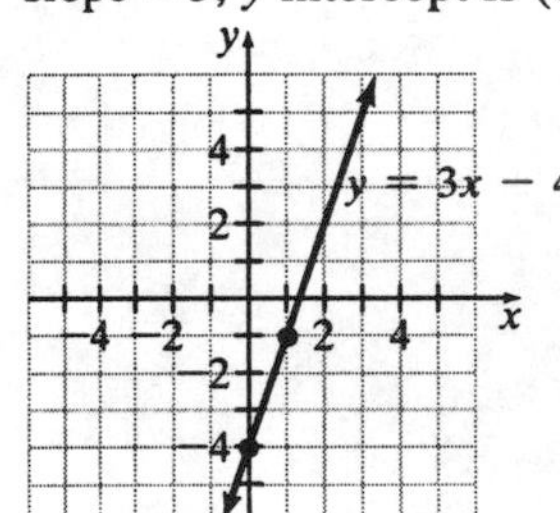

17. $3x - 2y = 8$
$$-2y = -3x + 8$$
$$y = \frac{3}{2}x - 4$$

Slope $= \frac{3}{2}$, y intercept is (0, –4).

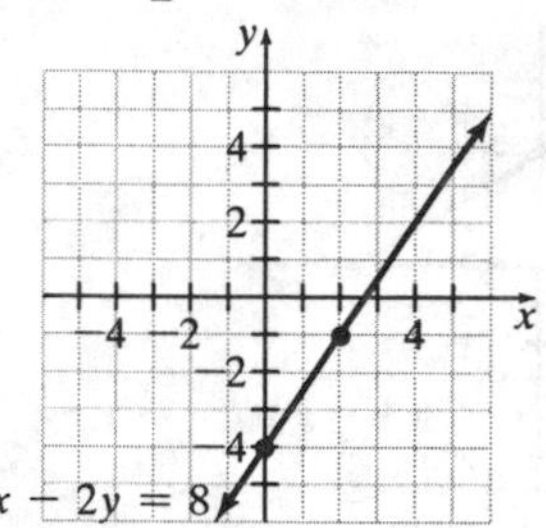

18. A function is a set of ordered pairs in which each first component corresponds to exactly one second component.

19. a. The relation is not a function; 1, a first component, is paired with more than 1 value.

b. Domain {1, 3, 5, 6}
Range {–4, 0, 2, 3, 5}

20. a. The graph is a function because it passes the vertical line test.

b. The graph is not a function because a vertical line can be drawn that intersects the graph at more than one point.

21. a.
$$f(2) = 2(2)^2 + 3(2) = 8 + 6 = 14$$

b.
$$f(-3) = 2(-3)^2 + 3(-3) = 18 - 9 = 9$$

22. Let $x = 0$, $y = f(0) = 2 \cdot 0 - 4 = -4$, (0, –4)
Let $x = 1$, $y = f(1) = 2 \cdot 1 - 4 = -2$, (1, –2)
Let $x = 2$, $y = f(2) = 2 \cdot 2 - 4 = 0$, (2, 0)

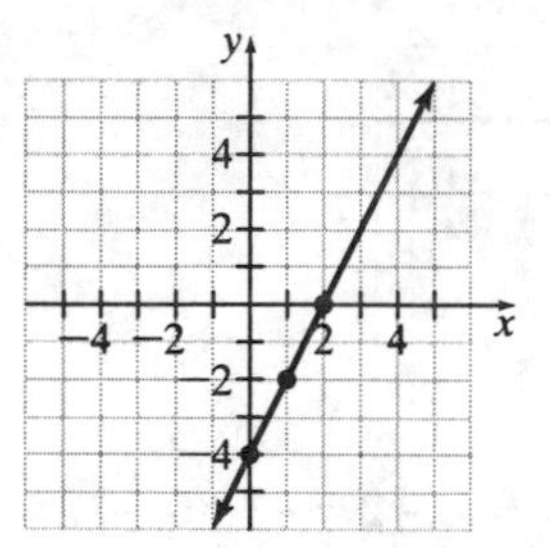

23.

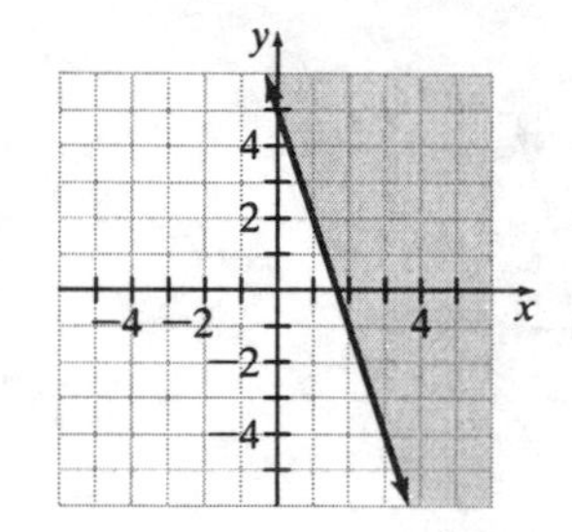

Check (0, 0): $y \geq -3x + 5$
$0 \geq -3 \cdot 0 + 5$
$0 \geq 5$ False

24.

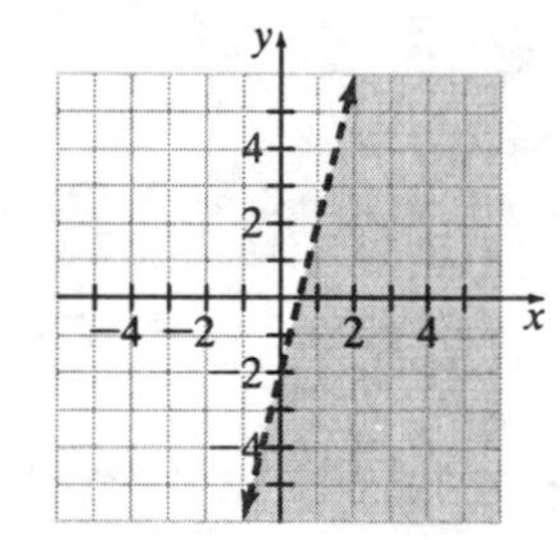

Check (0, 0): $y < 4x - 2$
$0 < 4 \cdot 0 - 2$
$0 < -2$ False

25. a.

i

700, 600, 500, 400, 300, 200, 100, 0

Income (dollars)

(10,000, 700)

0, 5000, 10,000 s

Sales (dollars)

b. $350

Cumulative Review Test

1. a. $\{1, 2, 3, \ldots\}$

b. $\{0, 1, 2, 3, \ldots\}$

2. a. Distributive Property

b. Commutative Property of Addition

3. $-3-(-2)=-3+2=-1$

4.
$$\begin{aligned}4^2+8\div(-2)&=16+8\div(-2)\\&=16+(-4)\\&=12\end{aligned}$$

5.
$$\begin{aligned}16\div(4-6)\cdot 5&=16\div(-2)\cdot 5\\&=-8\cdot 5\\&=-40\end{aligned}$$

6.
$$\begin{aligned}2x+5&=3(x-5)\\2x+5&=3x-15\\-x+5&=-15\\-x&=-20\\x&=20\end{aligned}$$

7.
$$\begin{aligned}3(x-2)-(x+4)&=2x-10\\3x-6-x-4&=2x-10\\2x-10&=2x-10\\0&=0\end{aligned}$$
All real numbers are solutions.

8.
$$\begin{aligned}2x-14&>5x+1\\-3x&>15\\x&<-5\end{aligned}$$

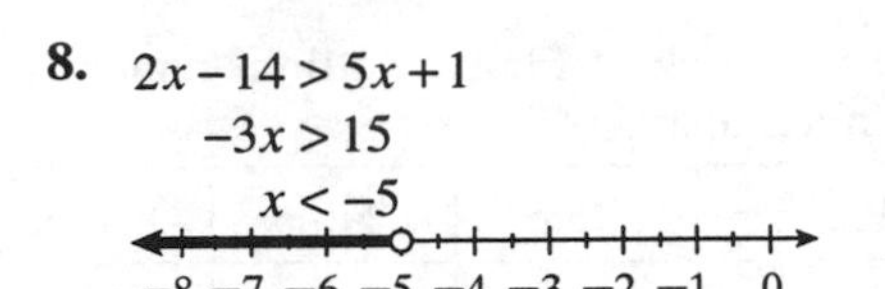

9.
$$\begin{aligned}v&=lwh\\\frac{v}{lh}&=\frac{lwh}{lh}\\\frac{v}{lh}&=w\end{aligned}$$

10.
$$\begin{aligned}\frac{3\text{ cans}}{\$1.25}&=\frac{8\text{ cans}}{x\text{ dollars}}\\\frac{3}{1.25}&=\frac{8}{x}\\3x&=10\\x&=\frac{10}{3}\approx 3.33\end{aligned}$$
8 cans sell for $3.33.

11. Let x = the number
$$\begin{aligned}11+2x&=19\\2x&=8\\x&=4\end{aligned}$$

12. Let x = width of rectangle.
Then $2x + 4$ = length of rectangle.
$$\begin{aligned}P&=2l+2w\\26&=2(2x+4)+2x\\26&=4x+8+2x\\26&=6x+8\\18&=6x\\3&=x\end{aligned}$$
The width is 3 feet and the length is $2(3)+4=10$ feet.

13. Let x = number of hours until the runners are 28 miles apart.

Runner	Rate	Time	Distance
First	6 mph	x	$6x$
Second	8 mph	x	$8x$

(Distance run by first runner) + (Distance run by second runner) = 28 miles

$$6x + 8x = 28$$
$$14x = 28$$
$$x = 2$$

It will take 2 hours.

14. Answers will vary. Sample answer: (0, 2), (2, 1), and (4, 0)

15. Let $x = 0,\ y = 3\cdot 0 - 5 = -5,\ (0, -5)$
Let $x = 1,\ y = 3\cdot 1 - 5 = -2,\ (1, -2)$
Let $x = 2,\ y = 3\cdot 2 - 5 = 1,\ (2, 1)$

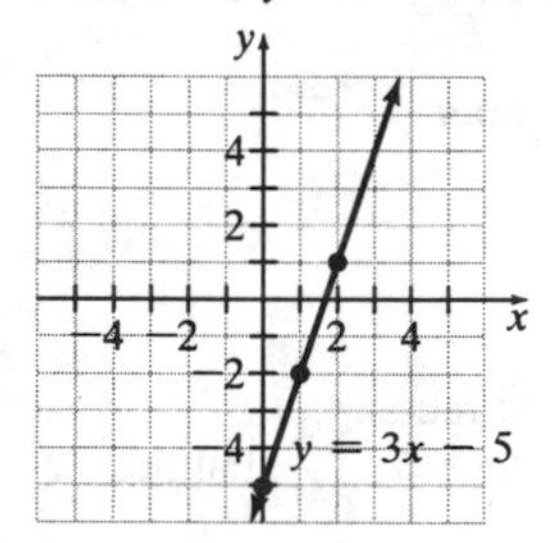

16. $2x + 6y = 12$

Let $x = 0$	Let $y = 0$
$2(0) + 6y = 12$	$2x + 6(0) = 12$
$6y = 12$	$2x = 12$
$y = 2$	$x = 6$

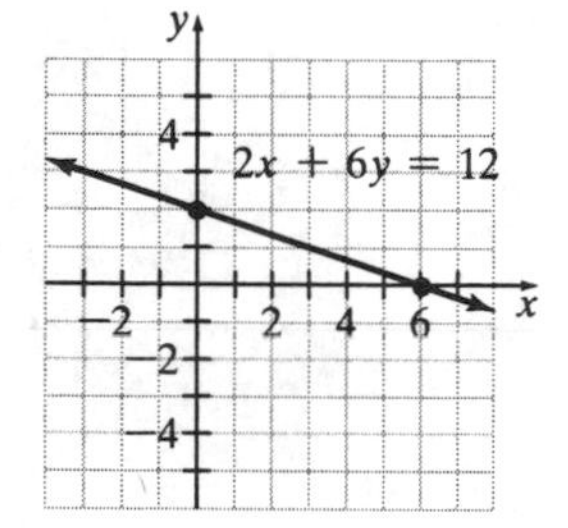

17. $3x + 5y = 12$
$$5y = -3x + 12$$
$$y = \frac{-3x}{5} + \frac{12}{5}$$
$$y = -\frac{3}{5}x + \frac{12}{5}$$

Slope $= -\frac{3}{5}$, y intercept is $\left(0, \frac{12}{5}\right)$

18. Slope $= \frac{2}{3}$, y intercept is (0, –3)

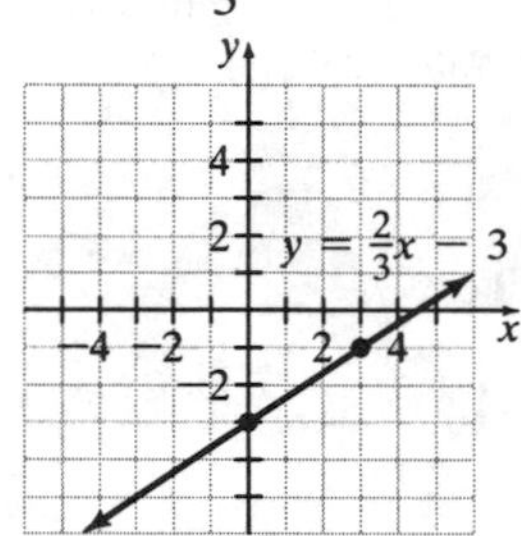

19. $y - 2 = 3(x - 5)$

20. **a.** The relation is not a function; it does not pass the vertical line test.

b. The relation is a function; each first component corresponds to exactly one second component.

Chapter 5

Exercise Set 5.1

1. The solution to a system of equations represents the ordered pairs that satisfy all the equations in the system.

3. Write the equations in slope-intercept form and compare their slopes and y-intercepts.

5. The point of intersection can only be estimated.

7. a. $y = 2x - 6$
$2 \stackrel{?}{=} 2(-1) - 6$
$2 = -8$ False
Since $(-1, 2)$ does not satisfy the first equation, it is not a solution to the system of equations.

b. $y = 2x - 6$ | $y = -x + 3$
$0 \stackrel{?}{=} 2(3) - 6$ | $0 \stackrel{?}{=} -3 + 3$
$0 = 0$ True | $0 = 0$ True
Since $(3, 0)$ satisfies both equations, it is a solution to the system of linear equations.

c. $y = 2x - 6$
$2 \stackrel{?}{=} 2(2) - 6$
$2 = -2$ False
Since $(2, 2)$ does not satisfy the first equation, it is not a solution to the system of equations.

9. a. $y = 2x - 3$ | $y = x + 5$
$13 \stackrel{?}{=} 2(8) - 3$ | $13 \stackrel{?}{=} 8 + 5$
$13 = 13$ True | $13 = 13$ True
Since $(8, 13)$ satisfies both equations, it is a solution to the system.

b. $y = 2x - 3$ | $y = x + 5$
$5 \stackrel{?}{=} 2(4) - 3$ | $5 \stackrel{?}{=} 4 + 5$
$5 = 5$ True | $5 = 9$ False
Since $(4, 5)$ does not satisfy both equations, it is not a solution to the system.

c. $y = 2x - 3$
$9 \stackrel{?}{=} 2(4) - 3$
$9 = 5$ False
Since $(4, 9)$ does not satisfy the first equation, it is not a solution to the system.

11. a. $3x - y = 6$ | $4x + y = 10$
$3(3) - 3 \stackrel{?}{=} 6$ | $4(3) + 3 \stackrel{?}{=} 10$
$6 = 6$ True | $15 = 10$ False
Since $(3, 3)$ does not satisfy both equations, it is not a solution to the system.

b. $3x - y = 6$
$3(2) - (-2) \stackrel{?}{=} 6$
$8 = 6$ False
Since $(2, -2)$ does not satisfy the first equation, it is not a solution to the system.

c. $3x - y = 6$ | $4x + y = 10$
$3(4) - 6 \stackrel{?}{=} 6$ | $4(4) + 6 \stackrel{?}{=} 10$
$6 = 6$ True | $22 = 10$ False
Since $(4, 6)$ does not satisfy both equations, it is not a solution to the system.

13. Solve the first equation for y.
$2x - 3y = 6$
$-3y = -2x + 6$
$y = \frac{2}{3}x - 2$
Notice that it is the same as the second equation. If the ordered pair satisfies the first equation, then it also satisfies the second equation.

a. $2x - 3y = 6$
$2(3) - 3(0) \stackrel{?}{=} 6$
$6 = 6$ True
Since $(3, 0)$ satisfies both equations, it is a solution to the system.

b. $2x - 3y = 6$

$2(3) - 3(-2) \stackrel{?}{=} 6$

$12 = 6$ False

Since (3, –2) does not satisfy the first equation, it is not a solution to the system.

c. $2x - 3y = 6$

$2(6) - 3(2) \stackrel{?}{=} 6$

$6 = 6$ True

Since (6, 2) satisfies both equations, it is a solution to the system.

15. a. $3x - 4y = 8$

$3(0) - 4(-2) \stackrel{?}{=} 8$

$8 = 8$ True

$2y = \frac{2}{3}x - 4$

$2(-2) \stackrel{?}{=} \frac{2}{3}(0) - 4$

$-4 = -4$ True

Since (0, –2) satisfies both equations, it is a solution to the system.

b. $3x - 4y = 8$

$3(1) - 4(-6) \stackrel{?}{=} 8$

$27 = 8$ False

Since (1, –6) does not satisfy the first equation, it is not a solution to the system.

c. $3x - 4y = 8$

$3\left(-\frac{1}{3}\right) - 4\left(-\frac{9}{4}\right) \stackrel{?}{=} 8$

$8 = 8$ True

$2y = \frac{2}{3}x - 4$

$2\left(-\frac{9}{4}\right) \stackrel{?}{=} \frac{2}{3}\left(-\frac{1}{3}\right) - 4$

$-\frac{9}{2} = -\frac{38}{9}$ False

Since $\left(-\frac{1}{3}, -\frac{9}{4}\right)$ does not satisfy both equations, it is not a solution to the system.

17. consistent—one solution

19. dependent—infinite number of solutions

21. consistent—one solution

23. inconsistent—no solution

25. Write each equation in slope-intercept form.

$y = 4x - 1$ $\qquad$ $2y = 4x - 6$

$y = 2x - 3$

Since the slopes of the lines are not the same, the lines intersect to produce one solution. This is a consistent system.

27. Write each equation in slope-intercept form.

$2y = 3x + 3$ $\qquad$ $y = \frac{3}{2}x - 2$

$y = \frac{3}{2}x + \frac{3}{2}$

Since the lines have the same slope, $\frac{3}{2}$, and different y-intercepts, the lines are parallel. There is no solution. This is an inconsistent system.

29. Write each equation in slope-intercept form.

$5x = y - 6$ $\qquad$ $3x = 4y + 5$

$y = 5x + 6$ $\qquad$ $4y = 3x - 5$

$y = \frac{3}{4}x - \frac{5}{4}$

Since the slopes of the lines are not the same, the lines intersect to produce one solution. This is a consistent system.

31. Write each equation in slope-intercept form.

$2x = 3y + 4$ $\qquad$ $6x - 9y = 12$

$2x - 4 = 3y$ $\qquad$ $-9y = -6x + 12$

$\frac{2}{3}x - \frac{4}{3} = y$ $\qquad$ $y = \frac{2}{3}x - \frac{4}{3}$

Since both equations are identical, the line is the same for both of them. There are an infinite number of solutions. This is a dependent system.

33. Write each equation in slope-intercept form.

$$3x+5y=-7 \qquad -3x-5y=-7$$
$$5y=-3x-7 \qquad -5y=3x-7$$
$$y=-\frac{3}{5}x-\frac{7}{5} \qquad y=-\frac{3}{5}x+\frac{7}{5}$$

Since the lines have the same slope and different y-intercepts, the lines are parallel. There is no solution. This is an inconsistent system.

35. Write each equation in slope-intercept form.

$$y=\frac{3}{2}x+\frac{1}{2} \qquad 3x-2y=-\frac{1}{2}$$
$$-2y=-3x-\frac{1}{2}$$
$$y=\frac{3}{2}x+\frac{1}{4}$$

Since the lines have the same slope and different y-intercepts, the lines are parallel. There is no solution. This is an inconsistent system.

37. Graph the equations $y=x+2$ and $y=-x+2$.

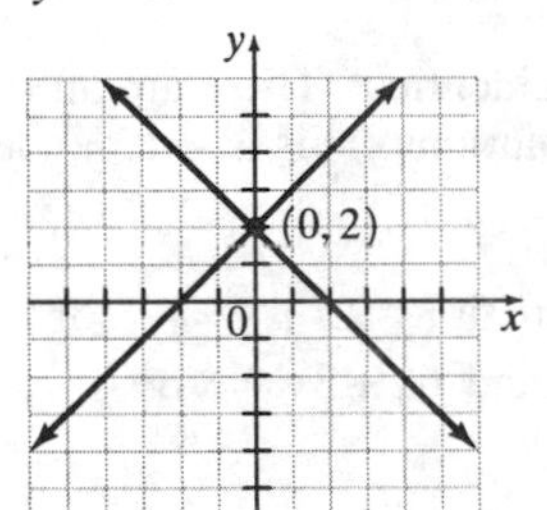

The lines intersect and the point of intersection is (0, 2). This is a consistent system.

39. Graph the equations $y=3x-6$ and $y=-x+6$.

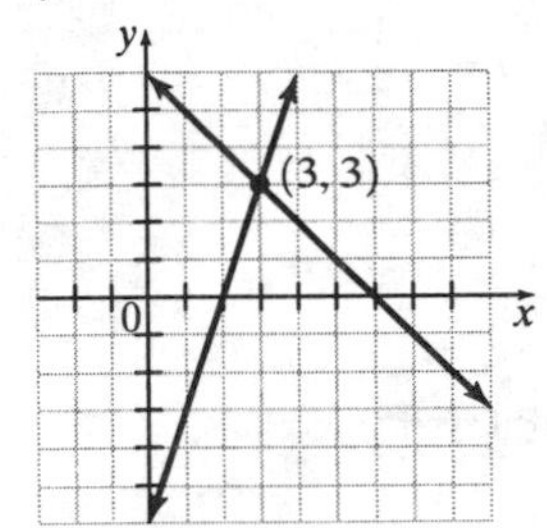

The lines intersect and the point of intersection is (3, 3). This is a consistent system.

41. Graph the equations $2x=4$ or $x=2$ and $y=-3$.

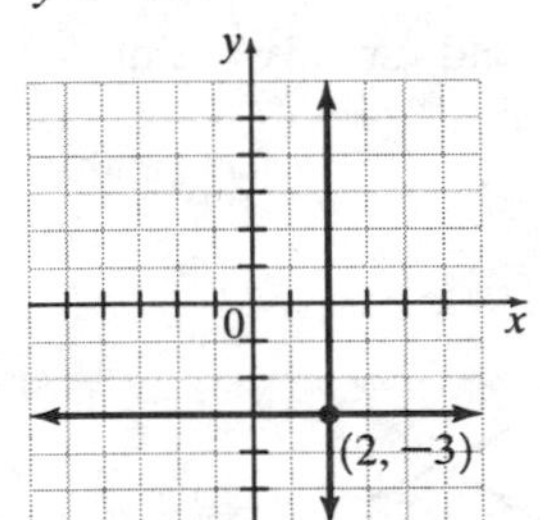

The lines intersect and the point of intersection is (2, –3). This is a consistent system.

43. Graph the equations $y=x+2$ and $x+y=4$ or $y=-x+4$.

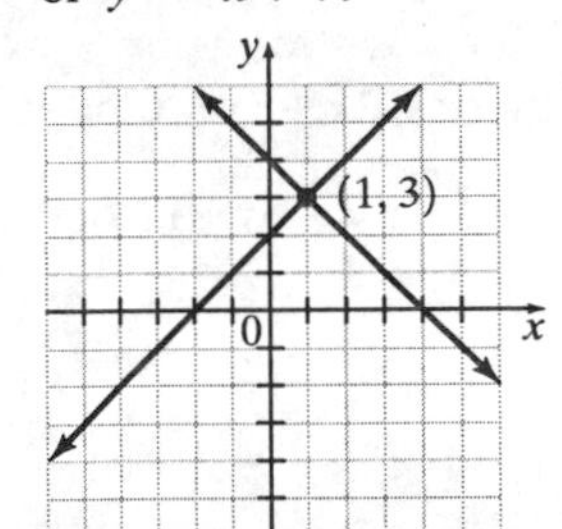

The lines intersect and the point of intersection is (1, 3). This is a consistent system.

45. Graph the equations $y=-\frac{1}{2}x+4$ and $x+2y=6$ or $y=-\frac{1}{2}x+3$.

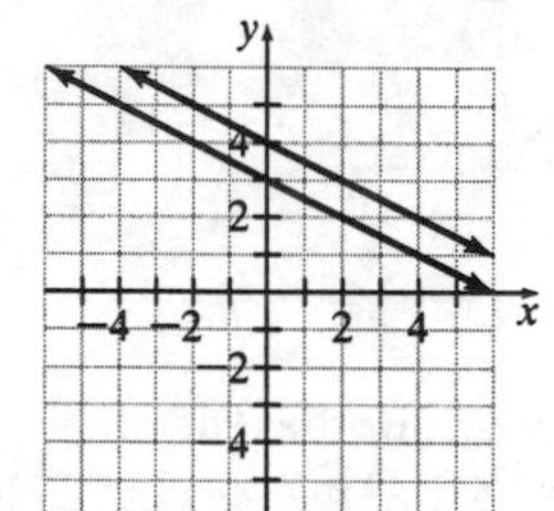

The lines are parallel. The system is inconsistent and there is no solution.

47. Graph the equations $x+2y=8$ or $y=-\frac{1}{2}x+4$ and $2x-3y=2$ or $y=\frac{2}{3}x-\frac{2}{3}$.

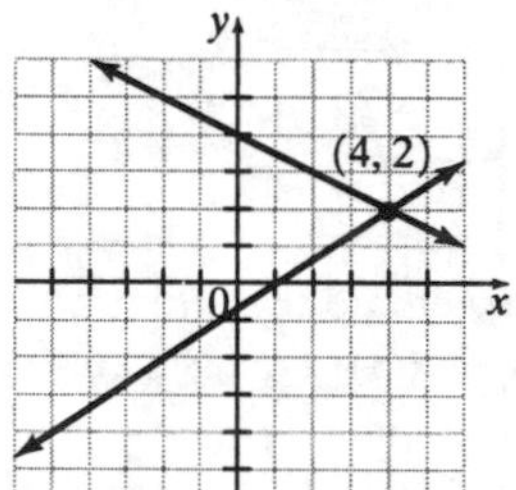

The lines intersect and the point of intersection is (4, 2). This is a consistent system.

49. Graph the equations $2x+3y=6$ or $y=-\frac{2}{3}x+2$ and $4x=-6y+12$ or $y=-\frac{2}{3}x+2$.

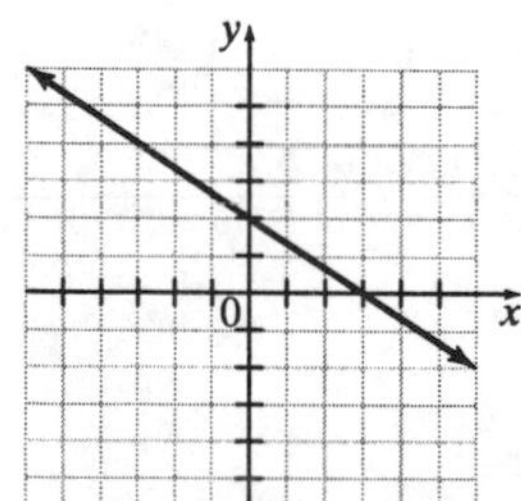

The lines are identical. There are an infinite number of solutions. This is a dependent system.

51. Graph the equations $y=3$ and $y=2x-3$.

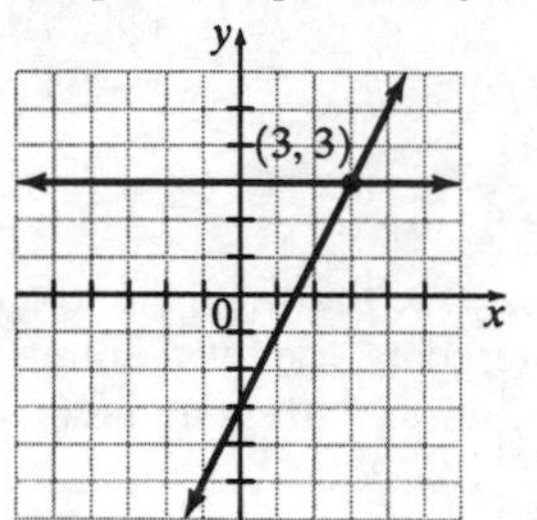

The lines intersect and the point of intersection is (3, 3). This is a consistent system.

53. Graph the equations $x-2y=4$ or $y=\frac{1}{2}x-2$ and $2x-4y=8$ or $y=\frac{1}{2}x-2$.

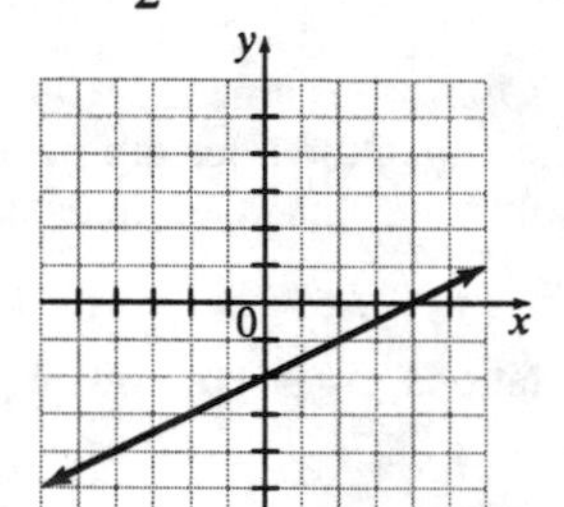

The lines are identical. There are an infinite number of solutions. This is a dependent system.

55. Graph the equations $2x+y=-2$ or $y=-2x-2$ and $6x+3y=6$ or $y=-2x+2$.

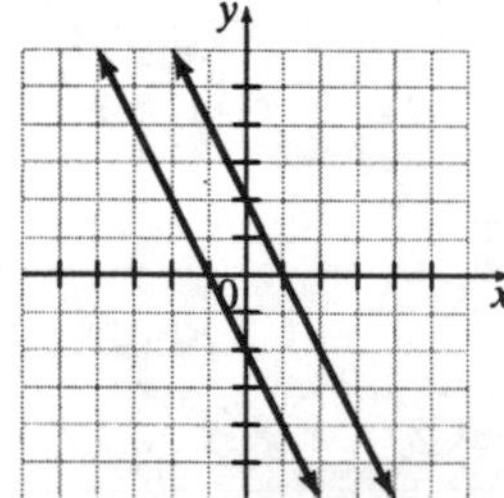

The lines are parallel. The system is inconsistent and there is no solution.

57. Graph the equations $4x - 3y = 6$ or $y = \frac{4}{3}x - 2$ and $2x + 4y = 14$ or $y = -\frac{1}{2}x + \frac{7}{2}$.

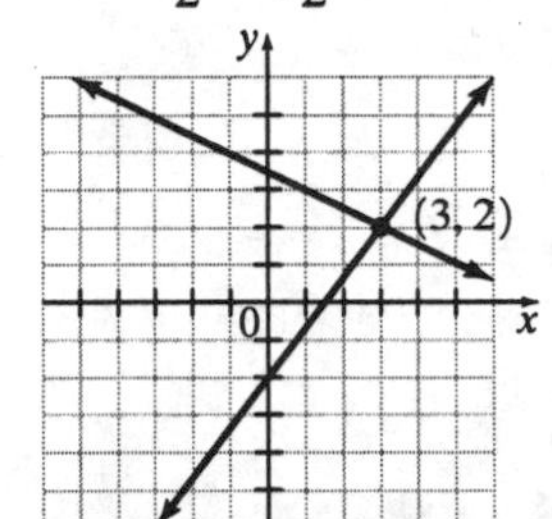

The lines intersect and the point of intersection is (3, 2). This is a consistent system.

59. Graph the equations $2x - 3y = 0$ or $y = \frac{2}{3}x$ and $x + 2y = 0$ or $y = -\frac{1}{2}x$.

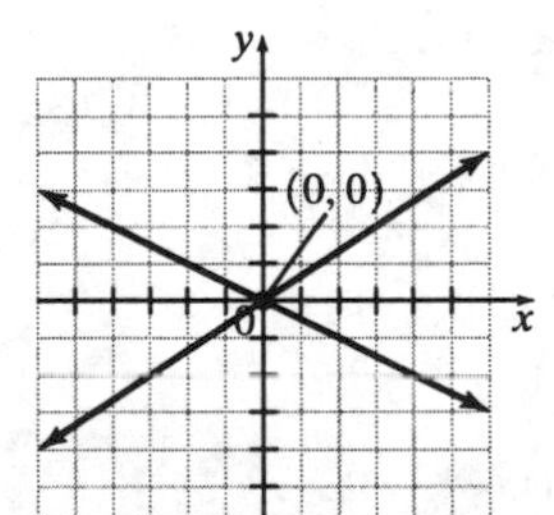

The lines intersect and the point of intersection is (0, 0). This is a consistent system.

61. Write each equation in slope-intercept form.

$$5x - 4y = 10 \qquad 12y = 15x - 20$$
$$-4y = -5x + 10 \qquad y = \frac{5}{4}x - \frac{5}{3}$$
$$y = \frac{5}{4}x - \frac{5}{2}$$

The lines are parallel because they have the same slope, $\frac{5}{4}$, and different y-intercepts.

63. The system has an infinite number of solutions. If the two lines have two points in common then they must be the same line.

65. The system has no solutions. Distinct parallel lines do not intersect.

67. $x = 4$, $y = 3$ has one solution, (4, 3).

69. (repair) $c = 600 + 650n$
(replacement) $c = 1800 + 450n$
Graph the equations and determine the intersection.

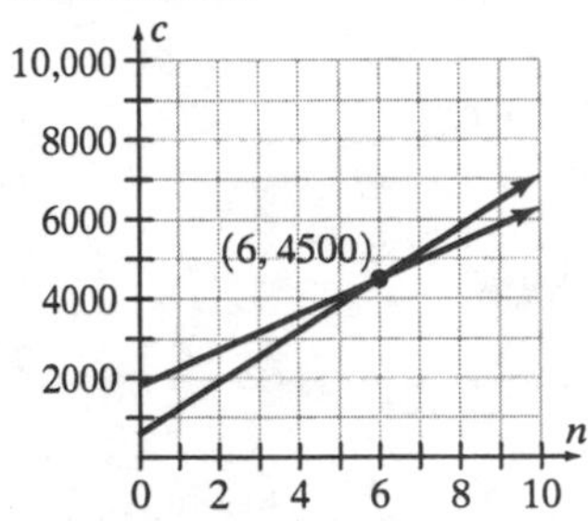

The solution is (6, 4500). Therefore, the total cost of repair equals the total cost of replacement at 6 years.

71. $c = 50 + 0.10s$
$c = 0.30s$
Graph the equations and determine the intersection.

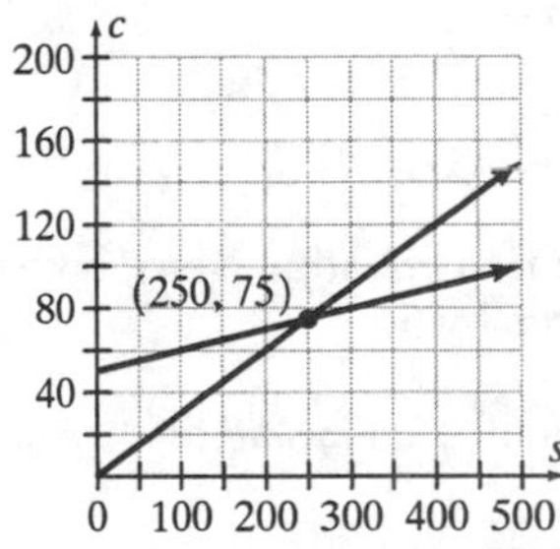

The solution (250, 75). Therefore, the cost is the same for 250 shares.

79. $3x-(x-6)+4(3-x)=3x-x+6+12-4x$
$=3x-x-4x+6+12$
$=-2x+18$

80. $2(x+3)-x=5x+2$
$2x+6-x=5x+2$
$x+6=5x+2$
$-4x=-4$
$x=1$

81. $A=P(1+rt)$
$1000=500(1+r\cdot 2)$
$1000=500+1000r$
$500=1000r$
$\frac{1}{2}=r$ or $r=\frac{1}{2}$

82. $A=\frac{1}{2}bh$
$2A=2\cdot\frac{1}{2}bh$
$2A=bh$
$\frac{2A}{b}=\frac{bh}{b}$
$\frac{2A}{b}=h$ or $h=\frac{2A}{b}$

Exercise Set 5.2

1. The x in the first equation, since both 6 and 9 are divisible by 3.

3. You will obtain a false statement, such as $3=0$.

5. $x+2y=5$
$2x-3y=3$
Solve the first equation for x, $x=5-2y$.
Substitute $5-2y$ for x in the second equation.
$2(5-2y)-3y=3$
$10-4y-3y=3$
$-7y=-7$
$y=1$
Substitute 1 for y in the equation $x=5-2y$.
$x=5-2(1)$
$x=3$
The solution is (3, 1).

7. $x+y=-2$
$x-y=0$
Solve the first equation for y, $y=-2-x$.
Substitute $-2-x$ for y in the second equation.
$x-(-2-x)=0$
$x+2+x=0$
$2x=-2$
$x=-1$
Substitute -1 for x in the equation $y=-2-x$.
$y=-2-(-1)$
$y=-2+1$
$y=-1$
The solution is (–1, –1).

9. $3x+y=3$
$3x+y+5=0$
Solve the first equation for y, $y=-3x+3$.
Substitute $-3x+3$ for y in the second equation.
$3x+y+5=0$
$3x-3x+3+5=0$
$8=0$ False
There is no solution.

11. $x=3$
$x+y+5=0$
Substitute 3 for x in the second equation.
$x+y+5=0$
$3+y+5=0$
$y=-8$
The solution is (3, –8).

13. $x - \frac{1}{2}y = 2$

$y = 2x - 4$

Substitute $2x - 4$ for y in the first equation.

$$x - \frac{1}{2}y = 2$$
$$x - \frac{1}{2}(2x - 4) = 2$$
$$x - x + 2 = 2$$
$$2 = 2$$

Since this is a true statement, there are an infinite number of solutions. This is a dependent system.

15. $2x + y = 9$

$y = 4x - 3$

Substitute $4x - 3$ for y in the first equation.

$$2x + y = 9$$
$$2x + 4x - 3 = 9$$
$$6x = 12$$
$$x = 2$$

Now substitute 2 for x in the second equation.

$$y = 4x - 3$$
$$y = 4(2) - 3$$
$$y = 8 - 3$$
$$y = 5$$

The solution is (2, 5).

17. $y = \frac{1}{3}x - 2$

$x - 3y = 6$

Substitute $\frac{1}{3}x - 2$ for y in the second equation.

$$x - 3y = 6$$
$$x - 3\left(\frac{1}{3}x - 2\right) = 6$$
$$x - x + 6 = 6$$
$$6 = 6$$

Since this is a true statement, there are an infinite number of solutions. This is a dependent system.

19. $2x + 3y = 7$

$6x - 2y = 10$

First solve the second equation for y.

$$\frac{1}{2}(6x - 2y) = \frac{1}{2}(10)$$
$$3x - y = 5$$
$$y = 3x - 5$$

Now substitute $3x - 5$ for y in the first equation.

$$2x + 3y = 7$$
$$2x + 3(3x - 5) = 7$$
$$2x + 9x - 15 = 7$$
$$11x = 22$$
$$x = 2$$

Finally substitute 2 for x in the equation $y = 3x - 5$.

$$y = 3(2) - 5$$
$$y = 6 - 5$$
$$y = 1$$

The solution is (2, 1).

21. $3x - y = 14$

$6x - 2y = 10$

First solve the first equation for y.

$$3x - y = 14$$
$$y = 3x - 14$$

Now substitute $3x - 14$ for y in the second equation.

$$6x - 2y = 10$$
$$6x - 2(3x - 14) = 10$$
$$6x - 6x + 28 = 10$$
$$28 = 10 \text{ False}$$

There is no solution.

23. $4x - 5y = -4$

$3x = 2y - 3$

First solve the second equation for x.

$$\frac{1}{3}(3x) = \frac{1}{3}(2y - 3)$$
$$x = \frac{2}{3}y - 1$$

Now substitute $\frac{2}{3}y - 1$ for x in the first equation.

$$4x - 5y = -4$$
$$4\left(\frac{2}{3}y - 1\right) - 5y = -4$$
$$\frac{8}{3}y - 4 - 5y = -4$$
$$\frac{8}{3}y - \frac{15}{3}y = 0$$
$$-\frac{7}{3}y = 0$$
$$y = 0$$

Finally substitute 0 for y in the equation $x = \frac{2}{3}y - 1$.

$$x = \frac{2}{3}(0) - 1$$
$$x = -1$$

The solution is (–1, 0).

25. $5x + 4y = -7$
$x - \frac{5}{3}y = -2$

First solve the second equation for x, $x = \frac{5}{3}y - 2$. Now substitute $\frac{5}{3}y - 2$ for x in the first equation.

$$5x + 4y = -7$$
$$5\left(\frac{5}{3}y - 2\right) + 4y = -7$$
$$\frac{25}{3}y - 10 + 4y = -7$$
$$\frac{25}{3}y + \frac{12}{3}y = 10 - 7$$
$$\frac{37}{3}y = 3$$
$$y = \frac{9}{37}$$

Finally, substitute $\frac{9}{37}$ for y in the equation $x = \frac{5}{3}y - 2$.

$$x = \frac{5}{3}\left(\frac{9}{37}\right) - 2$$
$$x = \frac{15}{37} - \frac{74}{37}$$
$$x = -\frac{59}{37}$$

The solution is $\left(-\frac{59}{37}, \frac{9}{37}\right)$.

27. $c = 1200 + 752n$
$c = 832n$

a. Substitute $832n$ for c in the first equation.

$$832n = 1200 + 752n$$
$$80n = 1200$$
$$n = 15$$

The mortgage plans will have the same total cost at 15 months.

b. $12 \text{ years} \cdot \left(\frac{12 \text{ months}}{1 \text{ year}}\right) = 144 \text{ months}$

$$c = 1200 + 752(144) = 109,488$$
$$c = 832(144) = 119,808$$

Yes, since \$109,488 is less than \$119,808, she should refinance.

29. $S = 300 + 3n$
$S = 400 + 2n$

Substitute $300 + 3n$ for S in the second equation.

$$300 + 3n = 400 + 2n$$
$$3n - 2n = 400 - 300$$
$$n = 100$$

The salaries will be the same for a sale of 100 tapes.

33. $\frac{25}{3.5} \approx 7.14$

The willow tree is about 7.14 years old.

34. $\frac{33}{1.6} = 20.625$

Steve would have to travel 21 days or more for the monthly pass to be worthwhile.

35. $4x - 8y = 16$
Let $x = 0$ and solve for y.
$4(0) - 8y = 16$
$y = -2$
Let $y = 0$ and solve for x.
$4x - 8(0) = 16$
$x = 4$
The intercepts are (0, –2) and (4, 0).

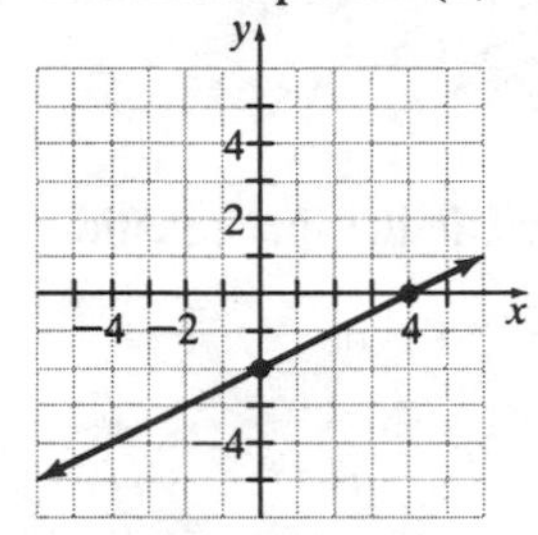

36. $3x - 5y = 8$
Write in slope-intercept form.
$-5y = -3x + 8$
$y = \frac{3}{5}x - \frac{8}{5}$
The slope is $\frac{3}{5}$ and the y intercept is $\left(0, -\frac{8}{5}\right)$.

Exercise Set 5.3

1. Multiply the top equation by 2. Now the top equation contains a $-2x$ and the bottom equation contains a $2x$. When the equations are added together, the variable x will be eliminated.

3. You will obtain a false statement, such as $0 = 6$

5.
$$\begin{aligned} x + y &= 6 \\ x - y &= 4 \\ \hline \text{Add: } 2x \quad &= 10 \\ x &= 5 \end{aligned}$$
Substitute 5 for x in the first equation.
$x + y = 6$
$5 + y = 6$
$y = 1$
The solution is (5, 1).

7.
$$\begin{aligned} -x + y &= 5 \\ x + y &= 1 \\ \hline \text{Add: } 2y &= 6 \\ y &= 3 \end{aligned}$$
Substitute 3 for y in the second equation.
$x + y = 1$
$x + 3 = 1$
$x = -2$
The solution is (–2, 3).

9.
$$\begin{aligned} x + 2y &= 15 \\ x - 2y &= -7 \\ \hline \text{Add: } 2x \quad &= 8 \\ x &= 4 \end{aligned}$$
Substitute 4 for x in the first equation.
$x + 2y = 15$
$4 + 2y = 15$
$2y = 11$
$y = \frac{11}{2}$
The solution is $\left(4, \frac{11}{2}\right)$

11. $4x + y = 6$
$-8x - 2y = 20$
Multiply the first equation by 2, then add to the second equation.
$2[4x + y = 6]$
gives
$$\begin{aligned} 8x + 2y &= 12 \\ -8x - 2y &= 20 \\ \hline 0 &= 32 \quad \text{False} \end{aligned}$$
There is no solution.

13. $-5x + y = 14$
$-3x + y = -2$
To eliminate y, multiply the first equation by –1 and then add.
$-1[-5x + y = 14]$
gives
$$\begin{aligned} 5x - y &= -14 \\ -3x + y &= -2 \\ \hline 2x \quad &= -16 \\ x &= -8 \end{aligned}$$

Substitute –8 for x in the first equation.

$$-5x + y = 14$$
$$-5(-8) + y = 14$$
$$40 + y = 14$$
$$y = -26$$

The solution is (–8, –26).

15. $3x + y = 10$
$3x - 2y = 16$

To eliminate x, multiply the first equation by –1 and then add.

$-1[3x + y = 10]$

gives

$$-3x - y = -10$$
$$3x - 2y = 16$$
$$-3y = 6$$
$$y = -2$$

Substitute –2 for y in the first equation.

$$3x + y = 10$$
$$3x + (-2) = 10$$
$$3x = 12$$
$$x = 4$$

The solution is (4, –2)

17. $4x - 3y = 8$
$2x + y = 14$

To eliminate y, multiply the second equation by 3, then add.

$3[2x + y = 14]$

gives

$$4x - 3y = 8$$
$$6x + 3y = 42$$
$$10x = 50$$
$$x = 5$$

Substitute 5 for x in the second equation.

$$2x + y = 14$$
$$2(5) + y = 14$$
$$10 + y = 14$$
$$y = 4$$

The solution is (5, 4).

19. $5x + 3y = 12$
$2x - 4y = 10$

To eliminate y, multiply the first equation by 4 and the second equation by 3 and then add.

$4[5x + 3y = 12]$
$3[2x - 4y = 10]$

gives

$$20x + 12y = 48$$
$$6x - 12y = 30$$
$$26x = 78$$
$$x = 3$$

Substitute 3 for x in the first equation.

$$5x + 3y = 12$$
$$5(3) + 3y = 12$$
$$15 + 3y = 12$$
$$3y = -3$$
$$y = -1$$

The solution is (3, –1).

21. $4x - 2y = 6$
$y = 2x - 3$

Align x- and y-terms on the left side.

$$4x - 2y = 6$$
$$-2x + y = -3$$

To eliminate x, multiply the second equation by 2 and then add.

$2[-2x + y = -3]$

gives

$$4x - 2y = 6$$
$$-4x + 2y = -6$$
$$0 = 0$$

Since this is a true statement, there are an infinite number of solutions. This is a dependent system.

23. $3x - 2y = -2$
$3y = 2x + 4$

Align x- and y-terms on the left side.

$$3x - 2y = -2$$
$$-2x + 3y = 4$$

To eliminate x, multiply the first equation by 2 and the second equation by 3 and then add.

$2[3x - 2y = -2]$
$3[-2x + 3y = 4]$

gives

$$\begin{aligned} 6x - 4y &= -4 \\ -6x + 9y &= 12 \\ \hline 5y &= 8 \\ y &= \frac{8}{5} \end{aligned}$$

Substitute $\frac{8}{5}$ for y in the first equation.

$$\begin{aligned} 3x - 2y &= -2 \\ 3x - 2\left(\frac{8}{5}\right) &= -2 \\ 3x - \frac{16}{5} &= -2 \\ 3x &= -\frac{10}{5} + \frac{16}{5} \\ 3x &= \frac{6}{5} \\ x &= \frac{2}{5} \end{aligned}$$

The solution is $\left(\frac{2}{5}, \frac{8}{5}\right)$.

25. $5x - 4y = 1$
$-10x + 8y = -4$
Multiply the first equation by 2, then add.
$2[5x - 4y = 1]$
gives

$$\begin{aligned} 10x - 8y &= 1 \\ -10x + 8y &= -4 \\ \hline 0 &= -3 \text{ False} \end{aligned}$$

There is no solution.

27. $3x - 5y = 0$
$2x + 3y = 0$
To eliminate y, multiply the first equation by 3 and the second equation by 5 and then add.
$3[3x - 5y = 0]$
$5[2x + 3y = 0]$
gives

$$\begin{aligned} 9x - 15y &= 0 \\ 10x + 15y &= 0 \\ \hline 19x \quad &= 0 \\ x &= 0 \end{aligned}$$

Substitute 0 for x in the second equation.

$$\begin{aligned} 2x + 3y &= 0 \\ 2(0) + 3y &= 0 \\ 3y &= 0 \\ y &= 0 \end{aligned}$$

The solution is (0, 0).

29. $-5x + 4y = -20$
$3x - 2y = 15$
To eliminate y, multiply the second equation by 2, then add.
$2[3x - 2y = 15]$
gives

$$\begin{aligned} -5x + 4y &= -20 \\ 6x - 4y &= 30 \\ \hline x \quad &= 10 \end{aligned}$$

Substitute 10 for x in the first equation.

$$\begin{aligned} -5x + 4y &= -20 \\ -5(10) + 4y &= -20 \\ -50 + 4y &= -20 \\ 4y &= 30 \\ y &= \frac{15}{2} \end{aligned}$$

The solution is $\left(10, \frac{15}{2}\right)$.

31. $7x - 3y = 4$
$2y = 4x - 6$
Align the x- and y-terms on the left side.
$7x - 3y = 4$
$-4x + 2y = -6$
To eliminate x, multiply the first equation by 4 and the second equation by 7 and then add.
$4[7x - 3y = 4]$
$7[-4x + 2y = -6]$
gives

$$\begin{aligned} 28x - 12y &= 16 \\ -28x + 14y &= -42 \\ \hline 2y &= -26 \\ y &= -13 \end{aligned}$$

Substitute −13 for y in the first equation.

$$\begin{aligned} 7x - 3y &= 4 \\ 7x - 3(-13) &= 4 \\ 7x + 39 &= 4 \\ 7x &= -35 \\ x &= -5 \end{aligned}$$

The solution is (−5, −13).

33. $4x+5y=0$
$3x=6y+4$
Align the x- and y-terms on the left side.
$4x+5y=0$
$3x-6y=4$
To eliminate y, multiply the first equation by 6 and the second equation by 5 and then add.
$6[4x+5y=0]$
$5[3x-6y=4]$
gives
$$\begin{aligned}24x+30y&=0\\ 15x-30y&=20\\ \hline 39x\quad\quad&=20\\ x&=\frac{20}{39}\end{aligned}$$
Substitute $\frac{20}{39}$ for x in the first equation.
$$\begin{aligned}4x+5y&=0\\ 4\left(\frac{20}{39}\right)+5y&=0\\ 5y&=-\frac{80}{39}\\ y&=-\frac{16}{39}\end{aligned}$$
The solution is $\left(\frac{20}{39}, -\frac{16}{39}\right)$.

35. $x-\frac{1}{2}y=4$
$3x+y=6$
To eliminate y, multiply the first equation by 2 and then add.
$2\left[x-\frac{1}{2}y=4\right]$
gives
$$\begin{aligned}2x-y&=8\\ 3x+y&=6\\ \hline 5x\quad&=14\\ x&=\frac{14}{5}\end{aligned}$$
Substitute $\frac{14}{5}$ for x in the second equation.
$$\begin{aligned}3x+y&=6\\ 3\left(\frac{14}{5}\right)+y&=6\\ \frac{42}{5}+y&=6\\ y&=6-\frac{42}{5}\\ y&=\frac{30}{5}-\frac{42}{5}\\ y&=-\frac{12}{5}\end{aligned}$$
The solution is $\left(\frac{14}{5}, -\frac{12}{5}\right)$.

37. Answers will vary.

39. a. $4x+2y=1000$
$2x+4y=800$
To eliminate x, multiply the first equation by −2.
$-2[4x+2y=1000]$
gives
$$\begin{aligned}-8x-4y&=-2000\\ 2x+4y&=\quad 800\\ \hline -6x\quad\quad&=-1200\\ x&=200\end{aligned}$$
Substitute 200 for x in the first equation.
$$\begin{aligned}4x+2y&=1000\\ 4(200)+2y&=1000\\ 800+2y&=1000\\ 2y&=200\\ y&=100\end{aligned}$$
The solution is (200, 100).

b. They will have the same solution. Dividing an equation by a nonzero number does not change the solutions.
$2x+y=500$
$2x+4y=800$
To eliminate x, multiply the first equation by −1 and then add.
$-1[2x+y=500]$
gives
$$\begin{aligned}-2x-y&=-500\\ 2x+4y&=800\\ \hline 3y&=300\\ y&=100\end{aligned}$$

Substitute 100 for y in the first equation.

$$\begin{aligned} 2x+y &= 500 \\ 2x+100 &= 500 \\ 2x &= 400 \\ x &= 200 \end{aligned}$$

The solution is (200, 100).

41. $\dfrac{x+2}{2}-\dfrac{y+4}{3}=4$

$\dfrac{x+y}{2}=\dfrac{1}{2}+\dfrac{x-y}{3}$

Start by writing each equation in standard form after clearing fractions.

For the first equation:

$$6\left[\frac{x+2}{2}-\frac{y+4}{3}=4\right]$$

$$\begin{aligned} 3(x+2)-2(y+4) &= 24 \\ 3x+6-2y-8 &= 24 \\ 3x-2y-2 &= 24 \\ 3x-2y &= 26 \end{aligned}$$

For the second equation:

$$6\left[\frac{x+y}{2}=\frac{1}{2}+\frac{x-y}{3}\right]$$

$$\begin{aligned} 3(x+y) &= 3+2(x-y) \\ 3x+3y &= 3+2x-2y \\ x+5y &= 3 \end{aligned}$$

The new system is:

$$\begin{aligned} 3x-2y &= 26 \\ x+5y &= 3 \end{aligned}$$

To eliminate x, multiply the second equation by −3 and then add.

$-3[x+5y=3]$

gives

$$\begin{aligned} 3x-2y &= 26 \\ -3x-15y &= -9 \\ \hline -17y &= 17 \\ y &= -1 \end{aligned}$$

Now, substitute −1 for y in the equation $x+5y=3$.

$$\begin{aligned} x+5(-1) &= 3 \\ x-5 &= 3 \\ x &= 8 \end{aligned}$$

The solution is (8, −1).

45. $5^3=5\cdot 5\cdot 5=125$

46. $$\begin{aligned} 2(2x-3) &= 2x+8 \\ 4x-6 &= 2x+8 \\ 4x-2x &= 8+6 \\ 2x &= 14 \\ x &= 7 \end{aligned}$$

47. $$\begin{aligned} 2x-4 &< 4x-2 \\ -4+2 &< 4x-2x \\ -2 &< 2x \\ -1 &< x \text{ or} \\ x &> -1 \end{aligned}$$

−1

48. $$\begin{aligned} f(x) &= 2x^2-4 \\ f(-3) &= 2(-3)^2-4 \\ &= 2(9)-4 \\ &= 18-4 \\ &= 14 \end{aligned}$$

Exercise Set 5.4

1. Let x and y be the integers, with x the larger integer.

$$\begin{aligned} x+y &= 29 \\ x &= y+3 \end{aligned}$$

Substitute $y+3$ for x in the first equation.

$$\begin{aligned} (y+3)+y &= 29 \\ 2y+3 &= 29 \\ 2y &= 26 \\ y &= \frac{26}{2}=13 \end{aligned}$$

$$\begin{aligned} x &= y+3 \\ x &= 13+3 \\ x &= 16 \end{aligned}$$

The integers are 13 and 16.

3. $$\begin{aligned} A+B &= 90 \\ B &= A+18 \end{aligned}$$

Substitute $A+18$ for B in the first equation.

$$\begin{aligned} A+(A+18) &= 90 \\ 2A+18 &= 90 \\ 2A &= 72 \\ A &= \frac{72}{2}=36 \end{aligned}$$

$B = A + 18$
$B = 36 + 18$
$B = 54$
The angles are $A = 36°$ and $B = 54°$.

5. $A + B = 180$
$A = B + 48$
Substitute $B + 48$ for A in the first equation.
$(B + 48) + B = 180$
$2B + 48 = 180$
$2B = 132$
$B = \frac{132}{2} = 66$
$A = B + 48$
$A = 66 + 48$
$A = 114$
The angles are $A = 114°$ and $B = 66°$.

7. Let w = the width of the screen
l = the height of the screen
The formula for the perimeter of the screen is
$P = 2w + 2h$.
$2w + 2l = 124$
$w = l + 8$
Substitute $l + 8$ for w in the first equation.
$2(l + 8) + 2l = 124$
$2l + 16 + 2l = 124$
$4l + 16 = 124$
$4l = 108$
$l = 27$
$w = l + 8$
$w = 27 + 8$
$w = 35$
The width of the screen is 35 inches and the height is 27 inches.

9. Let c = number of acres of corn
w = number of acres of wheat
$c + w = 100$
$450c + 430w = 44,400$
Solve the first equation for w.
$w = 100 - c$
Substitute $100 - c$ for w in the second equation.
$450c + 430(100 - c) = 44,400$
$450c + 43,000 - 430c = 44,400$
$20c + 43,000 = 44,400$
$20c = 1400$
$c = \frac{1400}{20} = 70$
$w = 100 - c$
$w = 100 - 70$
$w = 30$
She planted 70 acres of corn and 30 acres of wheat.

11. Let d = the number of shares of Disney stock
m = the number of shares of Microsoft stock
$d = 5m$
$26d + 85m = 10,750$
Substitute $5m$ for d in the second equation.
$26(5m) + 85m = 10,750$
$130m + 85m = 10,750$
$215m = 10,750$
$m = \frac{10,750}{215} = 50$
$d = 5m$
$d = 5(50)$
$d = 250$
She bought 250 shares of Disney stock and 50 shares of Microsoft stock.

13. Let k = the speed of the kayak
c = the speed of the current
$k + c = 4.7$
$\underline{k - c = 3.4}$
$2k \quad = 8.1$
$k = 4.05$
$k + c = 4.7$
$4.05 + c = 4.7$
$c = 0.65$
The speed of the kayak in still water is 4.05 miles per hour and the speed of the current is 0.65 miles per hour.

15. Let m = the cost of goods purchased (manufacturer's list price)
c = the total paid for the goods
Under plan A, $c = 50 + 0.85m$, while under plan B, $c = 100 + 0.80m$. Set the two costs equal.

$50 + 0.85m = 100 + 0.80m$
$0.05m = 50$
$m = 1000$
One would have to purchase goods with manufacturer's list prices totaling $1000 to pay the same amount under both plans.

17. a. Let n = the number of copies
c = the monthly cost
With the Kate Spence Company, $c = 18 + 0.02n$, while with Office Depot, $c = 25 + 0.015n$. Set the monthly costs equal
$18 + 0.02n = 25 + 0.015n$
$0.005n = 7$
$n = 1400$
When 1400 copies are made, the monthly costs of both plans are the same.

b. With Kate Spence Company:
$c = 18 + 0.02(2500)$
$c = 18 + 50$
$c = 68$
With Office Depot:
$c = 25 + 0.015(2500)$
$c = 25 + 37.5$
$c = 62.5$
When 2500 copies are made, it is less expensive to get the service contract from Office Depot.

19. a. Let y = the number of yards of carpet
c = the total cost of the carpet
At Carpet U.S.A., $c = 200 + 20y$, while at Tom Taylor's National Carpet Stores, $c = 260 + 16y$. Set the two costs equal.
$200 + 20y = 260 + 16y$
$4y = 60$
$y = 15$
The total cost of the carpet is the same at both stores when 15 square yards are purchased.

b. Carpet U.S.A.:
$c = 200 + 20(25)$
$= 200 + 500$
$= 700$
Tom Taylor's:
$c = 260 + 16(25)$
$= 260 + 400$
$= 660$
The total cost of 25 square yards of carpet is less at Tom Taylor's National Carpet Stores.

21. Let x = the amount giving 10% interest
y = the amount giving 8% interest
$x + y = 8000$
$0.10x + 0.08y = 750$
Eliminate the decimal numbers.
$x + y = 8000$
$10x + 8y = 75,000$
Multiply the first equation by –8.
$-8[x + y = 8000]$
$10x + 8y = 75,000$
gives
$-8x - 8y = -64,000$
$10x + 8y = 75,000$
$2x = 11,000$
$x = 5500$
$x + y = 8000$
$5500 + y = 8000$
$y = 2500$
They invested $5500 at 10%, and $2500 at 8%.

23. Let x = interest rate of the CD
y = interest rate of the money market account
$10,000x + 6000y = 740$
$x = y + 0.01$
Substitute $y + 0.01$ for x in the first equation.
$10,000(y + 0.01) + 6000y = 740$
$10,000y + 100 + 6000y = 740$
$16,000y = 640$
$y = 0.04$
$x = y + 0.01$
$x = 0.04 + 0.01$
$x = 0.05$
The interest rate of the CD is 5%, while the interest rate of the money market account is 4%.

25. Let e = the speed of Elizabeth's boat
m = the speed of Melissa's boat
If d is the distance that both boats travel, then $d = 3e$ and $d = 3.2m$.
$e = m + 4$
$3e = 3.2m$
Multiply the first equation by –3.
$-3[e = m + 4]$
$3e = 3.2m$
gives
$-3e = -3m - 12$
$3e = 3.2m$
$0 = 0.2m - 12$
$12 = 0.2m$
$60 = m$
$e = m + 4$
$e = 60 + 4$
$e = 64$
The speed of Elizabeth's boat is 64 miles per hour and the speed of Melissa's boat is 60 miles per hour.

27. Let j = the speed of John's crew
l = the speed of Leigh's crew
After 50 hours, John's crew has dug $50j$ miles and Leigh's crew has dug $50l$ miles. The total of these two distances is 30 miles.
$50j + 50l = 30$
$j = l + 0.1$
Substitute $l + 0.1$ for j in the first equation.
$50(l + 0.1) + 50l = 30$
$50l + 5 + 50l = 30$
$100l = 25$
$l = 0.25$
$j = l + 0.1$
$j = 0.25 + 0.1$
$j = 0.35$
John's crew digs 0.35 miles per hour and Leigh's crew digs 0.25 miles per hour.

29. Let a = time that Amanda jogs
d = time that Dolores jogs
The distance that Amanda jogs is $5a$ and the distance that Dolores jogs is $8d$.
$5a = 8d$
$a = d + 0.3$
Substitute $d + 0.3$ for a in the first equation.
$5(d + 0.3) = 8d$
$5d + 1.5 = 8d$
$1.5 = 3d$
$0.5 = d$
Dolores will catch up to Amanda when Dolores has been jogging for 0.5 hour.

31. Let x = amount of 25% solution
y = amount of 50% solution

Solution	Number of Liters	Concentration	Acid Content
25% solution	x	0.25	$0.25x$
50% solution	y	0.50	$0.50y$
Mixture	10	0.40	0.40(10)

$x + y = 10$
$0.25x + 0.50y = 4$
Solve the first equation for x.
$x = 10 - y$
Substitute $10 - y$ for x in the second equation.
$0.25(10 - y) + 0.50y = 4$
$2.5 - 0.25y + 0.50y = 4$
$0.25y = 1.5$
$y = 6$
$x = 10 - y$
$x = 10 - 6$
$x = 4$
She must use 4 liters of the 25% solution and 6 liters of the 50% solution.

33. Let x = the number of \$3 tiles
y = the number of \$5 tiles
$x + y = 380$
$3x + 5y = 1500$
Solve the first equation for x.
$x = 380 - y$
Substitute $380 - y$ for x in the second equation.
$3(380 - y) + 5y = 1500$
$1140 - 3y + 5y = 1500$
$2y = 360$
$y = 180$
She can purchase at most 180 of the \$5 tiles.

35. Let x = the amount of 5% butterfat milk
y = the amount of skim milk

Milk	Number of Gallons	Percentage butterfat	Butterfat Content
5% butterfat	x	0.05	$0.05x$
skim	y	0	0
Mixture	100	0.035	0.035(100)

$x + y = 100$
$0.05x = 3.5$
From the second equation, $x = 70$.
$x + y = 100$
$70 + y = 100$
$y = 30$
Wayne must use 70 gallons of 5% butterfat milk and 30 gallons of skim milk.

37. Let x = the amount of apple juice
y = the amount of apple drink

Liquid	Number of ounces	Cost per ounce	Total cost
Apple juice	x	12	$12x$
Apple drink	y	6	$6y$
Mixture	8	10	10(8)

$x + y = 8$
$12x + 6y = 80$
Solve the first equation for y.
$y = 8 - x$
Substitute $8 - x$ for y in the second equation.
$12x + 6(8 - x) = 80$
$12x + 48 - 6x = 80$
$6x = 32$
$x = \frac{32}{6} = 5\frac{1}{3}$
$y = 8 - x$
$y = 8 - 5\frac{1}{3}$
$y = \frac{24}{3} - \frac{16}{3}$
$y = \frac{8}{3} = 2\frac{2}{3}$
The cans should contain $5\frac{1}{3}$ ounces apple juice and $2\frac{2}{3}$ ounces apple drink.

39. Let t = the amount of time they jog
d = the distance to the school
The distance that Sean jogs is $9t$, while the distance that Meghan jogs is $5t$.
$9t = d$
$5t = d - 0.5$
Substitute $9t$ for d in the second equation.
$5t = 9t - 0.5$
$0.5 = 4t$
$0.125 = t$
$d = 9t$
$d = 9(0.125)$
$d = 1.125$
The school is 1.125 miles from their house.

43. **a.** $x + 4 = 4 + x$ illustrates the commutative property of addition.

b. $(3x)y = 3(xy)$ illustrates the associative property of multiplication.

c. $4(x + 2) = 4x + 8$ illustrates the distributive property.

44. $3x + 4 = -(x - 6)$
$3x + 4 = -x + 6$
$4x = 2$
$x = \frac{2}{4} = \frac{1}{2}$

45. Let l = the length of the rectangle
w = the width of the rectangle
$l = 2w + 2$
The formula for the perimeter of a rectangle is $P = 2l + 2w$.
$22 = 2l + 2w$
$22 = 2(2w + 2) + 2w$
$22 = 4w + 4 + 2w$
$18 = 6w$
$3 = w$
$l = 2w + 2$
$l = 2(3) + 2$
$l = 8$
The length of the rectangle is 8 feet and the width is 3 feet.

46. A graph is an illustration of the set of points that satisfy an equation.

Exercise Set 5.5

1. Yes, the solution to a system of linear inequalities contains all of the ordered pairs which satisfy both inequalities.

3. Yes, when the lines are parallel. One possible system is $x + y > 2$, $x + y < 1$.

5.

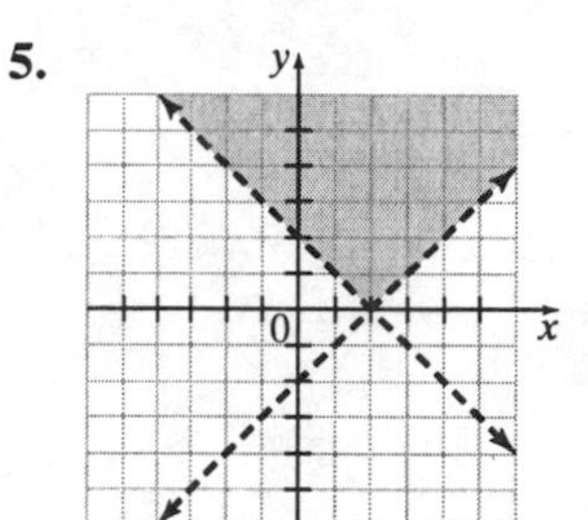

7.

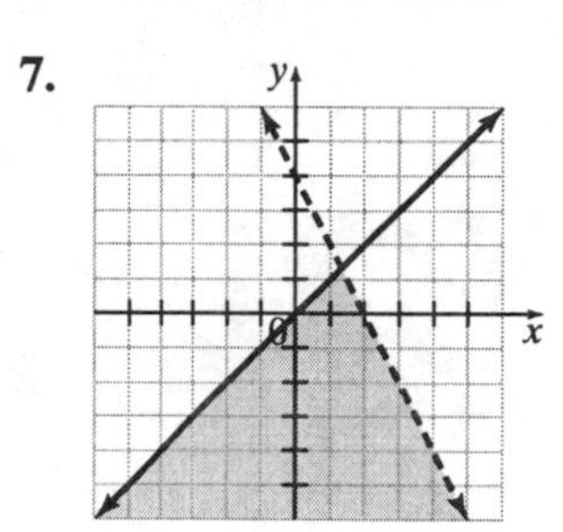

9.

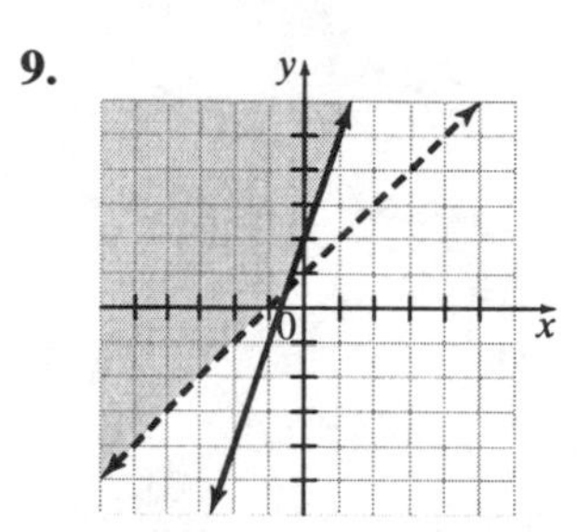

11.

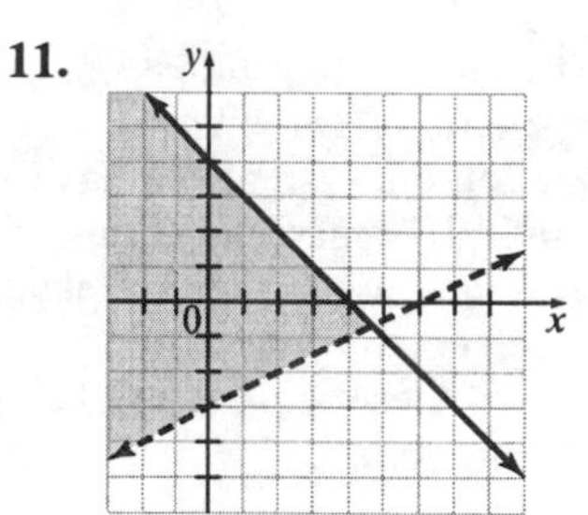

13.

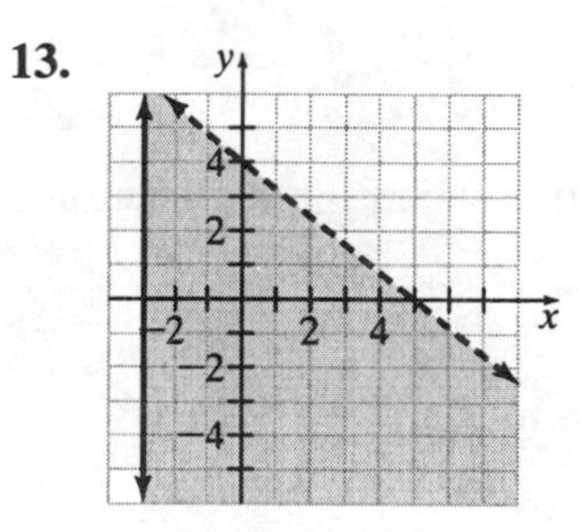

15.

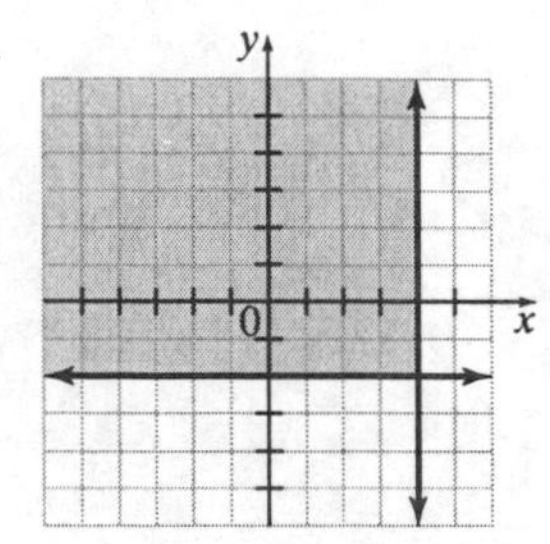

17.

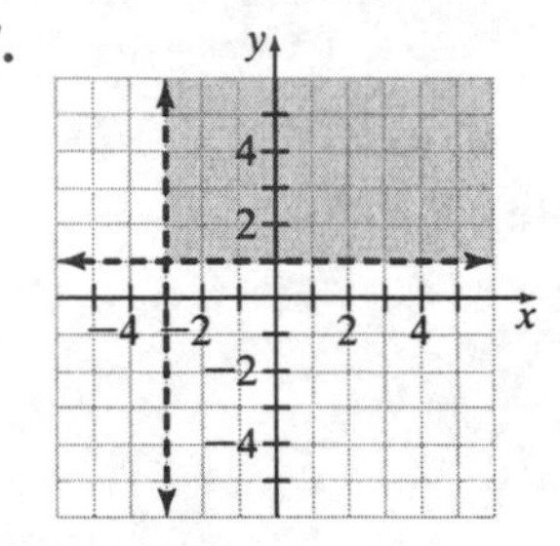

19.

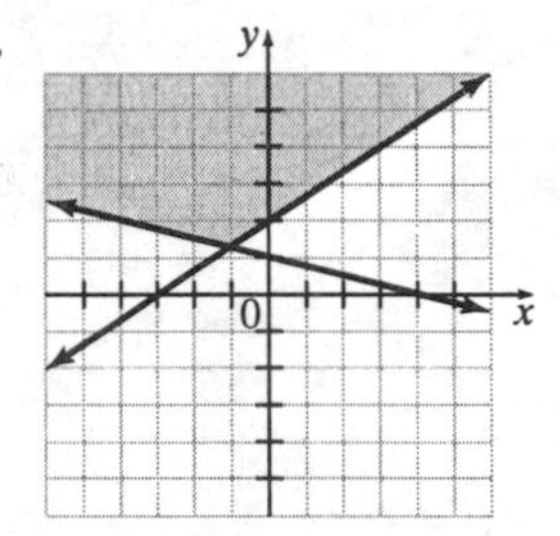

21. No, the system can have no solutions or infinitely many solutions. If the lines involved are parallel, the system will have no solutions or infinitely many solutions. If the lines intersect, they will divide the plane into 4 regions, each containing infinitely many points. One of these regions will be the solution to the system.

24. $6(x-2) < 4x-3+2x$
$6x-12 < 6x-3$
$-12 < -3$
Since this is a true statement, the solution is all real numbers.

0

25. $2x-5y=6$
$2x=5y+6$
$2x-6=5y$
$\frac{2x-6}{5}=y$
$y=\frac{2}{5}x-\frac{6}{5}$

26. $2x+y=4$
$y=-2x+4$
Let $x=0$
$y=-2(0)+4=4$
$(0, 4)$
Let $x=1$
$y=-2(1)+4=2$
$(1, 2)$
Let $x=2$
$y=-2(2)+4=0$
$(2, 0)$

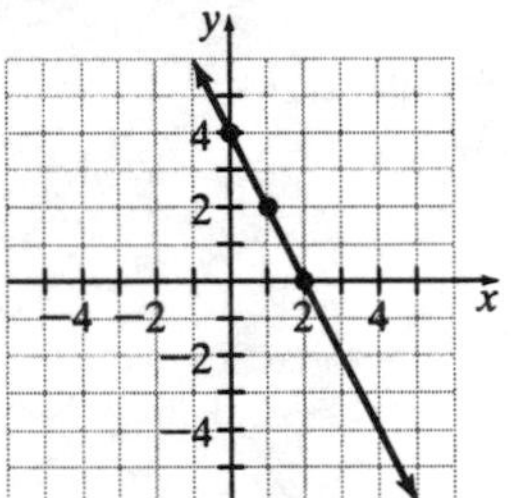

27. $m=\frac{y_2-y_1}{x_2-x_1}$
$m=\frac{-2-6}{3-(-4)}$
$m=\frac{-8}{7}$
$m=-\frac{8}{7}$

Review Exercises

1. a.

$y=3x-2$	$2x+3y=5$
$-2 \stackrel{?}{=} 3(0)-2$	$2(0)+3(-2) \stackrel{?}{=} 5$
$-2=-2$ True	$-6=5$ False

Since $(0, -2)$ does not satisfy both equations, it is not a solution to the system.

b. $y = 3x - 2$ $\qquad$ $2x + 3y = 5$
$4 \stackrel{?}{=} 3(2) - 2$ $\qquad$ $2(2) + 3(4) \stackrel{?}{=} 5$
$4 = 4$ True $\qquad$ $16 = 5$ False
Since (2, 4) does not satisfy both equations, it is not a solution to the system.

c. $y = 3x - 2$ $\qquad$ $2x + 3y = 5$
$1 \stackrel{?}{=} 3(1) - 2$ $\qquad$ $2(1) + 3(1) \stackrel{?}{=} 5$
$1 = 1$ True $\qquad$ $5 = 5$ True
Since (1, 1) satisfies both equations, it is a solution to the system.

2. a. $y = -x + 4$
$\frac{3}{2} \stackrel{?}{=} -\frac{5}{2} + 4$
$\frac{3}{2} = \frac{3}{2}$ True
$3x + 5y = 15$
$3\left(\frac{5}{2}\right) + 5\left(\frac{3}{2}\right) \stackrel{?}{=} 15$
$15 = 15$ True
Since $\left(\frac{5}{2}, \frac{3}{2}\right)$ satisfies both equations, it is a solution to the system.

b. $y = -x + 4$
$4 \stackrel{?}{=} -0 + 4$
$4 = 4$ True
$3x + 5y = 15$
$3(0) + 5(4) \stackrel{?}{=} 15$
$20 = 15$ False
Since (0, 4) does not satisfy both equations, it is not a solution to the system.

c. $y = -x + 4$
$\frac{3}{5} \stackrel{?}{=} -\frac{1}{2} + 4$
$\frac{3}{5} = \frac{7}{2}$ False
Since $\left(\frac{1}{2}, \frac{3}{5}\right)$ does not satisfy the first equation, it is not a solution to the system.

3. consistent, one solution

4. inconsistent, no solutions

5. dependent, infinite number of solutions

6. consistent, one solution

7. Write each equation in slope-intercept form.
$x + 2y = 8$ $\qquad$ $3x + 6y = 12$
$2y = -x + 8$ $\qquad$ $6y = -3x + 12$
$y = -\frac{1}{2}x + 4$ $\qquad$ $y = -\frac{1}{2}x + 2$
Since the slope of each line is $-\frac{1}{2}$ but the y-intercepts are different, the two lines are parallel. There is no solution. This is an inconsistent system.

8. Write each equation in slope-intercept form.
$y = -3x - 6$ is already in this form.
$2x + 5y = 8$
$5y = -2x + 8$
$y = -\frac{2}{5}x + \frac{8}{5}$
Since the slopes of the lines are different, the lines intersect to produce one solution. This is a consistent system.

9. Write each equation in slope-intercept form.
$y = \frac{1}{2}x - 4$ is already in this form.
$x - 2y = 8$
$-2y = -x + 8$
$y = \frac{1}{2}x - 4$
Since both equations are identical, the line is the same for both of them. There are an infinite number of solutions. This is a dependent system.

10. Write each equation in slope-intercept form.
$6x = 4y - 8$ $\qquad$ $4x = 6y + 8$
$6x + 8 = 4y$ $\qquad$ $4x - 8 = 6y$
$\frac{6x + 8}{4} = y$ $\qquad$ $\frac{4x - 8}{6} = y$
$\frac{3}{2}x + 2 = y$ $\qquad$ $\frac{2}{3}x - \frac{4}{3} = y$

Since the slopes of the lines are different, the lines intersect to produce one solution. This is a consistent system.

11. Graph $y = x - 5$ and $y = 2x - 8$.

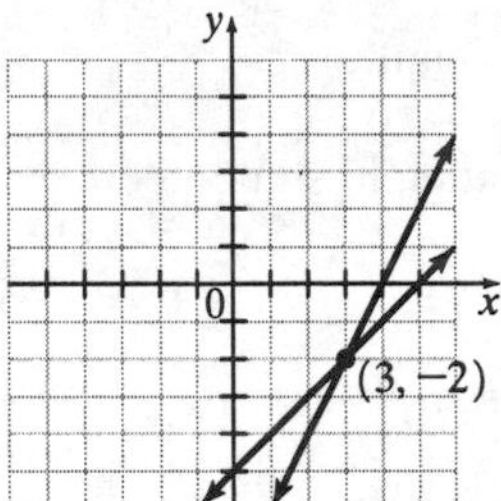

The lines intersect and the point of intersection is (3, –2). This is a consistent system.

12. Graph $x = -2$ and $y = 3$.

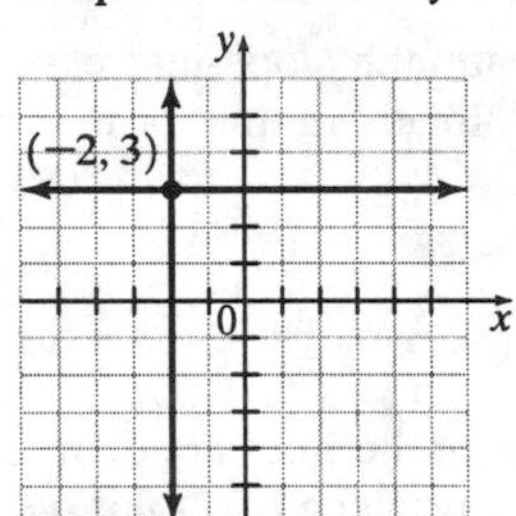

The lines intersect and the point of intersection is (–2, 3). This is a consistent system.

13. Graph $y = 3$ and $y = -2x + 5$.

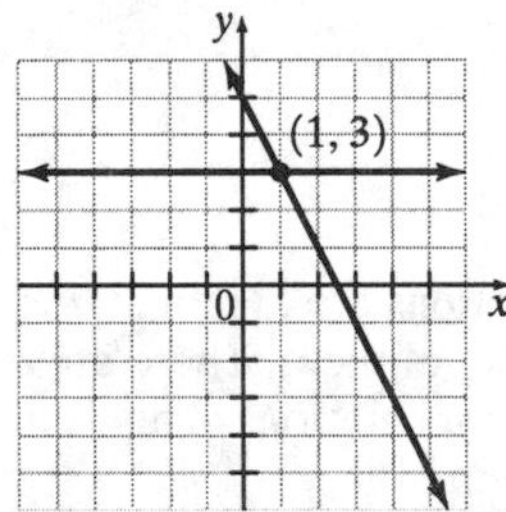

The lines intersect and the point of intersection is (1, 3). This is a consistent system.

14. Graph $x + 3y = 6$ and $y = 2$.

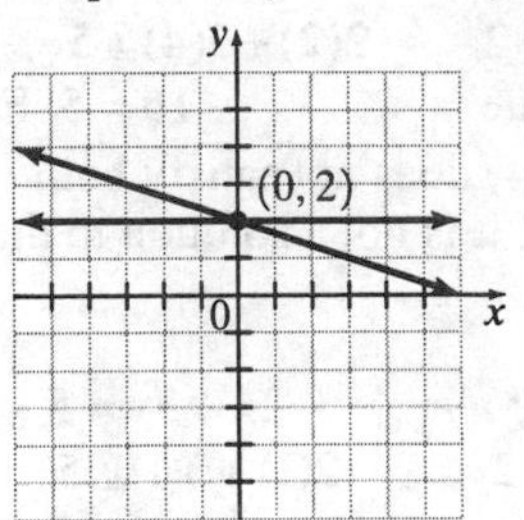

The lines intersect and the point of intersection is (0, 2). This is a consistent system.

15. Graph the equations $x + 2y = 8$ and $2x - y = -4$.

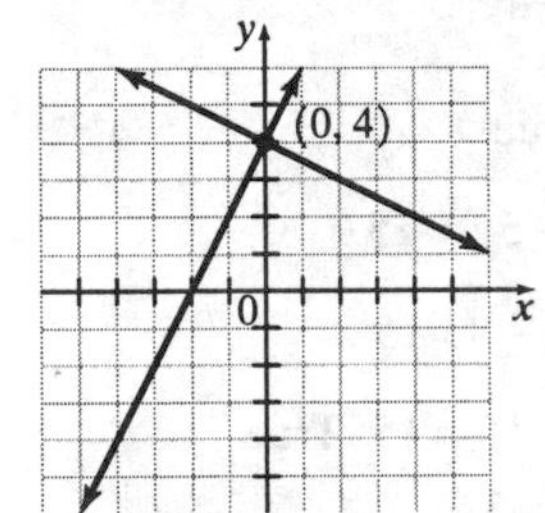

The lines intersect and the point of intersection is (0, 4). This is a consistent system.

16. Graph the equations $y = x - 3$ and $2x - 2y = 6$.

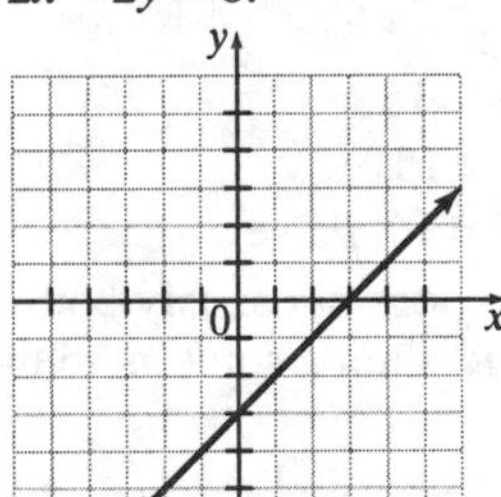

Both equations produce the same line. This is a dependent system. There are an infinite number of solutions.

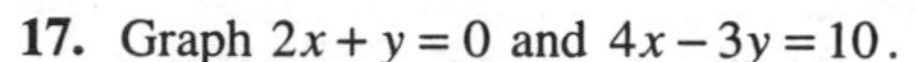

17. Graph $2x + y = 0$ and $4x - 3y = 10$.

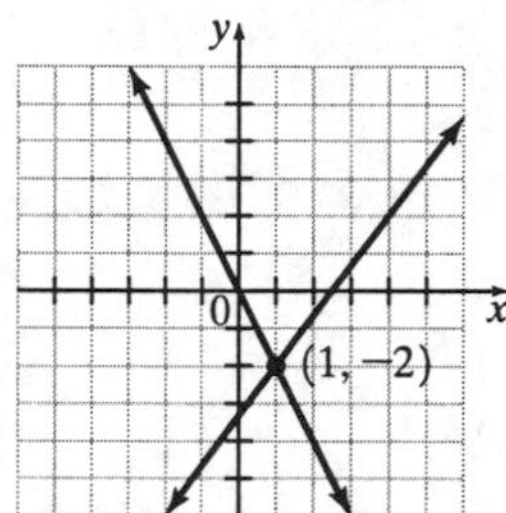

The lines intersect and the point of intersection is (1, –2). This is a consistent system.

18. Graph $x + 2y = 4$ and $\frac{1}{2}x + y = -2$.

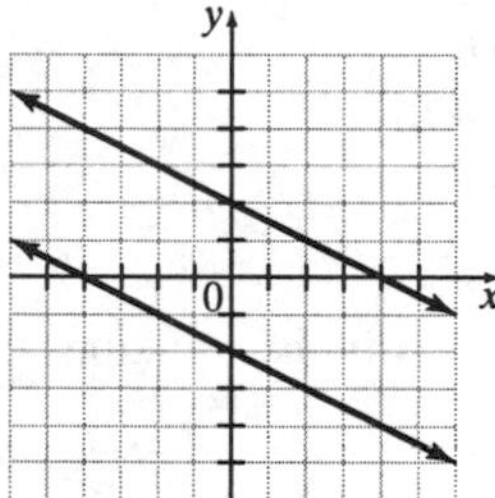

The lines are parallel and do not intersect. The system is inconsistent and there is no solution.

19. $y = 2x - 8$
$2x - 5y = 0$
Substitute $2x - 8$ for y in the second equation.

$$\begin{aligned} 2x - 5y &= 0 \\ 2x - 5(2x - 8) &= 0 \\ 2x - 10x + 40 &= 0 \\ -8x + 40 &= 0 \\ -8x &= -40 \\ x &= 5 \end{aligned}$$

Now, substitute 5 for x in the first equation.
$y = 2x - 8$
$y = 2(5) - 8$
$y = 10 - 8$
$y = 2$
The solution is (5, 2).

20. $x = 3y - 9$
$x + 2y = 1$
Substitute $3y - 9$ for x in the second equation.

$$\begin{aligned} x + 2y &= 1 \\ 3y - 9 + 2y &= 1 \\ 5y - 9 &= 1 \\ 5y &= 10 \\ y &= 2 \end{aligned}$$

Now substitute 2 for y in the first equation.
$x = 3y - 9$
$x = 3(2) - 9$
$x = 6 - 9$
$x = -3$
The solution is (–3, 2).

21. $2x - y = 6$
$x + 2y = 13$
Solve the second equation for x, $x = 13 - 2y$. Substitute $13 - 2y$ for x in the first equation.

$$\begin{aligned} 2x - y &= 6 \\ 2(13 - 2y) - y &= 6 \\ 26 - 4y - y &= 6 \\ -5y &= -20 \\ y &= 4 \end{aligned}$$

Substitute 4 for y in the equation $x = 13 - 2y$.
$x = 13 - 2(4)$
$x = 13 - 8$
$x = 5$
The solution is (5, 4).

22. $x = -3y$
$x + 4y = 6$
Substitute $-3y$ for x in the second equation.

$$\begin{aligned} x + 4y &= 6 \\ -3y + 4y &= 6 \\ y &= 6 \end{aligned}$$

Substitute 6 for y in the first equation.
$x = -3y$
$x = -3(6)$
$x = -18$
The solution is (–18, 6).

23. $4x-2y=10$
$y=2x+3$
Substitute $2x+3$ for y in the first equation.
$$4x-2y=10$$
$$4x-2(2x+3)=10$$
$$4x-4x-6=10$$
$$-6=10 \text{ False}$$
There is no solution.

24. $2x+4y=8$
$4x+8y=16$
Solve the first equation for x.
$$\frac{1}{2}(2x+4y)=\frac{1}{2}(8)$$
$$x+2y=4$$
$$x=4-2y$$
Substitute $4-2y$ for x in the second equation.
$$4x+8y=16$$
$$4(4-2y)+8y=16$$
$$16-8y+8y=16$$
$$16=16 \text{ True}$$
There are an infinite number of solutions.

25. $2x-3y=8$
$6x+5y=10$
Solve the first equation for x.
$$\frac{1}{2}(2x-3y)=\frac{1}{2}(8)$$
$$x-\frac{3}{2}y=4$$
$$x=\frac{3}{2}y+4$$
Substitute $\frac{3}{2}y+4$ for x in the second equation.
$$6x+5y=10$$
$$6\left(\frac{3}{2}y+4\right)+5y=10$$
$$9y+24+5y=10$$
$$14y=-14$$
$$y=-1$$
Substitute -1 for y in the equation $x=\frac{3}{2}y+4$.
$$x=\frac{3}{2}(-1)+4$$
$$x=-\frac{3}{2}+\frac{8}{2}$$
$$x=\frac{5}{2}$$
The solution is $\left(\frac{5}{2}, -1\right)$.

26. $4x-y=6$
$x+2y=8$
Solve the second equation for x, $x=8-2y$.
Substitute $8-2y$ for x in the first equation.
$$4x-y=6$$
$$4(8-2y)-y=6$$
$$32-8y-y=6$$
$$-9y=-26$$
$$y=\frac{26}{9}$$
Substitute $\frac{26}{9}$ for y in the equation $x=8-2y$.
$$x=8-2\left(\frac{26}{9}\right)$$
$$x=\frac{72}{9}-\frac{52}{9}$$
$$x=\frac{20}{9}$$
The solution is $\left(\frac{20}{9}, \frac{26}{9}\right)$.

27.
$$\begin{array}{r} x+y=6 \\ \underline{x-y=10} \\ 2x=16 \end{array}$$
$$x=8$$
$$x=\frac{16}{2}=8$$
Substitute 8 for x in the first equation.
$$x+y=6$$
$$8+y=6$$
$$y=-2$$
The solution is (8, –2).

28. $x+2y=-3$
$\underline{2x-2y=\ 6}$
$3x \quad\quad = 3$
$x=1$
Substitute 1 for x in the first equation.
$x+2y=-3$
$1+2y=-3$
$2y=-4$
$y=-2$
The solution is (1, –2).

29. $x+y=12$
$2x+y=5$
To eliminate y, multiply the first equation by –1 and then add.
$-1[x+y=12]$
gives
$-x-y=-12$
$\underline{2x+y=\ \ 5}$
$x \quad\quad =-7$
Substitute –7 for x in the first equation.
$x+y=12$
$-7+y=12$
$y=19$
The solution is (–7, 19).

30. $4x-3y=8$
$2x+5y=8$
To eliminate x, multiply the second equation by –2 and then add.
$-2[2x+5y=8]$
gives
$4x-3y=\ \ 8$
$\underline{-4x-10y=-16}$
$-13y=-8$
$y=\frac{8}{13}$
Substitute $\frac{8}{13}$ for y in the first equation.
$4x-3y=8$
$4x-3\left(\frac{8}{13}\right)=8$
$4x-\frac{24}{13}=8$
$4x=\frac{104}{13}+\frac{24}{13}$
$4x=\frac{128}{13}$
$x=\frac{32}{13}$
The solution is $\left(\frac{32}{13}, \frac{8}{13}\right)$.

31. $-2x+3y=15$
$3x+3y=10$
To eliminate y, multiply the second equation by –1 and then add.
$-1[3x+3y=10]$
gives
$-2x+3y=\ \ 15$
$\underline{-3x-3y=-10}$
$-5x \quad\quad = \ \ 5$
$x=-1$
Substitute –1 for x in the second equation.
$3x+3y=10$
$3(-1)+3y=10$
$-3+3y=10$
$3y=13$
$y=\frac{13}{3}$
The solution is $\left(-1, \frac{13}{3}\right)$.

32. $2x+y=9$
$-4x-2y=4$
Multiply the first equation by 2, and then add.
$2[2x+y=9]$
gives
$4x+2y=18$
$\underline{-4x-2y=\ 4}$
$0=22$ False
There is no solution.

33. $3x+4y=10$
$-6x-8y=-20$
To eliminate x, multiply the first equation by 2, and then add.
$2[3x+4y=10]$
gives
$$\begin{array}{r} 6x+8y=20 \\ \underline{-6x-8y=-20} \\ 0=0 \quad \text{True} \end{array}$$
There are an infinite number of solutions.

34. $2x-5y=12$
$3x-4y=-6$
To eliminate x, multiply the first equation by -3 and the second equation by 2 and then add.
$-3[2x-5y=12]$
$2[3x-4y=-6]$
gives
$$\begin{array}{r} -6x+15y=-36 \\ \underline{6x-8y=-12} \\ 7y=-48 \\ y=-\frac{48}{7} \end{array}$$
Now, substitute $-\frac{48}{7}$ for y in the first equation.
$$2x-5y=12$$
$$2x-5\left(-\frac{48}{7}\right)=12$$
$$2x+\frac{240}{7}=12$$
$$2x=\frac{84}{7}-\frac{240}{7}$$
$$2x=-\frac{156}{7}$$
$$x=-\frac{78}{7}$$
The solution is $\left(-\frac{78}{7}, -\frac{48}{7}\right)$.

35. Let x be the larger number and y be the smaller number.
$x+y=48$
$x=2y-3$
Substitute $2y-3$ for x in the first equation.
$$x+y=48$$
$$2y-3+y=48$$
$$3y=51$$
$$y=17$$
Substitute 17 for y in the second equation.
$x=2y-3$
$x=2(17)-3$
$x=31$
The numbers are 17 and 31.

36. Let x be the speed of the plane in still air and y be the speed of the wind.
$$\begin{array}{r} x+y=\ 600 \\ \underline{x-y=\ 530} \end{array}$$
Add: $2x \quad =1130$
$$x=565$$
Substitute 565 for x in the first equation.
$$x+y=600$$
$$565+y=600$$
$$y=35$$
The speed of the plane is 565 miles per hour and the speed of the wind is 35 miles per hour.

37. Let x be the number of miles traveled and c be the cost.
$c=20+0.5x$
$c=35+0.4x$
Substitute $20+0.5x$ for c in the second equation.
$$20+0.5x=35+0.4x$$
$$0.1x=15$$
$$x=150$$
The cost is the same for 150 miles of travel.

38. Let x be the amount invested at 4% and y be the amount invested at 6%.
$x+y=16,000$
$0.04x+0.06y=760$
Solve the first equation for x, $x=16,000-y$. Substitute $16,000-y$ for x in the second equation.

$0.04(16,000 - y) + 0.06y = 760$
$640 - 0.04y + 0.06y = 760$
$0.02y = 120$
$y = 6000$

Substitute 6000 for y in the equation $x = 16,000 - y$.
$x = 16,000 - 6000$
$x = 10,000$

She invested \$10,000 at 4% and \$6000 at 6%.

39. Let l be Liz's speed and m be Mary's speed. The distance that Liz traveled is $5l$ and the distance that Mary traveled is $5m$.
$m = l + 6$
$5l + 5m = 600$
Substitute $l + 6$ for m in the second equation.
$5l + 5(l + 6) = 600$
$5l + 5l + 30 = 600$
$10l = 570$
$l = 57$
Substitute 57 for l in the first equation.
$m = 57 + 6$
$m = 63$
Liz's speed was 57 miles per hour and Mary's speed was 63 miles per hour.

40. Let g be the pounds of Green Turf's grass seed and a be the pounds of Agway's grass seed.
$0.6g + 0.45a = 20.25$
$g + a = 40$
Solve the second equation for a, $a = 40 - g$.
Substitute $40 - g$ for a in the first equation.
$0.6g + 0.45(40 - g) = 20.25$
$0.6g + 18 - 0.45g = 20.25$
$0.15g = 2.25$
$g = 15$
Substitute 15 for g in the equation $a = 40 - g$.
$a = 40 - 15$
$a = 25$
There were 15 pounds of Green Turf's grass seed and 25 pounds of Agway's grass seed.

41. Let x be the amount of 30% acid solution and y be the amount of 50% acid solution.
$x + y = 6$
$0.3x + 0.5y = 0.4(6)$
To clear decimals, multiply the second equation by 10.
$x + y = 6$
$3x + 5y = 24$
Solve the first equation for y, $y = -x + 6$.
Substitute $-x + 6$ for y in the second equation.
$3x + 5y = 24$
$3x + 5(-x + 6) = 24$
$3x - 5x + 30 = 24$
$-2x = -6$
$x = 3$
Finally, substitute 3 for x in the equation $y = -x + 6$.
$y = -x + 6$
$y = -3 + 6$
$y = 3$
The chemist should combine 3 liters of each solution to produce the desired result.

42. $x + y > 2$
$2x - y \le 4$

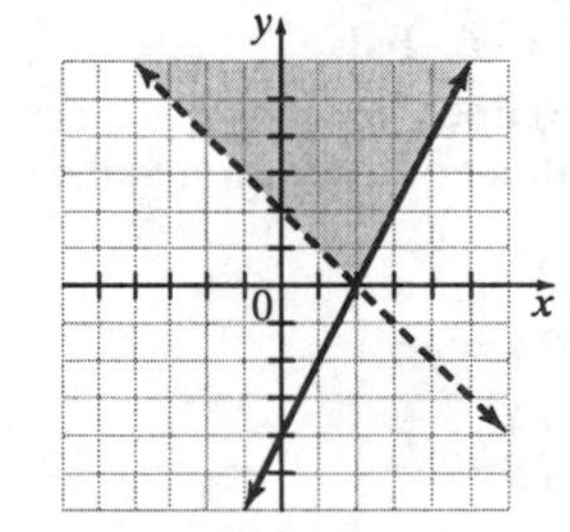

43. $2x - 3y \le 6$
$x + 4y > 4$

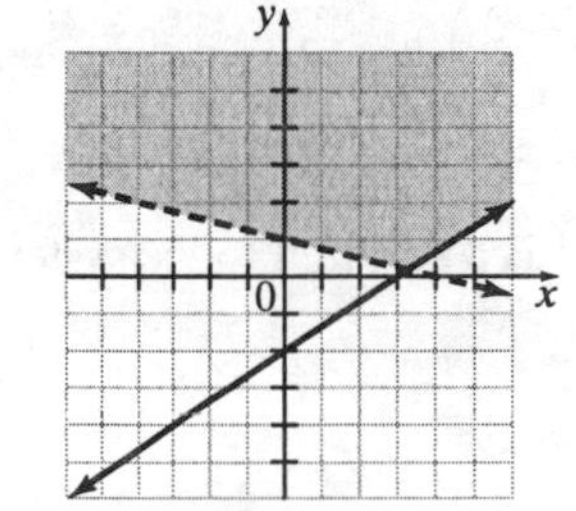

44. $2x - 6y > 6$
$x > -2$

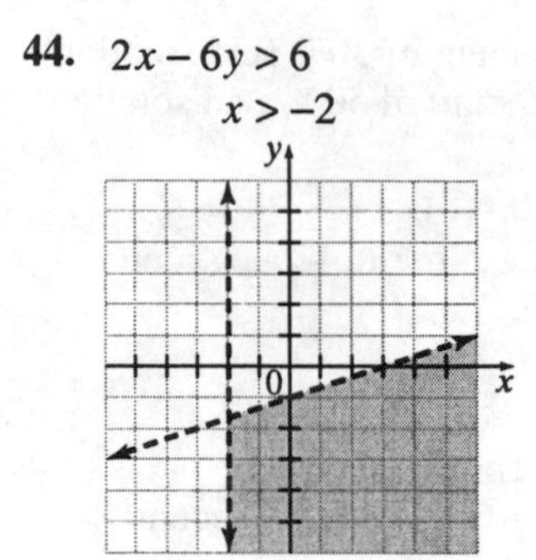

45. $x < 2$
$y \geq -3$

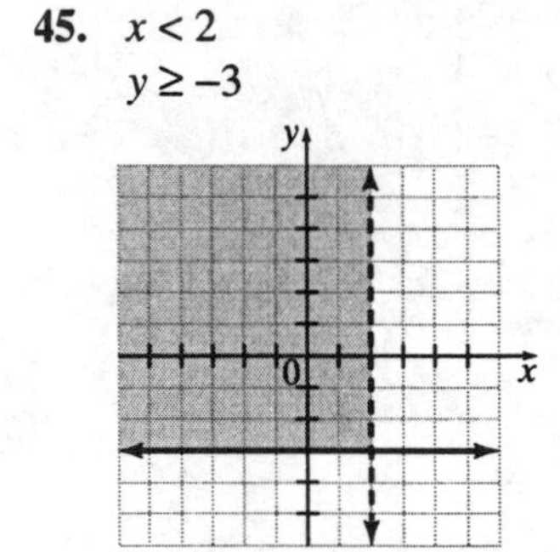

Practice Test

1. a. $x + 2y = -6$
$0 + 2(-6) \stackrel{?}{=} -6$
$-12 = -6$ False
Since (0, 6) does not satisfy the first equation, it is not a solution to the system.

b. $x + 2y = -6$
$-3 + 2\left(-\frac{3}{2}\right) \stackrel{?}{=} -6$
$-6 = -6$ True

$3x + 2y = -12$
$3(-3) + 2\left(-\frac{3}{2}\right) \stackrel{?}{=} -12$
$-12 = -12$ True
$\left(-3, -\frac{3}{2}\right)$ is a solution to the system.

c. $x + 2y = -6$
$2 + 2(-4) \stackrel{?}{=} -6$
$-6 = -6$ True
$3x + 2y = -12$
$3(2) + 2(-4) \stackrel{?}{=} -12$
$-2 = -12$ False
Since (2, –4) does not satisfy both equations, it is not a solution to the system.

2. The system is inconsistent, it has no solution.

3. The system is consistent; it has exactly one solution.

4. The system is dependent; it has an infinite number of solutions.

5. $3y = 6x - 9$ $\quad$ $2x - y = 6$
$y = 2x - 3$ $\quad$ $y = 2x - 6$
The lines have the same slope, but different y-intercepts, so they are parallel. Thus, the system of equations is inconsistent and has no solution.

6. $3x + 2y = 10$ $\quad$ $3x - 2y = 10$
$2y = -3x + 10$ $\quad$ $-2y = -3x + 10$
$y = -\frac{3}{2}x + 5$ $\quad$ $y = \frac{3}{2}x - 5$
The slopes of the lines are different. Thus, the system of equations is consistent and has one solution.

7. $4x = 6y - 12$ $\quad$ $2x - 3y = -6$
$6y = 4x + 12$ $\quad$ $-3y = -2x - 6$
$y = \frac{2}{3}x + 2$ $\quad$ $y = \frac{2}{3}x + 2$
The lines are the same. Thus, the system of equations is consistent and dependent, and it has infinite number of solutions.

8. a. You will obtain a false statement, such as $6 = 0$.

b. You will obtain a true statement, such as $0 = 0$.

9. $y = 3x - 2 \qquad y = -2x + 8$

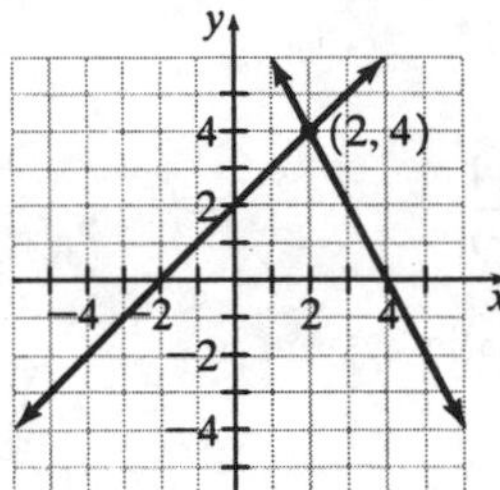

It appears that (2, 4) is the solution to the system.

$y = 3x - 2 \qquad\qquad y = -2x + 8$

$4 \stackrel{?}{=} 3(2) - 2 \qquad 4 \stackrel{?}{=} -2(2) + 8$

$4 = 4$ True $\qquad 4 = 4$ True

(2, 4) is the solution to the system.

10. $3x - 2y = -3 \qquad 3x + y = 6$

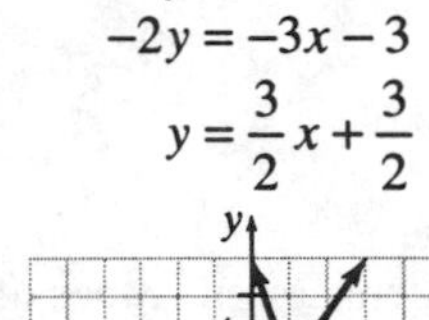

$-2y = -3x - 3 \qquad y = -3x + 6$

$y = \frac{3}{2}x + \frac{3}{2}$

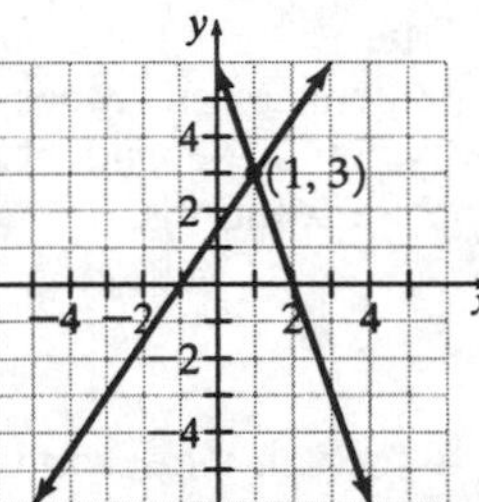

It appears that (1, 3) is the solution to the system

$3x - 2y = -3 \qquad\qquad 3x + y = 6$

$3(1) - 2(3) \stackrel{?}{=} -3 \qquad 3(1) + 3 \stackrel{?}{=} 6$

$-3 = -3$ True $\qquad 6 = 6$ True

(1, 3) is the solution to the system.

11. $y = 2x + 4 \qquad 4x - 2y = 6$

$-2y = -4x + 6$

$y = 2x - 3$

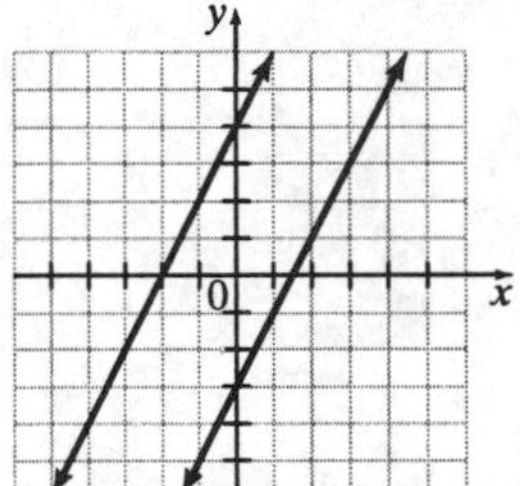

The lines are parallel, so the system has no solution.

12. $3x + y = 8$

$x - y = 6$

Solve the first equation for y,

$y = 8 - 3x$.

Substitute $8 - 3x$ for y in the second equation.

$x - (8 - 3x) = 6$

$x - 8 + 3x = 6$

$4x = 14$

$x = \frac{14}{4} = \frac{7}{2}$

Substitute $\frac{7}{2}$ for x in the equation $y = 8 - 3x$.

$y = 8 - 3x$

$y = 8 - 3\left(\frac{7}{2}\right)$

$y = \frac{16}{2} - \frac{21}{2}$

$y = -\frac{5}{2}$

The solution to the system is $\left(\frac{7}{2}, -\frac{5}{2}\right)$.

13. $4x - 3y = 9$

$2x + 4y = 10$

Solve the second equation for x since all the numbers are divisible by 2.

$2x = -4y + 10$

$x = -2y + 5$

Substitute $-2y + 5$ for x in the first equation.

$4(-2y + 5) - 3y = 9$

$-8y + 20 - 3y = 9$

$-11y = -11$

$y = 1$

Substitute 1 for y in the equation

$x = -2y + 5$.

$x = -2y + 5$

$x = -2(1) + 5$

$x = 3$

The solution to the system is (3, 1).

14. $y = 5x - 7$
$y = 3x + 5$
Both equations are solved for y. Substitute $5x - 7$ for y in the second equation.
$5x - 7 = 3x + 5$
$2x = 12$
$x = 6$
Substitute 6 for x in the first equation.
$y = 5x - 7$
$y = 5(6) - 7$
$y = 23$
The solution to the system is (6, 23).

15. $2x + y = 5$
$x + 3y = -10$
To eliminate x, multiply the second equation by –2.
$-2[x + 3y = -10]$
gives
$$\begin{array}{r} 2x + y = 5 \\ -2x - 6y = 20 \\ \hline -5y = 25 \\ y = -5 \end{array}$$
Substitute –5 for y in the first equation.
$2x + y = 5$
$2x + (-5) = 5$
$2x = 10$
$x = 5$
The solution to the system is (5, –5).

16. $3x + 2y = 12$
$-2x + 5y = 8$
To eliminate x, multiply the first equation by 2 and the second equation by 3.
$2[3x + 2y = 12]$
$3[-2x + 5y = 8]$
gives
$$\begin{array}{r} 6x + 4y = 24 \\ -6x + 15y = 24 \\ \hline 19y = 48 \\ y = \frac{48}{19} \end{array}$$
To eliminate y, multiply the first equation by 5 and the second equation by –2.
$5[3x + 2y = 12]$
$-2[-2x + 5y = 8]$
gives
$$\begin{array}{r} 15x + 10y = 60 \\ 4x - 10y = -16 \\ \hline 19x \quad\quad = 44 \\ x = \frac{44}{19} \end{array}$$
The solution to the system is $\left(\frac{44}{19}, \frac{48}{19}\right)$.

17. $5x - 10y = 20$
$x = 2y + 4$
Align the x- and y-terms on the left side of each equation.
$5x - 10y = 20$
$x - 2y = 4$
Multiply the second equation by –5.
$-5[x - 2y = 4]$
gives
$$\begin{array}{r} 5x - 10y = 20 \\ -5x + 10y = -20 \\ \hline 0 = 0 \text{ True} \end{array}$$
This is a true statement for all values of x and y. Thus, the system is dependent and has an infinite number of solutions.

18. $y = 3x - 4$
$2x + y = 6$
Substitute $3x - 4$ for y in the second equation.
$2x + (3x - 4) = 6$
$2x + 3x - 4 = 6$
$5x = 10$
$x = 2$
Substitute 2 for x in the first equation.
$y = 3x - 4$
$y = 3(2) - 4$
$y = 2$
The solution to the system is (2, 2).

19. $3x+5y=20$
$6x+3y=-12$
To eliminate x, multiply the first equation by –2.
$-2[3x+5y=20]$
gives

$$\begin{aligned} -6x-10y &= -40 \\ 6x+3y &= -12 \\ \hline -7y &= -52 \\ y &= \frac{52}{7} \end{aligned}$$

To eliminate y, multiply the first equation by –3 and the second equation by 5.
$-3[3x+5y=20]$
$5[6x+3y=-12]$
gives

$$\begin{aligned} -9x-15y &= -60 \\ 30x+15y &= -60 \\ \hline 21x \qquad &= -120 \\ x &= -\frac{120}{21} = -\frac{40}{7} \end{aligned}$$

The solution to the system is $\left(-\frac{40}{7}, \frac{52}{7}\right)$.

20. $4x-6y=8$
$3x+5y=10$
To eliminate x, multiply the first equation by –3 and the second equation by 4.
$-3[4x-6y=8]$
$4[3x+5y=10]$
gives

$$\begin{aligned} -12x+18y &= -24 \\ 12x+20y &= 40 \\ \hline 38y &= 16 \\ y &= \frac{16}{38} = \frac{8}{19} \end{aligned}$$

To eliminate y, multiply the first equation by 5 and the second equation by 6.
$5[4x-6y=8]$
$6[3x+5y=10]$
gives

$$\begin{aligned} 20x-30y &= 40 \\ 18x+30y &= 60 \\ \hline 38x \qquad &= 100 \\ x &= \frac{100}{38} = \frac{50}{19} \end{aligned}$$

The solution to the system is $\left(\frac{50}{19}, \frac{8}{19}\right)$.

21. Let m = the number of miles driven
c = the total cost per day
With Budget Rent A Car,
$c = 40 + 0.08m$, while with Hertz, $c = 45 + 0.03m$. Set the two costs equal.

$$\begin{aligned} 40+0.08m &= 45+0.03m \\ 0.05m &= 5 \\ m &= 100 \end{aligned}$$

The cars cost an equal amount when they are driven 100 miles per day.

22. Let l = amount of lemon candies
b = amount of butterscotch candies

Candy	Number of pounds	Cost per pound	Total cost
lemon	l	6	$6l$
butterscotch	b	4.5	$4.5b$
Mixture	20	5	5(20)

$l + b = 20$
$6l + 4.5b = 100$
Solve the first equation for b.
$l + b = 20$
$b = 20 - l$
Substitute $20 - l$ for b in the second equation.
$6l + 4.5(20 - l) = 100$
$6l + 90 - 4.5l = 100$
$1.5l = 10$
$l = \frac{10}{1.5} = 6\frac{2}{3}$
$b = 20 - l$
$b = 20 - 6\frac{2}{3}$
$b = 13\frac{1}{3}$
The mixture must contain $6\frac{2}{3}$ pounds of lemon candies and $13\frac{1}{3}$ pounds of butterscotch candies.

23. Let h = the speed of Dante Hull's boat
r = the speed of Deja Rocket's boat
In 3 hours, Dante's boat travels $3h$ miles, while in 3.2 hours, Deja's boat travels $3.2r$ miles.
The distance that the two boats travel are the same.
$3h = 3.2r$
$h = r + 4$
Substitute $r + 4$ for h in the first equation.
$3(r + 4) = 3.2r$
$3r + 12 = 3.2r$
$12 = 0.2r$
$60 = r$
Substitute 60 for r in the second equation.
$h = r + 4$
$h = 60 + 4$
$h = 64$
The speed of Dante Hull's boat is 64 miles per hour and the speed of Deja Rocket's boat is 60 miles per hour.

24. $2x + 4y < 8$
$x - 3y \geq 6$

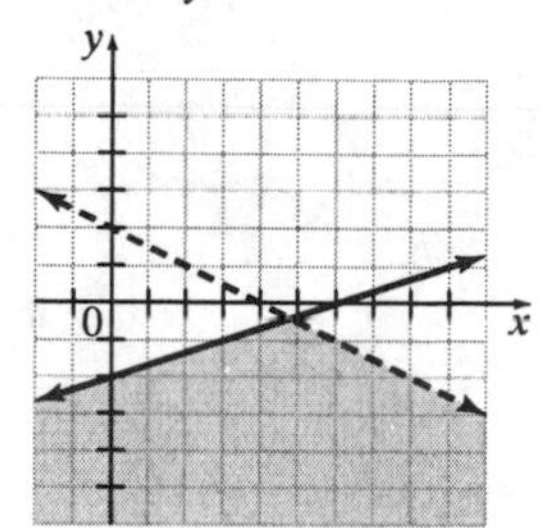

25. $x + 3y \geq 6$
$y < 3$

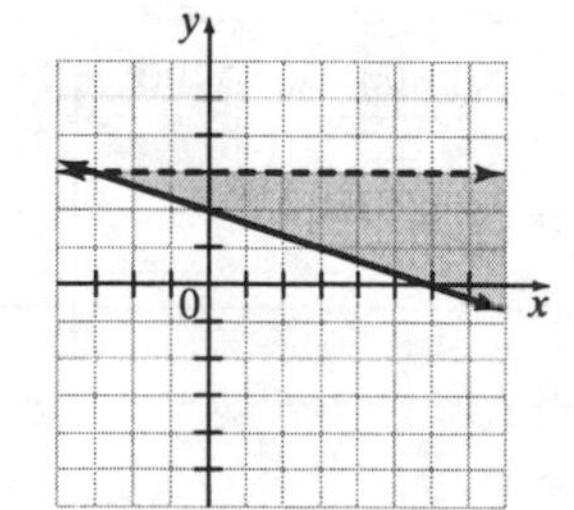

Cumulative Review Test

1. **a.** The most death occurred in 1900–1909.

   b. The fewest deaths occurred in 1990–1997.

   c. The most property damage occurred in 1990–1997.

2. In order, the test grades were 70, 78, 83, 92, and 99. The median test grade was 83.

3. $\frac{83+78+92+70+99}{5}=\frac{422}{5}=84.4$
   Her mean test score was 84.4.

4. $\frac{3}{8}+\frac{9}{10}=\frac{15}{40}+\frac{36}{40}=\frac{51}{40}$

5. **a.** 7

   b. $-6, -0.2, \frac{3}{5}, 7, 0, -\frac{5}{9}, 1.34$

   c. $\sqrt{7}, -\sqrt{2}$

   d. $-6, -0.2, \frac{3}{5}, \sqrt{7}, -\sqrt{2}, 7, 0, -\frac{5}{9}, 1.34$

6. $|-4|=4$
   $-|2|=-2$
   Since $4>-2$, $|-4|>-|-2|$.

7. $-64+74+(-192)=10+(-192)=-182$

8. $$\begin{aligned}-(2x+1)-3x^2 &= -[2(3)+1]-3(3)^2\\ &= -(6+1)-3(9)\\ &= -(7)-27\\ &= -7-27\\ &= -34\end{aligned}$$

9. $$\begin{aligned}&4-(3x-2)+2(x+3)\\ &= 4-3x+2+2x+6\\ &= (-3x+2x)+(4+2+6)\\ &= -x+12\end{aligned}$$

10. $$\begin{aligned}2(x-4)+2 &= 3x-4\\ 2x-8+2 &= 3x-4\\ 2x-6 &= 3x-4\\ -2 &= x\end{aligned}$$

11. $$\begin{aligned}\frac{20\text{ lb}}{8000\text{ sq ft}} &= \frac{x\text{ lb}}{25,000\text{ sq ft}}\\ 500,000 &= 8000x\\ 62.5 &= x\end{aligned}$$
    It takes 62.5 pounds of fertilizer.

12. $$\begin{aligned}3x-4 &\le x+6\\ 2x &\le 10\\ x &\le 5\end{aligned}$$
    (Number line: shaded to the left of a closed dot at 5.)

13. $$\begin{aligned}P &= 2l+2w\\ P-2l &= 2w\\ \frac{P-2l}{2} &= \frac{2w}{2}\\ w &= \frac{P-2l}{2}\end{aligned}$$

14. Let s = Maria's weekly sales
    w = Maria's weekly salary
    Under plan A, $w=0.12s$, while under plan B, $w=350+0.06s$. Set the two salaries equal.
    $$\begin{aligned}0.12s &= 350+0.06s\\ 0.06s &= 350\\ s &= \frac{350}{0.06}\approx 5833.33\end{aligned}$$
    Maria's weekly sales must be \$5833.33 for both plans to pay the same amount.

15. Let x = the measure of the smallest angle
    Then the other angles measure $x+20$ and $6x$. The sum of the measures of the angles in any triangle is 180°.
    $$\begin{aligned}x+(x+20)+6x &= 180\\ 8x+20 &= 180\\ 8x &= 160\\ x &= 20\end{aligned}$$
    $x+20=20+20=40$
    $6x=6(20)=120$
    The angles of the triangle measure 20°, 40°, and 120°.

16.

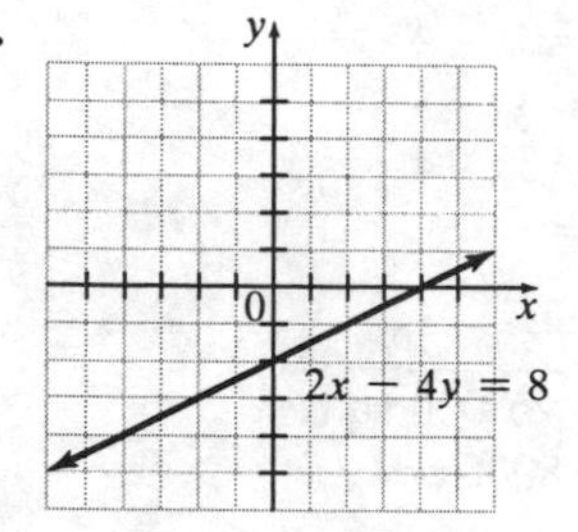

17.

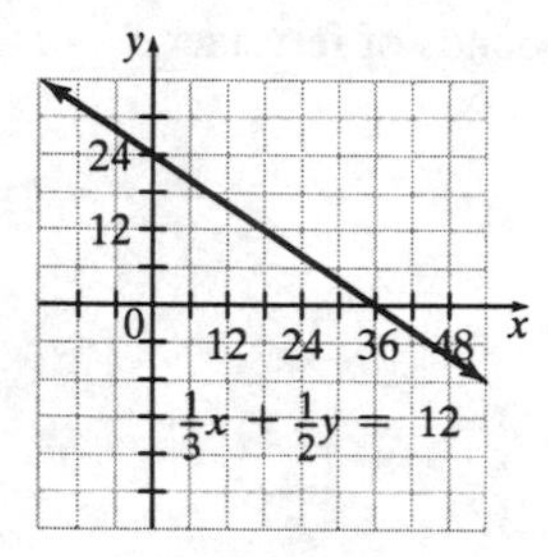

18. $3x - y = 6$

$-y = -3x + 6$

$y = 3x - 6$

$\frac{3}{2}x - 3 = \frac{1}{2}y$

$\frac{1}{2}y = \frac{3}{2}x - 3$

$y = 3x - 6$

The lines are the same, so the system has an infinite number of solutions.

19. Graph $2x + y = 5$ and $x - 2y = 0$.

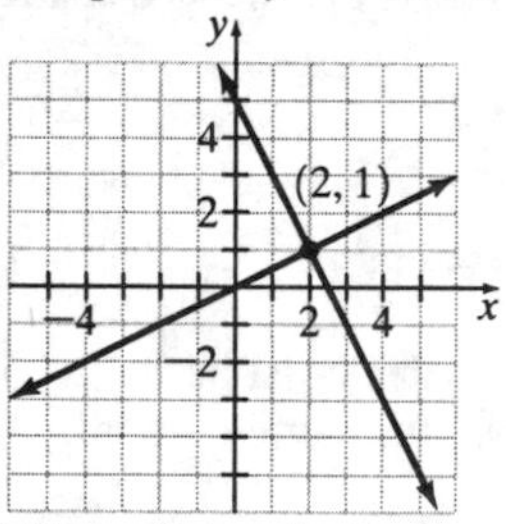

The solution is (2, 1).

20. $3x - 2y = 8$

$-6x + 3y = 2$

To eliminate x, multiply the first equation by 2.

$2[3x - 2y = 8]$

gives

$$\begin{array}{r} 6x - 4y = 16 \\ \underline{-6x + 3y = 2} \\ -y = 18 \\ y = -18 \end{array}$$

To eliminate y, multiply the first equation by 3 and the second equation by 2.

$3[3x - 2y = 8]$

$2[-6x + 3y = 2]$

gives

$$\begin{array}{r} 9x - 6y = 24 \\ \underline{-12x + 6y = 4} \\ -3x \quad = 28 \\ x = -\frac{28}{3} \end{array}$$

The solution to the system is $\left(-\frac{28}{3}, -18\right)$.

Chapter 6

Exercise Set 6.1

1. In the expression c^r, c is the base, r is the exponent.

3. a. $\frac{x^m}{x^n} = x^{m-n}, x \neq 0$

b. Answers will vary.

5. a. $\left(x^m\right)^n = x^{m \cdot n}$

b. Answers will vary.

7. $x^0 \neq 1$ when $x = 0$.

9. $x^2 \cdot x^4 = x^{2+4} = x^6$

11. $y \cdot y^2 = y^{1+2} = y^3$

13. $3^2 \cdot 3^3 = 3^{2+3} = 3^5 = 243$

15. $y^3 \cdot y^2 = y^{3+2} = y^5$

17. $z^3 \cdot z^5 = z^{3+5} = z^8$

19. $y^6 \cdot y = y^6 \cdot y^1 = y^{6+1} = y^7$

21. $\frac{2^2}{2} = \frac{2^2}{2^1} = 2^{2-1} = 2^1 = 2$

23. $\frac{x^{10}}{x^3} = x^{10-3} = x^7$

25. $\frac{5^4}{5^2} = 5^{4-2} = 5^2 = 25$

27. $\frac{y^2}{y} = \frac{y^2}{y^1} = y^{2-1} = y^1 = y$

29. $\frac{z^3}{z^3} = z^{3-3} = z^0 = 1$

31. $\frac{y^{12}}{y^9} = y^{12-9} = y^3$

33. $x^0 = 1$

35. $3x^0 = 3 \cdot 1 = 3$

37. $(3x)^0 = 3^0 x^0 = 1 \cdot 1 = 1$

39. $(-4x)^0 = (-4)^0 \cdot x^0 = 1 \cdot 1 = 1$

41. $\left(x^5\right)^2 = x^{5 \cdot 2} = x^{10}$

43. $\left(x^5\right)^5 = x^{5 \cdot 5} = x^{25}$

45. $\left(x^3\right)^1 = x^{3 \cdot 1} = x^3$

47. $\left(x^4\right)^3 = x^{4 \cdot 3} = x^{12}$

49. $\left(x^5\right)^3 = x^{5 \cdot 3} = x^{15}$

51. $(1.3x)^2 = (1.3)^2 x^2 = 1.69x^2$

53. $\left(-3x^3\right)^3 = (-3)^3 x^{3 \cdot 3} = (-27)x^9$

55. $\left(2x^2 y\right)^3 = 2^3 \cdot x^{2 \cdot 3} y^3 = 8x^6 y^3$

57. $\left(\frac{x}{4}\right)^3 = \frac{x^3}{4^3} = \frac{x^3}{64}$

59. $\left(\frac{y}{x}\right)^4 = \frac{y^4}{x^4}$

61. $\left(\frac{6}{x}\right)^3 = \frac{6^3}{x^3} = \frac{216}{x^3}$

63. $\left(\frac{3x}{y}\right)^3 = \frac{3^3x^3}{y^3} = \frac{27x^3}{y^3}$

65. $\left(\frac{3x}{5}\right)^2 = \frac{3^2x^2}{5^2} = \frac{9x^2}{25}$

67. $\left(\frac{2y^3}{x}\right)^4 = \frac{2^4y^{3\cdot4}}{x^4} = \frac{16y^{12}}{x^4}$

69. $\frac{x^5y}{xy^3} = \frac{x\cdot x^4\cdot y^4}{x\cdot y\cdot y^2} = \frac{x^4}{y^2}$

71. $\frac{10x^3y^8}{2xy^{10}} = \frac{2\cdot5\cdot x\cdot x^2\cdot y^8}{2\cdot x\cdot y^8\cdot y^2} = \frac{5x^2}{y^2}$

73. $\frac{2xy}{16x^3y^3} = \frac{2\cdot x\cdot y}{2\cdot8\cdot x\cdot x^2\cdot y\cdot y^2} = \frac{1}{8x^2y^2}$

75. $\frac{35x^4y^9}{15x^9y^{12}} = \frac{5\cdot7\cdot x^4\cdot y^9}{5\cdot3\cdot x^4\cdot x^5\cdot y^9\cdot y^3} = \frac{7}{3x^5y^3}$

77. $\frac{-36xy^7z}{12x^4y^5z} = -\frac{3\cdot12\cdot x\cdot y^5\cdot y^2\cdot z}{12\cdot x\cdot x^3\cdot y^5\cdot z} = -\frac{3y^2}{x^3}$

79.
$$-\frac{6x^2y^7z}{3x^5y^9z^6} = -\frac{2\cdot3\cdot x^2\cdot y^7\cdot z}{3\cdot x^2\cdot x^3\cdot y^7\cdot y^2\cdot z\cdot z^5}$$
$$= -\frac{2}{x^3y^2z^5}$$

81.
$$\left(\frac{4x^4}{2x^6}\right)^3 = \left(\frac{4}{2}\cdot\frac{x^4}{x^6}\right)^3$$
$$= \left(\frac{2}{x^2}\right)^3$$
$$= \frac{2^3}{x^{2\cdot3}}$$
$$= \frac{8}{x^6}$$

83.
$$\left(\frac{6y^6}{2y^3}\right)^3 = \left(\frac{6}{2}\cdot\frac{y^6}{y^3}\right)^3$$
$$= \left(3y^3\right)^3$$
$$= 3^3y^{3\cdot3}$$
$$= 27y^9$$

85. $\left(\frac{13x^9}{17x^5}\right)^0 = 1$

87.
$$\left(\frac{x^4y^3}{x^2y^5}\right)^2 = \left(\frac{x^4}{x^2}\cdot\frac{y^3}{y^5}\right)^2$$
$$= \left(\frac{x^2}{y^2}\right)^2$$
$$= \frac{x^{2\cdot2}}{y^{2\cdot2}}$$
$$= \frac{x^4}{y^4}$$

89.
$$\left(\frac{9y^2z^7}{18y^9z}\right)^4 = \left(\frac{9}{18}\cdot\frac{y^2}{y^9}\cdot\frac{z^7}{z}\right)^4$$
$$= \left(\frac{z^6}{2y^7}\right)^4$$
$$= \frac{z^{6\cdot4}}{2^4y^{7\cdot4}}$$
$$= \frac{z^{24}}{16y^{28}}$$

91.
$$\left(\frac{4xy^5}{y}\right)^3 = \left(4x\cdot\frac{y^5}{y}\right)^3$$
$$= (4xy^4)^3$$
$$= 4^3x^3y^{4\cdot3}$$
$$= 64x^3y^{12}$$

93. $\left(3xy^4\right)^2 = 3^2x^2y^{4\cdot2} = 9x^2y^8$

95. $\left(6xy^5\right)\left(3x^2y^4\right) = (6 \cdot 3)\left(x^{1+2}\right)\left(y^{5+4}\right)$
$= 18x^3y^9$

97. $\left(2x^4y^2\right)\left(5x^2y\right) = (2 \cdot 5)\left(x^{4+2}\right)\left(y^{2+1}\right)$
$= 10x^6y^3$

99. $(5xy)\left(2xy^6\right) = (5 \cdot 2)\left(x^{1+1}\right)\left(y^{1+6}\right) = 10x^2y^7$

101. $\left(2xy^3\right)\left(9x^4y^5\right)^0 = (2xy)^3 \cdot 1 = 2^3x^3y^3 = 8x^3y^3$

103. $\left(\dfrac{-x^3y^5}{xy^7}\right)^3 = \left(-\dfrac{x^3}{x} \cdot \dfrac{y^5}{y^7}\right)^3$
$= \left(\dfrac{-x^2}{y^2}\right)^3$
$= \dfrac{(-1)^3x^{2 \cdot 3}}{y^{2 \cdot 3}}$
$= -\dfrac{x^6}{y^6}$

105. $(-x)^2 = (-x)(-x) = x^2$

107. $\left(2.5x^3\right)^2 = 2.5^2 \cdot x^{3 \cdot 2} = 6.25x^6$

109. $\dfrac{x^7y^2}{xy^6} = \dfrac{x^7}{x} \cdot \dfrac{y^2}{y^6} = \dfrac{x^6}{y^4}$

111. $\left(\dfrac{-x^5}{y^2}\right)^3 = \dfrac{(-1)^3x^{5 \cdot 3}}{y^{2 \cdot 3}} = -\dfrac{x^{15}}{y^6}$

113. $\left(-6x^3y^2\right)^3 = (-6)^3x^{3 \cdot 3}y^{2 \cdot 3} = -216x^9y^6$

115. $\left(-2x^4y^2z\right)^3 = (-2)^3x^{4 \cdot 3}y^{2 \cdot 3}z^{1 \cdot 3} = -8x^{12}y^6z^3$

117. $\left(7x^2y^4\right)^2 = 7^2x^{2 \cdot 2}y^{4 \cdot 2} = 49x^4y^8$

119. $\left(4x^2y\right)\left(3xy^2\right)^3 = \left(4x^2y\right)\left(3^3x^3y^{2 \cdot 3}\right)$
$= \left(4x^2y\right)\left(27x^3y^6\right)$
$= 4 \cdot 27x^{2+3}y^{1+6}$
$= 108x^5y^7$

121. $\left(8.6x^2y^5\right)^2 = 8.6^2x^{2 \cdot 2}y^{5 \cdot 2} = 73.96x^4y^{10}$

123. $\left(x^7y^5\right)\left(xy^2\right)^4 = \left(x^7y^5\right)\left(x^{1 \cdot 4}y^{2 \cdot 4}\right)$
$= \left(x^7y^5\right)\left(x^4y^8\right)$
$= x^{7+4}y^{5+8}$
$= x^{11}y^{13}$

125. $\left(\dfrac{-x^4z^7}{x^2z^5}\right)^4 = \left(-1 \cdot \dfrac{x^4}{x^2} \cdot \dfrac{z^7}{z^5}\right)^4$
$= (-x^2z^2)^4$
$= (-1)^4x^{2 \cdot 4}z^{2 \cdot 4}$
$= x^8z^8$

127. $\dfrac{x+y}{x}$ cannot be simplified.

129. $\dfrac{x^2+2}{x}$ cannot be simplified.

131. $\dfrac{5yz^4}{yz^2} = 5y^{1-1} \cdot z^{4-2} = 5z^2$

133. $\dfrac{x}{x+1}$ cannot be simplified.

135. $x^2y = 3^2 \cdot 2 = 9 \cdot 2 = 18$

137. $(xy)^0 = (2 \cdot 4)^0 = 8^0 = 1$

139. The sign will be negative because a negative number with an odd number for an exponent will be negative. This is because $(-1)^m = -1$ when m is odd.

141. Area = Length × Width = $7x \cdot x = 7x^2$

143. Area = (Area of Shape 1) + (Area of Shape 2)

$$= x \cdot 3x + x \cdot 2y + xy + xy$$
$$= 3x^2 + 2xy + 2xy$$
$$= 3x^2 + 4xy$$

145.

$$\left(\frac{3x^4y^5}{6x^6y^8}\right)^3\left(\frac{9x^7y^8}{3x^3y^5}\right)^2 = \left(\frac{3}{6}\cdot\frac{x^4}{x^6}\cdot\frac{y^5}{y^8}\right)^3\left(\frac{9}{3}\cdot\frac{x^7}{x^3}\cdot\frac{y^8}{y^5}\right)^2$$
$$= \left(\frac{1}{2x^2y^3}\right)^3\left(\frac{3x^4y^3}{1}\right)^2$$
$$= \frac{1^3}{2^3x^{2\cdot3}y^{3\cdot3}}\cdot\frac{3^2x^{4\cdot2}y^{3\cdot2}}{1^2}$$
$$= \frac{9x^8y^6}{8x^6y^9}$$
$$= \frac{9x^2}{8y^3}$$

148.

$$2(x+4) - 3 = 5x + 4 - 3x + 1$$
$$2x + 8 - 3 = 2x + 5$$
$$2x + 5 = 2x + 5$$

All real numbers are solutions to this equation.

149. a.

$$P = 2l + 2w$$
$$26 = 2(x+5) + 2(x)$$
$$26 = 2x + 10 + 2x$$
$$26 = 4x + 10$$
$$16 = 4x$$
$$4 = x$$
$$x + 5 = 4 + 5 = 9$$

The sides are 4 and 9 inches long.

b.

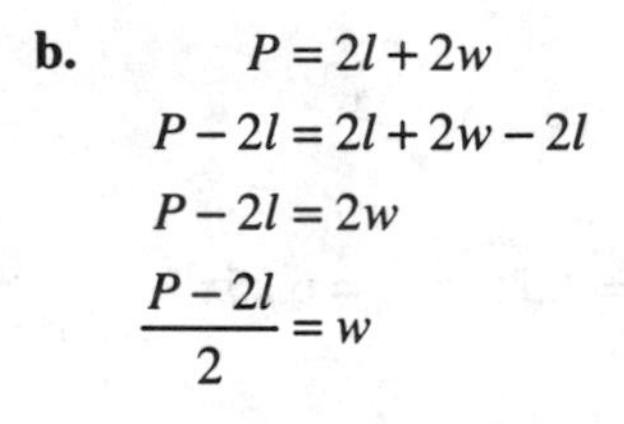

$$P = 2l + 2w$$
$$P - 2l = 2l + 2w - 2l$$
$$P - 2l = 2w$$
$$\frac{P - 2l}{2} = w$$

150.

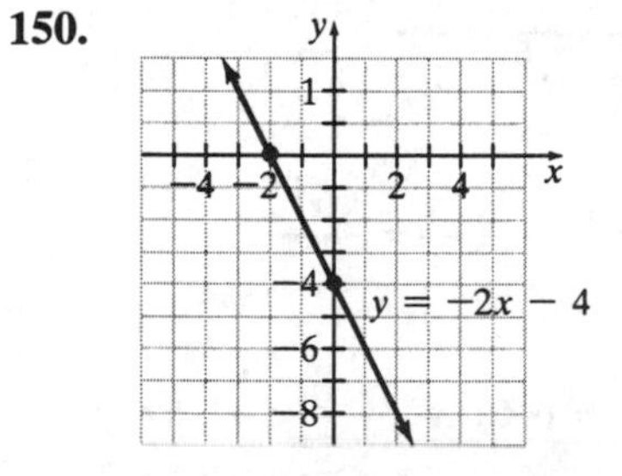

151. $m = \frac{x_2 - x_1}{y_2 - y_1}$
$= \frac{4-2}{-3-(-6)}$
$= \frac{2}{-3+6}$
$= \frac{2}{3}$

Exercise Set 6.2

1. Answers will vary.

3. No, it is not simplified because of the negative exponent.
$x^5y^{-3} = \frac{x^5}{y^3}$

5. The given simplification is not correct since $\left(y^4\right)^{-3} = y^{4\cdot(-3)} = y^{-12} = \frac{1}{y^{12}}$.

7. **a.** The numerator has one term, x^5y^2.
   b. The factors of the numerator are x^5 and y^2.

9. The sign of the exponent changes when a factor is moved from the numerator to the denominator of a fraction.

11. $x^{-4} = \frac{1}{x^4}$

13. $3^{-1} = \frac{1}{3}$

15. $\frac{1}{x^{-3}} = x^3$

17. $\frac{1}{x^{-1}} = x^1 = x$

19. $\frac{1}{4^{-2}} = 4^2 = 16$

21. $\left(x^{-2}\right)^3 = x^{-2\cdot 3} = x^{-6} = \frac{1}{x^6}$

23. $\left(y^{-5}\right)^4 = y^{-5\cdot 4} = y^{-20} = \frac{1}{y^{20}}$

25. $\left(x^4\right)^{-2} = x^{4(-2)} = x^{-8} = \frac{1}{x^8}$

27. $\left(2^{-3}\right)^{-2} = 2^{(-3)(-2)} = 2^6 = 64$

29. $y^4 \cdot y^{-2} = y^{4+(-2)} = y^2$

31. $x^7 \cdot x^{-5} = x^{7-5} = x^2$

33. $3^{-2} \cdot 3^4 = 3^{-2+4} = 3^2 = 9$

35. $\frac{r^4}{r^6} = r^{4-6} = r^{-2} = \frac{1}{r^2}$

37. $\frac{p^0}{p^{-3}} = p^{0-(-3)} = p^3$

39. $\frac{x^{-7}}{x^{-3}} = x^{-7-(-3)} = x^{-4} = \frac{1}{x^4}$

41. $\frac{3^2}{3^{-1}} = 3^{2-(-1)} = 3^3 = 27$

43. $3^{-3} = \frac{1}{3^3} = \frac{1}{27}$

45. $\frac{1}{z^{-9}} = z^9$

47. $\left(x^2\right)^{-5} = x^{2(-5)} = x^{-10} = \frac{1}{x^{10}}$

49. $\left(y^{-2}\right)^{-3} = y^{(-2)(-3)} = y^6$

51. $x^3 \cdot x^{-7} = x^{3-7} = x^{-4} = \frac{1}{x^4}$

53. $x^{-8} \cdot x^{-7} = x^{-8-7} = x^{-15} = \dfrac{1}{x^{15}}$

55. $\dfrac{x^{-3}}{x^5} = x^{-3-5} = x^{-8} = \dfrac{1}{x^8}$

57. $\dfrac{y^9}{y^{-1}} = y^{9-(-1)} = y^{10}$

59. $\dfrac{2^{-3}}{2^{-3}} = 2^{-3-(-3)} = 2^0 = 1$

61. $\left(2^{-1} + 3^{-1}\right)^0 = 1$

63. $\dfrac{1}{1^{-5}} = 1^{1-(-5)} = 1^6 = 1$

65. $\left(x^{-4}\right)^{-2} = x^{(-4)(-2)} = x^8$

67. $\left(x^0\right)^{-3} = (1)^{-3} = 1$

69. $2^{-3} \cdot 2 = 2^{-3+1} = 2^{-2} = \dfrac{1}{2^2} = \dfrac{1}{4}$

71. $6^{-4} \cdot 6^2 = 6^{-4+2} = 6^{-2} = \dfrac{1}{6^2} = \dfrac{1}{36}$

73. $\dfrac{x^{-1}}{x^{-4}} = x^{-1-(-4)} = x^3$

75. $\left(3^2\right)^{-1} = 3^{2(-1)} = 3^{-2} = \dfrac{1}{3^2} = \dfrac{1}{9}$

77. $\dfrac{5}{5^{-2}} = 5^{1-(-2)} = 5^3 = 125$

79. $\dfrac{3^{-4}}{3^{-2}} = 3^{-4-(-2)} = 3^{-2} = \dfrac{1}{3^2} = \dfrac{1}{9}$

81. $\dfrac{7^{-1}}{7^{-1}} = 7^{-1-(-1)} = 7^0 = 1$

83.
$$\begin{aligned}\left(6x^2\right)^{-2} &= 6^{-2}x^{2(-2)}\\ &= 6^{-2}x^{-4}\\ &= \frac{1}{6^2x^4}\\ &= \frac{1}{36x^4}\end{aligned}$$

85. $3x^{-2}y^2 = 3 \cdot \dfrac{1}{x^2} \cdot y^2 = \dfrac{3y^2}{x^2}$

87. $5x^{-5}y^{-2} = 5 \cdot \dfrac{1}{x^5} \cdot \dfrac{1}{y^2} = \dfrac{5}{x^5y^2}$

89. $\left(x^5y^{-3}\right)^{-3} = x^{5(-3)}y^{(-3)(-3)} = x^{-15}y^9 = \dfrac{y^9}{x^{15}}$

91. $\left(4y^{-2}\right)\left(5y^{-3}\right) = 4 \cdot 5 \cdot y^{-2} \cdot y^{-3} = 20y^{-5} = \dfrac{20}{y^5}$

93. $6x^4\left(-2x^{-2}\right) = 6 \cdot (-2) \cdot x^4 \cdot x^{-2} = -12x^2$

95.
$$\begin{aligned}\left(4x^2y\right)\left(3x^3y^{-1}\right) &= 4 \cdot 3 \cdot x^2 \cdot x^3 \cdot y \cdot y^{-1}\\ &= 12x^5\end{aligned}$$

97.
$$\begin{aligned}\left(2y^2\right)\left(4y^{-3}z^5\right) &= 2 \cdot 4 \cdot y^2 \cdot y^{-3} \cdot z^5\\ &= 8y^{-1}z^5\\ &= \frac{8z^5}{y}\end{aligned}$$

99. $\dfrac{8x^4}{4x^{-1}} = \dfrac{8}{4} \cdot x^{4-(-1)} = 2x^5$

101. $\dfrac{36x^{-4}}{9x^{-2}} = \dfrac{36}{4} \cdot \dfrac{x^{-4}}{x^{-2}} = 4 \cdot \dfrac{1}{x^2} = \dfrac{4}{x^2}$

103. $\dfrac{3x^4y^{-2}}{6y^3} = \dfrac{3}{6} \cdot x^4 \cdot \dfrac{y^{-2}}{y^3} = \dfrac{1}{2} \cdot x^4 \cdot \dfrac{1}{y^5} = \dfrac{x^4}{2y^5}$

105. $\dfrac{32x^4y^{-2}}{4x^{-2}y^0} = \left(\dfrac{32}{4}\right)x^{4-(-2)}y^{-2-0}$
$= 8x^6y^{-2}$
$= \dfrac{8x^6}{y^2}$

107. a. Yes, $a^{-1}b^{-1} = \dfrac{1}{a}\cdot\dfrac{1}{b} = \dfrac{1}{ab}$.

b. No, $a^{-1} + b^{-1} = \dfrac{1}{a} + \dfrac{1}{b} \neq \dfrac{1}{a+b}$.

109. $4^2 + 4^{-2} = 16 + \dfrac{1}{4^2} = 16 + \dfrac{1}{16} = 16\dfrac{1}{16}$

111. $5^2 + 5^{-2} = 25 + \dfrac{1}{5^2} = 25 + \dfrac{1}{25} = 25\dfrac{1}{25}$

113. The missing number is –2 since
$5^{-2} = \dfrac{1}{5^2} = \dfrac{1}{25}$.

115. $3^0 - 3^{-1} = 1 - \dfrac{1}{3^1} = 1 - \dfrac{1}{3} = \dfrac{2}{3}$

117. $2\cdot 4^{-1} + 4\cdot 3^{-1} = 2\left(\dfrac{1}{4^1}\right) + 4\left(\dfrac{1}{3^1}\right)$
$= 2\left(\dfrac{1}{4}\right) + 4\left(\dfrac{1}{3}\right)$
$= \dfrac{2}{4} + \dfrac{4}{3}$
$= \dfrac{6}{12} + \dfrac{16}{12}$
$= \dfrac{22}{12}$
$= \dfrac{11}{6}$

119. The missing coefficient is 2 since $2^3 = 8$.
The missing exponent is –3 since
$(x^{-3})^3 = x^{(-3)(3)} = x^{-9} = \dfrac{1}{x^9}$.

121. $-\left(-\dfrac{2}{5}\right)^{-1} = -\left[\dfrac{(-2)^{-1}}{5^{-1}}\right]$
$= -\left[\dfrac{5^1}{(-2)^1}\right]$
$= -\left(-\dfrac{5}{2}\right)$
$= \dfrac{5}{2}$

125. $2[6-(4-5)] \div 2 - 5^2 = 2[6-(-1)] \div 2 - 25$
$= 2(7) \div 2 - 25$
$= 14 \div 2 - 25$
$= 7 - 25$
$= -18$

126. $\dfrac{-3^2\cdot 4 \div 2}{\sqrt{9} - 2^2} = \dfrac{-9\cdot 4 \div 2}{3-4}$
$= \dfrac{-36 \div 2}{-1}$
$= \dfrac{-18}{-1}$
$= 18$

127. $\dfrac{x}{2.5} = \dfrac{8}{3}3x = 8(2.5)3x = 20x = \dfrac{20}{3} \approx 6.67$
You should use 6.67 ounces.

128. Let x = the larger integer, then
$37 - x$ = the smaller integer.
$x = 3(37 - x) + 1$
$x = 111 - 3x + 1$
$4x = 112$
$x = 28$
The numbers are 28 and 37 – 28 = 9.

Exercise Set 6.3

1. A number in scientific notation is written as a number greater than or equal to 1 and less than 10 that is multiplied by some power of 10.

3. a. Answers will vary.

b. 0.00469 in scientific notation is 4.69×10^{-3}

5. You will move the decimal point 6 places to the left.

7. The exponent will be negative when the number is less than 1.

9. The exponent will be negative since $0.00734 < 1$.

11. $0.000001 = 1\times10^{-6}$

13. $42,000 = 4.2\times10^{4}$

15. $450 = 4.5\times10^{2}$

17. $0.053 = 5.3\times10^{-2}$

19. $19,000 = 1.9\times10^{4}$

21. $0.00000186 = 1.86\times10^{-6}$

23. $0.00000914 = 9.14\times10^{-6}$

25. $110,100 = 1.101\times10^{5}$

27. $0.887 = 8.87\times10^{-1}$

29. $7.4\times10^{3} = 7400$

31. $4\times10^{7} = 40,000,000$

33. $2.13\times10^{-5} = 0.0000213$

35. $6.25\times10^{5} = 625,000$

37. $9\times10^{6} = 9,000,000$

39. $5.35\times10^{2} = 535$

41. $2.991\times10^{3} = 2991$

43. $1\times10^{4} = 10,000$

45. $$\begin{aligned}(4\times10^{2})(3\times10^{5}) &= (4\times3)(10^{2}\times10^{5})\\ &= 12\times10^{7}\\ &= 120,000,000\end{aligned}$$

47. $$\begin{aligned}(2.7\times10^{-1})(9\times10^{5}) &= (2.7\times9)(10^{-1}\times10^{5})\\ &= 24.3\times10^{4}\\ &= 243,000\end{aligned}$$

49. $$\begin{aligned}&(1.3\times10^{-8})(1.74\times10^{6})\\ &= (1.3\times1.74)(10^{-8}\times10^{6})\\ &= 2.262\times10^{-2}\\ &= 0.02262\end{aligned}$$

51. $$\frac{6.4\times10^{5}}{2\times10^{3}} = \left(\frac{6.4}{2}\right)\left(\frac{10^{5}}{10^{3}}\right) = 3.2\times10^{2} = 320$$

53. $$\frac{7.5\times10^{6}}{3\times10^{3}} = \left(\frac{7.5}{3}\right)\left(\frac{10^{6}}{10^{3}}\right) = 2.5\times10^{3} = 2500$$

55. $$\frac{4\times10^{5}}{2\times10^{4}} = \left(\frac{4}{2}\right)\left(\frac{10^{5}}{10^{4}}\right) = 2\times10^{1} = 20$$

57. $$\begin{aligned}(700,000)(6,000,000) &= (7\times10^{5})(6\times10^{6})\\ &= (7\times6)(10^{5}\times10^{6})\\ &= 42\times10^{11}\\ &= 4.2\times10^{12}\end{aligned}$$

59. $$\begin{aligned}(0.003)(0.00015) &= (3\times10^{-3})(1.5\times10^{-4})\\ &= (3\times1.5)(10^{-3}\times10^{-4})\\ &= 4.5\times10^{-7}\end{aligned}$$

61. $$\begin{aligned}\frac{1,400,000}{700} &= \frac{1.4\times10^{6}}{7\times10^{2}}\\ &= \left(\frac{1.4}{7}\right)\left(\frac{10^{6}}{10^{2}}\right)\\ &= 0.2\times10^{4}\\ &= 2\times10^{3}\end{aligned}$$

63. $\frac{0.00035}{0.000002} = \frac{3.5 \times 10^{-4}}{2.0 \times 10^{-6}}$
$= \left(\frac{3.5}{2.0}\right)\left(\frac{10^{-4}}{10^{-6}}\right)$
$= 1.75 \times 10^2$

65. 8.3×10^{-4}, 3.2×10^{-1}, 4.6, 4.8×10^5

67. a. 685 million dollars is $685,000,000 or $\$6.85 \times 10^8$.

b. $\frac{6.85 \times 10^8}{5} = \frac{6.85}{5} \times 10^8 = 1.37 \times 10^8$
A single plane cost $\$1.37 \times 10^8$.

69. a. $601,000,000 + 461,000,000 + 400,000,000 + 357,000,000 + 330,000,000$
$= 6.01 \times 10^8 + 4.61 \times 10^8 + 4 \times 10^8 + 3.57 \times 10^8 + 3.3 \times 10^8$
$= (6.01 + 4.61 + 4 + 3.57 + 3.3) \times 10^8$
$= 21.49 \times 10^8$
$= 2.149 \times 10^9$
The total gross ticket sales were $\$2.149 \times 10^9$.

b. $(6.01 \times 10^8) - (3.30 \times 10^8) = (6.01 - 3.30) \times 10^8 = 2.71 \times 10^8$
Titanic grossed $\$2.71 \times 10^8$ more than *Forrest Gump.*

71. Minimum volume $= (100,000 \text{ ft}^3/\text{sec})(60 \text{ sec}/\text{min})(60 \text{ min}/\text{hr})(24 \text{ hrs})$
$= (1 \times 10^5)(6 \times 10^1)(6 \times 10^1)(2.4 \times 10^1)\text{ft}^3$
$= (1 \times 6 \times 6 \times 2.4)(10^5 \times 10^1 \times 10^1 \times 10^1)\text{ft}^3$
$= 86.4 \times 10^8 \text{ ft}^3$
$= 8,640,000,000 \text{ ft}^3$

73. a. 14,300 kg = 1.43×10^4 kg

b. 8300 kg = 8.3×10^3 kg

c. $(1.43 \times 10^4) - (8.3 \times 10^3) = 14.3 \times 10^3 - 8.3 \times 10^3$
$= (14.3 - 8.3) \times 10^3$
$= 6 \times 10^3$
The 747 can carry 6×10^3 or 6000 kilograms more than the 767.

75. a. $1.6 \times 10^7 = 16,000,000$

b. $5.0 \times 10^8 = 500,000,000$

c. 16 million transistors are on a chip now.

d. 500 million

77. a. $\frac{\text{quarter}}{\text{dime}} = \frac{5.669}{2.264} \approx 2.504$

b. $\frac{\text{quarter}}{\text{dime}} = \frac{5.559 \times 10^{-2}}{2.220 \times 10^{-2}} = \frac{5.559}{2.220} \approx 2.504$

c. Yes, the quotients are the same.

d.
$$\begin{aligned}\left(1.128 \times 10^{-1}\right) - \left(2.220 \times 10^{-2}\right) &= \left(1.128 \times 10^{-1}\right) - \left(0.2220 \times 10^{-1}\right) \\ &= (1.128 - 0.2220) \times 10^{-1} \\ &= 0.906 \times 10^{-1} \\ &= 0.0906\end{aligned}$$
The weight is 0.0906 Newtons greater.

79. Andrew: $\$1.55 \times 10^{10}$
Bob: $\$6.20 \times 10^{8}$
$$\begin{aligned}\left(1.55 \times 10^{10}\right) + \left(6.20 \times 10^{8}\right) &= \left(1.55 \times 10^{10}\right) + \left(0.062 \times 10^{10}\right) \\ &= 1.612 \times 10^{10}\end{aligned}$$
The total loss was $\$1.612 \times 10^{10}$.

81. a.
$$\begin{aligned}2 \times \left(5.92 \times 10^{9}\right) &= 11.84 \times 10^{9} \\ &= 1.184 \times 10^{10}\end{aligned}$$
The world's population in 2052 will be about 1.184×10^{10}.

b. (53 years)(365 days/year) = 19,345 days = 1.9345×10^{4} days
Increase of 5.92×10^{9} people
$\frac{5.92 \times 10^{9}}{1.9345 \times 10^{4}} \approx 3.06 \times 10^{5} = 306{,}000$
About 306,000 people are added per day.

83. The unknown exponent is 11 since
$\frac{25 \times 10^{8}}{(5 \times 10^{-3})(5 \times 10^{11})} = \frac{25 \times 10^{8}}{25 \times 10^{8}} = 1.$

86.
$$\begin{aligned}4x^2 + 3x + \frac{x}{2} &= 4 \cdot 0^2 + 3 \cdot 0 + \frac{0}{2} \\ &= 0 + 0 + 0 \\ &= 0\end{aligned}$$

87. a. If $x = \frac{3}{2}$, $-x = -\frac{3}{2}$.

b. If $5x = 0$, then $x = 0$.

88. $2x - 3(x-2) = x+2$
$$2x - 3x + 6 = x + 2$$
$$-x + 6 = x + 2$$
$$4 = 2x$$
$$2 = x$$

89. $\left(-\frac{2x^5y^7}{8x^8y^3}\right)^3 = \left(\frac{-2}{8}\cdot\frac{x^5}{x^8}\cdot\frac{y^7}{y^3}\right)^3$
$$= \left(\frac{-1}{4}\cdot\frac{1}{x^3}\cdot y^4\right)^3$$
$$= \left(-\frac{y^4}{4x^3}\right)^3$$
$$= \frac{(-1)^3 y^{4\cdot 3}}{4^3 x^{3\cdot 3}}$$
$$= -\frac{y^{12}}{64x^9}$$

Exercise Set 6.4

1. A polynomial is an expression containing the sum of a finite number of terms of the form ax^n where a is a real number and n is a whole number.

3. a. The exponent on the variable is the degree of the term.

b. The degree of the polynomial is the same as the degree of the highest degree term in the polynomial.

5. $(3x+2)-(4x-6) = 3x + 2 - 4x + 6$
$$= -x + 8$$

7. Because the exponent on the variable in a constant term is 0.

9. a. Answers will vary.

b. $4x^3 + 5x - 7$ will be rewritten as $4x^3 + 0x^2 + 5x - 7$

11. No, it contains a fractional exponent.

13. Binomial

15. Monomial

17. Binomial

19. Monomial

21. Binomial

23. Polynomial

25. Trinomial

27. $2x + 3$, first

29. $x^2 - 2x - 4$, second

31. $3x^2 + x - 8$, second

33. Already in descending order, first

35. Already in descending order, third

37. $4x^3 - 3x^2 + x - 4$, third

39. $-2x^3 + 3x^2 + 5x - 6$, third

41. $(2x+3)+(x-5) = 2x + 3 + x - 5$
$$= 2x + x + 3 - 5$$
$$= 3x - 2$$

43. $(-4x+8)+(2x+3) = -4x + 8 + 2x + 3$
$$= -4x + 2x + 8 + 3$$
$$= -2x + 11$$

45. $(x+7)+(-6x-8) = x + 7 - 6x - 8$
$$= x - 6x + 7 - 8$$
$$= -5x - 1$$

47. $\left(x^2+2.6x-3\right)+(4x+3.8)=x^2+2.6x-3+4x+3.8$
$=x^2+2.6x+4x-3+3.8$
$=x^2+6.6x+0.8$

49. $(5x-7)+\left(2x^2+3x+12\right)=5x-7+2x^2+3x+12$
$=2x^2+5x+3x-7+12$
$=2x^2+8x+5$

51. $\left(2x^2-3x+5\right)+\left(-x^2+6x-8\right)=2x^2-3x+5-x^2+6x-8$
$=2x^2-x^2+6x-3x+5-8$
$=x^2+3x-3$

53. $\left(-x^2-4x+8\right)+\left(5x-2x^2+\frac{1}{2}\right)=-x^2-4x+8+5x-2x^2+\frac{1}{2}$
$=-x^2-2x^2-4x+5x+8+\frac{1}{2}$
$=-3x^2+x+\frac{17}{2}$

55. $\left(8x^2+4\right)+\left(-2.6x^2-5x\right)=8x^2+4-2.6x^2-5x$
$=8x^2-2.6x^2-5x+4$
$=5.4x^2-5x+4$

57. $\left(-7x^3-3x^2+4\right)+\left(4x+5x^3-7\right)=-7x^3-3x^2+4+4x+5x^3-7$
$=-7x^3+5x^3-3x^2+4x+4-7$
$=-2x^3-3x^2+4x-3$

59. $\left(3x^2+2xy+8\right)+\left(-x^2-5xy-3\right)=3x^2+2xy+8-x^2-5xy-3$
$=3x^2-x^2+2xy-5xy+8-3$
$=2x^2-3xy+5$

61. $\left(2x^2y+2x-3\right)+\left(3x^2y-5x+5\right)=2x^2y+2x-3+3x^2y-5x+5$
$=2x^2y+3x^2y+2x-5x-3+5$
$=5x^2y-3x+2$

63.
$$\begin{array}{r} 3x-6 \\ \underline{4x+5} \\ 7x-1 \end{array}$$

65.
$$\begin{array}{r} 5x^2-2x+4 \\ \underline{3x+1} \\ 5x^2+\quad x+5 \end{array}$$

67.
$$\begin{array}{r} -x^2-3x+3 \\ \underline{5x^2+5x-7} \\ 4x^2+2x-4 \end{array}$$

69.
$$\begin{array}{r} 2x^3+3x^2+6x\quad -9 \\ \underline{-4x^2\qquad\quad +7} \\ 2x^3-\ x^2+6x\quad -2 \end{array}$$

71.
$$\begin{array}{r} 6x^3-4x^2+x-9 \\ \underline{-x^3-3x^2-x+7} \\ 5x^3-7x^2\qquad -2 \end{array}$$

73.
$$\begin{aligned}(3x-4)-(2x+2) &= 3x-4-2x-2 \\ &= 3x-2x-4-2 \\ &= x-6\end{aligned}$$

75.
$$\begin{aligned}(-2x-3)-(-5x-7) &= -2x-3+5x+7 \\ &= -2x+5x-3+7 \\ &= 3x+4\end{aligned}$$

77.
$$\begin{aligned}(-x+8)-(3x+5) &= -x+8-3x-5 \\ &= -x-3x+8-5 \\ &= -4x+3\end{aligned}$$

79.
$$\begin{aligned}\left(9x^2+7x-5\right)-\left(3x^2+3.5\right) &= 9x^2+7x-5-3x^2-3.5 \\ &= 9x^2-3x^2+7x-5-3.5 \\ &= 6x^2+7x-8.5\end{aligned}$$

81.
$$\begin{aligned}\left(5x^2-x-1\right)-\left(-3x^2-2x-5\right) &= 5x^2-x-1+3x^2+2x+5 \\ &= 5x^2+3x^2-x+2x-1+5 \\ &= 8x^2+x+4\end{aligned}$$

83.
$$\begin{aligned}\left(-5x^2-2x\right)-\left(2x^2-7x+9\right) &= -5x^2-2x-2x^2+7x-9 \\ &= -5x^2-2x^2-2x+7x-9 \\ &= -7x^2+5x-9\end{aligned}$$

85.
$$\begin{aligned}\left(8x^3-2x^2-4x+5\right)-\left(5x^2+8\right) &= 8x^3-2x^2-4x+5-5x^2-8 \\ &= 8x^3-2x^2-5x^2-4x-8+5 \\ &= 8x^3-7x^2-4x-3\end{aligned}$$

87.
$$\begin{aligned}\left(2x^3-4x^2+5x-7\right)-\left(3x+\frac{3}{5}x^2-5\right) &= 2x^3-4x^2+5x-7-3x-\frac{3}{5}x^2+5 \\ &= 2x^3-4x^2-\frac{3}{5}x^2+5x-3x-7+5 \\ &= 2x^3-\frac{23}{5}x^2+2x-2\end{aligned}$$

89. $(8x+2)-(5x+4) = 8x+2-5x-4$
$= 8x-5x+2-4$
$= 3x-2$

91. $(2x^2-4x+8)-(5x-6) = 2x^2-4x+8-5x+6$
$= 2x^2-4x-5x+8+6$
$= 2x^2-9x+14$

93. $(3x^3+5x^2+9x-7)-(4x^3-6x^2) = 3x^3+5x^2+9x-7-4x^3+6x^2$
$= 3x^3-4x^3+5x^2+6x^2+9x-7$
$= -x^3+11x^2+9x-7$

95.
$$\begin{array}{r} 6x+5 \\ \underline{-(3x-3)} \end{array} \quad \text{or} \quad \begin{array}{r} 6x+5 \\ \underline{-\ 3x+3} \\ 3x+8 \end{array}$$

97.
$$\begin{array}{r} -5x+3 \\ \underline{-(-9x-4)} \end{array} \quad \text{or} \quad \begin{array}{r} -5x+3 \\ \underline{9x+4} \\ 4x+7 \end{array}$$

99.
$$\begin{array}{r} 7x^2-3x-4 \\ \underline{-(6x^2 \qquad -1)} \end{array} \quad \text{or} \quad \begin{array}{r} 7x^2-3x-4 \\ \underline{-\ 6x^2+0x+1} \\ x^2-3x-3 \end{array}$$

101.
$$\begin{array}{r} x-6 \\ \underline{-(-4x^2+6x)} \end{array} \quad \text{or} \quad \begin{array}{r} x-6 \\ \underline{4x^2-6x} \\ 4x^2-5x-6 \end{array}$$

103.
$$\begin{array}{r} 4x^3-6x^2+7x-9 \\ \underline{-(x^2+6x-7)} \end{array} \quad \text{or} \quad \begin{array}{r} 4x^3-6x^2+7x-9 \\ \underline{-x^2-6x+7} \\ 4x^3-7x^2+x-2 \end{array}$$

105. Answers will vary.

107. Answers will vary.

109. Sometimes

111. Sometimes

113. Answers will vary; one example is:
x^5+x^4+x

115. No, all three terms must have degree 5 or 0.

117. $a^2+2ab+b^2$

119. $4x^2+3xy$

121. $\left(3x^2-6x+3\right)-\left(2x^2-x-6\right)-\left(x^2+7x-9\right)=3x^2-6x+3-2x^2+x+6-x^2-7x+9$
$$=\left(3x^2-2x^2-x^2\right)+(-6x+x-7x)+(3+6+9)$$
$$=-12x+18$$

123. $4\left(x^2+2x-3\right)-6\left(2-4x-x^2\right)-2x(x+2)=4x^2+8x-12-12+24x+6x^2-2x^2-4x$
$$=\left(4x^2+6x^2-2x^2\right)+(8x+24x-4x)+(-12-12)$$
$$=8x^2+28x-24$$

125. $|-4|<|-6|$ since $|-4|=4$ and $|-6|=6$.

126. False

127. True

128. False

129. False

130. $\left(\dfrac{3x^4y^5}{6x^7y^4}\right)^3=\left(\dfrac{3}{6}\cdot\dfrac{x^4}{x^7}\cdot\dfrac{y^5}{y^4}\right)^3=\left(\dfrac{1}{2}\cdot x^{4-7}\cdot y^{5-4}\right)^3$
$$=\left(\frac{1}{2}\cdot x^{-3}\cdot y\right)^3=\left(\frac{y}{2x^3}\right)^3$$
$$=\frac{y^3}{2^3x^{3\cdot 3}}$$
$$=\frac{y^3}{8x^9}$$

Exercise Set 6.5

1. The distributive property is used when multiplying a monomial by a polynomial.

3. First, Outer, Inner, Last

5. Yes, FOIL is simply a way to remember the procedure.

7. $(a+b)^2=a^2+2ab+b^2$
$(a-b)^2=a^2-2ab+b^2$

9. No, $(x+5)^2 = x^2 + 10x + 25$

11. Answers will vary.

13. Answers will vary.

15. $x^2 \cdot 3xy = 3x^{2+1}y = 3x^3y$

17. $5x^3y^5(4x^2y) = (5 \cdot 4)x^{3+2}y^{5+1} = 20x^5y^6$

19. $4x^4y^6(-7x^2y^9) = -4 \cdot 7x^{4+2}y^{6+9}$
$= -28x^6y^{15}$

21. $9xy^6 \cdot 6x^5y^8 = 9 \cdot 6x^{1+5}y^{6+8}$
$= 54x^6y^{14}$

23. $(6x^2y)\left(\frac{1}{2}x^4\right) = 6 \cdot \frac{1}{2}x^{2+4}y$
$= 3x^6y$

25. $(3.3x^4)(1.8x^4y^3) = (3.3 \cdot 1.8)x^{4+4}y^3$
$= 5.94x^8y^3$

27. $3(x+4) = 3 \cdot x + 3(4) = 3x + 12$

29. $-3x(2x-2) = -3x(2x) - 3x(-2)$
$= -6x^2 + 6x$

31. $-2(8y+5) = (-2)8y + (-2)(5)$
$= -16y - 10$

33. $-2x(x^2 - 2x + 5) = (-2x)(x^2) + (-2x)(-2x) + (-2x)(5)$
$= -2x^3 + 4x^2 - 10x$

35. $5x(-4x^2 + 6x - 4) = 5x(-4x^2) + 5x(6x) + 5x(-4)$
$= -20x^3 + 30x^2 - 20x$

37. $0.5x^2(x^3 - 6x^2 - 1) = 0.5x^2(x^3) + 0.5x^2(-6x^2) + 0.5x^2(-1)$
$= 0.5x^5 - 3x^4 - 0.5x^2$

39. $0.3x(2xy + 5x - 6y) = (0.3x)(2xy) + (0.3x)(5x) + (0.3x)(-6y)$
$= 0.6x^2y + 1.5x^2 - 1.8xy$

41. $(x - y - 3)y = x \cdot y + (-y)y + (-3)y$
$= xy - y^2 - 3y$

43. $(x+3)(x+4) = x \cdot x + x \cdot 4 + 3 \cdot x + 3 \cdot 4$
$= x^2 + 4x + 3x + 12$
$= x^2 + 7x + 12$

45. $(2x+5)(3x-6) = (2x)(3x) + 2x(-6) + 5 \cdot 3x + 5(-6)$
$= 6x^2 - 12x + 15x - 30$
$= 6x^2 + 3x - 30$

47. $(2x-4)(2x+4) = (2x)(2x)+(2x)(4)+(-4)(2x)+(-4)(4)$
$= 4x^2+8x-8x-16$
$= 4x^2-16$

49. $(5-3x)(6+2x) = 5\cdot 6+5(2x)+(-3x)(6)+(-3x)(2x)$
$= 30+10x-18x-6x^2$
$= 30-8x-6x^2$
$= -6x^2-8x+30$

51. $(6x-1)(-2x+5) = 6x(-2x)+(6x)5+(-1)(-2x)+(-1)5$
$= -12x^2+30x+2x-5$
$= -12x^2+32x-5$

53. $(x-2)(4x-2) = 4x\cdot x-2\cdot x-2\cdot 4x+(-2)(-2)$
$= 4x^2-2x-8x+4$
$= 4x^2-10x+4$

55. $(3x-6)(4x-2) = (3x)(4x)+(3x)(-2)+(-6)(4x)+(-6)(-2)$
$= 12x^2-6x-24x+12$
$= 12x^2-30x+12$

57. $(x-1)(x+1) = x\cdot x+x\cdot 1+(-1)\cdot x+(-1)\cdot 1$
$= x^2+x-x-1$
$= x^2-1$

59. $(2x-3)(2x-3) = 2x\cdot 2x-3\cdot 2x-3\cdot 2x+(-3)(-3)$
$= 4x^2-6x-6x+9$
$= 4x^2-12x+9$

61. $(4x-4)(7-x) = 4x\cdot 7-4x\cdot x-28+(-4)(-x)$
$= 28x-4x^2-28+4x$
$= -4x^2+32x-28$

63. $(2x+3)(4-2x) = (2x)4+(2x)(-2x)+3\cdot 4+3(-2x)$
$= 8x-4x^2+12-6x$
$= -4x^2+2x+12$

65. $(x+y)(x-y) = x\cdot x+x(-y)+y\cdot x+y\cdot y$
$= x^2-xy+xy-y^2$
$= x^2-y^2$

67. $(2x-3y)(3x+2y)=(2x)(3x)+(2x)(2y)+(-3y)(3x)+(-3y)(2y)$
$=6x^2+4xy-9xy-6y^2$
$=6x^2-5xy-6y^2$

69. $(3x+y)(2+2x)=3x\cdot 2+2x\cdot 3x+2\cdot y+2x\cdot y$
$=6x+6x^2+2y+2xy$
$=6x^2+6x+2xy+2y$

71. $(x+0.6)(x+0.3)=x\cdot x+x(0.3)+(0.6)x+(0.6)(0.3)$
$=x^2+0.3x+0.6x+0.18$
$=x^2+0.9x+0.18$

73. $(2y-4)\left(\frac{1}{2}x-1\right)=2y\cdot\left(\frac{1}{2}x\right)-1\cdot 2y-4\left(\frac{1}{2}x\right)+(-4)(-1)$
$=xy-2y-2x+4$
$=xy-2x-2y+4$

75. $(x+5)(x-5)=x^2-5^2$
$=x^2-25$

77. $(3x-3)(3x+3)=(3x)^2-3^2=9x-9$

79. $(x+y)^2=(x)^2+2(x)(y)+(y)^2$
$=x^2+2xy+y^2$

81. $(x-0.2)^2=(x)^2-2(x)(0.2)+(0.2)^2$
$=x^2-0.4x+0.04$

83. $(4x+5)(4x+5)=(4x)^2+2(4x)(5)+(5)^2=16x^2+40x+25$

85. $(0.4x+y)^2=(0.4x)^2+2(0.4x)(y)+(y)^2$
$=0.16x^2+0.8xy+y^2$

87. $(x+3)\left(2x^2+4x-1\right)=x\left(2x^2+4x-1\right)+3\left(2x^2+4x-1\right)$
$=2x^3+4x^2-x+6x^2+12x-3$
$=2x^3+10x^2+11x-3$

89. $(3x+2)\left(4x^2-x+5\right)=(3x)\left(4x^2-x+5\right)+2\left(4x^2-x+5\right)$
$=12x^3-3x^2+15x+8x^2-2x+10$
$=12x^3+5x^2+13x+10$

91. $\left(-2x^2-4x+1\right)(7x-3)=-2x^2(7x-3)-4x(7x-3)+1(7x-3)$
$=-14x^3+6x^2-28x^2+12x+7x-3$
$=-14x^3-22x^2+19x-3$

93. $(-2x+6)\left(2x^2+3x-4\right)=-2x\left(2x^2+3x-4\right)+6\left(2x^2+3x-4\right)$
$=-4x^3-6x^2+8x+12x^2+18x-24$
$=-4x^3+6x^2+26x-24$

95. $\left(3x^2-2x+4\right)\left(2x^2+3x+1\right)=3x^2\left(2x^2+3x+1\right)-2x\left(2x^2+3x+1\right)+4\left(2x^2+3x+1\right)$
$=6x^4+9x^3+3x^2-4x^3-6x^2-2x+8x^2+12x+4$
$=6x^4+5x^3+5x^2+10x+4$

97. $\left(x^2-x+3\right)\left(x^2-2x\right)=x^2\left(x^2-2x\right)-x\left(x^2-2x\right)+3\left(x^2-2x\right)$
$=x^4-2x^3-x^3+2x^2+3x^2-6x$
$=x^4-3x^3+5x^2-6x$

99. $(a+b)\left(a^2-ab+b^2\right)=a\left(a^2-ab+b^2\right)+b\left(a^2-ab+b^2\right)$
$=a^3-a^2b+ab^2+a^2b-ab^2+b^3$
$=a^3+b^3$

101. Yes, it will always be a monomial.

103. No, it could have 2 or 4 terms.

105. The missing exponents are 6, 3, and 1 since
$3x^2(2x^6-5x^3+3x^1)=3x^2(2x^6)-3x^2(5x^3)+3x^2(3x^1)$
$=6x^8-15x^5+9x^3$

107. a. $A=(x+2)(2x+1)$
$=x(2x)+x\cdot 1+2\cdot 2x+2\cdot 1$
$=2x^2+x+4x+2$
$=2x^2+5x+2$

b. If $x=4$, $A=2\cdot 4^2+5\cdot 4+2=54$.
The area is 54 square feet.

c. For the rectangle to be a square, all sides must have the same length. Thus, $x+2=2x+1$.

$$\begin{aligned} x+2&=2x+1 \\ 2&=x+1 \\ 1&=x \end{aligned}$$

The rectangle is a square when $x = 1$ foot.

109. a. $a+b$

b. $a+b$

c. Yes

d. $(a+b)^2$

e. Area of small square = $a \cdot a$
Area of larger square = $b \cdot b$
Area of each rectangle = $a \cdot b$

$$\begin{aligned} (a+b)^2 &= a \cdot a + b \cdot b + 2(a \cdot b) \\ &= a^2+2ab+b^2 \end{aligned}$$

111. $(2x^3-6x^2+5x-3)(3x^3-6x+4)$

$$\begin{aligned} &= 2x^3(3x^3-6x+4)-6x^2(3x^3-6x+4)+5x(3x^3-6x+4)-3(3x^3-6x+4) \\ &= 2x^3(3x^3)+2x^3(-6x)+2x^3(4)-6x^2(3x^3)-6x^2(-6x)-6x^2(4)+5x(3x^3)+5x(-6x) \\ &\quad +5x(4)-3(3x^3)-3(-6x)-3(4) \\ &= 6x^6-12x^4+8x^3-18x^5+36x^3-24x^2+15x^4-30x^2+20x-9x^3+18x-12 \\ &= 6x^6-18x^5+3x^4+35x^3-54x^2+38x-12 \end{aligned}$$

113. Let x equal the maximum distance. Then

$$\begin{aligned} 2+1.5(x-1)&=20 \\ 2+1.5x-1.5&=20 \\ 1.5x+0.5&=20 \\ 1.5x&=19.5 \\ x&=\frac{19.5}{1.5}=13 \end{aligned}$$

Ingrid can go a maximum of 13 miles.

114. $\left(\frac{4x^8y^5}{8x^8y^6}\right)^4 = \left(\frac{4}{8}\cdot\frac{x^8}{x^8}\cdot\frac{y^5}{y^6}\right)^4$
$= \left(\frac{1}{2}\cdot\frac{1}{y}\right)^4$
$= \left(\frac{1}{2y}\right)^4$
$= \frac{1}{2^4y^4}$
$= \frac{1}{16y^4}$

115. a. $-6^3 = -(6^3) = -216$

b. $6^{-3} = \frac{1}{6^3} = \frac{1}{216}$

116. $(-x^2 - 6x + 5) - (5x^2 - 4x - 3) = -x^2 - 6x + 5 - 5x^2 + 4x + 3$
$= -x^2 - 5x^2 - 6x + 4x + 5 + 3$
$= -6x^2 - 2x + 8$

Exercise Set 6.6

1. To divide a polynomial by a monomial, divide each term in the polynomial by the monomial.

3. $\frac{y+5}{y} = \frac{y}{y} + \frac{5}{y} = 1 + \frac{5}{y}$

5. Terms should be listed in descending order.

7. $\frac{x^3+5x-1}{x+2} = \frac{x^3+0x^2+5x-1}{x+2}$

9. $(x+5)(x-3) - 2 = x^2 + 2x - 15 - 2$
$= x^2 + 2x - 17$

11. $\frac{x^2-2x-15}{x+3} = x - 5$ or $\frac{x^2-2x-15}{x-5} = x + 3$

13. $\frac{2x^2+5x+3}{2x+3} = x + 1$ or $\frac{2x^2+5x+3}{x+1} = 2x + 3$

15. $\frac{4x^2-9}{2x+3} = 2x - 3$ or $\frac{4x^2-9}{2x-3} = 2x + 3$

17. $\frac{2x+4}{2} = \frac{2x}{2} + \frac{4}{2} = x + 2$

19. $\frac{2x+6}{2} = \frac{2x}{2} + \frac{6}{2} = x + 3$

21. $\frac{3x+8}{2} = \frac{3x}{2} + \frac{8}{2} = \frac{3}{2}x + 4$

23. $\frac{-6x+4}{2} = \frac{-6x}{2} + \frac{4}{2} = -3x + 2$

25. $\frac{-9x-3}{-3} = 3x + 1$

27. $\frac{2x+16}{4} = \frac{2x}{4} + \frac{16}{4}$
$= \frac{x}{2} + 4$

29. $\dfrac{4-12x}{-3} = \dfrac{(-1)(4-12x)}{(-1)(-3)}$
$= \dfrac{-4+12x}{3}$
$= -\dfrac{4}{3} + \dfrac{12x}{3}$
$= -\dfrac{4}{3} + 4x$

31. $(3x^2+6x-9) \div 3x^2 = \dfrac{3x^2+6x-9}{3x^2}$
$= \dfrac{3x^2}{3x^2} + \dfrac{6x}{3x^2} + \dfrac{-9}{3x^2}$
$= 1 + \dfrac{2}{x} - \dfrac{3}{x^2}$

33. $\dfrac{-4x^5+6x+8}{2x^2} = \dfrac{-4x^5}{2x^2} + \dfrac{6x}{2x^2} + \dfrac{8}{2x^2}$
$= -2x^3 + \dfrac{3}{x} + \dfrac{4}{x^2}$

35. $(x^6+4x^4-3) \div x^3 = \dfrac{x^6+4x^4-3}{x^3}$
$= \dfrac{x^6}{x^3} + \dfrac{4x^4}{x^3} + \dfrac{-3}{x^3}$
$= x^3 + 4x - \dfrac{3}{x^3}$

37. $\dfrac{6x^5-4x^4+12x^3-5x^2}{2x^3} = \dfrac{6x^5}{2x^3} - \dfrac{4x^4}{2x^3} + \dfrac{12x^3}{2x^3} - \dfrac{5x^2}{2x^3}$
$= 3x^2 - 2x + 6 - \dfrac{5}{2x}$

39. $\dfrac{4x^3+6x^2-8}{-4} = \dfrac{(-1)\left(4x^3+6x^2-8\right)}{(-1)(-4x)}$
$= \dfrac{-4x^3-6x^2+8}{4x}$
$= \dfrac{-4x^3}{4x} - \dfrac{6x^2}{4x} + \dfrac{8}{4x}$
$= -x^2 - \dfrac{3}{2}x + \dfrac{2}{x}$

41. $$\frac{9x^6+3x^4-10x^2-9}{3x^2}=\frac{9x^6}{3x^2}+\frac{3x^4}{3x^2}-\frac{10x^2}{3x^2}-\frac{9}{3x^2}$$
$$=3x^4+x^2-\frac{10}{3}-\frac{3}{x^2}$$

43. $$\begin{array}{r} x+3 \\ x+1\overline{)x^2+4x+3} \\ \underline{x^2+x} \\ 3x+3 \\ \underline{3x+3} \\ 0 \end{array}$$

$$\frac{x^2+4x+3}{x+1}=x+3$$

45. $$\begin{array}{r} 2x+3 \\ x+5\overline{)2x^2+13x+15} \\ \underline{2x^2+10x} \\ 3x+15 \\ \underline{3x+15} \\ 0 \end{array}$$

$$\frac{2x^2+13x+15}{x+5}=2x+3$$

47. $$\begin{array}{r} 2x+4 \\ 3x+2\overline{)6x^2+16x+8} \\ \underline{6x^2+4x} \\ 12x+8 \\ \underline{12x+8} \\ 0 \end{array}$$

$$\frac{6x^2+16x+8}{3x+2}=2x+4$$

49. $$\frac{x^2-9}{-3+x}=\frac{x^2+0x-9}{x-3}$$

$$\begin{array}{r} x+3 \\ x-3\overline{)x^2+0x-9} \\ \underline{x^2-3x} \\ 3x-9 \\ \underline{3x-9} \\ 0 \end{array}$$

$$\frac{x^2-9}{-3+x}=x+3$$

51. $$\begin{array}{r} x+5 \\ 2x-3\overline{)2x^2+7x-18} \\ \underline{2x^2-3x} \\ 10x-18 \\ \underline{10x-15} \\ -3 \end{array}$$

$$(2x^2+7x-18)\div(2x-3)=x+5-\frac{3}{2x-3}$$

53. $$(4x^2-9)\div(2x-3)$$
$$=(4x^2+0x-9)\div(2x-3)$$

$$\begin{array}{r} 2x+3 \\ 2x-3\overline{)4x^2+0x-9} \\ \underline{4x^2-6x} \\ 6x-9 \\ \underline{6x-9} \\ 0 \end{array}$$

$$\frac{4x^2-9}{2x-3}=2x+3$$

55. $\dfrac{6x+8x^2-25}{4x+9}=\dfrac{8x^2+6x-25}{4x+9}$

$$\begin{array}{r|l} & 2x-3 \\ \hline 4x+9 & 8x^2+6x-25 \\ & \underline{8x^2+18x} \\ & \quad -12x-25 \\ & \quad \underline{-12x-27} \\ & \qquad\quad 2 \end{array}$$

$\dfrac{6x+8x^2-25}{4x+9}=2x-3+\dfrac{2}{4x+9}$

57. $\dfrac{6x+8x^2-12}{2x+3}=\dfrac{8x^2+6x-12}{2x+3}$

$$\begin{array}{r|l} & 4x-3 \\ \hline 2x+3 & 8x^2+6x-12 \\ & \underline{8x^2+12x} \\ & \quad -6x-12 \\ & \quad \underline{-6x-9} \\ & \qquad\; -3 \end{array}$$

$\dfrac{6x+8x^2-12}{2x+3}=4x-3-\dfrac{3}{2x+3}$

59.
$$\begin{array}{r|l} & 3x^2-5 \\ \hline x+6 & 3x^3+18x^2-5x-30 \\ & \underline{3x^3+18x^2} \\ & \qquad -5x-30 \\ & \qquad \underline{-5x-30} \\ & \qquad\qquad 0 \end{array}$$

$\dfrac{3x^3+18x^2-5x-30}{x+6}=3x^2-5$

61. $\dfrac{2x^3-4x^2+12}{x-2}=\dfrac{2x^3-4x^2+0x+12}{x-2}$

$$\begin{array}{r|l} & 2x \\ \hline x-2 & 2x^3-4x^2+0x+12 \\ & \underline{2x^3-4x^2} \\ & \qquad\qquad 12 \end{array}$$

$\dfrac{2x^3-4x^2+12}{x-2}=2x^2+\dfrac{12}{x-2}$

63. $\left(x^3-8\right)\div(x-3)$

$=(x^3+0x^2+0x-8)\div(x-3)$

$$\begin{array}{r|l} & x^2+3x+9 \\ \hline x-3 & x^3+0x^2+0x-8 \\ & \underline{x^3-3x^2} \\ & \quad 3x^2+0x \\ & \quad \underline{3x^2-9x} \\ & \qquad 9x-8 \\ & \qquad \underline{9x-27} \\ & \qquad\quad 19 \end{array}$$

$(x^3-8)\div(x-3)=x^2+3x+9+\dfrac{19}{x-3}$

65. $\dfrac{x^3-27}{x-3}=\dfrac{x^3+0x^2+0x-27}{x-3}$

$$\begin{array}{r|l} & x^2+3x+9 \\ \hline x-3 & x^3+0x^2+0x-27 \\ & \underline{x^3-3x^2} \\ & \quad 3x^2+0x \\ & \quad \underline{3x^2-9x} \\ & \qquad 9x-27 \\ & \qquad \underline{9x-27} \\ & \qquad\quad 0 \end{array}$$

$\dfrac{x^3-27}{x-3}=x^2+3x+9$

67. $\dfrac{4x^3-5x}{2x-1}=\dfrac{4x^3+0x^2-5x+0}{2x-1}$

$$\begin{array}{r}
2x^2+x-2\\
2x-1\overline{\big)\,4x^3+0x^2-5x+0}\\
\underline{4x^3-2x^2}\\
2x^2-5x\\
\underline{2x^2-x}\\
-4x+0\\
\underline{4x+2}\\
-2
\end{array}$$

$$\frac{4x^3-5x}{2x-1}=2x^2+x-2-\frac{2}{2x-1}$$

69.

$$\begin{array}{r}
-x^2-7x-5\\
x-1\overline{\big)\,-x^3-6x^2+2x-3}\\
\underline{-x^3+x^2}\\
-7x^2+2x\\
\underline{-7x^2+7x}\\
-5x-3\\
\underline{-5x+5}\\
-8
\end{array}$$

$$\frac{-x^3-6x^2+2x-3}{x-1}=-x^2-7x-5-\frac{8}{x-1}$$

71. No, $\dfrac{2x+1}{x^2}=\dfrac{2x}{x^2}+\dfrac{1}{x^2}=\dfrac{2}{x}+\dfrac{1}{x^2}$

73. $(x+4)(2x+3)+4=2x^2+3x+8x+12+4$
$=2x^2+11x+16$

75. First Degree

77. It has to be $4x$ since that is what must be multiplied with $4x^3$ in the quotient to get $16x^4$ in the dividend.

79. When dividing by $2x^2$, each exponent will decrease by two. So, the shaded areas must be 5, 3, 2, 1, respectively.

81. $\dfrac{4x^3-4x+6}{2x+3}=\dfrac{4x^3+0x^2-4x+6}{2x+3}$

$$\begin{array}{r}
2x^2-3x+\frac{5}{2}\\
2x+3\overline{\big)\,4x^3+0x^2-4x+6}\\
\underline{4x^3+6x^2}\\
-6x^2-4x\\
\underline{-6x^2-9x}\\
5x+6\\
\underline{5x+\frac{15}{2}}\\
-\frac{3}{2}
\end{array}$$

$$\frac{4x^3-4x+6}{2x+3}=2x^2-3x+\frac{5}{2}-\frac{3}{2(2x+3)}$$

83.

$$\begin{array}{r}
-3x+3\\
-x-3\overline{\big)\,3x^2+6x-10}\\
\underline{3x^2+9x}\\
-3x-10\\
\underline{-3x-9}\\
-1
\end{array}$$

$$\frac{3x^2+6x-10}{-x-3}=-3x+3+\frac{1}{x+3}$$

86. a. 2 is a natural number.

b. 2 and 0 are whole numbers.

c. $2, -5, 0, \dfrac{2}{5}, -6.3$, and $-\dfrac{23}{34}$ are rational numbers.

d. $\sqrt{7}$ and $\sqrt{3}$ are irrational numbers.

e. All of the numbers are real numbers.

87. a. $\dfrac{0}{1}=0$

b. $\dfrac{1}{0}$ is undefined

88. Evaluate expressions in parentheses first, then exponents, followed by multiplications and divisions from left to right, and finally additions and subtractions from left to right.

89. $2(x+3)+2x = x+4$
$2x+6+2x = x+4$
$4x+6 = x+4$
$4x = x-2$
$3x = -2$
$x = -\frac{2}{3}$

90.

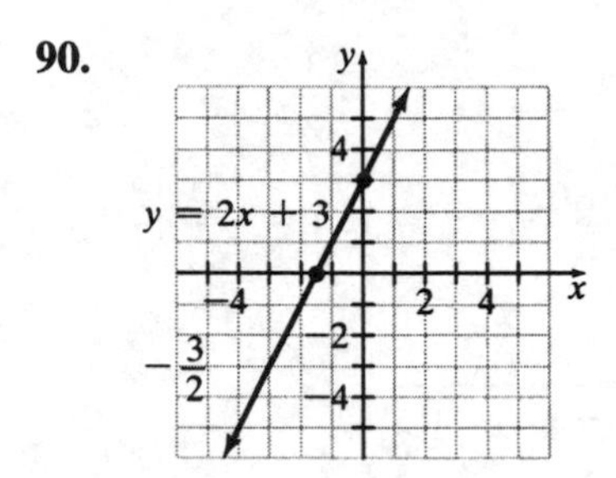

91.

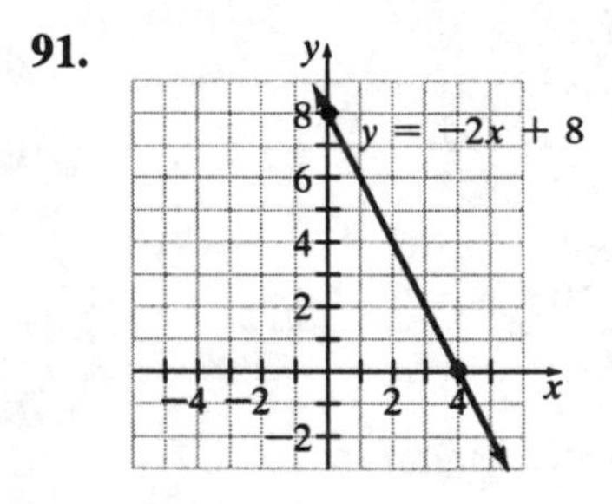

Review Exercises

1. $x^4 \cdot x^2 = x^{4+2} = x^6$

2. $x^2 \cdot x^4 = x^{2+4} = x^6$

3. $3^2 \cdot 3^3 = 3^{2+3} = 3^5 = 243$

4. $2^4 \cdot 2 = 2^{4+1} = 2^5 = 32$

5. $\frac{x^4}{x} = x^{4-1} = x^3$

6. $\frac{x^6}{x^6} = x^{6-6} = x^0 = 1$

7. $\frac{5^5}{5^3} = 5^{5-3} = 5^2 = 25$

8. $\frac{4^5}{4^3} = 4^{5-3} = 4^2 = 16$

9. $\frac{x^6}{x^8} = \frac{1}{x^{8-6}} = \frac{1}{x^2}$

10. $\frac{y^4}{y} = y^{4-1} = y^3$

11. $x^0 = 1$

12. $4x^0 = 4 \cdot 1 = 4$

13. $(3x)^0 = 1$

14. $4^0 = 1$

15. $(5x)^2 = 5^2x^2 = 25x^2$

16. $(3x)^3 = 3^3x^3 = 27x^3$

17. $(-4y)^2 = (-4)^2y^2 = 16y^2$

18. $(-3x)^3 = (-3)^3x^3 = -27x^3$

19. $\left(2x^2\right)^4 = 2^4x^{2\cdot4} = 16x^8$

20. $\left(-x^4\right)^3 = (-1)^3x^{4\cdot3} = -x^{12}$

21. $\left(-x^3\right)^4 = (-1)^4x^{3\cdot4} = x^{12}$

22. $\left(\frac{2x^3}{y}\right)^2 = \frac{2^2x^{3\cdot2}}{y^2} = \frac{4x^6}{y^2}$

23. $\left(\frac{8y^2}{2x}\right)^2 = \left(\frac{4y^2}{x}\right)^2$
$= \frac{4^2 y^{2\cdot 2}}{x^2}$
$= \frac{16y^4}{x^2}$

24. $6x^2 \cdot 4x^3 = 6 \cdot 4x^{2+3}$
$= 24x^5$

25. $\frac{16x^2y}{4xy^2} = \frac{16}{4} \cdot \frac{x^2}{x} \cdot \frac{y}{y^2}$
$= 4x\frac{1}{y}$
$= \frac{4x}{y}$

26. $2x(3xy^3)^2 = 2x(3^2x^2y^{3\cdot 2})$
$= 2x(9x^2y^6)$
$= 2 \cdot 9x^{1+2}y^6$
$= 18x^3y^6$

27. $\left(\frac{9x^2y}{3xy}\right)^2 = \left(\frac{9}{3} \cdot \frac{x^2}{x} \cdot \frac{y}{y}\right)^2$
$= (3x)^2$
$= 3^2x^2$
$= 9x^2$

28. $(2x^2y)^3(3xy^4) = (2^3x^{2\cdot 3}y^3)(3xy^4)$
$= (8x^6y^3)(3xy^4)$
$= 8 \cdot 3x^{6+1}y^{3+4}$
$= 24x^7y^7$

29. $4x^2y^3\left(2x^3y^4\right)^2 = 4x^2y^3\left(2^2x^{3\cdot 2}y^{4\cdot 2}\right)$
$= 4x^2y^3\left(4x^6y^8\right)$
$= 4 \cdot 4x^{2+6}y^{3+8}$
$= 16x^8y^{11}$

30. $4y^2\left(2x^2y^3\right)^2 = 4y^2\left(2^2x^{2\cdot 2}y^{3\cdot 2}\right)$
$= 4y^2\left(4x^4y^6\right)$
$= 4 \cdot 4x^4y^{2+6}$
$= 16x^4y^8$

31. $\left(\frac{8x^4y^3}{2xy^5}\right)^2 = \left(\frac{8}{2} \cdot \frac{x^4}{x} \cdot \frac{y^3}{y^5}\right)^2$
$= \left(4x^3 \cdot \frac{1}{y^2}\right)^2$
$= \left(\frac{4x^3}{y^2}\right)^2$
$= \frac{4^2x^{3\cdot 2}}{y^{2\cdot 2}}$
$= \frac{16x^6}{y^4}$

32. $\left(\frac{21x^4y^3}{7y^2}\right)^3 = \left(\frac{21}{7} \cdot x^4 \cdot \frac{y^3}{y^2}\right)^3 = \left(3x^4y\right)^3$
$= 3^3x^{4\cdot 3}y^3$
$= 27x^{12}y^3$

33. $x^{-3} = \frac{1}{x^3}$

34. $3^{-3} = \frac{1}{3^3} = \frac{1}{27}$

35. $5^{-2} = \frac{1}{5^2} = \frac{1}{25}$

36. $\frac{1}{x^{-3}} = x^3$

37. $\frac{1}{x^{-7}} = x^7$

38. $\frac{1}{3^{-2}} = 3^2 = 9$

39. $y^5 \cdot y^{-8} = y^{5-8} = y^{-3} = \frac{1}{y^3}$

40. $x^{-2} \cdot x^{-3} = x^{-2-3} = x^{-5} = \frac{1}{x^5}$

41. $x^4 \cdot x^{-7} = x^{4-7} = x^{-3} = \frac{1}{x^3}$

42. $x^{-2} \cdot x^{-2} = x^{-2+(-2)} = x^{-4} = \frac{1}{x^4}$

43. $\frac{x^2}{x^{-3}} = x^{2-(-3)} = x^5$

44. $\frac{x^5}{x^{-2}} = x^{5-(-2)} = x^7$

45. $\frac{x^{-3}}{x^3} = \frac{1}{x^{3+3}} = \frac{1}{x^6}$

46. $$\begin{aligned}(3x^4)^{-2} &= 3^{-2}x^{4(-2)}\\ &= 3^{-2}x^{-8}\\ &= \frac{1}{3^2x^8}\\ &= \frac{1}{9x^8}\end{aligned}$$

47. $$\begin{aligned}(4x^{-3}y)^{-3} &= 4^{-3}x^{(-3)(-3)}y^{-3}\\ &= 4^{-3}x^9y^{-3}\\ &= \frac{x^9}{4^3y^3}\\ &= \frac{x^9}{64y^3}\end{aligned}$$

48. $$\begin{aligned}(-2x^{-2}y)^2 &= (-2)^2x^{-2\cdot 2}y^{1\cdot 2}\\ &= 4x^{-4}y^2\\ &= \frac{4y^2}{x^4}\end{aligned}$$

49. $5y^{-2} \cdot 2y^4 = 5 \cdot 2y^{-2+4} = 10y^2$

50. $$\begin{aligned}(3x^{-2}y)^3 &= 3^3x^{(-2)3}y^3\\ &= 27x^{-6}y^3\\ &= \frac{27y^3}{x^6}\end{aligned}$$

51. $$\begin{aligned}(4x^{-2}y^3)^{-2} &= 4^{-2}x^{(-2)(-2)}y^{3(-2)}\\ &= 4^{-2}x^4y^{-6}\\ &= \frac{x^4}{4^2y^6}\\ &= \frac{x^4}{16y^6}\end{aligned}$$

52. $2x(3x^{-2}) = 2 \cdot 3x^{1-2} = 6x^{-1} = \frac{6}{x}$

53. $(5x^{-2}y)(2x^4y) = 5 \cdot 2x^{-2+4}y^{1+1} = 10x^2y^2$

54. $$\begin{aligned}4x^5(6x^{-7}y^2) &= 4 \cdot 6x^{5-7}y^2\\ &= 24x^{-2}y^2\\ &= \frac{24y^2}{x^2}\end{aligned}$$

55. $$\begin{aligned}4y^{-2}(3x^2y) &= 4 \cdot 3x^2y^{-2+1}\\ &= 12x^2y^{-1}\\ &= \frac{12x^2}{y}\end{aligned}$$

56. $\frac{6xy^4}{2xy^{-1}} = \frac{6}{2} \cdot \frac{x}{x} \cdot \frac{y^4}{y^{-1}} = 3y^5$

57. $$\begin{aligned}\frac{9x^{-2}y^3}{3xy^2} &= \frac{9}{3} \cdot \frac{x^{-2}}{x} \cdot \frac{y^3}{y^2}\\ &= 3 \cdot \frac{1}{x^3} \cdot y\\ &= \frac{3y}{x^3}\end{aligned}$$

58. $\frac{49x^2y^{-3}}{7x^{-3}y} = \frac{49}{7} \cdot \frac{x^2}{x^{-3}} \cdot \frac{y^{-3}}{y}$
$= \left(\frac{49}{7}\right)x^{2-(-3)}y^{-3-1}$
$= 7x^5y^{-4}$
$= \frac{7x^5}{y^4}$

59. $\frac{36x^4y^7}{9x^5y^{-3}} = \frac{36}{9} \cdot \frac{x^4}{x^5} \cdot \frac{y^7}{y^{-3}}$
$= 4 \cdot \frac{1}{x} \cdot y^{10}$
$= \frac{4y^{10}}{x}$

60. $\frac{4x^5y^{-2}}{8x^7y^3} = \frac{4}{8} \cdot \frac{x^5}{x^7} \cdot \frac{y^{-2}}{y^3}$
$= \frac{1}{2} \cdot \frac{1}{x^2} \cdot \frac{1}{y^5}$
$= \frac{1}{2x^2y^5}$

61. $364,000 = 3.64 \times 10^5$

62. $0.153 = 1.53 \times 10^{-1}$

63. $0.00763 = 7.63 \times 10^{-3}$

64. $47,000 = 4.7 \times 10^4$

65. $1,370,000 = 1.37 \times 10^6$

66. $0.000314 = 3.14 \times 10^{-4}$

67. $4.2 \times 10^{-3} = 0.0042$

68. $6.52 \times 10^{-4} = 0.000652$

69. $9.7 \times 10^5 = 970,000$

70. $4.38 \times 10^{-6} = 0.00000438$

71. $9.14 \times 10^{-1} = 0.914$

72. $1.103 \times 10^7 = 11,030,000$

73. $\left(1.7 \times 10^5\right)\left(3.4 \times 10^{-4}\right)$
$= (1.7 \times 3.4)\left(10^5 \times 10^{-4}\right)$
$= 5.78 \times 10^1$
$= 57.8$

74. $\left(4.2 \times 10^{-3}\right)\left(3 \times 10^5\right) = (4.2 \times 3)\left(10^{-3} \times 10^5\right)$
$= 12.6 \times 10^2$
$= 1260$

75. $\left(3.5 \times 10^{-2}\right)\left(7.0 \times 10^3\right)$
$= (3.5 \times 7.0)\left(10^{-2} \times 10^3\right)$
$= 24.5 \times 10^1$
$= 245$

76. $\frac{6.8 \times 10^3}{2 \times 10^{-2}} = \left(\frac{6.8}{2}\right)\left(\frac{10^3}{10^{-2}}\right)$
$= 3.4 \times 10^5$
$= 340,000$

77. $\frac{6.5 \times 10^4}{2.0 \times 10^6} = \left(\frac{6.5}{2.0}\right)\left(\frac{10^4}{10^6}\right)$
$= 3.25 \times 10^{-2}$
$= 0.0325$

78. $\frac{15 \times 10^{-3}}{5 \times 10^2} = \left(\frac{15}{5}\right)\left(\frac{10^{-3}}{10^2}\right)$
$= 3 \times 10^{-5}$
$= 0.00003$

79. $(60,000)(20,000) = \left(6 \times 10^4\right)\left(2 \times 10^4\right)$
$= (6 \times 2)\left(10^4 \times 10^4\right)$
$= 12 \times 10^8$
$= 1.2 \times 10^9$

80. $(12,000)(400,000)$
$= (1.2\times 10^4)(4.0\times 10^5)$
$= (1.2\times 4.0)\times(10^4\times 10^5)$
$= 4.8\times 10^9$

81. $(0.00023)(40,000) = (2.3\times 10^{-4})(4\times 10^4)$
$= (2.3\times 4)(10^{-4}\times 10^4)$
$= 9.2\times 10^0$

82. $\dfrac{250}{500,000} = \dfrac{2.5\times 10^2}{5.0\times 10^5}$
$= \left(\dfrac{2.5}{5.0}\right)\left(\dfrac{10^2}{10^5}\right)$
$= 0.5\times 10^{-3}$
$= 5.0\times 10^{-4}$

83. $\dfrac{0.000068}{0.02} = \dfrac{6.8\times 10^{-5}}{2\times 10^{-2}}$
$= \left(\dfrac{6.8}{2}\right)\left(\dfrac{10^{-5}}{10^{-2}}\right)$
$= 3.4\times 10^{-3}$

84. $\dfrac{850,000}{0.025} = \dfrac{8.5\times 10^5}{2.5\times 10^{-2}}$
$= \left(\dfrac{8.50}{2.50}\right)\left(\dfrac{10^5}{10^{-2}}\right)$
$= 3.40\times 10^7$

85. **a.** \$97.2 billion is \$97,200,000,000 or $\$9.72\times 10^{10}$

b. $(9.72\times 10^{10}) - (5.1\times 10^9) = (9.72\times 10^{10}) - (0.51\times 10^{10})$
$= (9.72 - 0.51)\times 10^{10}$
$= 9.21\times 10^{10}$

There is $\$9.21\times 10^{10}$ or \$92,100,000,000 more in the CREF Stock Fund.

86. Distance $= (520)(5.87\times 10^{12}) = (5.2\times 10^2)(5.87\times 10^{12})$
$= (5.2\times 5.87)(10^2\times 10^{12})$
$= 30.524\times 10^{14}$
$= 3.0524\times 10^{15}$

The distance is 3.0524×10^{15} miles.

87. Not a polynomial

88. Monomial, zero degree

89. $x^2 + 3x - 4$, trinomial, second degree

90. $4x^2 - x - 3$, trinomial, second degree

91. $13x^3 - 4$, Binomial, third degree

92. Not a polynomial

93. $-4x^2 + x$, binomial, second degree

94. Not a polynomial

95. $2x^3 + 4x^2 - 3x - 7$, polynomial, third degree

96. $$\begin{aligned}(x+3)+(2x+4) &= x+3+2x+4\\ &= x+2x+3+4\\ &= 3x+7\end{aligned}$$

97. $$\begin{aligned}(4x-6)+(3x+15) &= 4x-6+3x+15\\ &= 4x+3x-6+15\\ &= 7x+9\end{aligned}$$

98. $$\begin{aligned}(-x-10)+(-2x+5) &= -x-10-2x+5\\ &= -x-2x-10+5\\ &= -3x-5\end{aligned}$$

99. $$\begin{aligned}\left(-x^2+6x-7\right)+\left(-2x^2+4x-8\right) &= -x^2+6x-7-2x^2+4x-8\\ &= -x^2-2x^2+6x+4x-7-8\\ &= -3x^2+10x-15\end{aligned}$$

100. $$\begin{aligned}\left(12x^2+4x-8\right)+\left(-x^2-6x+5\right) &= 12x^2+4x-8-x^2-6x+5\\ &= 12x^2-x^2+4x-6x-8+5\\ &= 11x^2-2x-3\end{aligned}$$

101. $$\begin{aligned}(2x-4.3)-(x+2.4) &= 2x-4.3-x-2.4\\ &= 2x-x-4.3-2.4\\ &= x-6.7\end{aligned}$$

102. $$\begin{aligned}(-4x+8)-(-2x+6) &= -4x+8+2x-6\\ &= -4x+2x+8-6\\ &= -2x+2\end{aligned}$$

103. $$\begin{aligned}\left(4x^2-9x\right)-(3x+15) &= 4x^2-9x-3x-15\\ &= 4x^2-12x-15\end{aligned}$$

104. $\left(6x^2-6x+1\right)-(12x+5)=6x^2-6x+1-12x-5$
$=6x^2-6x-12x+1-5$
$=6x^2-18x-4$

105. $\left(-2x^2+8x-7\right)-\left(3x^2+12\right)=-2x^2+8x-7-3x^2-12$
$=-2x^2-3x^2+8x-7-12$
$=-5x^2+8x-19$

106. $\left(x^2+7x-3\right)-\left(x^2+3x-5\right)=x^2+7x-3-x^2-3x+5$
$=x^2-x^2+7x-3x-3+5$
$=4x+2$

107. $\frac{1}{2}x(14x+20)=\frac{1}{2}x(14x)+\frac{1}{2}x(20)$
$=7x^2+10x$

108. $-3x(5x+4)=-3x\cdot 5x+(-3x)4$
$=-15x^2-12x$

109. $3x\left(2x^2-4x+7\right)=3x\left(2x^2\right)+3x(-4x)+3x(7)$
$=6x^3-12x^2+21x$

110. $-x\left(3x^2-6x-1\right)=(-x)\left(3x^2\right)+(-x)(-6x)+(-x)(-1)$
$=-3x^3+6x^2+x$

111. $-4x\left(-6x^2+4x-2\right)=(-4x)\left(-6x^2\right)+(-4x)(4x)+(-4x)(-2)$
$=24x^3-16x^2+8x$

112. $(x+4)(x+5)=x\cdot x+x\cdot 5+4\cdot x+4\cdot 5$
$=x^2+5x+4x+20$
$=x^2+9x+20$

113. $(3x+6)(-4x+1)=3x(-4x)+3x(1)+6(-4x)+6(1)$
$=-12x^2+3x-24x+6$
$=-12x^2-21x+6$

114. $(-2x+6)^2=(-2x)^2+2(-2x)(6)+(6)^2$
$=4x^2-24x+36$

115. $\begin{aligned}(6-2x)(2+3x) &= 6\cdot 2+6\cdot 3x+(-2x)(2)+(-2x)(3x)\\ &= 12+18x-4x-6x^2\\ &= 12+14x-6x^2\\ &= -6x^2+14x+12\end{aligned}$

116. $\begin{aligned}(x+4)(x-4) &= (x)^2-(4)^2\\ &= x^2-16\end{aligned}$

117. $\begin{aligned}(3x+1)\left(x^2+2x+4\right) &= 3x\left(x^2+2x+4\right)+1\left(x^2+2x+4\right)\\ &= 3x^3+6x^2+12x+x^2+2x+4\\ &= 3x^3+7x^2+14x+4\end{aligned}$

118. $\begin{aligned}(x-1)\left(3x^2+4x-6\right) &= x\left(3x^2+4x-6\right)-1\left(3x^2+4x-6\right)\\ &= 3x^3+4x^2-6x-3x^2-4x+6\\ &= 3x^3+x^2-10x+6\end{aligned}$

119. $\begin{aligned}(-4x+2)\left(3x^2-x+7\right) &= -4x\left(3x^2-x+7\right)+2\left(3x^2-x+7\right)\\ &= -12x^3+4x^2-28x+6x^2-2x+14\\ &= -12x^3+10x^2-30x+14\end{aligned}$

120. $\begin{aligned}\frac{2x+4}{2} &= \frac{2x}{2}+\frac{4}{2}\\ &= x+2\end{aligned}$

121. $\begin{aligned}\frac{10x+12}{2} &= \frac{10x}{2}+\frac{12}{2}\\ &= 5x+6\end{aligned}$

122. $\begin{aligned}\frac{8x^2+4x}{x} &= \frac{8x^2}{x}+\frac{4x}{x}\\ &= 8x+4\end{aligned}$

123. $\begin{aligned}\frac{6x^2+9x-4}{3} &= \frac{6x^2}{3}+\frac{9x}{3}-\frac{4}{3}\\ &= 2x^2+3x-\frac{4}{3}\end{aligned}$

124. $\begin{aligned}\frac{4x^2-8x-6}{2x} &= \frac{4x^2}{2x}-\frac{8x}{2x}-\frac{6}{2x}\\ &= 2x-4-\frac{3}{x}\end{aligned}$

125. $$\frac{8x^5-4x^4+3x^2-2}{2x}=\frac{8x^5}{2x}-\frac{4x^4}{2x}+\frac{3x^2}{2x}-\frac{2}{2x}$$
$$=4x^4-2x^3+\frac{3}{2}x-\frac{1}{x}$$

126. $$\frac{16x-4}{-2}=\frac{(-1)(16x-4)}{(-1)(-2)}$$
$$=\frac{-16x+4}{2}$$
$$=\frac{-16x}{2}+\frac{4}{2}$$
$$=-8x+2$$

127. $$\frac{5x^2-6x+15}{3x}=\frac{5x^2}{3x}-\frac{6x}{3x}+\frac{15}{3x}$$
$$=\frac{5x}{3}x-2+\frac{5}{x}$$

128. $$\frac{5x^3+10x+2}{2x^2}=\frac{5x^3}{2x^2}+\frac{10x}{2x^2}+\frac{2}{2x^2}$$
$$=\frac{5x}{2}+\frac{5}{x}+\frac{1}{x^2}$$

129. $$\begin{array}{r} x+4 \\ x-3\overline{)\,x^2+x-12} \\ \underline{x^2-3x} \\ 4x-12 \\ \underline{4x-12} \\ 0 \end{array}$$

$$\frac{x^2+x-12}{x-3}=x+4$$

130. $$\begin{array}{r} 4x+2 \\ 2x-5\overline{)\,8x^2-16x-10} \\ \underline{8x^2-20x} \\ 4x-10 \\ \underline{4x-10} \\ 0 \end{array}$$

$$\frac{8x^2-16x-10}{2x-5}=4x+2$$

131. $$\begin{array}{r} 5x-2 \\ x+6\overline{)\,5x^2+28x-10} \\ \underline{5x^2+30x} \\ -2x-10 \\ \underline{-2x-12} \\ 2 \end{array}$$

$$\frac{5x^2+28x-10}{x+6}=5x-2+\frac{2}{x+6}$$

132. $$\begin{array}{r} 2x^2+3x-4 \\ 2x+3\overline{)\,4x^3+12x^2+x-12} \\ \underline{4x^3+6x^2} \\ 6x^2+x \\ \underline{6x^2+9x} \\ -8x-12 \\ \underline{-8x-12} \\ 0 \end{array}$$

$$\frac{4x^3+12x^2+x-12}{2x+3}=2x^2+3x-4$$

133. $$\begin{array}{r} 2x-3 \\ 2x-3\overline{)\,4x^2-12x+9} \\ \underline{4x^2-6x} \\ -6x+9 \\ \underline{-6x+9} \\ 0 \end{array}$$

$$\frac{4x^2-12x+9}{2x-3}=2x-3$$

Practice Test

1. $$4x^3\cdot 3x^2=4\cdot 3x^{3+2}$$
$$=12x^5$$

2. $\left(3xy^2\right)^3 = 3^3x^3y^{2\cdot 3}$
$= 27x^3y^6$

3. $\dfrac{8x^4}{2x} = \dfrac{8}{2}x^{4-1}$
$= 4x^3$

4. $\left(\dfrac{3x^2y}{6xy^3}\right)^3 = \left(\dfrac{3}{6}\cdot\dfrac{x^2}{x}\cdot\dfrac{y}{y^3}\right)^3$
$= \left(\dfrac{1}{2}\cdot x\cdot\dfrac{1}{y^2}\right)^3$
$= \left(\dfrac{x}{2y^2}\right)^3$
$= \dfrac{x^3}{2^3y^{2\cdot 3}}$
$= \dfrac{x^3}{8y^6}$

5. $\left(2x^3y^{-2}\right)^{-2} = 2^{-2}x^{3(-2)}y^{(-2)(-2)}$
$= 2^{-2}x^{-6}y^4$
$= \dfrac{y^4}{2^2x^6}$
$= \dfrac{y^4}{4x^6}$

6. $\dfrac{15x^4y^2}{45x^{-1}y} = \dfrac{15}{45}x^{4-(-1)}y^{2-1}$
$= \dfrac{1}{3}x^5y$
$= \dfrac{x^5y}{3}$

7. $\left(4x^0\right)\left(3x^2\right)^0 = (4\cdot 1)\cdot 1$
$= 4$

8. $(42{,}000)(30{,}000) = \left(4.2\times 10^4\right)\left(3.0\times 10^4\right)$
$= (4.2\times 3.0)\left(10^4\times 10^4\right)$
$= 12.60\times 10^8$
$= 1.26\times 10^9$

9. $\dfrac{0.0008}{4000} = \dfrac{8.0\times 10^{-4}}{4.0\times 10^3}$
$= \left(\dfrac{8.0}{4.0}\right)\left(\dfrac{10^{-4}}{10^3}\right)$
$= 2.0\times 10^{-7}$

10. $x^2 - 4 + 6x = x^2 + 6x - 4$
trinomial

11. y, monomial

12. $x^{-2} + 4$, not a polynomial

13. $-5 + 6x^3 - 2x^2 + 5x = 6x^3 - 2x^2 + 5x - 5$,
third degree

14. $(2x+4) + \left(3x^2 - 5x - 3\right)$
$= 2x + 4 + 3x^2 - 5x - 3$
$= 3x^2 + 2x - 5x + 4 - 3$
$= 3x^2 - 3x + 1$

15. $\left(x^2 - 4x + 7\right) - \left(3x^2 - 8x + 7\right)$
$= x^2 - 4x + 7 - 3x^2 + 8x - 7$
$= x^2 - 3x^2 - 4x + 8x + 7 - 7$
$= -2x^2 + 4x$

16. $\left(4x^2 - 5\right) - \left(x^2 + x - 8\right)$
$= 4x^2 - 5 - x^2 - x + 8$
$= 4x^2 - x^2 - x - 5 + 8$
$= 3x^2 - x + 3$

17. $-4x(2x+4) = -4x(2x) - 4x(4)$
$= -8x^2 - 16x$

18. $(4x+7)(2x-3) = (4x)(2x)+(4x)(-3)+7(2x)+7(-3)$
$= 8x^2 - 12x + 14x - 21$
$= 8x^2 + 2x - 21$

19. $(6-4x)(5+3x) = 6\cdot 5 + 6(3x) + (-4x)\cdot 5 + (-4x)(3x)$
$= 30 + 18x - 20x - 12x^2$
$= -12x^2 - 2x + 30$

20. $(3x-5)(2x^2+4x-5) = 3x(2x^2+4x-5) - 5(2x^2+4x-5)$
$= 3x\cdot 2x^2 + 3x\cdot 4x - 3x\cdot 5 - 5\cdot 2x^2 - 5\cdot 4x - 5(-5)$
$= 6x^3 + 12x^2 - 15x - 10x^2 - 20x + 25$
$= 6x^3 + 2x^2 - 35x + 25$

21. $\dfrac{16x^2+8x-4}{4} = \dfrac{16x^2}{4} + \dfrac{8x}{4} - \dfrac{4}{4}$
$= 4x^2 + 2x - 1$

22. $\dfrac{3x^2-6x+5}{-3x} = \dfrac{(-1)(3x^2-6x+5)}{(-1)(-3x)}$
$= \dfrac{-3x^2+6x-5}{3x}$
$= -\dfrac{3x^2}{3x} + \dfrac{6x}{3x} - \dfrac{5}{3x}$
$= -x + 2 - \dfrac{5}{3x}$

23.
$$\begin{array}{r} 4x+5 \\ 2x-3\overline{)\ 8x^2-2x-15} \\ \underline{8x^2-12x} \\ 10x-15 \\ \underline{10x-15} \\ 0 \end{array}$$

$\dfrac{8x^2-2x-15}{2x-3} = 4x+5$

24.
$$\begin{array}{r} 3x-2 \\ 4x+5\overline{)12x^2+7x-12} \\ \underline{12x^2+15x} \\ -8x-12 \\ \underline{-8x-10} \\ -2 \end{array}$$

$\dfrac{12x^2+7x-12}{4x+5} = 3x - 2 - \dfrac{2}{4x+5}$

25. a. $5730 = 5.73\times 10^3$

b. $\dfrac{4.46\times 10^9}{5.73\times 10^3} = \left(\dfrac{4.46}{5.73}\right)\left(\dfrac{10^9}{10^3}\right)$
$\approx 0.778\times 10^6$
$\approx 7.78\times 10^5$

Cumulative Review Test

1. $4+8\div 2+3 = 4+4+3$
$= 11$

2. $3x+5 = 4(x-2)$
$3x+5 = 4x-8$
$13 = x$

3. $2(x+3)+3x-5=4x-2$
$2x+6+3x-5=4x-2$
$5x+1=4x-2$
$x=-3$

4. $3x-11<5x-2$
$-11+2<5x-3x$
$-9<2x$
$-\frac{9}{2}<x$
$x>-\frac{9}{2}$

$-\frac{9}{2}$

5. $5x-2=y-7$
$y-7=5x-2$
$y=5x-2+7$
$y=5x+5$

6. $m=\frac{y_2-y_1}{x_2-x_1}=\frac{1-3}{5-1}=\frac{-2}{4}=-\frac{1}{2}$

7. The relation is a function.
Domain: {–4, 1, 2, 5}
Range: {–5, 5, 7, 13}

8. $3x+2y=10$
$x-y=5$
Solve the second equation for x.
$x=5+y$
Substitute $5+y$ for x in the first equation.
$3(5+y)+2y=10$
$15+3y+2y=10$
$5y=-5$
$y=-1$
Substitute –1 for y in the second equation.
$x-y=5$
$x-(-1)=5$
$x+1=5$
$x=4$
The solution is (4, –1).

9.

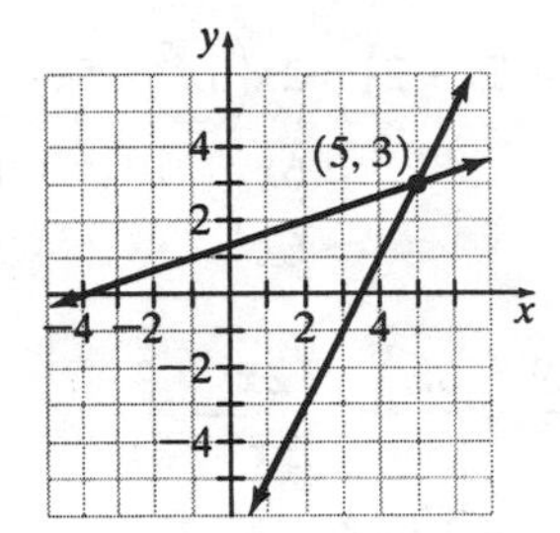

10. $\left(3x^2y^4\right)^3\left(5x^2y\right)=\left(3^3x^{2\cdot 3}y^{4\cdot 3}\right)\left(5x^2y\right)$
$=\left(27x^6y^{12}\right)\left(5x^2y\right)$
$=27\cdot 5x^{6+2}y^{12+1}$
$=135x^8y^{13}$

11. $-5x+2-7x^2=-7x^2-5x+2$
second degree

12. $\left(x^2+4x-3\right)+\left(2x^2+5x+1\right)$
$=x^2+4x-3+2x^2+5x+1$
$=x^2+2x^2+4x+5x-3+1$
$=3x^2+9x-2$

13. $\left(8x^2-3x+2\right)-\left(4x^2-3x+3\right)$
$=8x^2-3x+2-4x^2+3x-3$
$=8x^2-4x^2-3x+3x+2-3$
$=4x^2-1$

14. $(3y-5)(2y+3)$
$=3y\cdot 2y+3\cdot 3y-5\cdot 2y-5\cdot 3$
$=6y^2+9y-10y-15$
$=6y^2-y-15$

15. $(2x-1)(3x^2-5x+2) = 2x(3x^2-5x+2) - 1(3x^2-5x+2)$
$= 6x^3 - 10x^2 + 4x - 3x^2 + 5x - 2$
$= 6x^3 - 13x^2 + 9x - 2$

16. $\dfrac{6x^2+12x-9}{3x} = \dfrac{6x^2}{3x} + \dfrac{12x}{3x} - \dfrac{9}{3x}$
$= 2x + 4 - \dfrac{3}{x}$

17.
$$\begin{array}{r}
2x+5 \\
3x-2\overline{\big)\,6x^2+11x-10} \\
\underline{6x^2-4x} \\
15x-10 \\
\underline{15x-10} \\
0
\end{array}$$

$\dfrac{6x^2+11x-10}{3x-2} = 2x+5$

18. $\dfrac{x}{8} = \dfrac{1.25}{3}$
$3x = 8(1.25)$
$3x = 10$
$x = \dfrac{10}{3} \approx 3.33$
Eight cans of soup cost \$3.33.

19. Let x = the width of the rectangle. Then $3x - 2$ = the length of the rectangle.
$P = 2l + 2w$
$28 = 2(3x-2) + 2x$
$28 = 6x - 4 + 2x$
$32 = 8x$
$4 = x$
The width of the rectangle is 4 feet and the length is $3(4) - 2 = 10$ feet.

20. Let b = Bob's average speed. Then $b + 7$ = Nick's average speed.
$d = r \cdot t$. Both Bob and Nick drove for 0.5 hour and the total distance they covered was 60 miles.
$0.5b + 0.5(b+7) = 60$
$0.5b + 0.5b + 3.5 = 60$
$b = 56.5$
Bob's average speed was 56.5 miles per hour and Nick's average speed was $56.5 + 7 = 63.5$ miles per hour.

Chapter 7

Exercise Set 7.1

1. A prime number is an integer greater than 1 that has exactly two factors, itself and 1.

3. The number 1 is not a prime number. It is called unit.

5. The greatest common factor of two or more numbers is the greatest number that divides into all the numbers.

7. A factoring problem may be checked by multiplying the factors.

9. $48 = 6 \cdot 8$
$= 2 \cdot 3 \cdot 2 \cdot 4$
$= 2 \cdot 3 \cdot 2 \cdot 2 \cdot 2$
$= 2 \cdot 2 \cdot 2 \cdot 2 \cdot 3$
$= 2^4 \cdot 3$

11. $90 = 9 \cdot 10$
$= 3 \cdot 3 \cdot 2 \cdot 5$
$= 2 \cdot 3^2 \cdot 5$

13. $200 = 20 \cdot 10$
$= 2 \cdot 10 \cdot 2 \cdot 5$
$= 2 \cdot 2 \cdot 5 \cdot 2 \cdot 5$
$= 2^3 \cdot 5^2$

15. $20 = 2^2 \cdot 5$, $24 = 2^3 \cdot 3$, so the greatest common factor is 2^2 or 4.

17. $70 = 2 \cdot 5 \cdot 7$, $98 = 2 \cdot 7^2$, so the greatest common factor is $2 \cdot 7$ or 14.

19. $72 = 2^3 \cdot 3^2$, $90 = 2 \cdot 3^2 \cdot 5$, so the greatest common factor is $2 \cdot 3^2$ or 18.

21. The greatest common factor is x.

23. The greatest common factor is $3x$.

25. The greatest common factor is 1.

27. The greatest common factor is xy.

29. The greatest common factor is x^3y^5.

31. The greatest common factor is 5.

33. The greatest common factor is x^2y^2.

35. The greatest common factor is x.

37. The greatest common factor is $x + 3$.

39. The greatest common factor is $2x - 3$.

41. The greatest common factor is $3x - 4$.

43. The greatest common factor is $x - 4$.

45. The greatest common factor is 5.
$5x + 10 = 5 \cdot x + 5 \cdot 2$
$= 5(x + 2)$

47. The greatest common factor is 5.
$15x - 5 = 5 \cdot 3x - 5 \cdot 1$
$= 5(3x - 1)$

49. $13x + 15$ cannot be factored.

51. The greatest common factor is $3x$.
$9x^2 - 12x = 3x \cdot 3x - 3x \cdot 4$
$= 3x(3x - 4)$

53. The greatest common factor is $2p$.
$26p^2 - 8p = 2p \cdot 13p - 2p \cdot 4$
$= 2p(13p - 4)$

55. The greatest common factor is $2x$.
$4x^3 - 10x = 2x \cdot 2x^2 - 2x \cdot 5$
$= 2x(2x^2 - 5)$

57. The greatest common factor is $12x^8$.
$36x^{12} + 24x^8 = 12x^8 \cdot 3x^4 + 12x^8 \cdot 2$
$= 12x^8(3x^4 + 2)$

59. The greatest common factor is $3y^3$.
$$\begin{aligned} 24y^{15} - 9y^3 &= 3y^3 \cdot 8y^{12} - 3y^3 \cdot 3 \\ &= 3y^3(8y^{12} - 3) \end{aligned}$$

61. The greatest common factor is x.
$$\begin{aligned} x + 3xy^2 &= x \cdot 1 + x \cdot 3y^2 \\ &= x(1 + 3y^2) \end{aligned}$$

63. $6x + 5y$ cannot be factored.

65. The greatest common factor is $4xy$.
$$\begin{aligned} 16xy^2z + 4x^3y &= 4xy \cdot 4yz + 4xy \cdot x^2 \\ &= 4xy(4yz + x^2) \end{aligned}$$

67. The greatest common factor is $2xy^2$.
$$\begin{aligned} 34x^2y^2 + 16xy^4 &= 2xy^2 \cdot 17x + 2xy^2 \cdot 8y^2 \\ &= 2xy^2(17x + 8y^2) \end{aligned}$$

69. The greatest common factor is $25x^2yz$.
$$\begin{aligned} 25x^2yz^3 + 25x^3yz &= 25x^2yz \cdot z^2 + 25x^2yz \cdot x \\ &= 25x^2yz(z^2 + x) \end{aligned}$$

71. The greatest common factor is y^2z^3.
$$\begin{aligned} 13y^5z^3 - 11xy^2z^5 &= y^2z^3 \cdot 13y^3 - y^2z^3 \cdot 11xz^2 \\ &= y^2z^3(13y^3 - 11xz^2) \end{aligned}$$

73. The greatest common factor is 3.
$$\begin{aligned} 3x^2 + 6x + 9 &= 3 \cdot x^2 + 3 \cdot 2x + 3 \cdot 3 \\ &= 3(x^2 + 2x + 3) \end{aligned}$$

75. The greatest common factor is 3.
$$\begin{aligned} 9x^2 + 18x + 3 &= 3 \cdot 3x^2 + 3 \cdot 6x + 3 \cdot 1 \\ &= 3(3x^2 + 6x + 1) \end{aligned}$$

77. The greatest common factor is $4x$.
$$\begin{aligned} 4x^3 - 8x^2 + 12x &= 4x \cdot x^2 - 4x \cdot 2x + 4x \cdot 3 \\ &= 4x(x^2 - 2x + 3) \end{aligned}$$

79. $35x^2 - 16x + 10$ cannot be factored.

81. The greatest common factor is 3.

$$\begin{aligned}15p^2 - 6p + 9 &= 3 \cdot 5p^2 - 3 \cdot 2p + 3 \cdot 3 \\ &= 3(5p^2 - 2p + 3)\end{aligned}$$

83. The greatest common factor is $4x^3$.

$$\begin{aligned}24x^6 + 8x^4 - 4x^3 &= 4x^3 \cdot 6x^3 + 4x^3 \cdot 2x - 4x^3 \cdot 1 \\ &= 4x^3(6x^3 + 2x - 1)\end{aligned}$$

85. The greatest common factor is xy.

$$\begin{aligned}8x^2y + 12xy^2 + 9xy &= xy \cdot 8x + xy \cdot 12y + xy \cdot 9 \\ &= xy(8x + 12y + 9)\end{aligned}$$

87. The greatest common factor is $x + 4$.

$$x(x+4) + 3(x+4) = (x+3)(x+4)$$

89. $7x(4x-3) - 4(4x+3)$ cannot be factored.

91. The greatest common factor is $2x + 1$.

$$4x(2x+1) + 1(2x+1) = (4x+1)(2x+1)$$

93. The greatest common factor is $2x + 1$.

$$\begin{aligned}4x(2x+1) + 2x + 1 &= 4x(2x+1) + 1(2x+1) \\ &= (4x+1)(2x+1)\end{aligned}$$

95. $3\star + 6 = 3 \cdot \star + 3 \cdot 2 = 3(\star + 2)$

97.

$$\begin{aligned}35\Delta^3 - 7\Delta^2 + 14\Delta &= 7\Delta \cdot 5\Delta^2 - 7\Delta \cdot \Delta + 7\Delta \cdot 2 \\ &= 7\Delta(5\Delta^2 - \Delta + 2)\end{aligned}$$

99. The greatest common factor is $2(x-3)$.

$$\begin{aligned}4x^2(x-3)^3 - 6x(x-3)^2 + 4(x-3) &= 2(x-3) \cdot 2x^2(x-3)^2 - 2(x-3) \cdot 3x(x-3) - 2(x-3) \cdot 2 \\ &= 2(x-3)\left[2x^2(x-3)^2 - 3x(x-3) + 2\right]\end{aligned}$$

101. First factor $x^{1/3}$ from terms.

$$x^{7/3} + 5x^{4/3} + 6x^{1/3} = x^{1/3}(x^2 + 5x + 6)$$

103.

$$\begin{aligned}x^2 + 2x + 3x + 6 &= x \cdot x + x \cdot 2 + 3 \cdot x + 3 \cdot 2 \\ &= x(x+2) + 3(x+2) \\ &= (x+3)(x+2)\end{aligned}$$

105. $2x-(x-5)+4(3-x)=2x-x+5+12-4x$
$=x-4x+17$
$=-3x+17$

106. $4+3(x-8)=x-4(x+2)$
$4+3x-24=x-4x-8$
$3x-20=-3x-8$
$6x=12$
$x=2$

107. $4x-5y=20$
$-5y=-4x+20$
$y=-\dfrac{-4x+20}{-5}$
$y=\dfrac{4}{5}x-4$

108. a. $2x+3y=12$
$x=0\quad y=4\quad y=0\quad x=6$
(0, 4) (6, 0)

b.

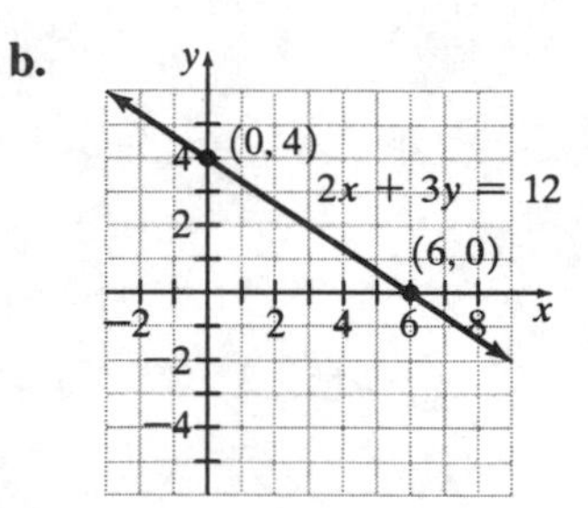

109. $\left(\dfrac{3x^2y^3}{2x^5y^2}\right)^2=\left(\dfrac{3y}{2x^3}\right)^2=\dfrac{(3y)^2}{(2x^3)^2}=\dfrac{3^2y^2}{2^2(x^3)^2}=\dfrac{9y^2}{4x^6}$

Exercise Set 7.2

1. The first step in any factoring by grouping problem is to factor out a common factor, if one exists.

3. If you multiply $(x-2)(x+4)$ by the FOIL method, you get the polynomial $x^2+4x-2x-8$

5. Answers will vary.

7. $x^2+3x+2x+6=x(x+3)+2(x+3)$
$=(x+2)(x+3)$

9. $x^2+3x+4x+12=x(x+3)+4(x+3)$
$=(x+4)(x+3)$

11. $x^2+2x+5x+10=x(x+2)+5(x+2)$
$=(x+5)(x+2)$

13. $x^2+3x-5x-15 = x(x+3)-5(x+3)$
$= (x-5)(x+3)$

15. $4x^2-14x+14x-49$
$= 2x(2x-7)+7(2x-7)$
$= (2x+7)(2x-7)$

17. $3x^2+9x+x+3 = 3x(x+3)+1(x+3)$
$= (3x+1)(x+3)$

19. $6x^2+3x-2x-1 = 3x(2x+1)-1(2x+1)$
$= (3x-1)(2x+1)$

21. $8x^2+32x+x+4 = 8x(x+4)+1(x+4)$
$= (8x+1)(x+4)$

23. $3x^2-2x+3x-2 = x(3x-2)+1(3x-2)$
$= (x+1)(3x-2)$

25. $2x^2-4x-3x+6 = 2x(x-2)-3(x-2)$
$= (2x-3)(x-2)$

27. $3x^2-9x+15x-15 = 3x(x-3)+5(x-3)$
$= (3x+5)(x-3)$

29. $x^2+2xy-3xy-6y^2$
$= x(x+2y)-3y(x+2y)$
$= (x-3y)(x+2y)$

31. $3x^2+2xy-9xy-6y^2$
$= x(3x+2y)-3y(3x+2y)$
$= (x-3y)(3x+2y)$

33. $10x^2-12xy-25xy+30y^2$
$= 2x(5x-6y)-5y(5x-6y)$
$= (2x-5y)(5x-6y)$

35. $x^2+bx+ax+ab = x(x+b)+a(x+b)$
$= (x+a)(x+b)$

37. $xy+4x-2y-8 = x(y+4)-2(y+4)$
$= (x-2)(y+4)$

39. $a^2+3a+ab+3b = a(a+3)+b(a+3)$
$= (a+b)(a+3)$

41. $xy-x+5y-5 = x(y-1)+5(y-1)$
$= (x+5)(y-1)$

43. $12+8y-3x-2xy = 4(3+2y)-x(3+2y)$
$= (4-x)(3+2y)$

45. $z^3+3z^2+z+3 = z^2(z+3)+1(z+3)$
$= (z^2+1)(z+3)$

47. $x^3+4x^2-3x-12 = x^2(x+4)-3(x+4)$
$= (x^2-3)(x+4)$

49. $2x^2-12x+8x-48$
$= 2\cdot x^2-2\cdot 6x+2\cdot 4x-2\cdot 24$
$= 2(x^2-6x+4x-24)$
$= 2[x(x-6)+4(x-6)]$
$= 2(x+4)(x-6)$

51.
$4x^2+8x+8x+16$
$= 4\cdot x^2+4\cdot 2x+4\cdot 2x+4\cdot 4$
$= 4(x^2+2x+2x+4)$
$= 4[x(x+2)+2(x+2)]$
$= 4(x+2)(x+2)$
$= 4(x+2)^2$

53. $6x^3+9x^2-2x^2-3x$
$= x\cdot 6x^2+x\cdot 9x-x\cdot 2x-x\cdot 3$
$= x(6x^2+9x-2x-3)$
$= x[3x(2x+3)-1(2x+3)]$
$= x(3x-1)(2x+3)$

55. $x^3 - 3x^2y + 2x^2y - 6xy^2$
$= x \cdot x^2 - x \cdot 3xy + x \cdot 2xy - x \cdot 6y^2$
$= x\left(x^2 - 3xy + 2xy - 6y^2\right)$
$= x[x(x-3y) + 2y(x-3y)]$
$= x(x+2y)(x-3y)$

57. $2x + 4y + 8 + xy = 2x + xy + 4y + 8$
$= x(2+y) + 4(y+2)$
$= (x+4)(y+2)$

59. $6x + 5y + xy + 30 = 6x + xy + 5y + 30$
$= x(6+y) + 5(y+6)$
$= (x+5)(y+6)$

61. $ax + by + ay + bx = ax + ay + bx + by$
$= a(x+y) + b(x+y)$
$= (a+b)(x+y)$

63. $cd - 12 - 4d + 3c = cd - 4d + 3c - 12$
$= d(c-4) + 3(c-4)$
$= (d+3)(c-4)$

65. $ac - bd - ad + bc = ac - ad + bc - bd$
$= a(c-d) + b(c-d)$
$= (a+b)(c-d)$

67. Not *any* arrangement of the terms of a polynomial is factorable by grouping. $xy + 2x + 5y + 10$ is factorable but $xy + 10 + 2x + 5y$ is not factorable in this arrangement.

69. $\Delta^2 + 3\Delta - 5\Delta - 15 = \Delta(\Delta+3) - 5(\Delta+3)$
$= (\Delta-5)(\Delta+3)$

71. a. $3x^2 + 10x + 8 = 3x^2 + 6x + 4x + 8$

b. $3x^2 + 6x + 4x + 8 = 3x(x+2) + 4(x+2)$
$= (3x+4)(x+2)$

73. a. $2x^2 - 11x + 15 = 2x^2 - 6x - 5x + 15$

b. $2x^2 - 6x - 5x + 15$
$= 2x(x-3) - 5(x-3)$
$= (2x-5)(x-3)$

75. a. $4x^2 - 17x - 15 = 4x^2 - 20x + 3x - 15$

b. $4x^2 - 20x + 3x - 15$
$= 4x(x-5) + 3(x-5)$
$= (4x+3)(x-5)$

77. $\bigstar\odot + 3\bigstar + 2\odot + 6 = \bigstar(\odot+3) + 2(\odot+3)$
$= (\bigstar+2)(\odot+3)$

79. $5 - 3(2x-7) = 4(x+5) - 6$
$5 - 6x + 21 = 4x + 20 - 6$
$-6x + 26 = 4x + 14$
$12 = 10x$
$\frac{12}{10} = x$
$\frac{5}{6} = x$

80. Let w = the number of pounds of chocolate wafers
p = the number of pounds of peppermint candies
$w + p = 50$
$6.25w + 2.50p = 4.75(50)$
or
$w + p = 50$
$6.25w + 2.50p = 237.50$
Multiply the first equation by –2.5 and then add.
$-2.5[w + p = 50]$
gives
$-2.5w - 2.5p = -125$
$\underline{6.25w + 2.5p = 237.5}$
$3.75w \quad = 112.5$
$w = 30$
$w + p = 50$
$30 + p = 50$
$p = 20$
They should mix 30 pounds of chocolate wafers with 20 pounds of peppermint hard candies.

81. $\dfrac{15x^3-6x^2-9x+5}{3x}$
$=\dfrac{15x^3}{3x}-\dfrac{6x^2}{3x}-\dfrac{9x}{3x}+\dfrac{5}{3x}$
$=5x^2-2x-3+\dfrac{5}{3x}$

82.
$$\begin{array}{r|l} & \quad\quad\quad x+3 \\ x-3 & x^2 \quad\quad -9 \\ & \underline{x^2-3x} \\ & \quad\quad 3x-9 \\ & \quad\quad \underline{3x-9} \\ & \quad\quad\quad\quad 0 \end{array}$$

$\dfrac{x^2-9}{x-3}=x+3$

Exercise Set 7.3

1. Since 8000 is positive, both signs will be the same. Since 180 is positive, both signs will be positive.

3. Since –8000 is negative, one sign will be positive, the other will be negative.

5. Since 8000 is positive, both signs will be the same. Since –240 is negative, both signs will be negative.

7. The trinomial $x^2+3xy-18y^2$ is obtained by multiplying the factors using the FOIL method.

9. The trinomial $3x^2-3y^2$ is obtained by multiplying all the factors and combing like terms.

11. To determine the factors when factoring a trinomial of the form x^2+bx+c. First, find two numbers whose product is *c*, and whose sum is *b*. The factors are (*x* + first number) and
(*x* + second number).

13. $x^2-7x+12=(x-3)(x-4)$

15. $x^2+6x+8=(x+4)(x+2)$

17. $x^2+7x+12=(x+4)(x+3)$

19. x^2+5x-9 cannot be factored.

21. $y^2-16y+15=(y-1)(y-15)$

23. $x^2+x-6=(x+3)(x-2)$

25. $r^2-2r-15=(r-5)(r+3)$

27. $b^2-11b+18=(b-9)(b-2)$

29. $x^2-8x-15$ cannot be factored.

31. $a^2+12a+11=(a+1)(a+11)$

33. $x^2+13x-30=(x+15)(x-2)$

35. $x^2+4x+4=(x+2)(x+2)$
$=(x+2)^2$

37. $k^2+6k+9=(k+3)(k+3)$
$=(k+3)^2$

39. $x^2-10x+25=(x-5)(x-5)$
$=(x-5)^2$

41. $w^2-18w+45=(w-15)(w-3)$

43. $x^2+10x-39=(x+13)(x-3)$

45. $x^2-x-20=(x-5)(x+4)$

47. $y^2+9y+14=(y+7)(y+2)$

49. $x^2+12x-64=(x+16)(x-4)$

51. $x^2+14x-24$ cannot be factored

53. $x^2-2x-80=(x-10)(x+8)$

55. $x^2 - 18x + 65 = (x-5)(x-13)$

57. $x^2 - 8xy + 15y^2 = (x-3y)(x-5y)$

59. $x^2 - 4xy + 4y^2 = (x-2y)(x-2y)$
$= (x-2y)^2$

61. $x^2 + 8xy + 15y^2 = (x+3y)(x+5y)$

63. $7x^2 - 42x + 35 = 7(x^2 - 6x + 5)$
$= 7(x-5)(x-1)$

65. $5x^2 + 20x + 15 = 5(x^2 + 4x + 3)$
$= 5(x+1)(x+3)$

67. $2x^2 - 14x + 24 = 2(x^2 - 7x + 12)$
$= 2(x-4)(x-3)$

69. $x^3 - 3x^2 - 18x = x(x^2 - 3x - 18)$
$= x(x-6)(x+3)$

71. $2x^3 + 6x^2 - 56x = 2x(x^2 + 3x - 28)$
$= 2x(x+7)(x-4)$

73. $x^3 + 8x^2 + 16x = x(x^2 + 8x + 16)$
$= x(x+4)(x+4)$
$= x(x+4)^2$

75.

Sign of Coefficient of x-term	Sign of Constant of Trinomial	Signs of Constant Terms in the Binomial Factors
–	+	both negative
–	–	one positive and one negative
+	–	one positive and one negative
+	+	both positive

77. $x^2 + 5x + 4 = (x+1)(x+4)$

79. $x^2 + 11x + 28 = (x+7)(x+4)$

81. $x^2 + 0.6x + 0.08 = (x+0.4)(x+0.2)$

83. $x^2 + \frac{2}{5}x + \frac{1}{25} = \left(x + \frac{1}{5}\right)\left(x + \frac{1}{5}\right)$
$= \left(x + \frac{1}{5}\right)^2$

85. $-x^2 - 6x - 8 = -(x^2 + 6x + 8)$
$= -(x+2)(x+4)$

87. $x^2 + 5x - 300 = (x+20)(x-15)$

89.
$$\begin{aligned} 3(3x-4) &= 5x+12 \\ 9x-12 &= 5x+12 \\ 9x-5x &= 12+12 \\ 4x &= 24 \\ x &= \frac{24}{4} \\ x &= 6 \end{aligned}$$

90. Let x be the percent of acid in the mixture.

Solution	Strength	Liters	Amount
18%	0.18	4	0.72
26%	0.26	1	0.26
Mixture	$\frac{x}{100}$	5	$\frac{5x}{100}$

$$\begin{aligned} 0.72+0.26 &= \frac{5x}{100} \\ 0.98 &= \frac{x}{20} \\ 19.6 &= x \end{aligned}$$

The mixture is a 19.6% acid solution.

91.

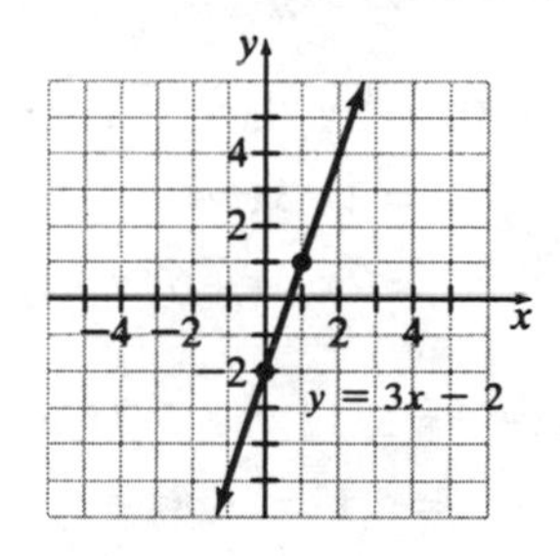

92.
$$\begin{aligned} (2x^2+5x-6)(x-2) &= 2x^2(x-2)+5x(x-2)-6(x-2) \\ &= 2x^3-4x^2+5x^2-10x-6x+12 \\ &= 2x^3+x^2-16x+12 \end{aligned}$$

93.
$$\begin{array}{r} 3x+2 \\ x-4 \overline{\smash{\big)}\, 3x^2-10x-10} \\ \underline{3x^2-12x} \\ 2x-10 \\ \underline{2x-8} \\ -2 \end{array}$$

$$\frac{3x^2-10x-10}{x-4} = 3x+2-\frac{2}{x-4}$$

94. $3x^2+5x-6x-10=x(3x+5)-2(3x+5)$
$=(x-2)(3x+5)$

Exercise Set 7.4

1. Factoring trinomials is the reverse process of multiplying binomials.

3. When factoring` a trinomial of the form ax^2+bx+c, the product of the constants in the binomial factors must equal the constant, c, of the trinomial.

5. $2x^2+11x+5=(2x+1)(x+5)$

7. $3x^2+14x+8=(3x+2)(x+4)$

9. $3x^2+4x+1=(3x+1)(x+1)$

11. $2x^2+13x+20=(2x+5)(x+4)$

13. $4x^2+4x-3=(2x-1)(2x+3)$

15. $5y^2-8y+3=(5y-3)(y-1)$

17. $5a^2-12a+6$ cannot be factored.

19. $6x^2+19x+3=(6x+1)(x+3)$

21. $5x^2+11x+4$ cannot be factored.

23. $5y^2-16y+3=(5y-1)(y-3)$

25. $7x^2+43x+6=(7x+1)(x+6)$

27. $7x^2-8x+1=(7x-1)(x-1)$

29. $3x^2-10x+7=(x-1)(3x-7)$

31. $5z^2-6z-8=(5z+4)(z-2)$

33. $4y^2+5y-6=(4y-3)(y+2)$

35. $10x^2-27x+5=(5x-1)(2x-5)$

37. $8x^2-2x-15=(4x+5)(2x-3)$

39. $6x^2+33x+15=3\cdot 2x^2+3\cdot 11x+3\cdot 5$
$=3\left(2x^2+11x+5\right)$
$=3(2x+1)(x+5)$

41. $6x^2+16x+10=2\cdot 3x^2+2\cdot 8x+2\cdot 5$
$=2\left(3x^2+8x+5\right)$
$=2(3x+5)(x+1)$

43. $6x^3-5x^2-4x=x\cdot 6x^2-x\cdot 5x-x\cdot 4$
$=x\left(6x^2-5x-4\right)$
$=x(2x+1)(3x-4)$

45. $12x^3+28x^2+8x$
$=4x\cdot 3x^2+4x\cdot 7x+4x\cdot 2$
$=4x\left(3x^2+7x+2\right)$
$=4x(3x+1)(x+2)$

47. $6x^3+4x^2-10x=2x\cdot 3x^2+2x\cdot 2x-2x\cdot 5$
$=2x\left(3x^2+2x-5\right)$
$=2x(3x+5)(x-1)$

49. $60x^2+40x+5=5\cdot 12x^2+5\cdot 8x+5\cdot 1$
$=5\left(12x^2+8x+1\right)$
$=5(6x+1)(2x+1)$

51. $2x^2+5xy+2y^2=(2x+y)(x+2y)$

53. $2x^2-7xy+3y^2=(2x-y)(x-3y)$

55. $12x^2+10xy-8y^2$
$=2\cdot 6x^2+2\cdot 5xy-2\cdot 4y^2$
$=2\left(6x^2+5xy-4y^2\right)$
$=2(2x-y)(3x+4y)$

57. $6x^2 - 15xy - 36y^2 = 3\cdot 2x^2 - 3\cdot 5xy - 3\cdot 12y^2$
$= 3(2x^2 - 5xy - 12y^2)$
$= 3(2x + 3y)(x - 4y)$

59. $4x^2 - 27x - 7$. This polynomial was obtained by multiplying the factors.

61. $10x^2 + 35x + 15$. This polynomial was obtained by multiplying the factors.

63. $2x^4 - x^3 - 3x^2$. This polynomial was obtained by multiplying the factors.

65. a. The second factor can be found by dividing the trinomial by the binomial.

b.
$$\begin{array}{r} 6x+11 \\ 3x+10\overline{)\ 18x^2+93x+110} \\ \underline{18x^2+60x} \\ 33x+110 \\ \underline{33x+110} \\ 0 \end{array}$$

The other factor is $6x + 11$.

67. The other factor is $15x - 32$. The product of the two first terms must equal $105x^2$, and the product of the constants must equal 480.

69. $8x^2 - 99x + 36 = (8x - 3)(x - 12)$

71. $16x^2 - 62x - 45 = (8x + 5)(2x - 9)$

73. $72x^2 + 417x - 420$
$= 3\cdot 24x^2 + 3\cdot 139x - 3\cdot 140$
$= 3(24x^2 + 139x - 140)$
$= 3(8x - 7)(3x + 20)$

75. The third factor is $x + 4$. The product of the three first terms must equal $2x^3$, and the product of the constants must equal -36.

76. $-x^2 - 4(y + 3) + 2y^2$
$= -(-3)^2 - 4(-5 + 3) + 2(-5)^2$
$= -9 - 4(-2) + 2(25)$
$= -9 + 8 + 50$
$= 49$

77. $\dfrac{500}{3.75} \approx 133.33$
His average speed was about 133.33 miles per hour.

78. a. $f(2) = 2^2 + 2 + 5 = 4 + 2 + 5 = 11$

b. $f(-4) = (-4)^2 + (-4) + 5$
$= 16 - 4 + 5$
$= 17$

79. $36x^4y^3 - 12xy^2 + 24x^5y^6$
$= 12xy^2\cdot 3x^3y - 12xy^2\cdot 1 + 12xy^2\cdot 2x^4y^4$
$= 12xy^2(3x^3y - 1 + 2x^4y^4)$

80. $x^2 - 15x + 54 = (x - 9)(x - 6)$

Exercise Set 7.5

1. a. $a^2 - b^2 = (a + b)(a - b)$

b. Answers will vary.

3. a. $a^3 - b^3 = (a - b)(a^2 + ab + b^2)$

b. Answers will vary.

5. No, there is no special formula for factoring the sum of two squares.

7. $y^2 - 16 = y^2 - 4^2 = (y + 4)(y - 4)$

9. $y^2 - 100 = y^2 - 10^2$
$= (y + 10)(y - 10)$

11. $x^2 - 49 = x^2 - 7^2$
$= (x + 7)(x - 7)$

13. $x^2 - y^2 = (x+y)(x-y)$

15. $9y^2 - 25z^2 = (3y)^2 - (5z)^2$
$= (3y+5z)(3y-5z)$

17. $64a^2 - 36b^2 = 4(16a^2 - 9b^2)$
$= 4[(4a)^2 - (3b)^2]$
$= 4(4a+3b)(4a-3b)$

19. $25x^2 - 36 = (5x)^2 - 6^2$
$= (5x+6)(5x-6)$

21. $z^4 - 81x^2 = (z^2)^2 - (9x)^2$
$= (z^2+9x)(z^2-9x)$

23. $9x^4 - 81y^2 = 9(x^4 - 9y^2)$
$= 9[(x^2)^2 - (3y)^2]$
$= 9(x^2+3y)(x^2-3y)$

25. $36m^4 - 49n^2 = (6m^2)^2 - (7n)^2$
$= (6m^2+7n)(6m^2-7n)$

27. $20x^2 - 180 = 20(x^2 - 9)$
$= 20(x^2 - 3^2)$
$= 20(x+3)(x-3)$

29. $16x^2 - 100y^4 = 4(4x^2 - 25y^4)$
$= 4[(2x)^2 - (5y^2)^2]$
$= 4(2x+5y^2)(2x-5y^2)$

31. $x^3 + y^3 = (x+y)(x^2 - xy + y^2)$

33. $a^3 - b^3 = (a-b)(a^2 + ab + b^2)$

35. $x^3 + 8 = x^3 + 2^3$
$= (x+2)(x^2 - 2x + 4)$

37. $x^3 - 27 = x^3 - 3^3$
$= (x-3)(x^2 + 3x + 9)$

39. $a^3 + 1 = a^3 + 1^3$
$= (a+1)(a^2 - a + 1)$

41. $27x^3 - 1 = (3x)^3 - 1^3$
$= (3x-1)(9x^2 + 3x + 1)$

43. $27a^3 - 125 = (3a)^3 - 5^3$
$= (3a-5)(9a^2 + 15a + 25)$

45. $27 - 8y^3 = 3^3 - (2y)^3$
$= (3-2y)(9 + 6y + 4y^2)$

47. $27x^3 + 64y^3 = (3x)^3 + (4y)^3$
$= (3x+4y)(9x^2 - 12xy + 16y^2)$

49. $8x^2 + 16x + 8 = 8(x^2 + 2x + 1)$
$= 8(x+1)^2$

51. $a^2b - 9b = b(a^2 - 9)$
$= b(a^2 - 3^2)$
$= b(a+3)(a-3)$

53. $3x^2 + 12x + 12 = 3(x^2 + 4x + 4)$
$= 3(x+2)^2$

55. $5x^2 - 10x - 15 = 5(x^2 - 2x - 3)$
$= 5(x-3)(x+1)$

57. $3xy - 6x + 9y - 18 = 3(xy - 2x + 3y - 6)$
$= 3[x(y-2) + 3(y-2)]$
$= 3(x+3)(y-2)$

59. $2x^2 - 50 = 2(x^2 - 25)$
$= 2(x^2 - 5^2)$
$= 2(x+5)(x-5)$

61. $3x^2y - 27y = 3y(x^2 - 9)$
$= 3y(x^2 - 3^2)$
$= 3y(x+3)(x-3)$

63. $3x^3y^2 + 3y^2 = 3y^2(x^3 + 1)$
$= 3y^2(x^3 + 1^3)$
$= 3y^2(x+1)(x^2 - x + 1)$

65. $2x^3 - 16 = 2(x^3 - 8)$
$= 2(x^3 - 2^3)$
$= 2(x-2)(x^2 + 2x + 4)$

67. $6x^2 - 4x + 24x - 16$
$= 2(3x^2 - 2x + 12x - 8)$
$= 2[x(3x-2) + 4(3x-2)]$
$= 2(x+4)(3x-2)$

69. $3x^3 - 10x^2 - 8x = x(3x^2 - 10x - 8)$
$= x(3x+2)(x-4)$

71. $4x^2 + 5x - 6 = (x+2)(4x-3)$

73. $25b^2 - 100 = 25(b^2 - 4)$
$= 25(b^2 - 2^2)$
$= 25(b+2)(b-2)$

75. $a^5b^2 - 4a^3b^4 = a^3b^2(a^2 - 4b^2)$
$= a^3b^2[a^2 - (2b)^2]$
$= a^3b^2(a+2b)(a-2b)$

77. $3x^4 - 18x^3 + 27x^2 = 3x^2(x^2 - 6x + 9)$
$= 3x^2(x-3)(x-3)$
$= 3x^2(x-3)^2$

79. $x^3 + 25x = x(x^2 + 25)$

81. $y^4 - 16 = (y^2)^2 - 4^2$
$= (y^2 + 4)(y^2 - 4)$
$= (y^2 + 4)(y^2 - 2^2)$
$= (y^2 + 4)(y+2)(y-2)$

83. $60a^2 - 25ab - 10b^2 = 5(12a^2 - 5ab - 2b^2)$
$= 5(3a - 2b)(4a + b)$

85. $2ab + 4a - 3b - 6 = 2a(b+2) - 3(b+2)$
$= (2a-3)(b+2)$

87. $9 - 9y^4 = 9(1 - y^4)$
$= 9\left[1^2 - (y^2)^2\right]$
$= 9(1 + y^2)(1 - y^2)$
$= 9(1 + y^2)(1^2 - y^2)$
$= 9(1 + y^2)(1 + y)(1 - y)$

89. You cannot divide both sides of the equation by $(a - b)$, because it equals 0.

91. $2◆^6 + 4◆^4✹^2 = 2◆^4(◆^2 + 2✹)$

93. $x^6 + 1 = (x^2)^3 + 1^3$
$= (x^2 + 1)(x^4 - x^2 + 1)$

95. $x^2 - 6x + 9 - 4y^2 = (x-3)^2 - (2y)^2$
$= (x - 3 + 2y)(x - 3 - 2y)$

97. $x^6 - y^6 = (x^3)^2 - (y^3)^2$
$= (x^3 + y^3)(x^3 - y^3)$
$= [(x+y)(x^2 - xy + y^2)][(x-y)(x^2 + xy + y^2)]$
$= (x+y)(x-y)(x^2 - xy + y^2)(x^2 + xy + y^2)$

98. $3x - 2(x+4) \geq 2x - 9$
$3x - 2x - 8 \geq 2x - 9$
$x - 8 \geq 2x - 9$
$1 \geq x$

[Number line: shaded to the left from a closed dot at 1]

99. $A = \frac{1}{2}h(b+d)$
$2A = bh + dh$
$2A - bh = dh$
$\frac{2A - bh}{h} = d$
$d = \frac{2A - bh}{h}$

100. $m = \frac{3-(-6)}{-2-4} = \frac{3+6}{-6} = \frac{9}{-6} = -\frac{3}{2}$

101. $\left(\frac{4x^4y}{6xy^5}\right)^3 = \left(\frac{4}{6} \cdot \frac{x^4}{x} \cdot \frac{y}{y^5}\right)^3$
$= \left(\frac{2}{3} \cdot x^3 \cdot \frac{1}{y^4}\right)^3$
$= \left(\frac{2x^3}{3y^4}\right)^3$
$= \frac{2^3 x^{3 \cdot 3}}{3^3 y^{4 \cdot 3}}$
$= \frac{8x^9}{27y^{12}}$

102. $x^{-2}x^{-3} = x^{-2-3}$
$= x^{-5}$
$= \frac{1}{x^5}$

Exercise Set 7.6

3. The standard form of a quadratic equation is $ax^2 + bx + c = 0$.

5. a. The zero-factor property may only be used when one side of the equation is equal to 0.

b. $(x+1)(x-2) = 4$
$x^2 - 2x + x - 2 = 4$
$x^2 - x - 6 = 0$
$(x-3)(x+2) = 0$
$x - 3 = 0$ or $x + 2 = 0$
$x = 3$ $\quad$ $x = -2$

7. $x(x+1) = 0$
$x = 0$ or $x + 1 = 0$
$x = 0$ $\quad$ $x = -1$

9. $7x(x-8) = 0$
$x = 0$ or $x - 8 = 0$
$x = 8$

11. $(2x+5)(x-3) = 0$
$2x + 5 = 0$ or $x - 3 = 0$
$2x = -5$ $\quad$ $x = 3$
$x = -\frac{5}{2}$

13. $x^2 - 16 = 0$
$(x+4)(x-4) = 0$
$x + 4 = 0$ or $x - 4 = 0$
$x = -4$ $\quad$ $x = 4$

15. $x^2 - 12x = 0$
$x(x-12) = 0$
$x = 0$ or $x - 12 = 0$
$x = 12$

17. $9x^2 + 27x = 0$
$9x(x+3) = 0$
$x = 0$ or $x + 3 = 0$
$x = -3$

19. $x^2 + x - 20 = 0$
$(x+5)(x-4) = 0$
$x + 5 = 0$ or $x - 4 = 0$
$x = -5$ $\quad x = 4$

21. $x^2 + 12x = -20$
$x^2 + 12x + 20 = 0$
$(x+10)(x+2) = 0$
$x + 10 = 0$ or $x + 2 = 0$
$x = -10$ $\quad x = -2$

23. $z^2 + 3z = 18$
$z^2 + 3z - 18 = 0$
$(z+6)(z-3) = 0$
$z + 6 = 0$ or $z - 3 = 0$
$z = -6$ $\quad z = 3$

25. $3x^2 + 6x - 24 = 0$
$3(x^2 + 2x - 8) = 0$
$3(x+4)(x-2) = 0$
$x + 4 = 0$ or $x - 2 = 0$
$x = -4$ $\quad x = 2$

27. $x^2 + 19x = 42$
$x^2 + 19x - 42 = 0$
$(x-2)(x+21) = 0$
$x - 2 = 0$ or $x + 21 = 0$
$x = 2$ $\quad x = -21$

29. $2y^2 + 22y + 60 = 0$
$2(y^2 + 11y + 30) = 0$
$2(y+6)(y+5) = 0$
$y + 6 = 0$ or $y + 5 = 0$
$y = -6$ $\quad y = -5$

31. $-2x - 15 = -x^2$
$x^2 - 2x - 15 = 0$
$(x-5)(x+3) = 0$
$x - 5 = 0$ or $x + 3 = 0$
$x = 5$ $\quad x = -3$

33. $-x^2 + 29x + 30 = 0$
$x^2 - 29x - 30 = 0$
$(x-30)(x+1) = 0$
$x - 30 = 0$ or $x + 1 = 0$
$x = 30$ $\quad x = -1$

35. $x^2 - 3x - 18 = 0$
$(x-6)(x+3) = 0$
$x - 6 = 0$ or $x + 3 = 0$
$x = 6$ $\quad x = -3$

37. $3p^2 = 22p - 7$
$3p^2 - 22p + 7 = 0$
$(3p-1)(p-7) = 0$
$3p - 1 = 0$ or $p - 7 = 0$
$3p = 1$ $\quad p = 7$
$p = \frac{1}{3}$

39. $3r^2 + r = 2$
$3r^2 + r - 2 = 0$
$(3r-2)(r+1) = 0$
$3r - 2 = 0$ or $r + 1 = 0$
$3r = 2$ $\quad r = -1$
$r = \frac{2}{3}$

41. $4x^2+4x-48=0$
$4(x^2+x-12)=0$
$4(x+4)(x-3)=0$
$x+4=0$ or $x-3=0$
$x=-4$ $x=3$

43. $8x^2+2x=3$
$8x^2+2x-3=0$
$(4x+3)(2x-1)=0$
$4x+3=0$ or $2x-1=0$
$4x=-3$ $2x=1$
$x=-\frac{3}{4}$ $x=\frac{1}{2}$

45. $2x^2-10x=-12$
$2x^2-10x+12=0$
$2(x^2-5x+6)=0$
$2(x-2)(x-3)=0$
$x-2=0$ or $x-3=0$
$x=2$ $x=3$

47. $2x^2=32x$
$2x^2-32x=0$
$2x(x-16)=0$
$x=0$ or $x-16=0$
$x=16$

49. $x^2=100$
$x^2-100=0$
$x^2-10^2=0$
$(x+10)(x-10)=0$
$x+10=0$ or $x-10=0$
$x=-10$ $x=10$

51. $x^2=9$
$x^2-9=0$
$(x+3)(x-3)=0$
$x+3=0$ or $x-3=0$
$x=-3$ $x=3$

53. $-t^2=-81$
$-t^2-81=0$
$t^2-81=0$
$(t+9)(t-9)=0$
$t+9=0$ or $t-9=0$
$t=-9$ $t=9$

55. $P=x^2-15x-50$
$x^2-15x-50=400$
$x^2-15-450=0$
$(x-30)(x+15)=0$
$x-30=0$ or $x+15=0$
$x=30$ $x=-15$
Since x must be positive, she must sell 30 videos for a profit of $400.

57. a. $n^2+n=20$
$n^2+n-20=0$
$(n-4)(n+5)=0$
$n-4=0$ or $n+5=0$
$n=4$ $n=-5$
Since n must be positive, $n=4$.

b. $n^2+n=90$
$n^2+n-90=0$
$(n-9)(n+10)=0$
$n-9=0$ or $n+10=0$
$n=9$ $n=-10$
Since n must be positive, $n=9$.

59. 6 is the only solution, so the only factor is $x-6$, repeated twice.
$(x-6)^2=x^2-12x+36$
The equation is $x^2-12x+36=0$

61. Let a = length of a side of the original square. Then $a+4$ is the length of the new side.
$(a+4)(a+4)=49$
$a^2+8a+16=49$
$a^2+8a-33=0$
$(a-3)(a+11)=0$

$a-3=0$ or $a+11=0$
$a=3$ or $a=-11$
Since length must be positive, the original square had sides of length 3 meters.

63. $d=16t^2$
$256=16t^2$
$\frac{256}{16}=t^2$
$16=t^2$
$0=t^2-16$
$0=(t+4)(t-4)$
$t+4=0$ or $t-4=0$
$t=-4$ $t=4$
Since time must be positive, it would take the egg 4 seconds to hit the ground.

65. Let x be the smaller of the two positive integers. Then $x+4$ is the other integer.
$x(x+4)=117$
$x^2+4x+117=0$
$(x-9)(x+13)=0$
$x-9=0$ or $x+13=0$
$x=9$ $x=-13$
Since x must be positive, the two integers are 9 and $9+4=13$.

67. Let w = width. Then length = $4w$.
$A=lw$
$36=(4w)(w)$
$36=4w^2$
$0=4w^2-36$
$0=4(w^2-9)$
$0=4(w+3)(w-3)$
$w+3=0$ or $w-3=0$
$w=-3$ $w=3$
Since dimensions must be positive, the width
is 3 feet and the length is $4(3)=12$ feet.

69. Let w = width of the garden, l = length of the garden.
$w=\frac{2}{3}l$
$lw=150$
$l\left(\frac{2}{3}l\right)=150$
$2l^2=450$
$2l^2-450=0$
$2(l^2-225)=0$
$2(l^2-15^2)=0$
$2(l+15)(l-15)=0$
$l+15=0$ or $l-15=0$
$l=-15$ $l=15$
Since dimensions must be positive, the length is 15 feet and the width is $\frac{2}{3}(15)=10$ feet.

71. Let a be the length of the side of the second square. Then the length of the side of the first square is $b=\frac{3}{4}a$. The sum of the areas is a^2+b^2.
$a^2+b^2=100$
$a^2+\left(\frac{3}{4}a\right)^2=100$
$a^2+\frac{9}{16}a^2=100$
$\frac{25}{16}a^2=100$
$25a^2=1600$
$25a^2-1600=0$
$25(a^2-64)=0$
$25(a^2-8^2)=0$
$25(a+8)(a-8)=0$
$a+8=0$ or $a-8=0$
$a=-8$ $a=8$
Since dimensions are positive, the length of the side of the second square is 8 cm, and the length of the side of the first square is $\frac{3}{4}\cdot 8=6$ cm.

73. Solutions are 0, 3, and 5, so the factors are x, $x-3$, and $x-5$.

$$x(x-3)(x-5) = x\left(x^2 - 8x + 15\right)$$
$$= x^3 + 8x^2 + 15x$$

The equation is $x^3 + 8x^2 + 15x = 0$.

75. $x + y = 9$ and $x^2 + y^2 = 45$

$y = 9 - x$

$$x^2 + (9-x)^2 = 45$$
$$x^2 + 81 - 18x + x^2 = 45$$
$$2x^2 - 18x + 81 - 45 = 0$$
$$2\left(x^2 - 9x + 18\right) = 0$$
$$2(x-3)(x-6) = 0$$
$$x - 3 = 0 \text{ or } x - 6 = 0$$
$$x = 3 \qquad x = 6$$

When $x = 3$, $y = 6$ and when $x = 6$, $y = 3$. The numbers are 3 and 6.

79. The slope of a vertical line is undefined because you cannot divide by zero.

80. $(3x+2) - \left(x^2 - 4x + 6\right) = 3x + 2 - x^2 + 4x - 6$

$$= -x^2 + 3x + 4x + 2 - 6$$
$$= -x^2 + 7x - 4$$

81. $\left(3x^2 + 2x - 4\right)(2x-1) = 3x^2(2x-1) + 2x(2x-1) - 4(2x-1)$

$$= 6x^3 - 3x^2 + 4x^2 - 2x - 8x + 4$$
$$= 6x^3 + x^2 - 10x + 4$$

82.

$$\begin{array}{r} 2x - 3 \\ 3x - 5 \overline{)\ 6x^2 - 19x + 15} \\ \underline{6x^2 - 10x} \quad\quad \\ -9x + 15 \\ \underline{-9x + 15} \\ 0 \end{array}$$

$$\frac{6x^2 - 19x + 15}{3x - 5} = 2x - 3$$

83. $\dfrac{6x^2 - 19x + 15}{3x-5} = \dfrac{(3x-5)(2x-3)}{3x-5}$

$$= 2x - 3$$

Review Exercises

1. The greatest common factor is y^2.
2. The greatest common factor is $3p$.
3. The greatest common factor is $5a^2$.
4. The greatest common factor is $4x^2y^2$.
5. The greatest common factor is 1.
6. The greatest common factor is 1.
7. The greatest common factor is $x - 5$.

8. The greatest common factor is $x + 5$.

9. $3x-9=3(x-3)$

10. $12x+6=6(2x+1)$

11. $24y^2-4y=4y(6y-1)$

12. $55p^3-20p^2=5p^2(11p-4)$

13. $48a^2b-36ab^2=12ab(4a-3b)$

14. $6xy-12x^2y=6xy(1-2x)$

15. $60x^4y^2+6x^9y^3-18x^5y^2$
$=6x^4y^2\left(10+x^5y-3x\right)$

16. $24x^2-13y^2+6xy$ cannot be factored.

17. $14a^2b-7b-a^3$ cannot be factored.

18. $x(5x+3)-2(5x+3)=(x-2)(5x+3)$

19. $3x(x-1)-2(x-1)=(3x-2)(x-1)$

20. $2x(4x-3)+4x-3=2x(4x-3)+1(4x-3)$
$=(2x+1)(4x-3)$

21. $x^2+5x+2x+10=x(x+5)+2(x+5)$
$=(x+2)(x+5)$

22. $x^2-3x+4x-12=x(x-3)+4(x-3)$
$=(x+4)(x-3)$

23. $y^2-7y-7y+69=y(y-7)-7(y-7)$
$=(y-7)(y-7)$
$=(y-7)^2$

24. $3a^2-3ab-a+b=3a(a-b)-1(a-b)$
$=(3a-1)(a-b)$

25. $3xy+3x+2y+2=3x(y+1)+2(y+1)$
$=(3x+2)(y+1)$

26. $x^2+3x-2xy-6y=x(x+3)-2y(x+3)$
$=(x-2y)(x+3)$

27. $5x^2+20x-x-4=5x(x+4)-1(x+4)$
$=(5x-1)(x+4)$

28. $5x^2-xy+20xy-4y^2=x(5x-y)+4y(5x-y)$
$=(x+4y)(5x-y)$

29. $4x^2+12xy-5xy-15y^2=4x(x+3y)-5y(x+3y)$
$=(4x-5y)(x+3y)$

30. $6a^2-10ab-3ab+5b^2=2a(3a-5b)-b(3a-5b)$
$=(2a-b)(3a-5b)$

31. $ab-a+b-1=a(b-1)+1(b-1)$
$=(a+1)(b-1)$

32. $3x^2-9xy+2xy-6y^2=3x(x-3y)+2y(x-3y)$
$=(3x+2y)(x-3y)$

33. $7a^2 + 14ab - ab - 2b^2 = 7a(a+2b) - b(a+2b)$
$= (7a - b)(a + 2b)$

34. $6x^2 + 9x - 2x - 3 = 3x(2x+3) - 1(2x+3)$
$= (3x-1)(2x+3)$

35. $x^2 - 7x - 8 = (x+1)(x-8)$

36. $x^2 + 8x - 15$ cannot be factored.

37. $x^2 - 13x + 42 = (x-6)(x-7)$

38. $x^2 + x - 20 = (x-4)(x+5)$

39. $x^2 + 11x + 30 = (x+6)(x+5)$

40. $x^2 - 15x + 56 = (x-8)(x-7)$

41. $x^2 - 12x - 44$ cannot be factored.

42. $x^2 + 11x - 26$ cannot be factored.

43. $x^3 - 17x^2 + 72x = x(x^2 - 17x + 72)$
$= x(x-9)(x-8)$

44. $x^3 - 3x^2 - 40x = x(x^2 - 3x - 40)$
$= x(x-8)(x+5)$

45. $x^2 - 2xy - 15y^2 = (x-5y)(x+3y)$

46. $4x^3 + 32x^2y + 60xy^2 = 4x(x^2 + 8xy + 15y^2)$
$= 4x(x+3y)(x+5y)$

47. $2x^2 + 11x + 12 = (2x+3)(x+4)$

48. $3x^2 - 13x + 4 = (3x-1)(x-4)$

49. $4x^2 - 9x + 5 = (4x-5)(x-1)$

50. $5x^2 - 13x - 6 = (5x+2)(x-3)$

51. $9x^2 + 3x - 2 = (3x-1)(3x+2)$

52. $5x^2 - 32x + 12 = (5x-2)(x-6)$

53. $3x^2 + 13x + 8$ cannot be factored.

54. $6x^2 + 31x + 5 = (6x+1)(x+5)$

55. $5x^2 + 37x - 24 = (5x-3)(x+8)$

56. $6x^2 + 11x - 10 = (3x-2)(2x+5)$

57. $8x^2 - 18x - 35 = (4x+5)(2x-7)$

58. $9x^2 - 6x + 1 = (3x-1)(3x-1)$
$= (3x-1)^2$

59. $9x^3 - 12x^2 + 4x = x(9x^2 - 12x + 4)$
$= x(3x-2)(3x-2)$
$= x(3x-2)^2$

60. $18x^3 - 24x^2 - 10x = 2x(9x^2 - 12x - 5)$
$= 2x(3x+1)(3x-5)$

61. $16a^2 - 22ab - 3b^2 = (8a+b)(2a-3b)$

62. $4a^2 - 16ab + 15b^2 = (2a-3b)(2a-5b)$

63. $x^2 - 25 = x^2 - 5^2$
$= (x+5)(x-5)$

64. $x^2 - 100 = x^2 - 10^2$
$= (x+10)(x-10)$

65. $4x^2 - 16 = 4(x^2 - 4)$
$= 4(x^2 - 2^2)$
$= 4(x+2)(x-2)$

66. $81x^2 - 9y^2 = 9(9x^2 - y^2)$
$= 9[(3x)^2 - y^2]$
$= 9(3x+y)(3x-y)$

67. $49 - x^2 = 7^2 - x^2$
$= (7+x)(7-x)$

68. $64 - x^2 = 8^2 - x^2$
$= (8+x)(8-x)$

69. $16x^4 - 49y^2 = (4x^2)^2 - (7y)^2$
$= (4x^2 + 7y)(4x^2 - 7y)$

70. $100x^4 - 121y^4$
$= (10x^2)^2 - (11y^2)^2$
$= (10x^2 + 11y^2)(10x^2 - 11y^2)$

71. $x^3 - y^3 = (x-y)(x^2 + xy + y^2)$

72. $x^3 + y^3 = (x+y)(x^2 - xy + y^2)$

73. $x^3 - 1 = (x-1)(x^2 + x + 1)$

74. $x^3 + 8 = x^3 + 2^3$
$= (x+2)(x^2 - 2x + 4)$

75. $a^3 + 27 = a^3 + 3^3$
$= (a+3)(a^2 - 3a + 9)$

76. $x^3 - 8 = x^3 - 2^3$
$= (x-2)(x^2 + 2x + 4)$

77. $125a^3 + b^3 = (5a)^3 + b^3$
$= (5a+b)(25a^2 - 5ab + b^2)$

78. $27 - 8y^3 = 3^3 - (2y)^3$
$= (3-2y)(9 + 6y + 4y^2)$

79. $27x^4 - 75y^2 = 3(9x^4 - 25y^2)$
$= 3[(3x^2)^2 - (5y)^2]$
$= 3(3x^2 + 5y)(3x^2 - 5y)$

80. $2x^3 - 128y^3 = 2(x^3 - 64y^3)$
$= 2[x^3 - (4y)^3]$
$= 2(x-4y)(x^2 + 4xy + 16y^2)$

81. $x^2 - 16x + 60 = (x-6)(x-10)$

82. $3x^2 - 18x + 27 = 3(x^2 - 6x + 9)$
$= 3(x-3)^2$

83. $4a^2 - 64 = 4(a^2 - 16)$
$= 4(a^2 - 4^2)$
$= 4(a+4)(a-4)$

84. $4y^2 - 36 = 4(y^2 - 9)$
$= 4(y^2 - 3^2)$
$= 4(y+3)(y-3)$

85. $8x^2 + 16x - 24 = 8(x^2 + 2x - 3)$
$= 8(x+3)(x-1)$

86. $x^2 - 6x - 27 = (x-9)(x+3)$

87. $9x^2 + 3x - 2 = (3x-1)(3x+2)$

88. $4x^2 + 7x - 2 = (4x-1)(x+2)$

89. $8x^3-8=8(x^3-1)$
$=8(x^3-1^3)$
$=8(x-1)(x^2+x+1)$

90. $x^3y-27y=y(x^3-27)$
$=y(x^3-3^3)$
$=y(x-3)(x^2+3x+9)$

91. $a^2b-2ab-15b=b(a^2-2a-15)$
$=b(a+3)(a-5)$

92. $6x^3+30x^2+9x^2+45x$
$=3x(2x^2+10x+3x+15)$
$=3x[2x(x+5)+3(x+5)]$
$=3x(2x+3)(x+5)$

93. $x^2+5xy+6y^2=(x+2y)(x+3y)$

94. $2x^2-xy-10y^2=(2x-5y)(x+2y)$

95. $4x^2-20xy+25y^2=(2x-5y)(2x-5y)$
$=(2x-5y)^2$

96. $25a^2-49b^2=(5a)^2-(7b)^2$
$=(5a+7b)(5a-7b)$

97. $xy-7x+2y-14=x(y-7)+2(y-7)$
$=(x+2)(y-7)$

98. $16y^5-25y^7=y^5(16-25y^2)$
$=y^5[4^2-(5y)^2]$
$=y^5(4+5y)(4-5y)$

99. $2x^3+12x^2y+16xy^2=2x(x^2+6xy+8y^2)$
$=2x(x+2y)(x+4y)$

100. $6x^2+5xy-21y^2=(2x-3y)(3x+7y)$

101. $16x^4-8x^3-3x^2=x^2(16x^2-8x-3)$
$=x^2(4x+1)(4x-3)$

102. $a^4-1=(a^2)^2-1^2$
$=(a^2+1)(a^2-1)$
$=(a^2+1)(a+1)(a-1)$

103. $x(x+6)=0$
$x=0$ or $x+6=0$
$x=-6$

104. $(x-2)(x+8)=0$
$x-2=0$ or $x+8=0$
$x=2$ $\quad x=-8$

105. $(x+5)(4x-3)=0$
$x+5=0$ or $4x-3=0$
$x=-5$ $\quad 4x=3$
$x=\frac{3}{4}$

106. $x^2-3x=0$
$x(x-3)=0$
$x=0$ or $x-3=0$
$x=3$

107. $5x^2+20x=0$
$5x(x+4)=0$
$x=0$ or $x+4=0$
$x=-4$

108. $7x^2+14x=0$
$7x(x+2)=0$
$x=0$ or $x+2=0$
$x=-2$

109. $x^2+7x+10=0$
$(x+2)(x+5)=0$
$x+2=0$ or $x+5=0$
$x=-2$ $\quad$ $x=-5$

110. $x^2-12=-x$
$x^2+x-12=0$
$(x+4)(x-3)=0$
$x+4=0$ or $x-3=0$
$x=-4$ $\quad$ $x=3$

111. $x^2-3x=-2$
$x^2-3x+2=0$
$(x-1)(x-2)=0$
$x-1=0$ or $x-2=0$
$x=1$ $\quad$ $x=2$

112. $3x^2+15x+12=0$
$3\left(x^2+5x+4\right)=0$
$3(x+1)(x+4)=0$
$x+1=0$ or $x+4=0$
$x=-1$ $\quad$ $x=-4$

113. $x^2-6x+8=0$
$(x-4)(x-2)=0$
$x-4=0$ or $x-2=0$
$x=4$ $\quad$ $x=2$

114. $6x^2+6x-12=0$
$6\left(x^2+x-2\right)=0$
$6(x+2)(x-1)=0$
$x+2=0$ or $x-1=0$
$x=-2$ $\quad$ $x=1$

115. $8x^2-3=-10x$
$8x^2+10x-3=0$
$(4x-1)(2x+3)=0$
$4x-1=0$ or $2x+3=0$
$4x=1$ $\quad$ $2x=-3$
$x=\frac{1}{4}$ $\quad$ $x=-\frac{3}{2}$

116. $2x^2+15x=8$
$2x^2+15x-8=0$
$(2x-1)(x+8)=0$
$2x-1=0$ or $x+8=0$
$2x=1$ $\quad$ $x=-8$
$x=\frac{1}{2}$

117. $4x^2-16=0$
$4\left(x^2-4\right)=0$
$4\left(x^2-2^2\right)=0$
$4(x+2)(x-2)=0$
$x+2=0$ or $x-2=0$
$x=-2$ $\quad$ $x=2$

118. $49x^2-100=0$
$(7x)^2-10^2=0$
$(7x+10)(7x-10)=0$
$7x+10=0$ or $7x-10=0$
$7x=-10$ $\quad$ $7x=10$
$x=-\frac{10}{7}$ $\quad$ $x=\frac{10}{7}$

119. $8x^2-14x+3=0$
$(2x-3)(4x-1)=0$
$2x-3=0$ or $4x-1=0$
$2x=3$ $\quad$ $4x=1$
$x=\frac{3}{2}$ $\quad$ $x=\frac{1}{4}$

120. $-48x=-12x^2-45$
$12x^2-48x+45=0$
$3\left(4x^2-16x+15\right)=0$
$3(2x-3)(2x-5)=0$
$2x-3=0$ or $2x-5=0$
$2x=3$ $\quad$ $2x=5$
$x=\frac{3}{2}$ $\quad$ $x=\frac{5}{2}$

121. Let x be the smaller of the two integers. Then the other is $x + 1$.

$$x(x+1) = 156$$
$$x^2 + x = 156$$
$$x^2 + x - 156 = 0$$
$$(x+13)(x-12) = 0$$
$$x+13 = 0 \quad \text{or} \quad x-12 = 0$$
$$x = -13 \qquad x = 12$$

Since the integers must be positive, the smaller is 12 and the larger is 13.

122. Let x be the smaller integer. The larger is $x + 2$.

$$x(x+2) = 48$$
$$x^2 + 2x = 48$$
$$x^2 + 2x - 48 = 0$$
$$(x+8)(x-6) = 0$$
$$x+8 = 0 \quad \text{or} \quad x-6 = 0$$
$$x = -8 \qquad x = 6$$

Since the integers must be positive, they are 6 and 8.

123. Let x be the smaller integer. Then the larger is $2x+6$.

$$x(2x+6) = 56$$
$$2x^2 + 6x = 56$$
$$2x^2 + 6x - 56 = 0$$
$$2\left(x^2 + 3x - 28\right) = 0$$

124. Let w be the width of the rectangle. Then the length is $w + 2$.

$$w(w+2) = 63$$
$$w^2 + 2w = 63$$
$$w^2 + 2w - 63 = 0$$
$$(w+9)(w-7) = 0$$
$$w+9 = 0 \quad \text{or} \quad w-7 = 0$$
$$w = -9 \qquad w = 7$$

Since the width must be positive, it is 7 feet, and the length is 9 feet.

125. Let x be the length of a side of the original square. Then $x - 4$ is the length of a side of the smaller square.

$$(x-4)^2 = 25$$
$$x^2 - 8x + 16 = 25$$
$$x^2 - 8x + 16 - 25 = 0$$
$$x^2 - 8x - 9 = 0$$
$$(x-9)(x+1) = 0$$
$$x-9 = 0 \quad \text{or} \quad x+1 = 0$$
$$x = 9 \qquad x = -1$$

Since lengths must be positive, the length of a side of the original square is 9 inches.

126. $C = x^2 - 79x + 20$

$$100 = x^2 - 79x + 20$$
$$0 = x^2 - 79x - 80$$
$$0 = (x-80)(x+1)$$
$$x-80 = 0 \quad \text{or} \quad x+1 = 0$$
$$x = 80 \qquad x = -1$$

Since only a positive number of dozens of cookies can be made, the association can make 80 dozen cookies.

Practice Test

1. The greatest common factor is $3y^2$.

2. The greatest common factor is $3xy^2$.

3. $10x^3y^2 - 8xy = 2xy\left(5x^2y - 4\right)$

4. $8a^3b - 12a^2b^2 + 28a^2b = 4a^2b(2a - 3b + 7)$

5. $x^2 - 8x + 2x - 16 = x(x-8) + 2(x-8)$
$= (x+2)(x-8)$

6. $5x^2 - 15x + 2x - 6 = 5x(x-3) + 2(x-3)$
$= (5x+2)(x-3)$

7. $a^2 - 4ab - 5ab + 20b^2$
$= a(a-4b) - 5b(a-4b)$
$= (a-5b)(a-4b)$

8. $x^2 + 12x + 32 = (x+4)(x+8)$

9. $x^2 + 5x - 24 = (x+8)(x-3)$

10. $25a^2 - 5ab - 6b^2 = (5a-3b)(5a+2b)$

11. $4x^2 - 16x - 48 = 4(x^2 - 4x - 12)$
$= 4(x+2)(x-6)$

12. $2x^3 - 3x^2 + x = x(2x^2 - 3x + 1)$
$= x(2x-1)(x-1)$

13. $12x^2 - xy - 6y^2 = (3x+2y)(4x-3y)$

14. $x^2 - 9y^2 = x^2 - (3y)^2$
$= (x+3y)(x-3y)$

15. $x^3 + 27 = x^3 + 3^3$
$= (x+3)(x^2 - 3x + 9)$

16. $(5x-3)(x-1) = 0$
$5x - 3 = 0$ or $x - 1 = 0$
$5x = 3$ $\quad x = 1$
$x = \frac{3}{5}$

17. $x^2 - 6x = 0$
$x(x-6) = 0$
$x = 0$ or $x - 6 = 0$
$x = 6$

18. $x^2 = 64$
$x^2 - 64 = 0$
$x^2 - 8^2 = 0$
$(x+8)(x-8) = 0$
$x + 8 = 0$ or $x - 8 = 0$
$x = -8$ $\quad x = 8$

19. $x^2 - 14x + 49 = 0$
$(x-7)^2 = 0$
$x - 7 = 0$
$x = 7$

20. $x^2 + 6 = -5x$
$x^2 + 5x + 6 = 0$
$(x+2)(x+3) = 0$
$x - 2 = 0$ or $x + 3 = 0$
$x = -2$ $\quad x = -3$

21. $x^2 - 7x + 12 = 0$
$(x-3)(x-4) = 0$
$x - 3 = 0$ or $x - 4 = 0$
$x = 3$ $\quad x = 4$

22. Let x be the smaller of the two integers. Then $2x+1$ is the larger.
$x(2x+1) = 36$
$2x^2 + x - 36 = 0$
$(x-4)(2x+9) = 0$
$x - 4 = 0$ or $2x + 9 = 0$
$x = 4$ $\quad 2x = -9$
$x = -\frac{9}{2}$
Since x must be positive and an integer, the smaller integer is 4 and the larger is $2 \cdot 4 + 1 = 9$.

23. Let x be the smaller of the two consecutive odd integers. Then $x + 2$ is the larger.
$x(x+2) = 99$
$x^2 + 2x - 99 = 0$
$(x-9)(x+11) = 0$
$x - 9 = 0$ or $x + 11 = 0$
$x = 9$ $\quad x = -11$
Since x must be positive, then the smaller integer is 9 and the larger is 11.

24. Let w be the width of the rectangle. Then the length is $w + 2$.
$w(w+2) = 24$
$w^2 + 2w = 24$
$w^2 + 2w - 24 = 0$
$(w+6)(w-4) = 0$
$w + 6 = 0$ or $w - 4 = 0$
$w = -6$ $\quad w = 4$
Since the width is positive, it is 4 meters, and the length is 6 meters.

25. $d = 16t^2$

$$1600 = 16t^2$$
$$16t^2 - 1600 = 0$$
$$16(t^2 - 100) = 0$$
$$16(t^2 - 10^2) = 0$$
$$16(t+10)(t-10) = 0$$
$$t + 10 = 0 \quad \text{or} \quad t - 10 = 0$$
$$t = -10 \qquad t = 10$$

Since time must be positive, then it would take the object 10 seconds to fall 1600 feet to the ground.

Cumulative Review Test

1.
$$\begin{aligned} 3 - 7(x + 4x^2 - 21) &= 3 - 7[-5 + 4(-5)^2 - 21] \\ &= 3 - 7[-5 + 4(25) - 21] \\ &= 3 - 7(-5 + 100 - 21) \\ &= 3 - 7(74) \\ &= 3 - 518 \\ &= -515 \end{aligned}$$

2.
$$\begin{aligned} 5x^2 - 3y + 7(2 + y^2 - 4x) &= 5(3)^2 - 3(-2) + 7[2 + (-2)^2 - 4(3)] \\ &= 5(9) + 6 + 7(2 + 4 - 12) \\ &= 45 + 6 + 7(-6) \\ &= 51 - 42 \\ &= 9 \end{aligned}$$

3. Let x = the total estimated population in 2010. Then 13.2% of x is 39,408 thousand.

$$0.132x = 39{,}408{,}000$$
$$x = \frac{39{,}408{,}000}{0.132} \approx 298{,}545{,}455$$

The population is estimated to be about 298,545,455 in 2010.

4.
$$3x - 4 = 5(x - 7) + 2x$$
$$3x - 4 = 5x - 35 + 2x$$
$$3x - 4 = 7x - 35$$
$$31 = 4x$$
$$\frac{31}{4} = x$$

5.
$$3x - 5 \geq 10(6 - x)$$
$$3x - 5 \geq 60 - 10x$$
$$13x \geq 65$$
$$x \geq 5$$

6.
$$2x - 3y = 6$$
$$-3y = -2x + 6$$
$$y = \frac{-2x + 6}{-3}$$
$$y = \frac{2}{3}x - 2$$

7. Let t = the number of hours that Brooke has been skiing. Then Bob has been skiing for $\left(t + \frac{1}{4}\right)$ hours.

	rate	time	distance
Brooke	8 kph	t	$8t$
Bob	4 kph	$t+\frac{1}{4}$	$4\left(t+\frac{1}{4}\right)$

Brooke catches Bob when they have both gone the same distance.

$$8t = 4\left(t+\frac{1}{4}\right)$$
$$8t = 4t+1$$
$$4t = 1$$
$$t = \frac{1}{4}$$

It will take Brooke $\frac{1}{4}$ hour to catch Bob.

8. Let d = the number of days. The cost in Orlando is 1370 + 160d. The cost in New York City is 919 + 342d. The cost in Orlando is less than the cost in New York City when
1370 + 160d < 919 + 342d.

$$1370+160d < 919+342d$$
$$451 < 182d$$
$$\frac{451}{182} < d$$

Since $\frac{451}{182} \approx 2.5$, the conference will be less expensive in Orlando if it lasts for more than 2 days.

9. For the x-intercept, set $y = 0$.

$$4x+6y = 24$$
$$4x+6(0) = 24$$
$$4x = 24$$
$$x = 6$$

The x-intercept is (6, 0).
For the y-intercept, set $x = 0$.

$$4x+6y = 24$$
$$4(0)+6y = 24$$
$$6y = 24$$
$$y = 4$$

The y-intercept is (0, 4).

10. $m = \frac{-8-5}{6-(-3)} = \frac{-13}{6+3} = -\frac{13}{9}$

11. Use the point-slope form with $m = \frac{1}{2}$ and $(x_1,\ y_1) = (4,\ -3)$.

$$y - y_1 = m(x - x_1)$$
$$y-(-3) = \frac{1}{2}(x-4)$$
$$y+3 = \frac{1}{2}x-2$$
$$y = \frac{1}{2}x-5$$

12.

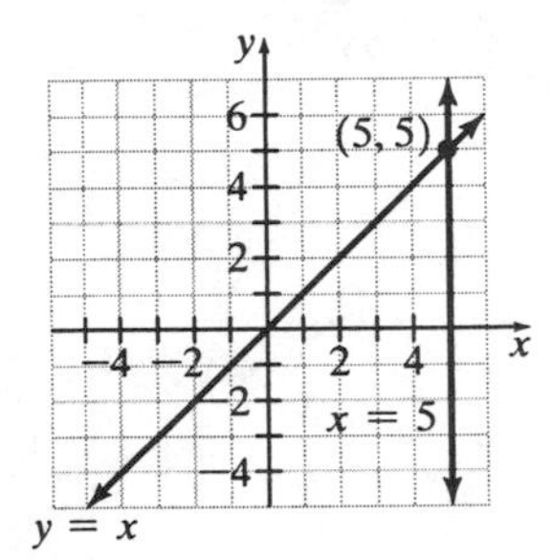

13. Let m = Mallory's current age. One year ago, Mallory's age was m – 1. At that time, Chris' age was (m – 1) + 6. Since Chris was twice Mallory's age, 2(m – 1) = (m – 1) + 6.

$$2m-2 = m+5$$
$$m = 7$$

Mallory is 7 years old.

14.

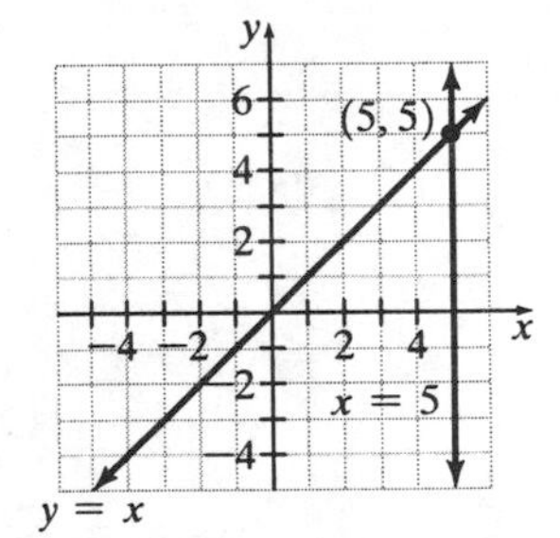

The solution is (5, 5).

15. $y = x + 1$
$y - 3x = -1$
Substitute $x + 1$ for y in the second equation.
$y - 3x = -1$
$x + 1 - 3x = -1$
$-2x = -2$
$x = 1$
Substitute 1 for x in the first equation.
$y = x + 1$
$y = 1 + 1$
$y = 2$
The solution is (1, 2).

16. $\left(\frac{3x}{2y}\right)^3 = \frac{3^3x^3}{2^3y^3} = \frac{27x^3}{8y^3}$

17. $(2x^{-3})^{-2}(4x^{-3}y^2)^3$
$= 2^{-2}x^{-3(-2)}4^3x^{-3(3)}y^{2(3)}$
$= 2^{-2}x^6 4^3x^{-9}y^6$
$= 2^{-2} \cdot 4^3x^{-3}y^6$
$= \frac{4^3y^6}{2^2x^3}$
$= \frac{64y^6}{4x^3}$
$= \frac{16y^6}{x^3}$

18. $(x^3 - x^2 + 6x - 5) - (2x^2 - 3x + 7)$
$= x^3 - x^2 + 6x - 5 - 2x^2 + 3x - 7$
$= x^3 - x^2 - 2x^2 + 6x + 3x - 5 - 7$
$= x^3 - 3x^2 + 9x - 12$

19. $x^2 - x - 110 = (x - 11)(x + 10)$

20. $5x^3 - 125x = 5x(x^2 - 25)$
$= 5x(x^2 - 5^2)$
$= 5x(x + 5)(x - 5)$

Chapter 8

Exercise Set 8.1

1. Answers will vary.

3. The value of the variable does not make the denominator equal to 0.

5. There is no factor common to both terms in the numerator of $\frac{3x+4y}{12xy}$.

7. The denominator cannot be 0.

9. $x-3=0$
$x \neq 3$

11. $-\frac{x+4}{4-x} = \frac{x+4}{-(4-x)}$
$= \frac{x+4}{-4+x}$
$= \frac{x+4}{x-4}$
$\frac{x+4}{x-4} \neq -1$
No

13. The expression is defined for all real numbers except $x = 0$.

15. $2x-6=0$
$2x=6$
$x=\frac{6}{2}$
The expression is defined for all real numbers except $x = 3$.

17. $x^2-4=0$
$(x-2)(x+2)=0$
The expression is defined for all real numbers except $x = 2$, $x = -2$.

19. $x^2+6x-16=0$
$(x-2)(x+8)=0$
The expression is defined for all real numbers except $x = 2$, $x = -8$.

21. $\frac{x}{x+xy} = \frac{x}{x(1+y)}$
$= \frac{1}{1+y}$

23. $\frac{5x+15}{x+3} = \frac{5(x+3)}{x+3}$
$= 5$

25. $\frac{x^3+6x^2+3x}{2x} = \frac{x\left(x^2+6x+3\right)}{2x}$
$= \frac{x^2+6x+3}{2}$

27. $\frac{x^2+2x+1}{x+1} = \frac{(x+1)^2}{x+1}$
$= x+1$

29. $\frac{x^2+2x}{x^2+4x+4} = \frac{x(x+2)}{(x+2)^2}$
$= \frac{x}{x+2}$

31. $\frac{x^2-x-6}{x^2-4} = \frac{(x-3)(x+2)}{(x-2)(x+2)}$
$= \frac{x-3}{x-2}$

33. $\frac{x^2-2x-3}{x^2-x-6} = \frac{(x+1)(x-3)}{(x+2)(x-3)}$
$= \frac{x+1}{x+2}$

35. $\frac{2x-3}{3-2x} = \frac{2x-3}{-(2x-3)}$
$= -1$

37. $\frac{x^2-2x-8}{4-x} = \frac{(x-4)(x+2)}{-(x-4)}$
$= -(x+2)$

39. $\frac{x^2+3x-18}{-2x^2+6x}=\frac{(x+6)(x-3)}{-2x(x-3)}$
$=-\frac{x+6}{2x}$

41. $\frac{2x^2+5x-3}{1-2x}=\frac{(2x-1)(x+3)}{-(2x-1)}$
$=-(x+3)$

43. $\frac{6x^2+x-2}{2x-1}=\frac{(2x-1)(3x+2)}{2x-1}$
$=3x+2$

45. $\frac{6x^2+7x-20}{2x+5}=\frac{(2x+5)(3x-4)}{2x+5}$
$=3x-4$

47. $\frac{6x^2-13x+6}{3x-2}=\frac{(3x-2)(2x-3)}{3x-2}$
$=2x-3$

49. $\frac{x^2-3x+4x-12}{x-3}=\frac{x(x-3)+4(x-3)}{x-3}$
$=\frac{(x+4)(x-3)}{x-3}$
$=x+4$

51. $\frac{2x^2-8x+3x-12}{2x^2+8x+3x+12}=\frac{2x(x-4)+3(x-4)}{2x(x+4)+3(x+4)}$
$=\frac{(x-4)(2x+3)}{(x+4)(2x+3)}$
$=\frac{x-4}{x+4}$

53. $\frac{x^3-8}{x-2}=\frac{(x-2)\left(x^2+2x+4\right)}{x-2}$
$=x^2+2x+4$

55. $\frac{5\Delta}{15}=\frac{5\Delta}{5\cdot 3}=\frac{\Delta}{3}$

57. $\frac{7\Delta}{14\Delta+21}=\frac{7\Delta}{7(2\Delta+3)}=\frac{\Delta}{2\Delta+3}$

59. $\frac{3\Delta-2}{2-3\Delta}=\frac{-(2-3\Delta)}{(2-3\Delta)}=-1$

61. $x^2-x-6=(x-3)(x+2)$
Denominator $= x+2$

63. $(x+3)(x+4)=x^2+7x+12$
Numerator $= x^2+7x+12$

65. a. $\frac{x+3}{x^2-2x+3x-6}=\frac{x+3}{x(x-2)+3(x-2)}$
$=\frac{x+3}{(x+3)(x-2)}$

$x\neq -3,\ x\neq 2$

b. $\frac{x+3}{(x+3)(x-2)}=\frac{1}{x-2}$

67. a. $\frac{x+5}{2x^3+7x^2-15x}=\frac{x+5}{x\left(2x^2+7x-15\right)}$
$=\frac{x+5}{x(2x-3)(x+5)}$

$x\neq 0,\ x\neq \frac{3}{2},\ x\neq -5$

b. $\frac{x+5}{x(2x-3)(x+5)}=\frac{1}{x(2x-3)}$

69. $\frac{\frac{1}{5}x^5-\frac{2}{3}x^4}{\frac{1}{5}x^5-\frac{2}{3}x^4}=1$

72. $z=\frac{x-y}{2}$
$2z=x-y$
$2z-x=-y$
$y=x-2z$

73. Let x = measure of the smallest angle.
Then the second angle = $x + 30$ and third angle = $3x + 10$.
angle 1 + angle 2 + angle 3 = 180°

$$x+(x+30)+(3x+10)=180$$
$$5x+40=180$$
$$5x=140$$
$$x=28$$
$x+30=28+30=58$

$3x+10=3(28)+10=84+10=94$. The three angles are 28°, 58°, and 94°.

74. $3x-4y=1$

$2x-2y=2$

Multiply the second equation by –2.

$-2[2x-2y=2]$

gives

$$\begin{aligned}3x-4y&=1\\ -4x+4y&=-4\\ \hline -x\quad&=-3\\ x&=3\end{aligned}$$

Substitute 3 for x in the first equation.

$$\begin{aligned}3x-4y&=1\\ 3(3)-4y&=1\\ 9-4y&=1\\ -4y&=-8\\ y&=2\end{aligned}$$

The solution is (3, 2).

75. $$\left(\frac{3x^6y^2}{9x^4y^3}\right)^2=\left(\frac{x^2}{3y}\right)^2=\frac{x^4}{9y^2}$$

76. $6x^2-4x-8-\left(-3x^2+6x+9\right)$

$=6x^2-4x-8+3x^2-6x-9$

$=9x^2-10x-17$

Exercise Set 8.2

1. Answers will vary.

3. $\dfrac{x+3}{x-4}\cdot\dfrac{\square}{x+3}=x+2$

Numerator must be

$(x+2)(x-4)=x^2-2x-8$

5. $\dfrac{x-4}{x+5}\cdot\dfrac{x+5}{\square}=\dfrac{1}{x+3}$

Denominator must be

$(x+3)(x-4)=x^2-x-12$

7. $$\frac{5x}{3y}\cdot\frac{y^2}{10}=\frac{5x}{3y}\cdot\frac{y^2}{5\cdot 2}=\frac{xy}{6}$$

9. $$\frac{16x^2}{y^4}\cdot\frac{5x^2}{y^2}=\frac{80x^4}{y^6}$$

11. $$\frac{6x^5y^3}{5z^3}\cdot\frac{6x^4}{5yz^4}=\frac{36x^9y^2}{25z^7}$$

13. $$\frac{3x-2}{3x+2}\cdot\frac{4x-1}{1-4x}=\frac{3x-2}{3x+2}\cdot\frac{4x-1}{-(4x-1)}=\frac{-3x+2}{3x+2}$$

15. $$\frac{x^2+7x+12}{x+4}\cdot\frac{1}{x+3}=\frac{(x+4)(x+3)}{(x+4)(x+3)}=1$$

17. $$\frac{a}{a^2-b^2}\cdot\frac{a+b}{a^2+ab}=\frac{a(a+b)}{(a+b)(a-b)\cdot a(a+b)}=\frac{1}{(a-b)(a+b)}=\frac{1}{a^2-b^2}$$

19. $\frac{6x^2-14x-12}{6x+4}\cdot\frac{x+3}{2x^2-2x-12}=\frac{2\left(3x^2-7x-6\right)}{2(3x+2)}\cdot\frac{x+3}{2\left(x^2-x-6\right)}$
$=\frac{2(3x+2)(x-3)(x+3)}{2(3x+2)\cdot 2(x-3)(x+2)}$
$=\frac{x+3}{2(x+2)}$

21. $\frac{x+3}{x-3}\cdot\frac{x^3-27}{x^2+3x+9}=\frac{(x+3)(x-3)\left(x^2+3x+9\right)}{(x-3)\left(x^2+3x+9\right)}$
$=x+3$

23. $\frac{9x^3}{y^2}\div\frac{3x}{y^3}=\frac{9x^3}{y^2}\cdot\frac{y^3}{3x}$
$=\frac{3\cdot 3x^3}{y^2}\cdot\frac{y^3}{3x}$
$=3x^2y$

25. $\frac{25xy^2}{7z}\div\frac{5x^2y^2}{14z^2}=\frac{5\cdot 5xy^2}{7z}\cdot\frac{2\cdot 7z^2}{5x^2y^2}$
$=\frac{5\cdot 5xy^2}{7z}\cdot\frac{2\cdot 7z^2}{5x^2y^2}$
$=\frac{10z}{x}$

27. $\frac{xy}{7a^2b}\div\frac{6xy}{7}=\frac{xy}{7a^2b}\cdot\frac{7}{6xy}$
$=\frac{1}{6a^2b}$

29. $\frac{3x^2+6x}{x}\div\frac{2x+4}{x^2}=\frac{3x(x+2)}{x}\cdot\frac{x^2}{2(x+2)}$
$=\frac{3x^2}{2}$

31. $\frac{x^2+3x-18}{x}\div\frac{x-3}{1}=\frac{(x-3)(x+6)}{x}\cdot\frac{1}{(x-3)}$
$=\frac{x+6}{x}$

33. $\frac{x^2-12x+32}{x^2-6x-16} \div \frac{x^2-x-12}{x^2-5x-24} = \frac{x^2-12x+32}{x^2-6x-16} \cdot \frac{x^2-5x-24}{x^2-x-12}$
$= \frac{(x-8)(x-4)}{(x-8)(x+2)} \cdot \frac{(x-8)(x+3)}{(x-4)(x+3)}$
$= \frac{x-8}{x+2}$

35. $\frac{2x^2+9x+4}{x^2+7x+12} \div \frac{2x^2-x-1}{(x+3)^2} = \frac{(2x+1)(x+4)}{(x+3)(x+4)} \cdot \frac{(x+3)^2}{(2x+1)(x-1)}$
$= \frac{x+3}{x-1}$

37. $\frac{x^2-y^2}{x^2-2xy+y^2} \div \frac{x+y}{y-x} = \frac{(x-y)(x+y)}{(x-y)^2} \cdot \frac{-(x-y)}{x+y}$
$= -1$

39. $\frac{9x}{6y^2} \cdot \frac{24x^2y^4}{9x} = 4x^2y^2$

41. $\frac{45a^2b^3}{12c^3} \cdot \frac{4c}{9a^3b^5} = \frac{5}{3ab^2c^2}$

43. $\frac{-xy}{a} \div \frac{-2ax}{6y} = \frac{-xy}{a} \cdot \frac{6y}{-2ax}$
$= \frac{3y^2}{a^2}$

45. $\frac{80m^4}{49x^5y^7} \cdot \frac{14x^{12}y^5}{25m^5} = \frac{5 \cdot 16m^4}{7 \cdot 7x^5y^7} \cdot \frac{2 \cdot 7x^{12}y^5}{5 \cdot 5m^5} = \frac{32x^7}{35y^2m}$

47. $(3x+5) \cdot \frac{1}{6x+10} = (3x+5) \cdot \frac{1}{2(3x+5)}$
$= \frac{1}{2}$

49. $\frac{1}{7x^2y} \div \frac{1}{21x^3y} = \frac{1}{7x^2y} \cdot \frac{3 \cdot 7x^3y}{1}$
$= 3x$

51. $\frac{6\Delta^2}{12} \cdot \frac{12}{36\Delta^5} = \frac{6\Delta^2}{12} \cdot \frac{12}{6 \cdot 6\Delta^5} = \frac{1}{6\Delta^3}$

53. $\dfrac{\Delta - ☺}{9\Delta - 9☺} \div \dfrac{\Delta^2 - ☺^2}{\Delta^2 + 2\Delta ☺ + ☺^2}$

$= \dfrac{\Delta - ☺}{9(\Delta - ☺)} \cdot \dfrac{(\Delta + ☺)^2}{(\Delta + ☺)(\Delta - ☺)} = \dfrac{\Delta + ☺}{9(\Delta - ☺)}$

55. $(x+2)(x+1) = x^2 + x + 2x + 2$

$= x^2 + 3x + 2$

Numerator is $x^2 + 3x + 2$.

57. $(x-5)(x+2) = x^2 + 2x - 5x - 10$

$= x^2 - 3x - 10$

Numerator is $x^2 - 3x - 10$.

59. $\left(x^2 - 4\right)(x-1) \div (x+2)$

$(x-2)(x+2)(x-1) \cdot \dfrac{1}{x+2} = x^2 - x - 2x + 2$

$= x^2 - 3x + 2$

Numerator is $x^2 - 3x + 2$.

61. $\left(\dfrac{x+2}{x^2 - 4x - 12} \cdot \dfrac{x^2 - 9x + 18}{x-2}\right) \div \dfrac{x^2 + 5x + 6}{x^2 - 4} = \dfrac{x+2}{(x-6)(x+2)} \cdot \dfrac{(x-3)(x-6)}{x-2} \cdot \dfrac{(x+2)(x-2)}{(x+2)(x+3)}$

$= \dfrac{x-3}{x+3}$

63. $\left(\dfrac{x^2 + 4x + 3}{x^2 - 6x - 16}\right) \div \left(\dfrac{x^2 + 5x + 6}{x^2 - 9x + 8} \cdot \dfrac{x^2 - 1}{x^2 + 4x + 4}\right) = \dfrac{x^2 + 4x + 3^2}{x^2 - 6x - 16} \cdot \dfrac{x^2 - 9x + 8}{x^2 + 5x + 6} \cdot \dfrac{x^2 + 4x + 4}{x^2 - 1}$

$= \dfrac{(x+1)(x+3)}{(x-8)(x+2)} \cdot \dfrac{(x-8)(x-1)}{(x+2)(x+3)} \cdot \dfrac{(x+2)^2}{(x-1)(x+1)}$

$= 1$

68. $\left(4x^3y^2z^4\right)\left(5xy^3z^7\right) = 4 \cdot 5 \cdot x^3xy^2y^3z^4z^7$

$= 20x^4y^5z^{11}$

69.
$$
\begin{array}{r}
2x^2+x-2 \\
2x-1\overline{\big)\,4x^3+0x^2-5x+0} \\
\underline{4x^3-2x^2} \qquad\qquad \\
2x^2-5x \qquad \\
\underline{2x^2-\ x} \qquad \\
-4x+0 \\
\underline{-4x+2} \\
-2
\end{array}
$$

$$\frac{4x^3-5x}{2x-1}=2x^2+x-2-\frac{2}{2x-1}$$

70. $3x^2-9x-30=3(x^2-3x-10)$
$=3(x-5)(x+2)$

71. $3x^2-9x-30=0$
$3(x^2-3x-10)=0$
$3(x-5)(x+2)=0$
$x-5=0$ or $x+2=0$
$x=5$ $\quad x=-2$

Exercise Set 8.3

1. Answers will vary.

3. Answers will vary.

5. $\frac{2}{x+5}+\frac{3}{5}$
The only factor (other than 1) of the first denominator is $x+5$. The only factor (other than 1) of the second denominator is 5. The LCD is therefore $5(x+5)$.

7. $\frac{6}{x-3}+\frac{1}{x}-\frac{1}{3}$
The only factor (other than 1) of the first denominator is $x-3$. The only factor (other than 1) of the second denominator is x. The only factor (other than 1) of the third denominator is 3. The LCD is therefore $3x(x-3)$.

9. a. The negative sign in $-(3x^2-4x+5)$ was not distributed.

b.
$$\frac{6x-2}{x^2-4x+3}-\frac{3x^2-4x+5}{x^2-4x+3}=\frac{6x-2-(3x^2-4x+5)}{x^2-4x+3}$$
$$=\frac{6x-2-3x^2+4x-5}{x^2-4x+3}$$
$$\neq\frac{6x-2-3x^2-4x+5}{x^2-4x+3}$$

11. $\frac{x-1}{8}+\frac{2x}{8}=\frac{x-1+2x}{8}$
$=\frac{3x-1}{8}$

13. $\frac{5x+3}{7}-\frac{x}{7}=\frac{5x+3-x}{7}$
$=\frac{4x+3}{7}$

15. $\frac{2}{x}+\frac{x+4}{x}=\frac{2+x+4}{x}$
$=\frac{x+6}{x}$

17. $\frac{x-7}{x}+\frac{x+7}{x}=\frac{x-7+x+7}{x}$
$=\frac{2x}{x}$
$=2$

19. $\frac{x}{x-1}+\frac{4x+7}{x-1}=\frac{x+4x+7}{x-1}$
$=\frac{5x+7}{x-1}$

21. $\frac{9x+7}{6x^2}-\frac{3x+4}{6x^2}=\frac{9x+7-(3x+4)}{6x^2}$
$=\frac{9x+7-3x-4}{6x^2}$
$=\frac{6x+3}{6x^2}$
$=\frac{3(2x+1)}{3\left(2x^2\right)}$
$=\frac{2x+1}{2x^2}$

23. $\frac{5x+4}{x^2-x-12}+\frac{-4x-1}{x^2-x-12}=\frac{5x+4-4x-1}{x^2-x-12}$
$=\frac{x+3}{(x+3)(x-4)}$
$=\frac{1}{x-4}$

25. $\frac{x+4}{3x+2}-\frac{x+4}{3x+2}=\frac{x+4-(x+4)}{3x+2}$
$=0$

27. $\frac{3x-5}{x+7}-\frac{6x+5}{x+7}=\frac{3x-5-(6x+5)}{x+7}$
$=\frac{3x-5-6x-5}{x+7}$
$=\frac{-3x-10}{x+7}$

29. $\frac{x^2+4x+3}{x+2}-\frac{5x+9}{x+2}=\frac{x^2+4x+3-(5x+9)}{x+2}$
$=\frac{x^2+4x+3-5x-9}{x+2}$
$=\frac{x^2-x-6}{x+2}$
$=\frac{(x-3)(x+2)}{x+2}$
$=x-3$

31. $\frac{-2x+5}{5x-10}+\frac{2(x-5)}{5x-10}=\frac{-2x+5+2(x-5)}{5x-10}$
$=\frac{-2x+5+2x-10}{5x-10}$
$=\frac{-5}{5(x-2)}$
$=-\frac{1}{x-2}$

33. $\frac{x^2-2x-3}{x^2-x-6}+\frac{x-3}{x^2-x-6}=\frac{x^2-2x-3+x-3}{x^2-x-6}$
$=\frac{x^2-x-6}{x^2-x-6}$
$=1$

35. $\frac{-x-7}{2x-9}-\frac{-3x-16}{2x-9}=\frac{-x-7-(-3x-16)}{2x-9}$
$=\frac{-x-7+3x+16}{2x-9}$
$=\frac{2x+9}{2x-9}$

37. $\frac{x^2+2x}{(x+6)(x-3)}-\frac{15}{(x+6)(x-3)}=\frac{x^2+2x-15}{(x+6)(x-3)}$
$=\frac{(x+5)(x-3)}{(x+6)(x-3)}$
$=\frac{x+5}{x+6}$

39. $\frac{3x^2-7x}{4x^2-8x}+\frac{x}{4x^2-8x}=\frac{3x^2-7x+x}{4x^2-8x}$
$=\frac{3x^2-6x}{4x^2-8x}$
$=\frac{3x(x-2)}{4x(x-2)}$
$=\frac{3}{4}$

41. $\frac{3x^2-4x+4}{3x^2+7x+2}-\frac{10x+9}{3x^2+7x+2}=\frac{3x^2-4x+4-(10x+9)}{3x^2+7x+2}$
$=\frac{3x^2-4x+4-10x-9}{3x^2+7x+2}$
$=\frac{3x^2-14x-5}{3x^2+7x+2}$
$=\frac{(3x+1)(x-5)}{(3x+1)(x+2)}$
$=\frac{x-5}{x+2}$

43. $\dfrac{x^2+3x-6}{x^2-5x+4}-\dfrac{-2x^2+4x-4}{x^2-5x+4}=\dfrac{x^2+3x-6-(-2x^2+4x-4)}{x^2-5x+4}$

$=\dfrac{x^2+3x-6+2x^2-4x+4}{x^2-5x+4}$

$=\dfrac{3x^2-x-2}{x^2-5x+4}$

$=\dfrac{(3x+2)(x-1)}{(x-4)(x-1)}$

$=\dfrac{3x+2}{x-4}$

45. $\dfrac{5x^2+40x+8}{x^2-64}+\dfrac{x^2+9x}{x^2-64}=\dfrac{5x^2+40x+8+x^2+9x}{x^2-64}$

$=\dfrac{6x^2+49x+8}{x^2-64}$

$=\dfrac{(6x+1)(x+8)}{(x-8)(x+8)}$

$=\dfrac{6x+1}{x-8}$

47. $\dfrac{x}{7}+\dfrac{x+4}{7}$

Least common denominator = 7

49. $\dfrac{1}{3x}+\dfrac{1}{3}$

Least common denominator = $3x$

51. $\dfrac{3}{5x}+\dfrac{7}{4}$

Least common denominator $=5x\cdot 4=20x$

53. $\dfrac{4}{x}+\dfrac{3}{x^3}$

Least common denominator $=x^3$

55. $\dfrac{x+3}{6x+5}+x=\dfrac{x+3}{6x+5}+\dfrac{x}{1}$

Least common denominator $=6x+5$

57. $\dfrac{x}{x+1}+\dfrac{4}{x^2}$

Least common denominator $=x^2(x+1)$

59. $\dfrac{x+1}{12x^2y}-\dfrac{7}{9x^3}=\dfrac{x+1}{3\cdot 4x^2y}-\dfrac{7}{3^2x^3}$

Least common denominator

$=4\cdot 3^2\cdot x^3\cdot y=36x^3y$

61. $\dfrac{x^2-7}{24x}-\dfrac{x+3}{9(x+5)}=\dfrac{x^2-7}{3\cdot 8x}-\dfrac{x+3}{3^2(x+5)}$

Least common denominator

$=8\cdot 3^2x(x+5)=72x(x+5)$

63. $\dfrac{5x-2}{x^2+x}-\dfrac{x^2}{x}=\dfrac{5x-2}{x(x+1)}-\dfrac{x^2}{x}$

Least common denominator $=x(x+1)$

65. $\dfrac{21}{24x^2y}+\dfrac{x+4}{15xy^3}=\dfrac{21}{3\cdot 8x^2y}+\dfrac{x+4}{3\cdot 5xy^3}$

Least common denominator

$=8\cdot 3\cdot 5x^2y^3=120x^2y^3$

67. $\frac{3}{3x+12}+\frac{3x+6}{2x+4}=\frac{3}{3(x+4)}+\frac{3x+6}{2(x+2)}$
Least common denominator
$=3\cdot 2(x+4)(x+2)=6(x+4)(x+2)$

69. $\frac{9x+4}{x+6}-\frac{3x-6}{x+5}$
Least common denominator $=(x+6)(x+5)$

71. $\frac{x-2}{x^2-5x-24}+\frac{3}{x^2+11x+24}$
$=\frac{x-2}{(x-8)(x+3)}+\frac{3}{(x+8)(x+3)}$
Least common denominator
$=(x-8)(x+3)(x+8)$

73. $\frac{7}{x+4}-\frac{x+5}{x^2-3x-4}$
$=\frac{7}{x+4}-\frac{x+5}{(x-4)(x+1)}$
Least common denominator
$=(x+4)(x-4)(x+1)$

75. $\frac{2x}{x^2+6x+5}-\frac{5x^2}{x^2+4x+3}$
$=\frac{2x}{(x+5)(x+1)}-\frac{5x^2}{(x+3)(x+1)}$
Least common denominator
$=(x+5)(x+1)(x+3)$

77. $\frac{3x-5}{x^2-6x+9}+\frac{3}{x-3}$
$=\frac{3x+5}{(x-3)^2}+\frac{3}{x-3}$
Least common denominator $=(x-3)^2$

79. $\frac{8x^2}{x^2-7x+6}+x-3$
$=\frac{8x^2}{(x-6)(x-1)}+\frac{x-3}{1}$
Least common denominator $=(x-6)(x-1)$

81. $\frac{x}{3x^2+16x-12}+\frac{6}{3x^2+17x-6}$
$=\frac{x}{(3x-2)(x+6)}+\frac{6}{(3x-1)(x+6)}$
Least common denominator
$=(3x-2)(x+6)(3x-1)$

83. $\frac{2x-3}{4x^2+4x+1}+\frac{x^2-4}{8x^2+10x+3}$
$=\frac{2x-3}{(2x+1)^2}+\frac{x^2-4}{(2x+1)(4x+3)}$
Least common denominator
$=(2x+1)^2(4x+3)$

85. $x^2-6x+3+\square=2x^2-5x-6$
$\square=2x^2-5x-6-(x^2-6x+3)$
$=2x^2-5x-6-x^2+6x-3$
$=x^2+x-9$
Sum of numerators must be $2x^2-5x-6$

87. $-x^2-4x+3+\square=5x-7$
$\square=5x-7-(-x^2-4x+3)$
$=5x-7+x^2+4x-3$
$=x^2+9x-10$
Sum of numerator must be $5x-7$

89. $\dfrac{3x}{x^2-x-20}-\dfrac{2x}{-x^2+x+20}=\dfrac{3x}{x^2-x-20}+\dfrac{2x}{-1(-x^2+x+20)}$

$=\dfrac{3x}{x^2-x-20}+\dfrac{2x}{x^2-x-20}$

$=\dfrac{3x}{(x-5)(x+4)}+\dfrac{2x}{(x-5)(x+4)}$

Least common denominator $=(x-5)(x+4)$

91. $\dfrac{5}{8\Delta^2 ☺^2}+\dfrac{6}{5\Delta^4 ☺^5}$

Least common denominator $=8\cdot 5\Delta^4 ☺^5=40\Delta^4 ☺^5$

93. $\dfrac{6}{\Delta+3}-\dfrac{\Delta+5}{\Delta^2-4\Delta+3}=\dfrac{6}{\Delta+3}-\dfrac{\Delta+5}{(\Delta-3)(\Delta-1)}$

Least common denominator $=(\Delta+3)(\Delta-3)(\Delta-1)$

95. $\dfrac{x^2-8x+2}{x+7}+\dfrac{2x^2-5x}{x+7}-\dfrac{3x^2+7x+6}{x+7}=\dfrac{x^2-8x+2+(2x^2-5x)-(3x^2+7x+6)}{x+7}$

$=\dfrac{x^2-8x+2+2x^2-5x-3x^2-7x-6}{x+7}$

$=\dfrac{-20x-4}{x+7}$

97. $\dfrac{12}{x-3}-\dfrac{5}{x^2-9}+\dfrac{7}{x+3}=\dfrac{12}{x-3}-\dfrac{5}{(x+3)(x-3)}+\dfrac{7}{x+3}$

Least common denominator $=(x+3)(x-3)$

99. $\dfrac{4}{x^2-4}-\dfrac{11}{3x^2+5x-2}+\dfrac{5}{3x^2-7x+2}$

$=\dfrac{4}{(x+2)(x-2)}-\dfrac{11}{(3x-1)(x+2)}+\dfrac{5}{(3x-1)(x-2)}$

Least common denominator $(x+2)(x-2)(3x-1)$

101. $4\dfrac{3}{5}-2\dfrac{5}{9}=\dfrac{23}{5}-\dfrac{23}{9}$

$=\dfrac{207}{45}-\dfrac{115}{45}$

$=\dfrac{92}{45}$

$=2\dfrac{2}{45}$

102.
$$\begin{aligned} 6x+4 &= -(x+2)-3x+4 \\ 6x+4 &= -x-2-3x+4 \\ 6x+4 &= -4x+2 \\ 6x+4x &= 2-4 \\ 10x &= -2 \\ x &= \frac{-2}{10} = -\frac{1}{5} \end{aligned}$$

103.
$$\begin{aligned} \frac{6}{128} &= \frac{x}{48} \\ 128x &= 6 \cdot 48 \\ 128x &= 288 \\ x &= \frac{288}{128} \\ x &= \frac{9}{4} = 2\frac{1}{4} \end{aligned}$$
You should use 2.25 ounces of concentrate.

104. Let h = the number of hours played. The cost under Plan 1 is $C = 125 + 2.5h$ while the cost under Plan 2 is $C = 300$.

a. Set the two costs equal
$$\begin{aligned} 125+2.5h &= 300 \\ 2.5h &= 175 \\ h &= 70 \end{aligned}$$
If Malcolm plays 70 hours in a year, the cost of the two plans is equal.

b. $4 \cdot 52 = 208$
4 hours per week is 208 hours in a year. The cost for Plan 1 would be $125 + 2.5(208) = 645$.
Thus, he should use Plan 2.

105.

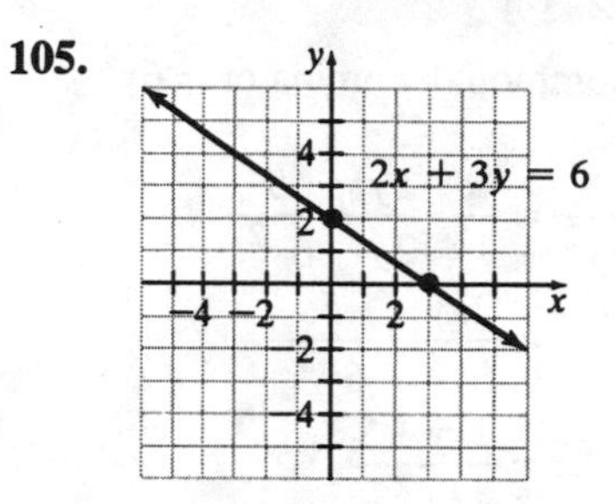

106.
$$\begin{aligned} m &= \frac{y_2-y_1}{x_2-x_1} \\ &= \frac{5-(-2)}{4-(-3)} \\ &= \frac{5+2}{4+3} \\ &= \frac{7}{7} \\ &= 1 \end{aligned}$$

Exercise Set 8.4

1. For each fraction, divide the LCD by the denominator.

3. a. Answers will vary.

b.
$$\begin{aligned} \frac{x}{x^2-x-6}+\frac{3}{x^2-4} &= \frac{x}{(x-3)(x+2)}+\frac{3}{(x-2)(x+2)} \\ &= \frac{x(x-2)}{(x-3)(x+2)(x-2)}+\frac{3(x-3)}{(x-2)(x+2)(x-3)} \\ &= \frac{x^2-2x+3x-9}{(x-3)(x+2)(x-2)} \\ &= \frac{x^2+x-9}{(x-3)(x+2)(x-2)} \end{aligned}$$

5. a. $\frac{x}{3y}+\frac{5}{6y^2}$

$6y^2=2\cdot 3\cdot y^2$

Least common denominator $=6y^2$

b. $\frac{x}{3y}+\frac{5}{6y^2}=\frac{x}{3y}\cdot\frac{2y}{2y}+\frac{5}{6y^2}$

$=\frac{2xy}{6y^2}+\frac{5}{6y^2}$

$=\frac{2xy+5}{6y^2}$

c. Yes. After factoring out the common factors, the reduced form would be the same.

7. $\frac{1}{3x}+\frac{5}{x}=\frac{1}{3x}+\frac{5\cdot 3}{3x}$

$=\frac{1+15}{3x}$

$=\frac{16}{3x}$

9. $\frac{4}{x^2}+\frac{3}{2x}=\frac{4\cdot 2}{2x^2}+\frac{3x}{2x^2}$

$=\frac{3x+8}{2x^2}$

11. $3+\frac{5}{x}=\frac{3x}{x}+\frac{5}{x}=\frac{3x+5}{x}$

13. $\frac{2}{x^2}+\frac{3}{5x}=\frac{2\cdot 5}{5x^2}+\frac{3x}{5x^2}=\frac{3x+10}{5x^2}$

15. $\frac{7}{4x^2y}+\frac{3}{5xy^2}=\frac{7\cdot 5y}{4x^2y\cdot 5y}+\frac{3\cdot 4x}{5xy^2\cdot 4x}$

$=\frac{35y}{20x^2y^2}+\frac{12x}{20x^2y^2}$

$=\frac{35y+12x}{20x^2y^2}$

17. $y+\frac{x}{y}=\frac{y\cdot y}{y}+\frac{x}{y}=\frac{y^2+x}{y}$

19. $\frac{3a-1}{2a}+\frac{2}{3a}=\frac{(3a-1)3}{2a\cdot 3}+\frac{2\cdot 2}{3a\cdot 2}$

$=\frac{9a-3+4}{6a}$

$=\frac{9a+1}{6a}$

21. $\frac{4x}{y}+\frac{2y}{xy}=\frac{4x\cdot x}{xy}+\frac{2y}{xy}$

$=\frac{4x^2+2y}{xy}$

23. $\frac{6}{b}-\frac{4}{5a^2}=\frac{6\cdot 5a^2}{b\cdot 5a^2}-\frac{4\cdot b}{5a^2\cdot b}$

$=\frac{30a^2-4b}{5a^2b}$

25. $\frac{3}{x}+\frac{7}{x-4}=\frac{3(x-4)}{x(x-4)}+\frac{7x}{x(x-4)}$

$=\frac{3x-12+7x}{x(x-4)}$

$=\frac{10x-12}{x(x-4)}$

27. $\frac{9}{p+3}+\frac{2}{p}=\frac{9p}{p(p+3)}+\frac{2(p+3)}{p(p+3)}$

$=\frac{9p+2p+6}{p(p+3)}$

$=\frac{11p+6}{p(p+3)}$

29. $\frac{3}{4x}-\frac{x}{3x+5}=\frac{3(3x+5)}{4x(3x+5)}-\frac{x(4x)}{4x(3x+5)}$

$=\frac{9x+15-4x^2}{4x(3x+5)}$

$=\frac{-4x^2+9x+15}{4x(3x+5)}$

31. $\frac{4}{p-3}+\frac{2}{3-p}=\frac{4}{p-3}-\frac{2}{p-3}$
$=\frac{2}{p-3}$

33. $\frac{9}{x+7}-\frac{5}{-x-7}=\frac{9}{x+7}+\frac{5}{x+7}$
$=\frac{14}{x+7}$

35. $\frac{3}{a-2}+\frac{a}{2a-4}=\frac{3\cdot 2}{2(a-2)}+\frac{a}{2(a-2)}$
$=\frac{a+6}{2(a-2)}$

37. $\frac{x+5}{x-5}-\frac{x-5}{x+5}=\frac{(x+5)^2}{(x-5)(x+5)}-\frac{(x-5)^2}{(x-5)(x+5)}$
$=\frac{x^2+10x+25-\left(x^2-10x+25\right)}{(x-5)(x+5)}$
$=\frac{x^2+10x+25-x^2+10x-25}{(x-5)(x+5)}$
$=\frac{20x}{(x-5)(x+5)}$

39. $\frac{x}{x^2-9}+\frac{4}{x+3}=\frac{x}{(x+3)(x-3)}+\frac{4(x-3)}{(x+3)(x-3)}$
$=\frac{x+4x-12}{(x+3)(x-3)}$
$=\frac{5x-12}{(x+3)(x-3)}$

41. $\frac{x+2}{x^2-4}-\frac{2}{x+2}=\frac{x+2}{(x+2)(x-2)}-\frac{2(x-2)}{(x+2)(x-2)}$
$=\frac{x+2-(2x-4)}{(x+2)(x-2)}$
$=\frac{x+2-2x+4}{(x+2)(x-2)}$
$=\frac{-x+6}{(x+2)(x-2)}$

43.
$$\frac{2x+3}{x^2-7x+12}-\frac{2}{x-3}=\frac{2x+3}{(x-3)(x-4)}-\frac{2(x-4)}{(x-3)(x-4)}$$
$$=\frac{2x+3-(2x-8)}{(x-3)(x-4)}$$
$$=\frac{2x+3-2x+8}{(x-3)(x-4)}$$
$$=\frac{11}{(x-3)(x-4)}$$

45.
$$\frac{x^2}{x^2+2x-8}-\frac{x-4}{x+4}=\frac{x^2}{(x+4)(x-2)}-\frac{(x-4)(x-2)}{(x+4)(x-2)}$$
$$=\frac{x^2-\left(x^2-6x+8\right)}{(x+4)(x-2)}$$
$$=\frac{x^2-x^2+6x-8}{(x+4)(x-2)}$$
$$=\frac{6x-8}{(x+4)(x-2)}$$

47.
$$\frac{x-3}{x^2+10x+25}+\frac{x-3}{x+5}=\frac{x-3}{(x+5)^2}+\frac{(x-3)(x+5)}{(x+5)^2}$$
$$=\frac{x-3+x^2+2x-15}{(x+5)^2}$$
$$=\frac{x^2+3x-18}{(x+5)^2}$$

49.
$$\frac{5}{a^2-9a+8}-\frac{3}{a^2-6a-16}=\frac{5}{(a-8)(a-1)}-\frac{3}{(a-8)(a+2)}$$
$$=\frac{5(a+2)}{(a-8)(a-1)(a+2)}-\frac{3(a-1)}{(a-8)(a-1)(a+2)}$$
$$=\frac{5a+10-(3a-3)}{(a-8)(a-1)(a+2)}$$
$$=\frac{5a+10-3a+3}{(a-8)(a-1)(a+2)}$$
$$=\frac{2a+13}{(a-8)(a-1)(a+2)}$$

51. $$\frac{1}{x^2-4}+\frac{3}{x^2+5x+6}=\frac{1}{(x-2)(x+2)}+\frac{3}{(x+3)(x+2)}$$
$$=\frac{x+3}{(x-2)(x+2)(x+3)}+\frac{3(x-2)}{(x-2)(x+2)(x+3)}$$
$$=\frac{x+3+3x-6}{(x-2)(x+2)(x+3)}$$
$$=\frac{4x-3}{(x-2)(x+2)(x+3)}$$

53. $$\frac{x}{2x^2+7x+3}-\frac{3}{3x^2+7x-6}=\frac{x}{(2x+1)(x+3)}-\frac{3}{(3x-2)(x+3)}$$
$$=\frac{x(3x-2)}{(2x+1)(3x-2)(x+3)}-\frac{3(2x+1)}{(2x+1)(3x-2)(x+3)}$$
$$=\frac{3x^2-2x-(6x+3)}{(2x+1)(3x-2)(x+3)}$$
$$=\frac{3x^2-2x-6x-3}{(2x+1)(3x-2)(x+3)}$$
$$=\frac{3x^2-8x-3}{(2x+1)(3x-2)(x+3)}$$

55. $$\frac{x}{4x^2+11x+6}-\frac{2}{8x^2+2x-3}=\frac{x}{(4x+3)(x+2)}-\frac{2}{(4x+3)(2x-1)}$$
$$=\frac{x(2x-1)}{(4x+3)(x+2)(2x-1)}-\frac{2(x+2)}{(4x+3)(x+2)(2x-1)}$$
$$=\frac{2x^2-x-(2x+4)}{(4x+3)(x+2)(2x-1)}$$
$$=\frac{2x^2-x-2x-4}{(4x+3)(x+2)(2x-1)}$$
$$=\frac{2x^2-3x-4}{(4x+3)(x+2)(2x-1)}$$

57. $$\frac{3x+12}{x^2+x-12}-\frac{2}{x-3}=\frac{3(x+4)}{(x-3)(x+4)}-\frac{2}{x-3}$$
$$=\frac{3}{x-3}-\frac{2}{x-3}$$
$$=\frac{3-2}{x-3}$$
$$=\frac{1}{x-3}$$

59. $\frac{3}{x}+4$ is defined for all real numbers except $x=0$.

61. $\frac{5}{x-3}+\frac{7}{x+6}$ is defined for all real numbers except $x = 3$ and $x = -6$.

63. $$\begin{aligned}\frac{3}{\Delta-2}-\frac{1}{2-\Delta}&=\frac{3}{\Delta-2}+\frac{1}{\Delta-2}\\&=\frac{3+1}{\Delta-2}\\&=\frac{4}{\Delta-2}\end{aligned}$$

65. $\frac{5}{a+b}+\frac{3}{a}$
$a + b = 0$ when $a = -b$. The expression is defined for all real numbers except $a = 0$ and $a = -b$.

67. $$\begin{aligned}\frac{x}{x^2-4}+\frac{3x}{x+2}+\frac{3x^2-5x}{4-x^2}&=\frac{x}{(x+2)(x-2)}+\frac{3x(x-2)}{(x+2)(x-2)}-\frac{3x^2-5x}{(x+2)(x-2)}\\&=\frac{x+3x^2-6x-(3x^2-5x)}{(x+2)(x-2)}\\&=\frac{x+3x^2-6x-3x^2+5x}{(x+2)(x-2)}\\&=\frac{0}{(x+2)(x-2)}\\&=0\end{aligned}$$

69. $$\begin{aligned}\frac{x+6}{4-x^2}-\frac{x+3}{x+2}+\frac{x-3}{2-x}&=\frac{x+6}{(2-x)(2+x)}-\frac{(x+3)(2-x)}{(2+x)(2-x)}+\frac{(x-3)(2+x)}{(2-x)(2+x)}\\&=\frac{x+6-(-x^2-x+6)+(x^2-x-6)}{(2-x)(2+x)}\\&=\frac{x+6+x^2+x-6+x^2-x-6}{(2-x)(2+x)}\\&=\frac{2x^2+x-6}{(2-x)(2+x)}\\&=\frac{(2x-3)(x+2)}{(2-x)(2+x)}\\&=\frac{2x-3}{2-x}\end{aligned}$$

71. $$\frac{2}{x^2-x-6}+\frac{3}{x^2-2x-3}+\frac{1}{x^2+3x+2}=\frac{2}{(x+2)(x-3)}+\frac{3}{(x-3)(x+1)}+\frac{1}{(x+2)(x+1)}$$
$$=\frac{2(x+1)}{(x+2)(x-3)(x+1)}+\frac{3(x+2)}{(x+2)(x-3)(x+1)}+\frac{x-3}{(x+2)(x-3)(x+1)}$$
$$=\frac{2x+2+3x+6+x-3}{(x+2)(x-3)(x+1)}$$
$$=\frac{6x+5}{(x+2)(x-3)(x+1)}$$

74. $$\frac{18 \text{ counts}}{2 \text{ minutes}}\cdot\left(2\frac{1}{4}\text{ hours}\right)$$
$$\frac{18 \text{ counts}}{2 \text{ minutes}}(135 \text{ minutes}) = 1215 \text{ counts}$$
The counter will be at 1215.

75. $$\begin{aligned} 3(x-2)+2 &< 4(x+1) \\ 3x-6+2 &< 4x+4 \\ 3x-4 &< 4x+4 \\ -8 &< x \\ x &> -8 \end{aligned}$$

−8

76. $$\begin{array}{r} 4x-3 \\ 2x+3\overline{)\ 8x^2+6x-13} \\ \underline{8x^2+12x} \\ -6x-13 \\ \underline{-6x-\ 9} \\ -4 \end{array}$$

$$(8x^2+6x-13)\div(2x+3)=4x-3-\frac{4}{2x+3}$$

77. $$\frac{x^2+xy-6y^2}{x^2-xy-2y^2}\cdot\frac{y^2-x^2}{x^2+2xy-3y^2}=\frac{(x+3y)(x-2y)}{(x+y)(x-2y)}\cdot\frac{-1(x-y)(x+y)}{(x+3y)(x-y)}$$
$$=-1$$

Exercise Set 8.5

1. A complex fraction is a fraction whose numerator or denominator (or both) contains a fraction.

3. a. $$\frac{\frac{x-1}{5}}{\frac{2}{x^2+5x+6}}$$
Numerator, $\frac{x-1}{5}$

Denominator $\frac{2}{x^2+5x+6}$

b. $\dfrac{\frac{1}{2y}+x}{\frac{3}{y}+x}$

Numerator, $\dfrac{1}{2y}+x$

Denominator, $\dfrac{3}{y}+x$

5. $$\frac{3+\frac{2}{3}}{1+\frac{1}{3}}=\frac{\left(3+\frac{2}{3}\right)3}{\left(1+\frac{1}{3}\right)3}=\frac{9+2}{3+1}=\frac{11}{4}$$

7. $$\frac{2+\frac{3}{8}}{1+\frac{1}{3}}=\frac{\left(2+\frac{3}{8}\right)8\cdot 3}{\left(1+\frac{1}{3}\right)8\cdot 3}=\frac{2\cdot 8\cdot 3+3\cdot 3}{8\cdot 3+8}=\frac{48+9}{24+8}=\frac{57}{32}$$

9. $$\frac{\frac{2}{7}-\frac{1}{4}}{6-\frac{2}{3}}=\frac{\left(\frac{2}{7}-\frac{1}{4}\right)3\cdot 4\cdot 7}{\left(6-\frac{2}{3}\right)3\cdot 4\cdot 7}=\frac{2\cdot 3\cdot 4-1\cdot 3\cdot 7}{6\cdot 3\cdot 4\cdot 7-2\cdot 4\cdot 7}=\frac{24-21}{504-56}=\frac{3}{448}$$

11. $$\frac{\frac{xy^2}{6}}{\frac{3}{y}}=\frac{xy^2}{6}\cdot\frac{y}{3}=\frac{xy^3}{18}$$

13. $$\frac{\frac{8x^2y}{3z^3}}{\frac{4xy}{9z^5}}=\frac{8x^2y}{3z^3}\cdot\frac{9z^5}{4xy}=6xz^2$$

15. $$\frac{a-\frac{a}{b}}{\frac{1+a}{b}}=\frac{\left(a-\frac{a}{b}\right)b}{\left(\frac{1+a}{b}\right)b}=\frac{ab-a}{1+a}$$

17. $$\frac{\frac{9}{x}+\frac{3}{x^2}}{3+\frac{1}{x}}=\frac{\left(\frac{9}{x}+\frac{3}{x^2}\right)x^2}{\left(3+\frac{1}{x}\right)x^2}=\frac{9x+3}{3x^2+x}=\frac{3(3x+1)}{x(3x+1)}=\frac{3}{x}$$

19. $$\frac{5-\frac{1}{x}}{4-\frac{1}{x}}=\frac{\left(5-\frac{1}{x}\right)x}{\left(4-\frac{1}{x}\right)x}=\frac{5x-1}{4x-1}$$

21. $$\frac{\frac{y}{x}+\frac{x}{y}}{\frac{x-y}{x}}=\frac{\left(\frac{y}{x}+\frac{x}{y}\right)xy}{\left(\frac{x-y}{x}\right)xy}=\frac{y^2+x^2}{y(x-y)}$$

23. $$\frac{\frac{a^2}{b}-b}{\frac{b^2}{a}-a}=\frac{\left(\frac{a^2}{b}-b\right)ab}{\left(\frac{b^2}{a}-a\right)ab}=\frac{\left(a^2-b^2\right)a}{\left(b^2-a^2\right)b}=-\frac{a}{b}$$

25. $$\frac{-\frac{a}{b}+3}{\frac{a}{b}-3}=\frac{\left(-\frac{a}{b}+3\right)b}{\left(\frac{a}{b}-3\right)b}=\frac{-a+3b}{a-3b}=\frac{-1(a-3b)}{a-3b}=-1$$

27. $$\frac{\frac{a^2-b^2}{a}}{\frac{a+b}{a^3}}=\frac{a^2-b^2}{a}\cdot\frac{a^3}{a+b}=\frac{(a+b)(a-b)}{a}\cdot\frac{a^3}{a+b}=a^2(a-b)$$

29. $$\frac{\frac{1}{a}-\frac{1}{b}}{\frac{1}{ab}}=\frac{\left(\frac{1}{a}-\frac{1}{b}\right)ab}{\left(\frac{1}{ab}\right)ab}=\frac{b-a}{1}=b-a$$

31. $$\frac{\frac{a}{b}+\frac{1}{a}}{\frac{b}{a}+\frac{1}{a}}=\frac{\left(\frac{a}{b}+\frac{1}{a}\right)ab}{\left(\frac{b}{a}+\frac{1}{a}\right)ab}=\frac{a^2+b}{b^2+b}=\frac{a^2+b}{b(b+1)}$$

33. $$\frac{\frac{1}{x}+\frac{1}{y}}{\frac{1}{x}-\frac{1}{y}}=\frac{\left(\frac{1}{x}+\frac{1}{y}\right)xy}{\left(\frac{1}{x}-\frac{1}{y}\right)xy}=\frac{y+x}{y-x}$$

35. $$\frac{\frac{3}{a}+\frac{3}{a^2}}{\frac{3}{b}+\frac{3}{b^2}}=\frac{\left(\frac{3}{a}+\frac{3}{a^2}\right)a^2b^2}{\left(\frac{3}{b}+\frac{3}{b^2}\right)a^2b^2}=\frac{3ab^2+3b^2}{3a^2b+3a^2}=\frac{3b^2(a+1)}{3a^2(b+1)}=\frac{ab^2+b^2}{a^2(b+1)}$$

37. a. Answers will vary.

b. c. $$\frac{5+\frac{3}{7}}{\frac{1}{9}-4}=\frac{\frac{35}{7}+\frac{3}{7}}{\frac{1}{9}-\frac{36}{9}}=\frac{\frac{38}{7}}{-\frac{35}{9}}=\frac{38}{7}\cdot\left(-\frac{9}{35}\right)=-\frac{342}{245}$$

$$\frac{5+\frac{3}{7}}{\frac{1}{9}-4}=\frac{\left(5+\frac{3}{7}\right)7\cdot 9}{\left(\frac{1}{9}-4\right)7\cdot 9}=\frac{5\cdot 7\cdot 9+3\cdot 9}{7-4\cdot 7\cdot 9}=\frac{315+27}{7-252}=-\frac{342}{245}$$

39. a. Answers will vary.

b. c. $$\frac{\frac{x-y}{x+y}+\frac{3}{x+y}}{2-\frac{7}{x+y}}=\frac{\frac{x-y+3}{x+y}}{\frac{2(x+y)}{x+y}-\frac{7}{x+y}}$$
$$=\frac{\frac{x-y+3}{x+y}}{\frac{2x+2y-7}{x+y}}$$
$$=\frac{x-y+3}{x+y}\cdot\frac{x+y}{2x+2y-7}$$
$$=\frac{x-y+3}{2x+2y-7}$$

$$\frac{\frac{x-y}{x+y}+\frac{3}{x+y}}{2-\frac{7}{x+y}}=\frac{\frac{x-y+3}{x+y}}{2-\frac{7}{x+y}}$$
$$=\frac{\left(\frac{x-y+3}{x+y}\right)(x+y)}{\left(2-\frac{7}{x+y}\right)(x+y)}$$
$$=\frac{x-y+3}{2(x+y)-7}$$
$$=\frac{x-y+3}{2x+2y-7}$$

41. a. $$\frac{\frac{5}{12x}}{\frac{8}{x^2}-\frac{4}{3x}}$$

b. $$\frac{\frac{5}{12x}}{\frac{8}{x^2}-\frac{4}{3x}}=\frac{\frac{5}{12x}}{\frac{24-4x}{3x^2}}=\frac{5}{12x}\cdot\frac{3x^2}{(24-4x)}$$
$$=\frac{5x}{4(24-4x)}$$
$$=\frac{5x}{96-16x}$$

43. $$\frac{x^{-1}+y^{-1}}{2}=\frac{\frac{1}{x}+\frac{1}{y}}{2}$$
$$=\frac{\left(\frac{1}{x}+\frac{1}{y}\right)xy}{2xy}$$
$$=\frac{y+x}{2xy}$$

45. $$\frac{x^{-1}+y^{-1}}{x^{-1}y^{-1}}=\frac{\frac{1}{x}+\frac{1}{y}}{\frac{1}{xy}}$$
$$=\frac{\left(\frac{1}{x}+\frac{1}{y}\right)xy}{\left(\frac{1}{xy}\right)xy}$$
$$=\frac{y+x}{1}$$
$$=x+y$$

47. a. $$E=\frac{\frac{1}{2}\left(\frac{2}{3}\right)}{\frac{2}{3}+\frac{1}{2}}$$
$$=\frac{\frac{2}{6}}{\frac{4+3}{6}}$$
$$=\frac{2}{6}\cdot\frac{6}{7}$$
$$=\frac{2}{7}$$

b. $$E=\frac{\frac{1}{2}\left(\frac{4}{5}\right)}{\frac{4}{5}+\frac{1}{2}}$$
$$=\frac{\frac{4}{10}}{\frac{8+5}{10}}$$
$$=\frac{4}{10}\cdot\frac{10}{13}$$
$$=\frac{4}{13}$$

49. $$\frac{\frac{a}{b}+b-\frac{1}{a}}{\frac{a}{b^2}-\frac{b}{a}+\frac{1}{a^2}}=\frac{\frac{a^2+b^2a-b}{ba}}{\frac{a^3-ab^3+b^2}{a^2b^2}}$$
$$=\frac{a^2+b^2a-b}{ba}\cdot\frac{a^2b^2}{a^3-ab^3+b^2}$$
$$=\frac{(a^2+b^2a-b)ab}{a^3-ab^3+b^2}$$
$$=\frac{a^3b+a^2b^3-ab^2}{a^3-ab^3+b^2}$$

51. $2x-8(5-x)=9x-3(x+2)$
$2x-40+8x=9x-3x-6$
$10x-40=6x-6$
$4x=34$
$x=\frac{34}{4}=\frac{17}{2}$

52. $2x+y=7$
$3x-2y=21$
Multiply the first equation by 2 and add.
$2[2x+y=7]$ gives
$4x+2y=14$
$3x-2y=21$
$7x \quad =35$
$x=5$
Substitute 5 for x in the first equation.
$2x+y=7$
$2(5)+y=7$
$10+y=7$
$y=-3$
The solution is (5, –3).

53. A polynomial is a sum of terms of the form ax^n where a is a real number and n is a whole number.

54.
$$\frac{x}{3x^2+17x-6}-\frac{2}{x^2+3x-18}=\frac{x}{(3x-1)(x+6)}-\frac{2}{(x+6)(x-3)}$$
$$=\frac{x(x-3)}{(3x+1)(x+6)(x-3)}-\frac{2(3x-1)}{(3x-1)(x+6)(x-3)}$$
$$=\frac{x^2-3x-(6x-2)}{(3x-1)(x+6)(x-3)}$$
$$=\frac{x^2-3x-6x+2}{(3x-1)(x+6)(x-3)}$$
$$=\frac{x^2-9x+2}{(3x-1)(x+6)(x-3)}$$

Exercise Set 8.6

1. a. Answers will vary.

b.
$$\frac{1}{x-1}-\frac{1}{x+1}=\frac{3x}{x^2-1}$$
$$\frac{1}{x-1}-\frac{1}{x+1}=\frac{3x}{(x-1)(x+1)}$$

Multiply both sides of the equation by the least common denominator, $(x-1)(x+1)$.

$$(x-1)(x+1)\left(\frac{1}{x-1}-\frac{1}{x+1}\right)=\left(\frac{3x}{(x-1)(x+1)}\right)(x-1)(x+1)$$

$$(x-1)(x+1)\left(\frac{1}{x-1}\right)-(x-1)(x+1)\left(\frac{1}{x+1}\right)=3x$$

$$x+1-(x-1)=3x$$

$$x+1-x+1=3x$$

$$2=3x$$

$$\frac{2}{3}=x$$

3. a. The problem on the left is an expression to be simplified while the problem on the right is an equation to be solved.

b. Left: Write the fractions with the LCD, $12(x-1)$, then combine numerators.
Right: Multiply both sides of the equation by the LCD, $12(x-1)$, then solve.

c. Left:

$$\frac{x}{3}-\frac{x}{4}+\frac{1}{x-1}=\frac{x\cdot 4(x-1)}{3\cdot 4(x-1)}-\frac{x\cdot 3(x-1)}{4\cdot 3(x-1)}+\frac{1\cdot 3\cdot 4}{(x-1)3\cdot 4}$$

$$=\frac{4x(x-1)-3x(x-1)+12}{3\cdot 4(x-1)}$$

$$=\frac{4x^2-4x-3x^2+3x+12}{12(x-1)}$$

$$=\frac{x^2-x+12}{12(x-1)}$$

Right:

$$\frac{x}{3}-\frac{x}{4}=\frac{1}{x-1}$$

$$12(x-1)\left(\frac{x}{3}-\frac{x}{4}\right)=\left(\frac{1}{x-1}\right)12(x-1)$$

$$12(x-1)\left(\frac{x}{3}\right)-12(x-1)\left(\frac{x}{4}\right)=12$$

$$4x(x-1)-3x(x-1)=12$$

$$4x^2-4x-3x^2+3x=12$$

$$x^2-x-12=0$$

$$(x-4)(x+3)=0$$

$$x-4=0 \quad \text{or} \quad x+3=0$$

$$x=4 \qquad\qquad x=-3$$

5. You must check for extraneous when there is a variable in the denominator.

7. No, because there is no variable in any denominator.

9. $\frac{4}{7}=\frac{x}{21}$

$4(21)=7x$

$84=7x$

$12=x$

Check: $\frac{4}{7}=\frac{x}{21}$

$\frac{4}{7}=\frac{12}{21}$

$\frac{4}{7}=\frac{4}{7}$ True

11. $\frac{6}{13}=\frac{24}{x}$

$6x=13(24)$

$6x=312$

$x=52$

Check: $\frac{6}{13}=\frac{24}{x}$

$\frac{6}{13}=\frac{24}{52}$

$\frac{6}{13}=\frac{6}{13}$ True

13. $\frac{12}{10}=\frac{z}{25}$

$12(25)=10z$

$300=10z$

$30=z$

Check: $\frac{12}{10}=\frac{z}{25}$

$\frac{12}{10}=\frac{30}{25}$

$\frac{6}{5}=\frac{6}{5}$ True

15. $\frac{-2}{6}=\frac{3c}{9}$

$-2(9)=6(3c)$

$-18=18c$

$-1=c$

Check: $\frac{-2}{6}=\frac{3c}{9}$

$\frac{-2}{6}=\frac{3(-1)}{9}$

$-\frac{1}{3}=-\frac{1}{3}$ True

17. $\frac{x+4}{7}=\frac{3}{7}$

$7(x+4)=7(3)$

$7x+28=21$

$7x=-7$

$x=-1$

Check: $\frac{x+4}{7}=\frac{3}{7}$

$\frac{-1+4}{7}=\frac{3}{7}$

$\frac{3}{7}=\frac{3}{7}$ True

19. $\frac{4x+5}{6}=\frac{7}{2}$

$2(4x+5)=7(6)$

$8x+10=42$

$8x=32$

$x=4$

Check: $\frac{4x+5}{6}=\frac{7}{2}$

$\frac{4(4)+5}{6}=\frac{7}{2}$

$\frac{21}{6}=\frac{7}{2}$

$\frac{7}{2}=\frac{7}{2}$ True

21. $6-\frac{x}{4}=\frac{x}{8}$

$8\left(6-\frac{x}{4}\right)=8\left(\frac{x}{8}\right)$

$48-2x=x$

$48=3x$

$16=x$

Check: $6-\frac{x}{4}=\frac{x}{8}$

$6-\frac{16}{4}=\frac{16}{8}$

$6-4=2$

$2=2$ True

23. $\frac{x}{3}-\frac{3x}{4}=\frac{1}{12}$

$12\left(\frac{x}{3}-\frac{3x}{4}\right)=\left(\frac{1}{12}\right)12$

$12\left(\frac{x}{3}\right)-12\left(\frac{3x}{4}\right)=1$

$4x-9x=1$

$-5x=1$

$x=-\frac{1}{5}$

Check: $\frac{x}{3}-\frac{3x}{4}=\frac{1}{12}$

$\frac{\left(-\frac{1}{5}\right)}{3}-\frac{3\left(-\frac{1}{5}\right)}{4}=\frac{1}{12}$

$\frac{5\left(-\frac{1}{5}\right)}{5\cdot 3}-\frac{3\left(-\frac{1}{5}\right)5}{4\cdot 5}=\frac{1}{12}$

$\frac{-1}{5\cdot 3}-\frac{-3}{4\cdot 5}=\frac{1}{3\cdot 4}$

Multiply both sides of the equation by the least common denominator, $3\cdot 4\cdot 5$.

$3\cdot 4\cdot 5\left(\frac{-1}{5\cdot 3}-\frac{-3}{4\cdot 5}\right)=\left(\frac{1}{3\cdot 4}\right)3\cdot 4\cdot 5$

$-4-(-9)=5$

$5=5$ True

25. $\frac{5}{2}-x=3x$

$2\left(\frac{5}{2}-x\right)=2(3x)$

$5-2x=6x$

$5=8x$

$\frac{5}{8}=x$

Check: $\frac{5}{2}-x=3x$

$\frac{5}{2}-\frac{5}{8}=3\left(\frac{5}{8}\right)$

$\frac{20}{8}-\frac{5}{8}=\frac{15}{8}$

$\frac{15}{8}=\frac{15}{8}$ True

27. $\frac{5}{3x}+\frac{3}{x}=1$

$3x\left(\frac{5}{3x}+\frac{3}{x}\right)=(1)3x$

$5+9=3x$

$14=3x$

$\frac{14}{3}=x$

Check: $\frac{5}{3x}+\frac{3}{x}=1$

$\frac{5}{3\left(\frac{14}{3}\right)}+\frac{3}{\frac{14}{3}}=1$

$\frac{5}{14}+\frac{3}{1}\left(\frac{3}{14}\right)=1$

$\frac{5}{14}+\frac{9}{14}=1$

$\frac{14}{14}=1$ True

29.
$$\frac{x-1}{x-5}=\frac{4}{x-5}$$
$$(x-5)\left(\frac{x-1}{x-5}\right)=\left(\frac{4}{x-5}\right)(x-5)$$
$$x-1=4$$
$$x=5$$
Check: $\dfrac{x-1}{x-5}=\dfrac{4}{x-5}$
$$\frac{5-1}{5-5}=\frac{4}{5-5}$$
$$\frac{4}{0}=\frac{4}{0}$$
Since $\frac{4}{0}$ is not a real number, 5 is an extraneous solution. This equation has no solution.

31.
$$\frac{3-5y}{4}=\frac{2-4y}{3}$$
$$3(3-5y)=4(2-4y)$$
$$9-15y=8-16y$$
$$y=-1$$
Check: $\dfrac{3-5y}{4}=\dfrac{2-4y}{3}$
$$\frac{3-5(-1)}{4}=\frac{2-4(-1)}{3}$$
$$\frac{3+5}{4}=\frac{2+4}{3}$$
$$\frac{8}{4}=\frac{6}{3}\text{ True}$$

33.
$$\frac{5}{-x-6}=\frac{2}{x}$$
$$5(x)=2(-x-6)$$
$$5x=-2x-12$$
$$7x=-12$$
$$x=-\frac{12}{7}$$
Check: $\dfrac{5}{-x-6}=\dfrac{2}{x}$
$$\frac{5}{-\left(-\frac{12}{7}\right)-6}=\frac{2}{-\frac{12}{7}}$$
$$\frac{5}{\frac{12}{7}-6}=2\left(-\frac{7}{12}\right)$$
$$\frac{5}{-\frac{30}{7}}=-\frac{7}{6}$$
$$5\left(-\frac{7}{30}\right)=-\frac{7}{6}\text{ True}$$

35.
$$\frac{2x-3}{x-4}=\frac{5}{x-4}$$
$$(x-4)\left(\frac{2x-3}{x-4}\right)=\left(\frac{5}{x-4}\right)(x-4)$$
$$2x-3=5$$
$$2x=8$$
$$x=4$$
Check: $\dfrac{2x-3}{x-4}=\dfrac{5}{x-4}$
$$\frac{2(4)-3}{4-4}=\frac{5}{4-4}$$
$$\frac{5}{0}=\frac{5}{0}$$
Since $\frac{5}{0}$ is not a real number, 4 is an extraneous solution. This equation has no solution.

37.
$$\frac{x+4}{x-3} = \frac{x+10}{x+2}$$
$$(x+4)(x+2) = (x-3)(x+10)$$
$$x^2 + 6x + 8 = x^2 + 7x - 30$$
$$38 = x$$

Check: $\frac{x+4}{x-3} = \frac{x+10}{x+2}$
$$\frac{38+4}{38-3} = \frac{38+10}{38+2}$$
$$\frac{42}{35} = \frac{48}{40}$$
$$\frac{6}{5} = \frac{6}{5} \text{ True}$$

39.
$$\frac{2x-1}{3} - \frac{3x}{4} = \frac{5}{6}$$
$$12\left(\frac{2x-1}{3} - \frac{3x}{4}\right) = \left(\frac{5}{6}\right)12$$
$$4(2x-1) - 3(3x) = 2(5)$$
$$8x - 4 - 9x = 10$$
$$-4 - x = 10$$
$$-4 = 10 + x$$
$$-14 = x$$

Check: $\frac{2x-1}{3} - \frac{3x}{4} = \frac{5}{6}$
$$\frac{2(-14)-1}{3} - \frac{3(-14)}{4} = \frac{5}{6}$$
$$\frac{-28-1}{3} - \frac{-42}{4} = \frac{5}{6}$$
$$-\frac{58}{6} + \frac{63}{6} = \frac{5}{6}$$
$$\frac{5}{6} = \frac{5}{6} \text{ True}$$

41.
$$x + \frac{20}{x} = -9$$
$$x\left(x + \frac{20}{x}\right) = -9x$$
$$x^2 + 20 = -9x$$
$$x^2 + 9x + 20 = 0$$
$$(x+4)(x+5) = 0$$
$$x + 4 = 0 \quad \text{or} \quad x + 5 = 0$$
$$x = -4 \qquad\qquad x = -5$$

Check $x = -4$: $x + \frac{20}{x} = -9$
$$-4 + \frac{20}{-4} = -9$$
$$-4 + (-5) = -9 \text{ True}$$

Check $x = -5$: $x + \frac{20}{x} = -9$
$$-5 + \frac{20}{-5} = -9$$
$$5 + (-4) = -9 \text{ True}$$

43.

$$\frac{3y-2}{y+1} = 4 - \frac{y+2}{y-1}$$
$$(y+1)(y-1)\left(\frac{3y-2}{y+1}\right) = \left(4 - \frac{y+2}{y-1}\right)(y+1)(y-1)$$
$$(y-1)(3y-2) = 4(y+1)(y-1) - \left(\frac{y+2}{y-1}\right)(y+1)(y-1)$$
$$3y^2 - 5y + 2 = 4\left(y^2 - 1\right) - (y+2)(y+1)$$
$$3y^2 - 5y + 2 = 4y^2 - 4 - (y^2 + 3y + 2)$$
$$3y^2 - 5y + 2 = 3y^2 - 3y - 6$$
$$-5y + 2 = -3y - 6$$
$$8 = 2y$$
$$4 = y$$

Check:
$$\frac{3y-2}{y+1} = 4 - \frac{y+2}{y-1}$$
$$\frac{3(4)-2}{4+1} = 4 - \frac{4+2}{4-1}$$
$$\frac{12-2}{5} = 4 - \frac{6}{3}$$
$$\frac{10}{5} = 4 - 2$$
$$2 = 2 \text{ True}$$

45.

$$\frac{1}{x+3} + \frac{1}{x-3} = \frac{-5}{x^2-9}$$
$$\frac{1}{x+3} + \frac{1}{x-3} = \frac{-5}{(x-3)(x+3)}$$
$$(x-3)(x+3)\left[\frac{1}{x+3} + \frac{1}{x-3}\right] = \left[\frac{-5}{(x-3)(x+3)}\right](x-3)(x+3)$$
$$(x-3)(x+3)\left(\frac{1}{x+3}\right) + (x-3)(x+3)\left(\frac{1}{x-3}\right) = -5$$
$$x - 3 + x + 3 = -5$$
$$2x = -5$$
$$x = -\frac{5}{2}$$

Check: $\frac{1}{x+3}+\frac{1}{x-3}=\frac{-5}{x^2-9}$

$$\frac{1}{-\frac{5}{2}+3}+\frac{1}{-\frac{5}{2}-3}=\frac{-5}{\left(-\frac{5}{2}\right)^2-9}$$

$$\frac{1}{\frac{1}{2}}-\frac{1}{-\frac{11}{2}}=\frac{-5}{\frac{25}{4}-9}$$

$$2-\frac{2}{11}=\frac{-5}{-\frac{11}{4}}$$

$$\frac{20}{11}=\frac{20}{11} \text{ True}$$

47. $a-\frac{a}{3}+\frac{a}{5}=26$

$$15\left(a-\frac{a}{3}+\frac{a}{5}\right)=15(26)$$

$$15a-5a+3a=390$$

$$13a=390$$

$$a=30$$

Check: $a-\frac{a}{3}+\frac{a}{5}=26$

$$30-\frac{30}{3}+\frac{30}{5}=26$$

$$30-10+6=26$$

$$26=26 \text{ True}$$

49.

$$\frac{3}{x-5}-\frac{4}{x+5}=\frac{11}{x^2-25}$$

$$\frac{3}{x-5}-\frac{4}{x+5}=\frac{11}{(x-5)(x+5)}$$

$$(x-5)(x+5)\left[\frac{3}{x-5}-\frac{4}{x+5}\right]=\left[\frac{11}{(x-5)(x+5)}\right](x-5)(x+5)$$

$$(x-5)(x+5)\left(\frac{3}{x-5}\right)-(x-5)(x+5)\left(\frac{4}{x+5}\right)=11$$

$$3(x+5)-4(x-5)=11$$

$$3x+15-4x+20=11$$

$$-x+35=11$$

$$24=x$$

Check: $\frac{3}{x-5}-\frac{4}{x+5}=\frac{11}{x^2-25}$

$$\frac{3}{24-5}-\frac{4}{24+5}=\frac{11}{24^2-25}$$
$$\frac{3}{19}-\frac{4}{29}=\frac{11}{576-25}$$
$$\frac{87}{551}-\frac{76}{551}=\frac{11}{551}$$
$$\frac{11}{551}=\frac{11}{551} \quad \text{True}$$

51.
$$\frac{y}{2y+2}+\frac{2y-16}{4y+4}=\frac{y-3}{y+1}$$
$$\frac{y}{2(y+1)}+\frac{2y-16}{4(y+1)}=\frac{y-3}{y+1}$$
$$4(y+1)\left[\frac{y}{2(y+1)}+\frac{2y-16}{4(y+1)}\right]=\left[\frac{y-3}{y+1}\right]4(y+1)$$
$$4(y+1)\left[\frac{y}{2(y+1)}\right]+4(y+1)\left[\frac{2y-16}{4(y+1)}\right]=4(y-3)$$
$$2y+2y-16=4y-12$$
$$4y-16=4y-12$$
$$-16=-12 \qquad \text{False}$$

Since this is a false statement, the equation has no solution.

53.
$$\frac{1}{y-1}+\frac{1}{2}=\frac{2}{y^2-1}$$
$$\frac{1}{y-1}+\frac{1}{2}=\frac{2}{(y+1)(y-1)}$$
$$2(y+1)(y-1)\left[\frac{1}{y-1}+\frac{1}{2}\right]=2(y+1)(y-1)\left[\frac{2}{(y+1)(y-1)}\right]$$
$$2(y+1)(y-1)\left(\frac{1}{y-1}\right)+2(y+1)(y-1)\left(\frac{1}{2}\right)=2(2)$$
$$2(y+1)+(y+1)(y-1)=4$$
$$2y+2+y^2-1=4$$
$$y^2+2y-3=0$$
$$(y+3)(y-1)=0$$

$y+3=0$ or $y-1=0$

$y=-3$ $\qquad$ $y=1$

Check: $y = -3$

$$\frac{1}{y-1}+\frac{1}{2}=\frac{2}{y^2-1}$$
$$\frac{1}{-3-1}+\frac{1}{2}=\frac{2}{(-3)^2-1}$$
$$\frac{1}{-4}+\frac{1}{2}=\frac{2}{8}$$
$$\frac{1}{4}=\frac{1}{4} \quad \text{True}$$

The solution to the equation is -3.

Check: $y = 1$

$$\frac{1}{y-1}+\frac{1}{2}=\frac{2}{y^2-1}$$
$$\frac{1}{1-1}+\frac{1}{2}=\frac{2}{1^2-1}$$
$$\frac{1}{0}+\frac{1}{2}=\frac{2}{0}$$

Since $\frac{1}{0}$ and $\frac{2}{0}$ are not real numbers, 1 is an extraneous solution.

55.

$$\frac{x-4}{x^2-2x}=\frac{-4}{x^2-4}$$
$$\frac{x-4}{x(x-2)}=\frac{-4}{(x-2)(x+2)}$$
$$x(x-2)(x+2)\left[\frac{x-4}{x(x-2)}\right]=\left[\frac{-4}{(x-2)(x+2)}\right](x)(x-2)(x+2)$$
$$(x-4)(x+2)=-4x$$
$$x^2-2x-8=-4x$$
$$x^2+2x-8=0$$
$$(x+4)(x-2)=0$$

$x+4=0$ or $x-2=0$

$x=-4$ $\qquad$ $x=2$

Check: $x=-4$

$$\frac{x-4}{x^2-2x}=\frac{-4}{x^2-4}$$
$$\frac{-4-4}{(-4)^2-2(-4)}=\frac{-4}{(-4)^2-4}$$
$$\frac{-8}{24}=\frac{-4}{12}$$
$$-\frac{1}{3}=-\frac{1}{3} \quad \text{True}$$

The solution to the equation is -4.

Check: $x = 2$

$$\frac{x-4}{x^2-2x}=\frac{-4}{x^2-4}$$
$$\frac{2-4}{2^2-2(2)}=\frac{-4}{2^2-4}$$
$$\frac{-2}{0}=\frac{-4}{0}$$

Since $\frac{-2}{0}$ and $\frac{-4}{0}$ are not real numbers, 2 is an extraneous solution.

57. The solution is 5. Since $3 = x-2$, $x = 5$.

59. The solution is 0. Since $x + x = 0$, $x = 0$.

61. x can be any real number.
$x-2+x-2=2x-4$.

63. $\frac{1}{p}+\frac{1}{q}=\frac{1}{f}$

$\frac{1}{30}+\frac{1}{q}=\frac{1}{10}$

$\frac{1}{q}=\frac{1}{10}-\frac{1}{30}$

$\frac{1}{q}=\frac{2}{30}$

$2q=30$

$q=15$

The image will appear 15 cm from the mirror.

65. No, it is impossible for both sides of the equation to be equal.

67. Let x be the number of minutes of internet access over 5 hours.
Plan 1: $7.95 + 0.15x$
Plan 2: 19.95

$7.95+0.15x=19.95$

$0.15x=12$

$x=80$

80 minutes $= \frac{80}{60}$ hours $=1\frac{1}{3}$ hours

Jake would have to use the internet more than $5+1\frac{1}{3}=6\frac{1}{3}$ hours.

68. Let x = measure of larger angle
y = measure of smaller angle

$x+y=180$

$\frac{1}{2}x-30=y$

Substitute $\frac{1}{2}x-30$ for y in the first equation.

$x+\frac{1}{2}x-30=180$

$\frac{3}{2}x=210$

$x=140$

$y=\frac{1}{2}x-30$

$y=\frac{1}{2}(140)-30$

$y=70-30$

$y=40$

The angles measure 40° and 140°.

69. $\frac{600 \text{ gallons}}{8 \text{ gallons / minute}}=75$ minutes

70. A linear equation has the form $ax+b=c$, $a\neq 0$, while a quadratic equation has the form $ax^2+bx+c=0,\ a\neq 0$.
Linear equation example: $3x+5=12$
Quadratic equation example:
$4x^2+3x-6=0$

Exercise Set 8.7

1. Some examples are:
$A=\frac{1}{2}bh,\ A=\frac{1}{2}h(b_1+b_2),\ V=\frac{1}{3}\pi r^2h,$
and $V=\frac{4}{3}\pi r^3$

3. It represents 1 complete task.

5. Let w = width, then
$\frac{2}{3}w+4$ = length
area = width · length

$$90 = w\left(\frac{2}{3}w+4\right)$$
$$90 = \frac{2w^2}{3}+4w$$
$$3(90) = 3\left(\frac{2w^2}{3}+4w\right)$$
$$270 = 2w^2+12w$$

$2w^2+12w-270=0$
$w^2+6w-135=0$
$(w+15)(w-9)=0$
$w+15=0$ or $w-9=0$
$w=-15$ $\quad w=9$
Since the width cannot be negative, $w = 9$
$l=\frac{2}{3}w+4$
$l=\frac{2}{3}(9)+4$
$l=6+4=10$
The length = 10 inches and the width = 9 inches.

7. Let x = height, then $x + 5$ = base
area $=\frac{1}{2}\cdot$ height · base

$$42=\frac{1}{2}x(x+5)$$
$$2(42)=2\left[\frac{1}{2}x(x+5)\right]$$
$$84=x(x+5)$$
$$84=x^2+5x$$
$$0=x^2+5x-84$$
$$0=(x-7)(x+12)$$

$x-7=0$ or $x+12=0$
$x=7$ $\quad x=-12$
Since the height cannot be negative, $x = 7$.
base = x + 5 = 7 + 15 = 12
The base is 12 cm and the height is 7 cm.

9. Let one number be x, then the other number is $10x$.

$$\frac{1}{x}-\frac{1}{10x}=3$$
$$10x\left(\frac{1}{x}-\frac{1}{10x}\right)=10x(3)$$
$$10-1=30x$$
$$9=30x$$
$$\frac{9}{30}=x$$
$$\frac{3}{10}=x$$
$$3=10x$$

The numbers are $\frac{3}{10}$ and 3.

11. Let x = amount by which the numerator was increased.

$$\frac{3+x}{4}=\frac{5}{2}$$
$$4\left(\frac{3+x}{4}\right)=\left(\frac{5}{2}\right)4$$
$$3+x=10$$
$$x=7$$

The numerator was increased by 7.

13. Let r = speed of Creole Queen paddle boat.
$t=\frac{d}{r}$
Time upstream = time downstream

$$\frac{4}{r-2}=\frac{6}{r+2}$$
$$4(r+2)=6(r-2)$$
$$4r+8=6r-12$$
$$20=2r$$
$$10=r$$

The boat's speed in still water is 10 mph.

15. $t=\frac{d}{r}$
Time pulling younger son + time pulling older son = $\frac{1}{2}$ hour.

$$\frac{d}{30}+\frac{d}{30}=\frac{1}{2}$$
$$\frac{2d}{30}=\frac{1}{2}$$
$$4d=30$$
$$d=\frac{30}{4}=7.5$$

Each son traveled 7.5 miles.

17. Let r be the speed of the propeller plane, then $4r$ is the speed of the jet.

$$\frac{d}{r}=t$$

time by jet + time by propeller plane = 6 hr

$$\frac{1600}{4r}+\frac{500}{r}=6$$
$$\frac{400}{r}+\frac{300}{r}=6$$
$$\frac{900}{r}=6$$
$$r=\frac{900}{6}=150$$

The speed of the propeller plane is 150 mph and the speed of the jet is 600 mph.

19. Let d = distance traveled by car, then $70 - d$ = distance traveled by train.

$$t=\frac{d}{r}$$

time traveled by car + time traveled by train = total time

$$\frac{d}{25}+\frac{70-d}{50}=1.6$$
$$50\left(\frac{d}{25}+\frac{70-d}{50}\right)=1.6(50)$$
$$2d+70-d=80$$
$$d=10$$
$$70-10=60$$

She travels 10 miles by car and 60 miles by train.

21. Let d = the distance flown with the wind, then $2800 - d$ = the time flown against the wind.

$$\text{time with wind }=\frac{d}{600}$$

$$\text{time against wind }=\frac{2800-d}{500}$$
$$\frac{d}{600}+\frac{2800-d}{500}=5$$
$$3000\left(\frac{d}{600}+\frac{2800-d}{500}\right)=5(3000)$$
$$5d+16,800-6d=15,000$$
$$d=1800$$
$$\text{time with wind }=\frac{1800}{600}=3$$
$$\text{time against wind }=\frac{1000}{500}=2$$

It flew 3 hours at 600 mph and 2 hours at 500 mph.

23. $\text{time at 30 m/s} = \frac{d}{30}$

$$\text{time at 40 m/s} = \frac{d}{40}$$
$$\frac{d}{30}=\frac{d}{40}+20$$
$$120\left(\frac{d}{30}\right)=\left(\frac{d}{40}+20\right)120$$
$$4d=3d+2400$$
$$d=2400$$

The lake is 2400 m wide.

25. $\text{Felicia's rate }=\frac{1}{6}$

$$\text{Reyrald's rate }=\frac{1}{8}$$
$$\frac{t}{6}+\frac{t}{8}=1$$
$$48\left(\frac{t}{6}+\frac{t}{8}\right)=48(1)$$
$$8t+6t=48$$
$$14t=48$$
$$t=3\frac{3}{7}$$

It will take them $3\frac{3}{7}$ hours.

27. input rate $= \frac{1}{2}$

output rate $= \frac{1}{3}$

$$\frac{t}{2} - \frac{t}{3} = 1$$
$$6\left(\frac{t}{2} - \frac{t}{3}\right) = 6(1)$$
$$3t - 2t = 6$$
$$t = 6$$

It will take 6 hours.

29. Jackie's rate $= \frac{1}{8}$

Tim's rate $= \frac{1}{t}$

In 3 hours, Jackie does $\frac{3}{8}$ of the exams and Tim does $\frac{3}{t}$.

$$\frac{3}{8} + \frac{3}{t} = 1$$
$$8t\left(\frac{3}{8} + \frac{3}{t}\right) = 8t$$
$$3t + 24 = 8t$$
$$24 = 5t$$
$$t = \frac{24}{5} = 4\frac{4}{5}$$

It would take Tim $4\frac{4}{5}$ hours.

31. Rate for first backhoe $= \frac{1}{12}$

Rate for second backhoe $= \frac{1}{15}$

Work done by first $= \frac{1}{12} \cdot 5 = \frac{5}{12}$

Work done by second $= \frac{1}{15} \cdot t = \frac{t}{15}$

$$\frac{5}{12} + \frac{t}{15} = 1$$
$$60\left(\frac{5}{12} + \frac{t}{15}\right) = 1 \cdot 60$$
$$25 + 4t = 60$$
$$4t = 35$$
$$t = \frac{35}{4} = 8\frac{3}{4}$$

It takes the smaller backhoe $8\frac{3}{4}$ days to finish the trench.

33. Ken's rate $= \frac{1}{4}$

Bettina's rate $= \frac{1}{6}$

$$\frac{t}{6} + \frac{t+3}{4} = 1$$
$$12\left(\frac{t}{6} + \frac{t+3}{4}\right) = 12$$
$$2t + 3(t+3) = 12$$
$$2t + 3t + 9 = 12$$
$$5t = 3$$
$$t = \frac{3}{5}$$

It will take them $\frac{3}{5}$ hour or 36 minutes longer.

35. Rate of first skimmer $= \frac{1}{60}$

Rate of second skimmer $= \frac{1}{50}$

Rate of transfer $= \frac{1}{30}$

$$\frac{t}{60} + \frac{t}{50} - \frac{t}{30} = 1$$
$$300\left(\frac{t}{60} + \frac{t}{50} - \frac{t}{30}\right) = 300$$
$$5t + 6t - 10t = 300$$
$$t = 300$$

It will take 300 hours to fill the tank.

37. Let x be the number.

$$\frac{4}{x}+5x=12$$
$$x\left(\frac{4}{x}+5x\right)=12x$$
$$4+5x^2=12x$$
$$5x^2-12x+4=0$$
$$(5x-2)(x-2)=0$$
$$5x-2=0 \quad \text{or} \quad x-2=0$$
$$x=\frac{2}{5} \qquad\qquad x=2$$

The numbers are 2 or $\frac{2}{5}$

39. Ed's rate $=\frac{1}{8}$

Samantha's rate $=\frac{1}{4}$

$$\frac{p}{4}-1=\frac{p}{8}$$
$$8\left(\frac{p}{4}-1\right)=8\left(\frac{p}{8}\right)$$
$$2p-8=p$$
$$p=8$$

Each must pick 8 pints.

41.
$$\begin{aligned}\frac{1}{2}(x+3)-(2x+6)&=\frac{1}{2}x+\frac{3}{2}-2x-6\\&=\frac{x}{2}-\frac{4x}{2}+\frac{3}{2}-\frac{12}{2}\\&=-\frac{3x}{2}-\frac{9}{2}\end{aligned}$$

42.

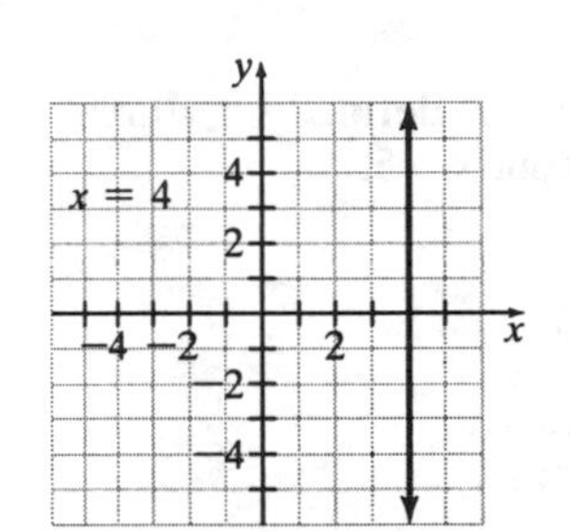

43.
$$\begin{aligned}\frac{x^2-14x+48}{x^2-5x-24}\div\frac{2x^2-13x+6}{2x^2+5x-3}&=\frac{x^2-14x+48}{x^2-5x-24}\cdot\frac{2x^2+5x-3}{2x^2-13x+6}\\&=\frac{(x-6)(x-8)}{(x+3)(x-8)}\cdot\frac{(2x-1)(x+3)}{(2x-1)(x-6)}\\&=1\end{aligned}$$

44. $$\frac{x}{6x^2-x-15}-\frac{5}{9x^2-12x-5}=\frac{x}{(2x+3)(3x-5)}-\frac{5}{(3x+1)(3x-5)}$$
$$=\frac{x(3x+1)}{(2x+3)(3x-5)(3x+1)}-\frac{5(2x+3)}{(2x+3)(3x-5)(3x+1)}$$
$$=\frac{3x^2+x-(10x+15)}{(2x+3)(3x-5)(3x+1)}$$
$$=\frac{3x^2+x-10x-15}{(2x+3)(3x-5)(3x+1)}$$
$$=\frac{3x^2-9x-15}{(2x+3)(3x-5)(3x+1)}$$

Review Exercises

1. $\frac{4}{3x-15}$
$3x-15=0$
$3(x-5)=0$
$x-5=0$
The expression is defined for all real numbers except $x=5$.

2. $\frac{2}{x^2-8x+15}$
$x^2-8x+15=0$
$(x-3)(x-5)=0$
$x-3=0, x-5=0$
The expression is defined for all real numbers except $x=3, x=5$.

3. $\frac{2}{5x^2+4x-1}$
$5x^2+4x-1=0$
$(5x-1)(x+1)=0$
$5x-1=0,\ x+1=0$
The expression is defined for all real numbers except $x=\frac{1}{5}$, $x=-1$.

4. $$\frac{y}{xy-y}=\frac{y}{y(x-1)}=\frac{1}{x-1}$$

5. $$\frac{x^3+4x^2+12x}{x}=\frac{x\left(x^2+4x+12\right)}{x}=x^2+4x+12$$

6. $$\frac{9x^2+6xy}{3x}=\frac{3x(3x+2y)}{3x}=3x+2y$$

7. $$\frac{x^2+2x-8}{x-2}=\frac{(x-2)(x+4)}{x-2}=x+4$$

8. $$\frac{x^2-25}{x-5}=\frac{(x-5)(x+5)}{x-5}=x+5$$

9. $$\frac{2x^2-7x+3}{3-x}=\frac{(2x-1)(x-3)}{-1(x-3)}=-(2x-1)$$

10. $$\frac{x^2-2x-24}{x^2+6x+8}=\frac{(x-6)(x+4)}{(x+2)(x+4)}=\frac{x-6}{x+2}$$

11. $$\frac{4x^2-11x-3}{4x^2-7x-2}=\frac{(4x+1)(x-3)}{(4x+1)(x-2)}=\frac{x-3}{x-2}$$

12. $\frac{2x^2-21x+40}{4x^2-4x-15}=\frac{(x-8)(2x-5)}{(2x+3)(2x-5)}$
$=\frac{x-8}{2x+3}$

13. $\frac{3a}{4b}\cdot\frac{2}{4a^2b}=\frac{3\cdot 2}{4\cdot 4}\cdot\frac{a}{a^2}\cdot\frac{1}{b\cdot b}$
$=\frac{3}{8ab^2}$

14. $\frac{15x^2y^3}{3z}\cdot\frac{6z^3}{5xy^3}=\frac{15\cdot 6}{15}\left(\frac{x^2}{x}\right)\left(\frac{y^3}{y^3}\right)\left(\frac{z^3}{z}\right)$
$=6xz^2$

15. $\frac{40a^3b^4}{7c^3}\cdot\frac{14c^5}{5a^5b}=\frac{40}{5}\cdot\frac{14}{7}\cdot\frac{a^3}{a^5}\cdot\frac{b^4}{b}\cdot\frac{c^5}{c^3}$
$=16\frac{1}{a^2}b^3c^2$
$=\frac{16b^3c^2}{a^2}$

16. $\frac{1}{x-4}\cdot\frac{4-x}{3}=\frac{1}{x-4}\cdot\frac{-1(x-4)}{3}$
$=-\frac{1}{3}$

17. $\frac{-x+3}{5}\cdot\frac{10x}{x-3}=\frac{-1(x-3)}{5}\cdot\frac{2\cdot 5x}{x-3}$
$=-2x$

18. $\frac{a-2}{a+3}\cdot\frac{a^2+4a+3}{a^2-a-2}=\frac{(a-2)(a+3)(a+1)}{(a+3)(a-2)(a+1)}$
$=1$

19. $\frac{8x^4}{y}\div\frac{x^4}{8y}=\frac{8x^4}{y}\cdot\frac{8y}{x^4}$
$=64$

20. $\frac{8xy^2}{z}\div\frac{x^4y^2}{4z^2}=\frac{8xy^2}{z}\cdot\frac{4z^2}{x^4y^2}$
$=\frac{32z}{x^3}$

21. $\frac{5a+5b}{a^2}\div\frac{a^2-b^2}{a^2}=\frac{5(a+b)}{a^2}\cdot\frac{a^2}{(a+b)(a-b)}$
$=\frac{5}{a-b}$

22. $\frac{1}{a^2+8a+15}\div\frac{3}{a+5}=\frac{1}{(a+5)(a+3)}\cdot\frac{a+5}{3}$
$=\frac{1}{3(a+3)}$

23. $(x+7)\div\frac{x^2+3x-28}{x-4}=(x+7)\cdot\frac{x-4}{(x-4)(x+7)}$
$=1$

24. $\frac{x^2+xy-2y^2}{4y}\div\frac{x+2y}{12y^2}=\frac{(x-y)(x+2y)}{4y}\cdot\frac{12y^2}{x+2y}$
$=3y(x-y)$

25. $\dfrac{x}{x+3}-\dfrac{3}{x+3}=\dfrac{x-3}{x+3}$

26. $\dfrac{3x}{x+7}+\dfrac{21}{x+7}=\dfrac{3x+21}{x+7}$
$=\dfrac{3(x+7)}{x+7}$
$=3$

27. $\dfrac{9x-4}{x+8}+\dfrac{76}{x+8}=\dfrac{9x-4+76}{x+8}$
$=\dfrac{9x+72}{x+8}$
$=\dfrac{9(x+8)}{x+8}$
$=9$

28. $\dfrac{7x-3}{x^2+7x-30}-\dfrac{3x+9}{x^2+7x-30}=\dfrac{7x-3-(3x+9)}{x^2+7x-30}$
$=\dfrac{7x-3-3x-9}{x^2+7x-30}$
$=\dfrac{4x-12}{x^2+7x-30}$
$=\dfrac{4(x-3)}{(x+10)(x-3)}$
$=\dfrac{4}{x+10}$

29. $\dfrac{6x^2-3x-21}{x+6}-\dfrac{2x^2-24x-3}{x+6}=\dfrac{6x^2-3x-21-\left(2x^2-24x-3\right)}{x+6}$
$=\dfrac{6x^2-3x-21-2x^2+24x+3}{x+6}$
$=\dfrac{4x^2+21x-18}{x+6}$
$\dfrac{(4x-3)(x+6)}{x+6}$
$=4x-3$

30. $\frac{6x^2-4x}{2x-3}-\frac{-3x+12}{2x-3}=\frac{6x^2+4x-(-3x+12)}{2x-3}$
$=\frac{6x^2-4x+3x-12}{2x-3}$
$=\frac{6x^2-x-12}{2x-3}$
$=\frac{(3x+4)(2x-3)}{2x-3}$
$=3x+4$

31. $\frac{a}{7}+\frac{4a}{3}$
Least common denominator $= 3\cdot 7 = 21$

32. $\frac{8}{x+3}+\frac{4}{x+3}$
Least common denominator $= x + 3$

33. $\frac{5}{4xy^3}-\frac{7}{10x^2y}$
Least common denominator $= 20x^2y^3$

34. $\frac{6}{x+1}-\frac{3x}{x}$
Least common denominator $= x(x+1)$

35. $\frac{4}{x+2}+\frac{6x-3}{x-3}$
Least common denominator $= (x+2)(x-3)$

36. $\frac{7x-12}{x^2+x}-\frac{4}{x+1}=\frac{7x-12}{x(x+1)}-\frac{4}{x+1}$
Least common denominator $= x(x + 1)$

37. $\frac{12x-9}{x-y}-\frac{5x+7}{x^2-y^2}=\frac{12x-9}{x-y}-\frac{5x+7}{(x+y)(x-y)}$
Least common denominator $= (x + y)(x - y)$

38. $\frac{4x^2}{x-7}+8x^2=\frac{4x^2}{x-y}+\frac{8x^2}{1}$
Least common denominator $= x - 7$

39. $\frac{19x-5}{x^2+2x-35}+\frac{3x-2}{x^2+9x+14}$
$=\frac{19x-5}{(x+7)(x-5)}+\frac{3x-2}{(x+7)(x+2)}$
Least common denominator
$=(x+7)(x-5)(x+2)$

40. $\frac{6}{3y}+\frac{y}{2y^2}=\frac{6}{3y}\cdot\frac{2y}{2y}+\frac{y}{2y^2}\cdot\frac{3}{3}$
$=\frac{12y}{6y^2}+\frac{3y}{6y^2}$
$=\frac{12y+3y}{6y^2}$
$=\frac{15y}{6y^2}$
$=\frac{5}{2y}$

41. $\frac{2x}{xy}+\frac{1}{5x}=\frac{2x}{xy}\cdot\frac{5}{5}+\frac{1}{5x}\cdot\frac{y}{y}$
$=\frac{10x}{5xy}+\frac{y}{5xy}$
$=\frac{10x+y}{5xy}$

42. $\frac{5x}{3xy}-\frac{4}{x^2}=\frac{5x}{3xy}\cdot\frac{x}{x}-\frac{4}{x^2}\cdot\frac{3y}{3y}$
$=\frac{5x^2}{3x^2y}-\frac{12y}{3x^2y}$
$=\frac{5x^2-12y}{3x^2y}$

43. $$6-\frac{2}{x+2}=6\left(\frac{x+2}{x+2}\right)-\frac{2}{x+2}=\frac{6x+12-2}{x+2}=\frac{6x+10}{x+2}$$

44. $$\frac{x-y}{y}-\frac{x+y}{x}=\frac{x-y}{y}\cdot\frac{x}{x}-\frac{x+y}{x}\cdot\frac{y}{y}=\frac{x(x-y)}{xy}-\frac{y(x+y)}{xy}=\frac{x^2-xy-xy-y^2}{xy}=\frac{x^2-2xy-y^2}{xy}$$

45. $$\frac{7}{x+4}+\frac{4}{x}=\frac{7}{x+4}\cdot\frac{x}{x}+\frac{4}{x}\cdot\frac{x+4}{x+4}=\frac{7x}{x(x+4)}+\frac{4(x+4)}{x(x+4)}=\frac{7x+4x+16}{x(x+4)}=\frac{11x+16}{x(x+4)}$$

46. $$\frac{2}{3x}-\frac{3}{3x-6}=\frac{2}{3x}-\frac{3}{3(x-2)}=\frac{2}{3x}\cdot\frac{x-2}{x-2}-\frac{3}{3(x-2)}\cdot\frac{x}{x}=\frac{2(x-2)}{3x(x-2)}-\frac{3x}{3x(x-2)}=\frac{2x-4-3x}{3x(x-2)}=\frac{-x-4}{3x(x-2)}$$

47. $$\frac{3}{x+2}+\frac{7}{(x+2)^2}=\frac{3}{x+2}\cdot\frac{x+2}{x+2}+\frac{7}{(x+2)^2}=\frac{3(x+2)}{(x+2)^2}+\frac{7}{(x+2)^2}=\frac{3x+6+7}{(x+2)^2}=\frac{3x+13}{(x+2)^2}$$

48. $$\frac{x+2}{x^2-x-6}+\frac{x-3}{x^2-8x+15}=\frac{x+2}{(x-3)(x+2)}+\frac{x-3}{(x-3)(x-5)}=\frac{1}{x-3}+\frac{1}{x-5}=\frac{1}{x-3}\cdot\frac{x-5}{x-5}+\frac{1}{x-5}\cdot\frac{x-3}{x-3}=\frac{x-5}{(x-3)(x-5)}+\frac{x-3}{(x-5)(x-3)}=\frac{x-5+x-3}{(x-5)(x-3)}=\frac{2x-8}{(x-5)(x-3)}$$

49. $$\frac{x+2}{x+5}-\frac{x-5}{x+2}=\frac{(x+2)(x+2)}{(x+5)(x+2)}-\frac{(x-5)(x+5)}{(x+2)(x+5)}$$
$$=\frac{x^2+4x+4-(x^2-25)}{(x+5)(x+2)}$$
$$=\frac{x^2+4x+4-x^2+25}{(x+5)(x+2)}$$
$$=\frac{4x+29}{(x+5)(x+2)}$$

50. $$3+\frac{x}{x-4}=\frac{3(x-4)}{x-4}+\frac{x}{x-4}$$
$$=\frac{3x-12+x}{x-4}$$
$$=\frac{4x-12}{x-4}$$

51. $$\frac{a+2}{b}\div\frac{a-2}{4b^2}=\frac{a+2}{b}\cdot\frac{4b^2}{a-2}$$
$$=\frac{4b(a+2)}{a-2}$$
$$=\frac{4ab+8b}{a-2}$$

52. $$\frac{x+3}{x^2-9}+\frac{2}{x+3}=\frac{x+3}{(x-3)(x+3)}+\frac{2}{x+3}$$
$$=\frac{x+3}{(x-3)(x+3)}+\frac{2(x-3)}{(x-3)(x+3)}$$
$$=\frac{x+3+2x-6}{(x-3)(x+3)}$$
$$=\frac{3x-3}{(x-3)(x+3)}$$

53. $$\frac{4x+4y}{x^2y}\cdot\frac{y^3}{8x}=\frac{4(x+y)y^2}{8x^3}$$
$$=\frac{(x+y)y^2}{2x^3}$$

54. $$\begin{aligned}\frac{4}{(x+2)(x-3)}-\frac{4}{(x-2)(x+2)}&=\frac{4(x-2)}{(x+2)(x-3)(x-2)}-\frac{4(x-3)}{(x+2)(x-3)(x-2)}\\&=\frac{4(x-2)-4(x-3)}{(x+2)(x-3)(x-2)}\\&=\frac{4x-8-4x+12}{(x+2)(x-3)(x-2)}\\&=\frac{4}{(x+2)(x-3)(x-2)}\end{aligned}$$

55. $$\begin{aligned}\frac{x+7}{x^2+9x+14}-\frac{x-10}{x^2-49}&=\frac{x+7}{(x+7)(x+2)}-\frac{x-10}{(x+7)(x-7)}\\&=\frac{x+7}{(x+7)(x+2)}\cdot\frac{(x-7)}{(x-7)}-\frac{x-10}{(x+7)(x-7)}\cdot\frac{(x+2)}{(x+2)}\\&=\frac{x^2-49-\left(x^2-8x-20\right)}{(x+7)(x-7)(x+2)}\\&=\frac{8x-29}{(x+7)(x-7)(x+2)}\end{aligned}$$

56. $$\begin{aligned}\frac{x-y}{x+y}\cdot\frac{xy+x^2}{x^2-y^2}&=\frac{x-y}{x+y}\cdot\frac{x(y+x)}{(x+y)(x-y)}\\&=\frac{x}{x+y}\end{aligned}$$

57. $$\begin{aligned}\frac{2x^2-18y^2}{15}\div\frac{(x-3y)^2}{3}&=\frac{2\left(x^2-9y^2\right)}{15}\cdot\frac{3}{(x-3y)^2}\\&=\frac{2(x-3y)(x+3y)}{3\cdot 5}\cdot\frac{3}{(x-3y)^2}\\&=\frac{2(x+3y)}{5(x-3y)}\\&=\frac{2x+6y}{5(x-3y)}\end{aligned}$$

58. $$\begin{aligned}\frac{a^2-9a+20}{a-4}\cdot\frac{a^2-8a+15}{a^2-10a+25}&=\frac{(a-4)(a-5)}{a-4}\cdot\frac{(a-5)(a-3)}{(a-5)^2}\\&=a-3\end{aligned}$$

59. $$\frac{a}{a^2-1}-\frac{2}{3a^2-2a-5}=\frac{a}{(a-1)(a+1)}-\frac{2}{(3a-5)(a+1)}$$
$$=\frac{a(3a-5)}{(a-1)(a+1)(3a-5)}-\frac{2(a-1)}{(a-1)(a+1)(3a-5)}$$
$$=\frac{a(3a-5)-2(a-1)}{(a-1)(a+1)(3a-5)}$$
$$=\frac{3a^2-5a-2a+2}{(a-1)(a+1)(3a-5)}$$
$$=\frac{3a^2-7a+2}{(a-1)(a+1)(3a-5)}$$

60. $$\frac{2x^2+6x-20}{x^2-2x}\div\frac{x^2+7x+10}{4x^2-16}=\frac{2\left(x^2+3x-10\right)}{x(x-2)}\cdot\frac{4\left(x^2-4\right)}{(x+5)(x+2)}$$
$$=\frac{2(x+5)(x-2)}{x(x-2)}\cdot\frac{4(x+2)(x-2)}{(x+5)(x+2)}$$
$$=\frac{8(x-2)}{x}$$
$$=\frac{8x-16}{x}$$

61. $$\frac{2+\frac{2}{3}}{\frac{6}{9}}=\frac{9\left(2+\frac{2}{3}\right)}{9\left(\frac{6}{9}\right)}$$
$$=\frac{18+6}{6}$$
$$=\frac{24}{6}$$
$$=4$$

62. $$\frac{1+\frac{5}{8}}{4-\frac{9}{16}}=\frac{16\left(1+\frac{5}{8}\right)}{16\left(4-\frac{9}{16}\right)}$$
$$=\frac{16+10}{64-9}$$
$$=\frac{26}{55}$$

63. $$\frac{\frac{12ab}{9c}}{\frac{4a}{c^2}}=\frac{12ab}{9c}\cdot\frac{c^2}{4a}$$
$$=\frac{bc}{3}$$

64. $$\frac{36x^4y^2}{\frac{9xy^5}{4z^2}}=\frac{36x^4y^2\cdot 4z^2}{\frac{9xy^5}{4z^2}\cdot 4z^2}$$
$$=\frac{144x^4y^2z^2}{9xy^5}$$
$$=\frac{16x^3z^2}{y^3}$$

65. $$\frac{a-\frac{a}{b}}{\frac{1+a}{b}}=\frac{\left(a-\frac{a}{b}\right)b}{\left(\frac{1+a}{b}\right)b}=\frac{ab-a}{1+a}$$

66. $$\frac{x+\frac{1}{y}}{y^2}=\frac{\left(x+\frac{1}{y}\right)y}{\left(y^2\right)y}$$
$$=\frac{xy+1}{y^3}$$

67. $\dfrac{\frac{4}{x}+\frac{2}{x^2}}{6-\frac{1}{x}}=\dfrac{\left(\frac{4}{x}+\frac{2}{x^2}\right)x^2}{\left(6-\frac{1}{x}\right)x^2}$
$=\dfrac{4x+2}{6x^2-x}$
$=\dfrac{4x+2}{x(6x-1)}$

68. $\dfrac{\frac{x}{x+y}}{\frac{x^2}{2x+2y}}=\dfrac{x}{x+y}\cdot\dfrac{2x+2y}{x^2}$
$=\dfrac{1}{x+y}\cdot\dfrac{2(x+y)}{x}$
$=\dfrac{2}{x}$

69. $\dfrac{\frac{3}{x}}{\frac{3}{x^2}}=\dfrac{3}{x}\cdot\dfrac{x^2}{3}=x$

70. $\dfrac{\frac{1}{a}+2}{\frac{1}{a}+\frac{1}{a}}=\dfrac{\frac{1}{a}+2}{\frac{2}{a}}$
$=\dfrac{\left(\frac{1}{a}+2\right)a}{\left(\frac{2}{a}\right)a}$
$=\dfrac{1+2a}{2}$

71. $\dfrac{\frac{1}{x^2}+\frac{1}{x}}{\frac{1}{x^2}-\frac{1}{x}}=\dfrac{x^2\left(\frac{1}{x^2}+\frac{1}{x}\right)}{x^2\left(\frac{1}{x^2}-\frac{1}{x}\right)}$
$=\dfrac{1+x}{1-x}$

72. $\dfrac{\frac{3x}{y}-x}{\frac{y}{x}-1}=\dfrac{\left(\frac{3x}{y}-x\right)xy}{\left(\frac{y}{x}-1\right)xy}$
$=\dfrac{3x^2-x^2y}{y^2-xy}$
$=\dfrac{3x^2-x^2y}{y(y-x)}$

73. $\dfrac{7}{35}=\dfrac{5}{x}$
$7x=5(35)$
$7x=175$
$x=25$

74. $\dfrac{3}{x}=\dfrac{9}{3}$
$3(3)=9x$
$9=9x$
$1=x$

75. $\dfrac{5}{9}=\dfrac{5}{x+3}$
$5(x+3)=5(9)$
$5x+15=45$
$5x=30$
$x=6$

76. $\dfrac{x}{6}=\dfrac{x-4}{2}$
$2x=6(x-4)$
$2x=6x-24$
$24=4x$
$6=x$

77. $\dfrac{3x+4}{5}=\dfrac{2x-8}{3}$
$3(3x+4)=5(2x-8)$
$9x+12=10x-40$
$52=x$

78. $\dfrac{x}{5}+\dfrac{x}{2}=-14$
$\dfrac{2x+5x}{10}=-14$
$7x=-140$
$x=-20$

79. $\dfrac{x+3}{2}=\dfrac{x}{2}$
$2(x+3)=2x$
$2x+3=2x$
$3=0$
No solution

80.
$$\frac{3}{x}-\frac{1}{6}=\frac{1}{x}$$
$$6x\left(\frac{3}{x}-\frac{1}{6}\right)=6x\left(\frac{1}{x}\right)$$
$$18-x=6$$
$$-x=-12$$
$$x=12$$

81.
$$\frac{1}{x-7}+\frac{1}{x+7}=\frac{1}{x^2-49}$$
$$\frac{1}{x-7}+\frac{1}{x+7}=\frac{1}{(x-7)(x+7)}$$
$$(x-7)(x+7)\left[\frac{1}{x-7}+\frac{1}{x+7}\right]=\left[\frac{1}{(x-7)(x+7)}\right](x-7)(x+7)$$
$$x+7+x-7=1$$
$$2x=1$$
$$x=\frac{1}{2}$$

82.
$$\frac{x-3}{x-2}+\frac{x+1}{x+3}=\frac{2x^2+x+1}{x^2+x-6}$$
$$\frac{x-3}{x-2}+\frac{x+1}{x+3}=\frac{2x^2+x+1}{(x-2)(x+3)}$$
$$(x-2)(x+3)\left(\frac{x-3}{x-2}+\frac{x+1}{x+3}\right)=(x-2)(x+3)\left[\frac{2x^2+x+1}{(x-2)(x+3)}\right]$$
$$(x+3)(x-3)+(x-2)(x+1)=2x^2+x+1$$
$$x^2-9+x^2-x-2=2x^2+x+1$$
$$2x^2-x-11=2x^2+x+1$$
$$-12=2x$$
$$-6=x$$

83.
$$\frac{a}{a^2-64}+\frac{4}{a+8}=\frac{3}{a-8}$$
$$\frac{a}{(a+8)(a-8)}+\frac{4}{a+8}=\frac{3}{a-8}$$
$$(a+8)(a-8)\left[\frac{a}{(a+8)(a-8)}+\frac{4}{a+8}\right]=(a+8)(a-8)\left(\frac{3}{a-8}\right)$$
$$a+4(a-8)=3(a+8)$$
$$a+4a-32=3a+24$$
$$5a-32=3a+24$$
$$2a=56$$
$$a=28$$

84. John and Amy rate $= \frac{1}{6}$

Paul and Cindy rate $= \frac{1}{5}$

$$\frac{t}{6} + \frac{t}{5} = 1$$
$$30\left(\frac{t}{6} + \frac{t}{5}\right) = 30(1)$$
$$5t + 6t = 30$$
$$11t = 30$$
$$t = \frac{30}{11} = 2\frac{8}{11}$$

It will take the 4 children $2\frac{8}{11}$ hours.

85. $\frac{3}{4}$-inch hose's rate $= \frac{1}{7}$

$\frac{5}{16}$-inch hose's rate $= \frac{1}{12}$

$$\frac{t}{7} - \frac{t}{12} = 1$$
$$\frac{12t - 7t}{84} = 1$$
$$5t = 84$$
$$t = \frac{84}{5} = 16\frac{4}{5}$$

It will take $16\frac{4}{5}$ hours to fill the pool.

86. Let x be one number then, $5x$ is the other number.

$$\frac{1}{x} + \frac{1}{5x} = 6$$
$$5x\left(\frac{1}{x} + \frac{1}{5x}\right) = 5x(6)$$
$$5 + 1 = 30x$$
$$6 = 30x$$
$$\frac{6}{30} = x$$
$$\frac{1}{5} = x$$

$$5x = 5\left(\frac{1}{5}\right) = 1$$

The numbers are $\frac{1}{5}$ and 1.

87. Let x = Robert's speed, then
$3.5 + x$ = Tran's speed

$$t = \frac{d}{r}$$

Robert's time = Tran's time

$$\frac{3}{x} = \frac{8}{3.5 + x}$$
$$3(3.5 + x) = 8x$$
$$10.5 + 3x = 8x$$
$$10.5 = 5x$$
$$2.1 = x$$

$x + 3.5 = 2.1 + 3.5 = 5.6$

Robert's speed is 2.1 mph and Tran's speed is 5.6 mph.

Practice Test

1. $\frac{x-5}{5-x} = \frac{-1(5-x)}{5-x} = -1$

2.
$$\frac{x^3 - 1}{x^2 - 1} = \frac{(x-1)(x^2 + x + 1)}{(x-1)(x+1)} = \frac{x^2 + x + 1}{x + 1}$$

3.
$$\frac{3x^2y}{4z^2} \cdot \frac{8xz^3}{9xy^4} = \frac{3x^2y}{4z^2} \cdot \frac{2 \cdot 4xz^3}{3 \cdot 3xy^4} = \frac{2x^2z}{3y^3}$$

4. $$\frac{a^2-9a+14}{a-2}\cdot\frac{a^2-4a-21}{(a-7)^2}=\frac{(a-7)(a-2)}{a-2}\cdot\frac{(a-7)(a+3)}{(a-7)^2}$$
$$=a+3$$

5. $$\frac{x^2-x-6}{x^2-9}\cdot\frac{x^2-6x+9}{x^2+4x+4}=\frac{(x-3)(x+2)}{(x-3)(x+3)}\cdot\frac{(x-3)(x-3)}{(x+2)(x+2)}$$
$$=\frac{(x-3)^2}{(x+3)(x+2)}$$
$$=\frac{x^2-6x+9}{(x+3)(x+2)}$$

6. $$\frac{x^2-1}{x+2}\cdot\frac{2x+4}{2-2x^2}=\frac{(x-1)(x+1)}{x+2}\cdot\frac{2(x+2)}{-2(x-1)(x+1)}$$
$$=-1$$

7. $$\frac{x^2-9y^2}{3x+6y}\div\frac{x+3y}{x+2y}=\frac{(x-3y)(x+3y)}{3(x+2y)}\cdot\frac{x+2y}{x+3y}$$
$$=\frac{x-3y}{3}$$

8. $$\frac{15}{y^2+2y-15}\div\frac{3}{y-3}=\frac{15}{(y-3)(y+5)}\cdot\frac{y-3}{3}$$
$$=\frac{5}{y+5}$$

9. $$\frac{x^2+x-20}{x-2}\div\frac{x^2-6x+8}{2-x}=\frac{(x+5)(x-4)}{x-2}\cdot\frac{-1(x-2)}{(x-4)(x-2)}$$
$$=\frac{-(x+5)}{x-2}$$
$$=-\frac{x+5}{x-2}$$

10. $$\frac{4x+3}{2y}+\frac{2x-5}{2y}=\frac{4x+3+2x-5}{2y}$$
$$=\frac{6x-2}{2y}$$
$$=\frac{2(3x-1)}{2y}$$
$$=\frac{3x-1}{y}$$

11. $\frac{7x^2-4}{x+3}-\frac{6x+7}{x+3}=\frac{7x^2-4-(6x+7)}{x+3}$
$=\frac{7x^2-4-6x-7}{x+3}$
$=\frac{7x^2-6x-11}{x+3}$

12. $\frac{5}{x}+\frac{3}{2x^2}=\frac{5}{x}\cdot\frac{2x}{2x}+\frac{3}{2x^2}$
$=\frac{10x}{2x^2}+\frac{3}{2x^2}$
$=\frac{10x+3}{2x^2}$

13. $\frac{4}{xy}-\frac{3}{xy^3}=\frac{4}{xy}\cdot\frac{y^2}{y^2}-\frac{3}{xy^3}$
$=\frac{4y^2}{xy^3}-\frac{3}{xy^3}$
$=\frac{4y^2-3}{xy^3}$

14. $3-\frac{6x}{x+4}=3\left(\frac{x+4}{x+4}\right)-\frac{6x}{x+4}$
$=\frac{3x+12}{x+4}-\frac{6x}{x+4}$
$=\frac{3x+12-6x}{x+4}$
$=\frac{-3x+12}{x+4}$

15.
$$\begin{aligned}\frac{x-5}{x^2-16}-\frac{x-2}{x^2+2x-8} &= \frac{x-5}{(x-4)(x+4)}-\frac{x-2}{(x-2)(x+4)}\\ &= \frac{x-5}{(x-4)(x+4)}-\frac{1}{x+4}\\ &= \frac{x-5}{(x-4)(x+4)}-\frac{1}{x+4}\cdot\frac{x-4}{x-4}\\ &= \frac{x-5-(x-4)}{(x-4)(x+4)}\\ &= \frac{x-5-x+4}{(x-4)(x+4)}\\ &= \frac{-1}{(x-4)(x+4)}\end{aligned}$$

16.
$$\begin{aligned}\frac{4+\frac{1}{2}}{3-\frac{1}{5}} &= \frac{10\left(4+\frac{1}{2}\right)}{10\left(3-\frac{1}{5}\right)}\\ &= \frac{40+5}{30-2}\\ &= \frac{45}{28}\end{aligned}$$

17.
$$\begin{aligned}\frac{x+\frac{x}{y}}{\frac{1}{x}} &= \left(x+\frac{x}{y}\right)\frac{x}{1}\\ &= \left(\frac{xy+x}{y}\right)\left(\frac{x}{1}\right)\\ &= \frac{yx^2+x^2}{y}\end{aligned}$$

18.
$$\begin{aligned}\frac{2+\frac{3}{x}}{\frac{2}{x}-5} &= \frac{x\left(2+\frac{3}{x}\right)}{x\left(\frac{2}{x}-5\right)}\\ &= \frac{2x+3}{2-5x}\end{aligned}$$

19.
$$\begin{aligned}6+\frac{2}{x} &= 7\\ x\left(6+\frac{2}{x}\right) &= 7x\\ 6x+2 &= 7x\\ 2 &= x\end{aligned}$$

20.
$$\frac{2x}{3}-\frac{x}{4}=x+1$$
$$12\left(\frac{2x}{3}-\frac{x}{4}\right)=12(x+1)$$
$$8x-3x=12x+12$$
$$5x=12x+12$$
$$-7x=12$$
$$x=-\frac{12}{7}$$

21.
$$\frac{x}{x-8}+\frac{6}{x-2}=\frac{x^2}{x^2-10x+16}$$
$$\frac{x}{x-8}+\frac{6}{x-2}=\frac{x^2}{(x-8)(x-2)}$$
$$(x-8)(x-2)\left(\frac{x}{x-8}+\frac{6}{x-2}\right)=(x-8)(x-2)\left[\frac{x^2}{(x-8)(x-2)}\right]$$
$$x(x-2)+6(x-8)=x^2$$
$$x^2-2x+6x-48=x^2$$
$$4x-48=0$$
$$4x=48$$
$$x=12$$

22.
$$\frac{t}{8}+\frac{t}{5}=1$$
$$\frac{5t+8t}{40}=1$$
$$13t=40$$
$$t=\frac{40}{13}=3\frac{1}{13}$$

It will take them $3\frac{1}{13}$ hours to level one acre together.

23. Let x be the number.
$$x+\frac{1}{x}=2$$
$$x\left(x+\frac{1}{x}\right)=x(2)$$
$$x^2+1=2x$$
$$x^2-2x+1=0$$
$$(x-1)(x-1)=0$$
$$x-1=0$$
$$x=1$$

The number is 1.

24. Let x = base, then $2x-3$ = height.

$$\text{area} = \frac{1}{2}\cdot \text{base}\cdot\text{height}$$
$$27 = \frac{1}{2}x(2x-3)$$
$$2(27) = 2\left[\frac{1}{2}x(2x-3)\right]$$
$$54 = x(2x-3)$$
$$54 = 2x^2-3x$$
$$0 = 2x^2-3x-54$$
$$0 = (2x+9)(x-6)$$
$$2x+9=0 \quad \text{or} \quad x-6=0$$
$$x=-\frac{9}{2} \qquad x=6$$

Since the base cannot be negative, the base is 6 inches and the height is $2(6)-3=9$ inches.

25. Let d = the distance she rollerblades, then $12-d$ is the distance she bicycles.

$$t=\frac{d}{r}$$
$$\frac{d}{4}+\frac{12-d}{10}=1.5$$
$$20\left(\frac{d}{4}+\frac{12-d}{10}\right)=20(1.5)$$
$$5d+2(12-d)=30$$
$$5d+24-2d=30$$
$$3d=6$$
$$d=2$$

She rollerblades for 2 miles.

Cumulative Review Test

1. $4x^2-7xy^2+3=4(-3)^2-7(-3)(-2)^2+3$
$$=4(9)-7(-3)(4)+3$$
$$=36+84+3$$
$$=123$$

2. $5z+4=-3(z-7)$
$$5z+4=-3z+21$$
$$5z+3z=21-4$$
$$8z=17$$
$$z=\frac{17}{8}$$

3. $\left(\frac{6x^2y^3}{2x^5y}\right)^3 = \left(\frac{3y^2}{x^3}\right)^3$
$= \frac{27y^6}{x^9}$

4. $P = 2E + 3R$
$P - 2E = 3R$
$R = \frac{P - 2E}{3}$

5. $(6x^2 - 3x - 5) - (-2x^2 - 8x - 9) = 6x^2 - 3x - 5 + 2x^2 + 8x + 9$
$= 6x^2 + 2x^2 - 3x + 8x - 5 + 9$
$= 8x^2 + 5x + 4$

6. $(4x^2 - 6x + 3)(3x - 5) = 4x^2(3x - 5) - 6x(3x - 5) + 3(3x - 5)$
$= 12x^3 - 20x^2 - 18x^2 + 30x + 9x - 15$
$= 12x^3 - 38x^2 + 39x - 15$

7. $6a^2 - 6a - 5a + 5 = 6a(a - 1) - 5(a - 1)$
$= (6a - 5)(a - 1)$

8. $13x^2 + 26x - 39 = 13(x^2 + 2x - 3)$
$= 13(x + 3)(x - 1)$

9. $4x - 3y = 12$
$-3y = -4x + 12$
$y = \frac{-4}{-3}x + \frac{12}{-3}$
$y = \frac{4}{3}x - 4$

10. $x - y = 5$
$2x - 3y = 14$
Solve the first equation for x.
$x = 5 + y$
Substitute $5 + y$ for x in the second equation.

$$2x - 3y = 14$$
$$2(5 + y) - 3y = 14$$
$$10 + 2y - 3y = 14$$
$$-y = 4$$
$$y = -4$$

Substitute -4 for y in the first equation.

$$x - (-4) = 5$$
$$x + 4 = 5$$
$$x = 1$$

The solution is $(1, -4)$.

11. $x + y > 5$
$y \le x - 4$

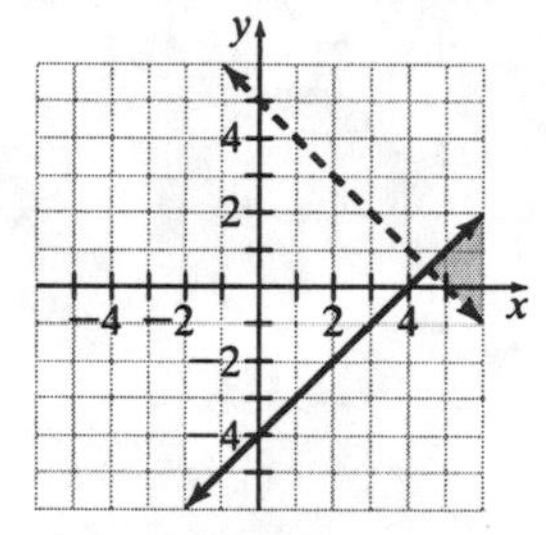

12.

$$2x^2 = 11x - 12$$
$$2x^2 - 11x + 12 = 0$$
$$(x - 4)(2x - 3) = 0$$
$$x - 4 = 0 \quad \text{or} \quad 2x - 3 = 0$$
$$x = 4 \qquad\qquad x = \tfrac{3}{2}$$

13.

$$\frac{x^2 - 9}{x^2 - x - 6} \cdot \frac{x^2 - 2x - 8}{2x^2 - 7x - 4} = \frac{(x-3)(x+3)}{(x-3)(x+2)} \cdot \frac{(x-4)(x+2)}{(2x+1)(x-4)}$$
$$= \frac{x+3}{2x+1}$$

14. $$\frac{x}{x+4}-\frac{3}{x-5}=\frac{x}{x+4}\cdot\frac{x-5}{x-5}-\frac{3}{x-5}\cdot\frac{x+4}{x+4}$$
$$=\frac{x(x-5)}{(x+4)(x-5)}-\frac{3(x+4)}{(x+4)(x-5)}$$
$$=\frac{x^2-5x-(3x+12)}{(x+4)(x-5)}$$
$$=\frac{x^2-5x-3x-12}{(x+4)(x-5)}$$
$$=\frac{x^2-8x-12}{(x+4)(x-5)}$$

15. $$\frac{4}{x^2-3x-10}+\frac{2}{x^2+5x+6}=\frac{4}{(x-5)(x+2)}+\frac{2}{(x+2)(x+3)}$$
$$=\frac{4(x+3)}{(x-5)(x+2)(x+3)}+\frac{2(x-5)}{(x-5)(x+2)(x+3)}$$
$$=\frac{4(x+3)+2(x-5)}{(x-5)(x+2)(x+3)}$$
$$=\frac{4x+12+2x-10}{(x-5)(x+2)(x+3)}$$
$$=\frac{6x+2}{(x-5)(x+2)(x+3)}$$

16. $$\frac{x}{9}-\frac{x}{6}=\frac{1}{12}$$
$$36\left(\frac{x}{9}-\frac{x}{6}\right)=36\left(\frac{1}{12}\right)$$
$$4x-6x=3$$
$$-2x=3$$
$$x=-\frac{3}{2}$$

17. $$\frac{7}{x+3}+\frac{5}{x+2}=\frac{5}{x^2+5x+6}$$
$$\frac{7}{x+3}+\frac{5}{x+2}=\frac{5}{(x+3)(x+2)}$$
$$(x+3)(x+2)\left(\frac{7}{x+3}+\frac{5}{x+2}\right)=(x+3)(x+2)\left[\frac{5}{(x+3)(x+2)}\right]$$
$$7(x+2)+5(x+3)=5$$
$$7x+14+5x+15=5$$
$$12x+29=5$$
$$12x=-24$$
$$x=-2$$

Check:

$$\frac{7}{x+3}+\frac{5}{x+2}=\frac{5}{x^2+5x+6}$$
$$\frac{7}{-2+3}+\frac{5}{-2+2}=\frac{5}{(-2)^2+5(-2)+6}$$
$$\frac{7}{1}+\frac{5}{0}=\frac{5}{0}$$

Since $\frac{5}{0}$ is not a real number, there is no solution.

18. Let x = the total medical bills. The cost under plan 1 is $0.10x$, while the cost under plan 2 is $100 + 0.05x$.

$$0.10x = 100 + 0.05x$$
$$0.05x = 100$$
$$x = 2000$$

The cost under both plans is the same for \$2000 in total medical bills.

19. Let x = pounds of sunflower seed and
y = pounds of premixed assorted seed mix

$$x + y = 50$$
$$0.50x + 0.15y = 14.50$$

Solve the first equation for y.

$$y = 50 - x$$

Substitute $50 - x$ for y in the second equation.

$$0.50x + 0.15(50 - x) = 14.50$$
$$0.50x + 7.5 - 0.15x = 14.50$$
$$0.35x = 7$$
$$x = 20$$
$$y = 50 - x = 50 - 20 = 30$$

He will have to use 20 pounds of sunflower seed and 30 pounds of premixed assorted seed mix.

20. Let d = distance on first leg, then the distance on the second leg is $12.75 - d$.

$$t = \frac{d}{r}$$

Time for first leg + time for second leg = total time

$$\frac{d}{6.5}+\frac{12.75-d}{9.5}=1.5$$
$$(9.5)(6.5)\left[\frac{d}{6.5}+\frac{12.75-d}{9.5}\right]=(9.5)(6.5)(1.5)$$
$$9.5d + 6.5(12.75 - d) = 92.625$$
$$9.5d + 82.875 - 6.5d = 92.625$$
$$3d = 9.75$$
$$d = 3.25$$
$$12.75 - d = 12.75 - 3.25 = 9.5$$

The distance traveled during the first leg of the race was 3.25 miles and the distance traveled in the second leg of the race was 9.5 miles.

Chapter 9

Exercise Set 9.1

1. The principal square root of a positive real number x is a positive number whose square equals x.

3. Answers will vary.

5. Answers will vary.

7. Yes, since $4^2 = 16$.

9. No, because the square root of a negative number is not a real number.

11. Yes, since $\sqrt{\frac{9}{25}} = \frac{3}{5}$ which is a rational number.

13. $\sqrt{0} = 0$ since $(0)^2 = 0$

15. $\sqrt{1} = 1$ since $1^2 = 1$

17. $\sqrt{81} = 9$
    Take the opposite of both sides to get
    $-\sqrt{81} = -9$

19. $\sqrt{400} = 20$ since $20^2 = 400$

21. $\sqrt{25} = 5$ since $(5)^2 = 25$
    Take the opposite of both sides to get
    $-\sqrt{25} = -5$

23. $\sqrt{144} = 12$ since $(12)^2 = 144$

25. $\sqrt{169} = 13$ since $(13)^2 = 169$

27. $\sqrt{1} = 1$ since $(1)^2 = 1$
    Take the opposite of both sides to get
    $-\sqrt{1} = -1$

29. $\sqrt{81} = 9$ since $9^2 = 81$

31. $\sqrt{121} = 11$ since $(11)^2 = 121$
    Take the opposite of both sides to get
    $-\sqrt{121} = -11$

33. $\sqrt{\frac{1}{4}} = \frac{1}{2}$ since $\left(\frac{1}{2}\right)^2 = \frac{1}{4}$

35. $\sqrt{\frac{16}{9}} = \frac{4}{3}$ since $\left(\frac{4}{3}\right)^2 = \frac{16}{9}$

37. $\sqrt{\frac{25}{30}} = \frac{5}{6}$ since $\left(\frac{5}{6}\right)^2 = \frac{25}{36}$
    Take the opposite of both sides to get
    $-\sqrt{\frac{25}{30}} = -\frac{5}{6}$

39. $\sqrt{\frac{36}{49}} = \frac{6}{7}$ since $\left(\frac{6}{7}\right)^2 = \frac{36}{49}$

41. $\sqrt{10} \approx 3.1622777$

43. $\sqrt{15} \approx 3.8729833$

45. $\sqrt{80} \approx 8.9442719$

47. $\sqrt{81} = 9$

49. $\sqrt{97} \approx 9.8488578$

51. $\sqrt{3} \approx 1.7320508$

53. True; since 49 is a perfect square, $\sqrt{49}$ is a rational number

55. True; since 25 is a perfect square, $\sqrt{25}$ is a rational number

57. False; since 9 is a perfect square, $\sqrt{9}$ is a rational number

59. True; $\sqrt{\frac{4}{9}} = \frac{2}{3}$ which is a rational number

61. True; since 125 is not a perfect square, $\sqrt{125}$ is an irrational number

63. True; $\sqrt{(18)^2} = \sqrt{324} = 18$ which is an integer

65. $\sqrt{3} = 3^{1/2}$

67. $\sqrt{17} = (17)^{1/2}$

69. $\sqrt{6y} = (6y)^{1/2}$

71. $\sqrt{12x^2} = \left(12x^2\right)^{1/2}$

73. $\sqrt{15ab^2} = (15ab^2)^{1/2}$

75. $\sqrt{50a^3} = (50a^3)^{1/2}$

77. Rational:

$7.24,\ \frac{5}{7},\ 0.666\ldots,\ 5,\ \sqrt{\frac{4}{49}},\ \frac{3}{7},\ -\sqrt{9}$

Irrational: $\sqrt{\frac{5}{16}}$

Imaginary: $\sqrt{-9},\ -\sqrt{-16}$

79. 6 and 7 since $6^2 = 36,\ 7^2 = 49,$ and $36 < 47 < 49$

81. **a.** Square 4.6 and compare it with 20.

b. $(4.6)^2 = 21.16 > 20$, so 4.6 is greater.

83. $-\sqrt{7},\ -\sqrt{4},\ -\frac{1}{2},\ 2.5,\ 3,\ \sqrt{16},\ 4.01,\ 12$

85.

$$\begin{aligned}\sqrt{4} &= 2\\ 6^{1/2} &\approx 2.45\\ -\sqrt{9} &= -3\\ -(25)^{1/2} &= -5\\ (30)^{1/2} &\approx 5.48\end{aligned}$$

$(-4)^{1/2}$, imaginary number

87. $\sqrt{0} = 0$ which is neither a positive nor negative number.
0 is a perfect square since $0^2 = 0$

a. Yes

b. No

c. No

d. Yes

e. No

89. **a.** Yes, $\sqrt{2^2} = \sqrt{4} = 2.$

b. Yes, $\sqrt{5^2} = \sqrt{25} = 5.$

c. $\sqrt{a^2} = a,\ a \geq 0$

91. $(x^3)^{1/2} = x^{3\cdot(1/2)} = x^{3/2}$

93. $x^{1/2}x^{5/2} = x^{(1/2)+(5/2)} = x^{6/2} = x^3$

95. $m = \frac{7-3}{6-(-5)} = \frac{4}{11}$

96.

$$\begin{aligned}f(x) &= x^2 - 4x - 5\\ f(-3) &= (-3)^2 - 4(-3) - 5\\ &= 9 + 12 - 5\\ &= 16\end{aligned}$$

97.
$$\frac{2x}{x^2-4}+\frac{1}{x-2}=\frac{2}{x+2}$$
$$\frac{2x}{(x-2)(x+2)}+\frac{1}{x-2}=\frac{2}{x+2}$$
$$(x-2)(x+2)\left[\frac{2x}{(x-2)(x+2)}+\frac{1}{(x-2)}\right]=\frac{2}{x+2}\cdot(x-2)(x+2)$$
$$2x+(x+2)=2(x-2)$$
$$2x+x+2=2x-4$$
$$3x+2=2x-4$$
$$3x=2x-6$$
$$x=-6$$

98.
$$\frac{4x}{x^2+6x+9}-\frac{2x}{x+3}=\frac{x+1}{x+3}$$
$$\frac{4x}{(x+3)^2}-\frac{2x}{x+3}=\frac{x+1}{x+3}$$
$$(x+3)^2\left[\frac{4x}{(x+3)^2}-\frac{2x}{x+3}\right]=\frac{x+1}{x+3}\cdot(x+3)^2$$
$$4x-2x(x+3)=(x+1)(x+3)$$
$$4x-2x^2-6x=x^2+4x+3$$
$$0=3x^2+6x+3$$
$$0=3(x^2+2x+1)$$
$$0=3(x+1)^2$$
$$0=x+1$$
$$x=-1$$

Exercise Set 9.2

1. Answers will vary.

3. The product rule cannot be used when radicands are negative.

5. a. Answers will vary.

b. $\sqrt{x^{13}}=\sqrt{x^{12}\cdot x}=\sqrt{x^{12}}\sqrt{x}=x^6\sqrt{x}$

7. a. There can be no perfect square factors or any exponents greater than 1 in the radicand.

b. $\sqrt{75x^5}=\sqrt{25x^4}\cdot\sqrt{3x}=5x^2\sqrt{3x}$

9. No; $\sqrt{32}=\sqrt{16\cdot 2}=\sqrt{16}\sqrt{2}=4\sqrt{2}$

11. No; $\sqrt{x^9}=\sqrt{x^8x}=\sqrt{x^8}\sqrt{x}=x^4\sqrt{x}$

13. $\sqrt{36}=6$ since $6^2=36$

15. $\sqrt{8} = \sqrt{4 \cdot 2}$
$= \sqrt{4} \cdot \sqrt{2}$
$= 2\sqrt{2}$

17. $\sqrt{96} = \sqrt{16 \cdot 6}$
$= \sqrt{16} \cdot \sqrt{6}$
$= 4\sqrt{6}$

19. $\sqrt{32} = \sqrt{16 \cdot 2}$
$= \sqrt{16} \cdot \sqrt{2}$
$= 4\sqrt{2}$

21. $\sqrt{90} = \sqrt{9 \cdot 10}$
$= \sqrt{9} \cdot \sqrt{10}$
$= 3\sqrt{10}$

23. $\sqrt{80} = \sqrt{16 \cdot 5}$
$= \sqrt{16} \cdot \sqrt{5}$
$= 4\sqrt{5}$

25. $\sqrt{72} = \sqrt{36 \cdot 2}$
$= \sqrt{36} \cdot \sqrt{2}$
$= 6\sqrt{2}$

27. $\sqrt{156} = \sqrt{4 \cdot 39}$
$= \sqrt{4} \cdot \sqrt{39}$
$= 2\sqrt{39}$

29. $\sqrt{256} = 16$ since $(16)^2 = 256$

31. $\sqrt{1600} = 40$ since $(40)^2 = 1600$

33. $\sqrt{x^8} = x^4$ since $(x^4)^2 = x^8$

35. $\sqrt{x^2y^4} = \sqrt{x^2}\sqrt{y^4} = xy^2$

37. $\sqrt{a^{12}b^9} = \sqrt{a^{12}b^8}\sqrt{b} = a^6b^4\sqrt{b}$

39. $\sqrt{a^2b^4c} = \sqrt{a^2b^4} \cdot \sqrt{c} = ab^2\sqrt{c}$

41. $\sqrt{2x^3} = \sqrt{x^2}\sqrt{2x} = x\sqrt{2x}$

43. $\sqrt{75a^3b^2} = \sqrt{25a^2b^2}\sqrt{3a} = 5ab\sqrt{3a}$

45. $\sqrt{300a^5b^{11}} = \sqrt{100a^4b^{10}}\sqrt{3ab}$
$= 10a^2b^5\sqrt{3ab}$

47. $\sqrt{243x^3y^4} = \sqrt{81x^2y^4}\sqrt{3x} = 9xy^2\sqrt{3x}$

49. $\sqrt{192a^2b^7c} = \sqrt{64a^2b^6}\sqrt{3bc} = 8ab^3\sqrt{3bc}$

51. $\sqrt{250x^4yz} = \sqrt{25x^4} \cdot \sqrt{10yz} = 5x^2\sqrt{10yz}$

53. $\sqrt{7} \cdot \sqrt{7} = \sqrt{7 \cdot 7} = \sqrt{49} = 7$

55. $\sqrt{18} \cdot \sqrt{3} = \sqrt{54}$
$= \sqrt{9} \cdot \sqrt{6}$
$= 3\sqrt{6}$

57. $\sqrt{48} \cdot \sqrt{15} = \sqrt{720}$
$= \sqrt{144} \cdot \sqrt{5}$
$= 12\sqrt{5}$

59. $\sqrt{3x}\sqrt{7x} = \sqrt{21x^2} = \sqrt{21}\sqrt{x^2} = x\sqrt{21}$

61. $\sqrt{4a^2}\sqrt{12ab^2} = \sqrt{48a^3b^2}$
$= \sqrt{16a^2b^2}\sqrt{3a}$
$= 4ab\sqrt{3a}$

63. $\sqrt{6xy^3}\sqrt{12x^2y} = \sqrt{72x^3y^4}$
$= \sqrt{36x^2y^4}\sqrt{2x}$
$= 6xy^2\sqrt{2x}$

65. $\sqrt{21xy}\sqrt{3x^3y^4} = \sqrt{63x^4y^5}$
$= \sqrt{9x^4y^4}\sqrt{7y}$
$= 3x^2y^2\sqrt{7y}$

67. $\sqrt{15xy^6}\sqrt{6xyz} = \sqrt{90x^2y^7z}$
$= \sqrt{9x^2y^6}\sqrt{10yz}$
$= 3xy^3\sqrt{10yz}$

69. $\sqrt{4a^2b^4}\sqrt{9a^4b^6} = \sqrt{36a^6b^{10}}$
$= 6a^3b^5$

71. $\left(\sqrt{2x}\right)^2 = (2x)^{(1/2)\cdot 2}$
$= (2x)^1$
$= 2x$

73. $\left(\sqrt{13x^4y^6}\right)^2 = (13x^4y^6)^{(1/2)\cdot 2}$
$= (13x^4y^6)^1$
$= 13x^4y^6$

75. $\left(\sqrt{5a}\right)^2\left(\sqrt{3a}\right)^2 = (5a)^{(1/2)\cdot 2}(3a)^{(1/2)\cdot 2}$
$= (5a)^1(3a)^1$
$= 5a \cdot 3a$
$= 15a^2$

77. Exponent on x is 4 because $\sqrt{x^4} = x^2$.

79. Exponent on x is 6 because $\sqrt{x^6} = x^3$; exponent on y is 5 because $\sqrt{y^5} = \sqrt{y^4}\sqrt{y} = y^2\sqrt{y}$.

81. Coefficient is 8 because $\sqrt{2} \cdot \sqrt{8} = \sqrt{16} = 4$.
Exponent on x is 12 because $\sqrt{x^{12}} \cdot \sqrt{x^3} = \sqrt{x^{15}} = x^7\sqrt{x}$.
Exponent on y is 7 because $\sqrt{y^5} \cdot \sqrt{y^7} = \sqrt{y^{12}} = y^6$.

83. a. $\left(\sqrt{13x^3}\right)^2 = \left(13x^3\right)^{(1/2)\cdot 2}$
$= \left(13x^3\right)^1$
$= 13x^3$

b. $\sqrt{\left(13x^3\right)^2} = \left(13x^3\right)^{2\cdot(1/2)}$
$= \left(13x^3\right)^1$
$= 13x^3$

c. Yes

85. $\sqrt{200\Delta^{11}} = \sqrt{100\Delta^{10}}\sqrt{2\Delta} = 10\Delta^5\sqrt{2\Delta}$

87. $\sqrt{5\Delta^{100}} \cdot \sqrt{5\nabla^{36}} = \sqrt{25\Delta^{100}\nabla^{36}} = 5\Delta^{50}\nabla^{18}$

89. $\sqrt{x^{2/6}} = \left(x^{2/6}\right)^{1/2} = x^{(2/6)(1/2)} = x^{1/6}$

91. $\sqrt{4x^{4/5}} = \sqrt{4}\sqrt{x^{4/5}}$
$= 2\left(x^{4/5}\right)^{1/2}$
$= 2x^{(4/5)(1/2)}$
$= 2x^{2/5}$

93. It is rational since $\sqrt{6.25} = 2.5$ and 2.5 is a terminating decimal number

95. a. $\sqrt{16} = 4$
The side has length 4 feet.

b. No, it is increased $\sqrt{2}$ or ≈ 1.414 times.

c. 4 times

97. a. Yes; a rational number is one that can be written in the form $\frac{a}{b}$ where a and b are integers and $b \neq 0$.
Let x and y be 2 rational numbers
$x = \frac{a}{b}$, a, b, integers $b \neq 0$
$y = \frac{c}{d}$, c, d integers $d \neq 0$
$xy = \frac{a}{b} \cdot \frac{c}{d} = \frac{ac}{bd}$, ac and bd are integers and $bd \neq 0$, hence xy is rational.

b. No, for example $\sqrt{2} \cdot \sqrt{2} = 2$

102. $3x+6y=9$

$$6y=-3x+9$$

$$\frac{6y}{6}=\frac{-3x+9}{6}$$

$$y=-\frac{1}{2}x+\frac{3}{2}$$

$$m=-\frac{1}{2},\ \left(0,\ \frac{3}{2}\right)$$

103. Graph the line $6x-5y=30$.
Since the inequality symbol is $\geq$, draw a solid line.

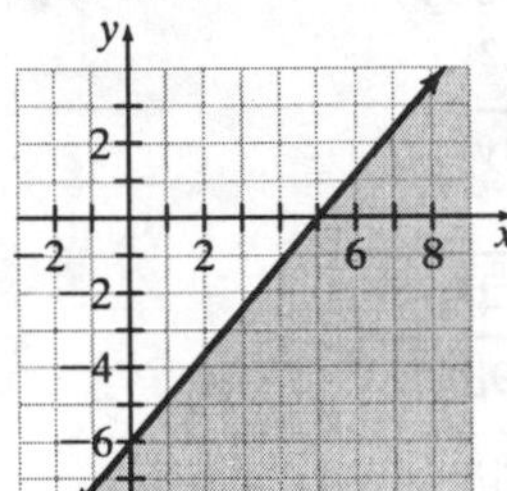

Check (0, 0):

$$6x-5y\geq 30$$

$$6(0)-5(0)\geq 30$$

$$0-0\geq 30$$

$$0\geq 30 \quad \text{False}$$

Since (0, 0) does not satisfy the inequality, shade the region that does not include (0, 0).

104. $3x-4y=6$
$5x-3y=5$
Multiply the first equation by 5 and the second equation by -3.

$$5[3x-4y=6]$$

$$-3[5x-3y=5]$$

givcs

$$15x-20y=30$$

$$-15x+9y=-15$$

$$-11y=15$$

$$y=-\frac{15}{11}$$

Substitute $-\frac{15}{11}$ for y in the first equation.

$$3x-4y=6$$

$$3x-4\left(-\frac{15}{11}\right)=6$$

$$3x+\frac{60}{11}=6$$

$$3x=\frac{6}{11}$$

$$x=\frac{2}{11}$$

The solution is $\left(\frac{2}{11},\ -\frac{15}{11}\right)$

105. $\dfrac{3x^2-16x-12}{3x^2-10x-8}\div\dfrac{x^2-7x+6}{3x^2-11x-4}$

$$=\frac{3x^2-16x-12}{3x^2-10x-8}\cdot\frac{3x^2-11x-4}{x^2-7x+6}$$

$$=\frac{(3x+2)(x-6)}{(3x+2)(x-4)}\cdot\frac{(3x+1)(x-4)}{(x-6)(x-1)}$$

$$=\frac{3x+1}{x-1}$$

Exercise Set 9.3

1. 1. No perfect square factors in any radicand.
2. No radicand contains a fraction.
3. No square roots in any denominator.

3. The radicand contains a fraction.

$$\sqrt{\frac{1}{3}}=\frac{\sqrt{1}}{\sqrt{3}}=\frac{1}{\sqrt{3}}=\frac{1}{\sqrt{3}}\cdot\frac{\sqrt{3}}{\sqrt{3}}=\frac{\sqrt{3}}{3}$$

5. Cannot be simplified: the radicand does not have a factor that is a perfect suqare, the radicand does not contain a fraction, and the denominator does not contain a square root.

7. The numerator and denominator have a common factor.

$$\frac{x^2\sqrt{2}}{x}=x\sqrt{2}$$

9. Cannot be simplified: the radicand does not have a factor that is a perfect suqare, the radicand does not contain a fraction, and the denominator does not contain a square root.

11. Answers will vary.

13. **a.** Answers will vary.

b. $\frac{a}{\sqrt{b}}=\frac{a}{\sqrt{b}}\cdot\frac{\sqrt{b}}{\sqrt{b}}=\frac{a\sqrt{b}}{b}$

15. $\sqrt{\frac{20}{5}}=\sqrt{4}=2$

17. $\sqrt{\frac{63}{7}}=\sqrt{9}=3$

19. $\frac{\sqrt{18}}{\sqrt{2}}=\sqrt{\frac{18}{2}}=\sqrt{9}=3$

21. $\sqrt{\frac{1}{36}}=\frac{\sqrt{1}}{\sqrt{36}}=\frac{1}{6}$

23. $\sqrt{\frac{81}{144}}=\frac{\sqrt{81}}{\sqrt{144}}=\frac{9}{12}=\frac{3}{4}$

25. $\frac{\sqrt{10}}{\sqrt{1000}}=\sqrt{\frac{10}{1000}}=\sqrt{\frac{1}{100}}=\frac{\sqrt{1}}{\sqrt{100}}=\frac{1}{10}$

27. $\sqrt{\frac{40x^3}{2x}}=\sqrt{20x^2}=\sqrt{4x^2}\cdot\sqrt{5}=2x\sqrt{5}$

29. $\sqrt{\frac{45x^2}{16x^2y^4}}=\sqrt{\frac{45}{16y^4}}$
$=\frac{\sqrt{45}}{\sqrt{16y^4}}$
$=\frac{\sqrt{9\cdot 5}}{4y^2}$
$=\frac{3\sqrt{5}}{4y^2}$

31. $\sqrt{\frac{16x^5y^3}{100x^7y}}=\sqrt{\frac{4y^2}{25x^2}}=\frac{\sqrt{4y^2}}{\sqrt{25x^2}}=\frac{2y}{5x}$

33. $\sqrt{\frac{24ab}{24a^5b^3}}=\sqrt{\frac{1}{a^4b^2}}=\frac{\sqrt{1}}{\sqrt{a^4b^2}}=\frac{1}{a^2b}$

35. $\frac{\sqrt{32x^5}}{\sqrt{8x}}=\sqrt{\frac{32x^5}{8x}}=\sqrt{4x^4}=2x^2$

37. $\frac{\sqrt{81x^5y}}{\sqrt{100xy^3}}=\sqrt{\frac{81x^5y}{100xy^3}}$
$=\sqrt{\frac{81x^4}{100y^2}}$
$=\frac{\sqrt{81x^4}}{\sqrt{100y^2}}$
$=\frac{9x^2}{10y}$

39. $\frac{\sqrt{45ab^6}}{\sqrt{9ab^4c^2}}=\sqrt{\frac{45ab^6}{9ab^4c^2}}$
$=\sqrt{\frac{5b^2}{c^2}}$
$=\frac{\sqrt{5b^2}}{\sqrt{c^2}}$
$=\frac{b\sqrt{5}}{c}$

41. $\frac{\sqrt{125a^6b^8}}{\sqrt{5a^2b^2}}=\sqrt{\frac{125a^6b^8}{5a^2b^2}}$
$=\sqrt{25a^4b^6}$
$=5a^2b^3$

43. $\frac{1}{\sqrt{5}}=\frac{1}{\sqrt{5}}\cdot\frac{\sqrt{5}}{\sqrt{5}}=\frac{\sqrt{5}}{\sqrt{25}}=\frac{\sqrt{5}}{5}$

45. $\frac{4}{\sqrt{8}} = \frac{4}{\sqrt{4}\sqrt{2}}$
$= \frac{4}{2\sqrt{2}}$
$= \frac{2}{\sqrt{2}}$
$= \frac{2}{\sqrt{2}} \cdot \frac{\sqrt{2}}{\sqrt{2}}$
$= \frac{2\sqrt{2}}{2}$
$= \sqrt{2}$

47. $\frac{6}{\sqrt{12}} = \frac{6}{\sqrt{4 \cdot 3}}$
$= \frac{6}{2\sqrt{3}}$
$= \frac{3}{\sqrt{3}}$
$= \frac{3}{\sqrt{3}} \cdot \frac{\sqrt{3}}{\sqrt{3}}$
$= \frac{3\sqrt{3}}{\sqrt{9}}$
$= \frac{3\sqrt{3}}{3}$
$= \sqrt{3}$

49. $\sqrt{\frac{2}{3}} = \frac{\sqrt{2}}{\sqrt{3}} = \frac{\sqrt{2}}{\sqrt{3}} \cdot \frac{\sqrt{3}}{\sqrt{3}} = \frac{\sqrt{6}}{\sqrt{9}} = \frac{\sqrt{6}}{3}$

51. $\sqrt{\frac{7}{21}} = \sqrt{\frac{1}{3}}$
$= \frac{\sqrt{1}}{\sqrt{3}}$
$= \frac{1}{\sqrt{3}}$
$= \frac{1}{\sqrt{3}} \cdot \frac{\sqrt{3}}{\sqrt{3}}$
$= \frac{\sqrt{3}}{\sqrt{9}}$
$= \frac{\sqrt{3}}{3}$

53. $\sqrt{\frac{x^2}{2}} = \frac{\sqrt{x^2}}{\sqrt{2}} = \frac{x}{\sqrt{2}} = \frac{x}{\sqrt{2}} \cdot \frac{\sqrt{2}}{\sqrt{2}} = \frac{x\sqrt{2}}{2}$

55. $\sqrt{\frac{a^2}{8}} = \frac{\sqrt{a^2}}{\sqrt{8}}$
$= \frac{a}{\sqrt{8}}$
$= \frac{a}{\sqrt{4 \cdot 2}}$
$= \frac{a}{2\sqrt{2}}$
$= \frac{a}{2\sqrt{2}} \cdot \frac{\sqrt{2}}{\sqrt{2}}$
$= \frac{a\sqrt{2}}{2 \cdot 2}$
$= \frac{a\sqrt{2}}{4}$

57. $\sqrt{\frac{x^4}{5}} = \frac{\sqrt{x^4}}{\sqrt{5}} = \frac{x^2}{\sqrt{5}} = \frac{x^2}{\sqrt{5}} \cdot \frac{\sqrt{5}}{\sqrt{5}} = \frac{x^2\sqrt{5}}{5}$

59. $\sqrt{\frac{a^8}{14b}} = \frac{\sqrt{a^8}}{\sqrt{14b}}$
$= \frac{a^4}{\sqrt{14b}}$
$= \frac{a^4}{\sqrt{14b}} \cdot \frac{\sqrt{14b}}{\sqrt{14b}}$
$= \frac{a^4\sqrt{14b}}{\sqrt{196b^2}}$
$= \frac{a^4\sqrt{14b}}{14b}$

61. $\sqrt{\frac{8x^4y^2}{32x^2y^3}} = \sqrt{\frac{x^2}{4y}}$
$= \frac{\sqrt{x^2}}{\sqrt{4y}}$
$= \frac{x}{\sqrt{4}\sqrt{y}}$
$= \frac{x}{2\sqrt{y}}$
$= \frac{x}{2\sqrt{y}} \cdot \frac{\sqrt{y}}{\sqrt{y}}$
$= \frac{x\sqrt{y}}{2y}$

63. $\sqrt{\frac{18yz}{75x^4y^5z^3}} = \sqrt{\frac{6}{25x^4y^4z^2}}$
$= \frac{\sqrt{6}}{\sqrt{25x^4y^4z^2}}$
$= \frac{\sqrt{6}}{5x^2y^2z}$

65. $\frac{\sqrt{90x^4y}}{\sqrt{2x^5y^5}} = \sqrt{\frac{90x^4y}{2x^5y^5}}$
$= \sqrt{\frac{45}{xy^4}}$
$= \frac{\sqrt{45}}{\sqrt{xy^4}}$
$= \frac{\sqrt{9}\sqrt{5}}{\sqrt{y^4}\sqrt{x}}$
$= \frac{3\sqrt{5}}{y^2\sqrt{x}}$
$= \frac{3\sqrt{5}}{y^2\sqrt{x}} \cdot \frac{\sqrt{x}}{\sqrt{x}}$
$= \frac{3\sqrt{5x}}{xy^2}$

67. Yes, $\frac{a}{b} \div \frac{c}{d} = \frac{a}{b} \cdot \frac{d}{c} = \frac{ad}{bc}$, which is a rational number since *b*, *c*, and *d* cannot be zero $\left(\frac{a}{b}, \frac{c}{d} \text{ both rational and } \frac{c}{d} \neq 0\right)$.

69. $\sqrt{5} + \sqrt{10} \approx 2.236 + 3.162 \approx 5.398 \approx 5.40$

71. $\sqrt{5} / \sqrt{10} \approx (2.236) \div (3.162) \approx 0.71$

73. $\sqrt{7} + \sqrt{21} \approx 2.646 + 4.583 \approx 7.229 \approx 7.23$

75. $\sqrt{7} / \sqrt{21} \approx (2.646) \div (4.583) \approx 0.58$

77. a. No

b. $2 \cdot \sqrt{5} = 2\sqrt{5}$ is twice as large as $\sqrt{5}$.

79. $\sqrt{(3^2) - 4 \cdot 1 \cdot 2} = \sqrt{9 - 8} = \sqrt{1} = 1$

81. $\sqrt{(-14)^2 - 4 \cdot 1 \cdot (-5)} = \sqrt{196 + 20}$
$= \sqrt{216}$
$= \sqrt{36 \cdot 6}$
$= 6\sqrt{6}$

83. $\sqrt{(4)^2 - 4 \cdot (-2)(7)} = \sqrt{16 + 56}$
$= \sqrt{72}$
$= \sqrt{36 \cdot 2}$
$= 6\sqrt{2}$

85. The missing expression is $64x^{10}$ since $\sqrt{\frac{64x^{10}}{4x^2}} = \sqrt{16x^8} = 4x^4$.

87. The missing expression is 2 since $\frac{1}{\sqrt{2}} = \frac{1}{\sqrt{2}} \cdot \frac{\sqrt{2}}{\sqrt{2}} = \frac{\sqrt{2}}{2}$.

89.

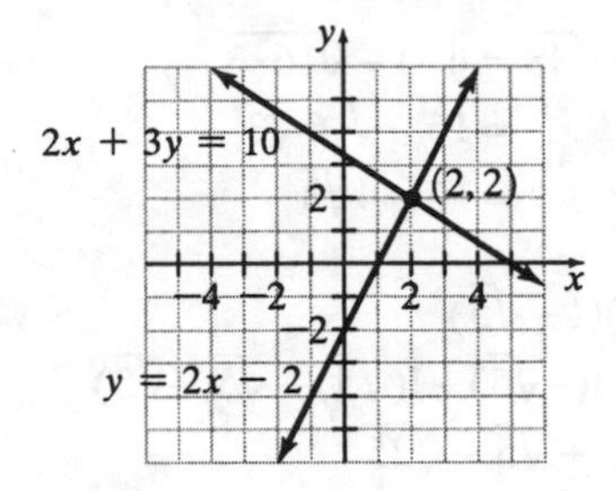

The solution is (2, 2).

90. $3x^2 - 12x - 96 = 3(x^2 - 4x - 32)$
$= 3(x+4)(x-8)$

91. $\frac{x-1}{x^2-1} = \frac{x-1}{(x+1)(x-1)} = \frac{1}{x+1}$

92.
$$x + \frac{24}{x} = 10$$
$$x\left(x + \frac{24}{x}\right) = 10 \cdot x$$
$$x^2 + 24 = 10x$$
$$x^2 - 10x + 24 = 0$$
$$(x-6)(x-4) = 0$$
$$x - 6 = 0 \text{ or } x - 4 = 0$$
$$x = 6 \qquad x = 4$$

Exercise Set 9.4

1. Like square roots are square roots having the same radicand. One example is $\sqrt{3},\ 5\sqrt{3}$.

3. Only like square roots can be added or subtracted.

5. a. $4+\sqrt{7}$

b. $5-\sqrt{11}$

c. $-3+2\sqrt{13}$

d. $\sqrt{5}+1$

7. a. Multiply the numerator and denominator by $b-\sqrt{c}$.

b.
$$\frac{a}{b+\sqrt{c}} = \frac{a}{b+\sqrt{c}} \cdot \frac{b-\sqrt{c}}{b-\sqrt{c}} = \frac{a(b-\sqrt{c})}{b^2-c} = \frac{ab-a\sqrt{c}}{b^2-c}$$

9. $6\sqrt{2} - 4\sqrt{2} = (6-4)\sqrt{2} = 2\sqrt{2}$

11. $7\sqrt{5} - 11\sqrt{5} = (7-11)\sqrt{5} = -4\sqrt{5}$

13. $\sqrt{7} + 4\sqrt{7} - 3\sqrt{7} + 6 = (1+4-3)\sqrt{7} + 6$
$= 2\sqrt{7} + 6$

15. $7\sqrt{x} + \sqrt{x} = (7+1)\sqrt{x} = 8\sqrt{x}$

17. $-\sqrt{y} + 3\sqrt{y} - 5\sqrt{y} = (-1+3-5)\sqrt{y} = -3\sqrt{y}$

19. $3\sqrt{y} - \sqrt{y} + 3 = (3-1)\sqrt{y} + 3 = 2\sqrt{y} + 3$

21. $\sqrt{x} + \sqrt{y} + x + 3\sqrt{y} = \sqrt{x} + (1+3)\sqrt{y} + x$
$= \sqrt{x} + 4\sqrt{y} + x$

23. $2 + 3\sqrt{y} - 6\sqrt{y} + 5 = (2+5) + (3-6)\sqrt{y}$
$= 7 - 3\sqrt{y}$

25. $-3\sqrt{7} + \sqrt{7} - 2\sqrt{x} - 7\sqrt{x}$
$= (-3+1)\sqrt{7} + (-2-7)\sqrt{x}$
$= -2\sqrt{7} - 9\sqrt{x}$

27. $\sqrt{12} + \sqrt{18} = \sqrt{4 \cdot 3} + \sqrt{9 \cdot 2}$
$= \sqrt{4}\sqrt{3} + \sqrt{9}\sqrt{2}$
$= 2\sqrt{3} + 3\sqrt{2}$

29. $\sqrt{300} - \sqrt{27} = \sqrt{100 \cdot 3} - \sqrt{9 \cdot 3}$
$= \sqrt{100}\sqrt{3} - \sqrt{9}\sqrt{3}$
$= 10\sqrt{3} - 3\sqrt{3}$
$= 7\sqrt{3}$

31. $\sqrt{75}+\sqrt{108}=\sqrt{25\cdot 3}+\sqrt{36\cdot 3}$
$=\sqrt{25}\sqrt{3}+\sqrt{36}\sqrt{3}$
$=5\sqrt{3}+6\sqrt{3}$
$=11\sqrt{3}$

33. $4\sqrt{50}-\sqrt{72}+\sqrt{8}$
$=4\sqrt{25\cdot 2}-\sqrt{36\cdot 2}+\sqrt{4\cdot 2}$
$=4\sqrt{25}\sqrt{2}-\sqrt{36}\sqrt{2}+\sqrt{4}\sqrt{2}$
$=4\cdot 5\sqrt{2}-6\sqrt{2}+2\sqrt{2}$
$=20\sqrt{2}-6\sqrt{2}+2\sqrt{2}$
$=16\sqrt{2}$

35. $-3\sqrt{125}+7\sqrt{75}=-3\sqrt{25\cdot 5}+7\sqrt{25\cdot 3}$
$=-3\sqrt{25}\sqrt{5}+7\sqrt{25}\sqrt{3}$
$=-3\cdot 5\sqrt{5}+7\cdot 5\sqrt{3}$
$=-15\sqrt{5}+35\sqrt{3}$

37. $2\sqrt{360}+4\sqrt{80}=2\sqrt{36\cdot 10}+4\sqrt{16\cdot 5}$
$=2\sqrt{36}\sqrt{10}+4\sqrt{16}\sqrt{5}$
$=2\cdot 6\sqrt{10}+4\cdot 4\sqrt{5}$
$=12\sqrt{10}+16\sqrt{5}$

39. $4\sqrt{16}-\sqrt{48}=4\cdot 4-\sqrt{16\cdot 3}$
$=16-\sqrt{16}\sqrt{3}$
$=16-4\sqrt{3}$

41. $(1+\sqrt{5})(1-\sqrt{5})$
$=1(1)+1(-\sqrt{5})+1(\sqrt{5})+\sqrt{5}(-\sqrt{5})$
$=1-\sqrt{5}+\sqrt{5}-\sqrt{25}$
$=1-5$
$=-4$

43. $(4-\sqrt{2})(4+\sqrt{2})$
$=4(4)+4(\sqrt{2})+4(-\sqrt{2})+\sqrt{2}(-\sqrt{2})$
$=16+4\sqrt{2}-4\sqrt{2}-2$
$=16-2$
$=14$

45. $(\sqrt{x}+3)(\sqrt{x}-3)$
$=\sqrt{x}(\sqrt{x})+\sqrt{x}(-3)+3\sqrt{x}+3(-3)$
$=x-3\sqrt{x}+3\sqrt{x}-9$
$=x-9$

47. $(\sqrt{6}+x)(\sqrt{6}-x)$
$=\sqrt{6}(\sqrt{6})+\sqrt{6}(-x)+\sqrt{6}(x)+x(-x)$
$=6-\sqrt{6}x+\sqrt{6}x-x^2$
$=6-x^2$

49. $(\sqrt{5x}+\sqrt{y})(\sqrt{5x}-\sqrt{y})=\sqrt{5x}(\sqrt{5x})+\sqrt{5x}(-\sqrt{y})+\sqrt{y}(\sqrt{5x})+\sqrt{y}(-\sqrt{y})$
$=5x-\sqrt{5x}\sqrt{y}+\sqrt{5x}\sqrt{y}-y$
$=5x-y$

51. $(2\sqrt{x}+3\sqrt{y})(2\sqrt{x}-3\sqrt{y})=(2\sqrt{x})(2\sqrt{x})+(2\sqrt{x})(-3\sqrt{y})+(3\sqrt{y})(2\sqrt{x})+(3\sqrt{y})(-3\sqrt{y})$
$=4x-6\sqrt{xy}+6\sqrt{xy}-9y$
$=4x-9y$

53. $\dfrac{3}{\sqrt{5}+2}=\dfrac{3}{\sqrt{5}+2}\cdot\dfrac{\sqrt{5}-2}{\sqrt{5}-2}$
$=\dfrac{3(\sqrt{5}-2)}{5-4}$
$=\dfrac{3\sqrt{5}-6}{1}$
$=3\sqrt{5}-6$

55. $\frac{2}{\sqrt{6}-1}=\frac{2}{\sqrt{6}-1}\cdot\frac{\sqrt{6}+1}{\sqrt{6}+1}$
$=\frac{2(\sqrt{6}+1)}{6-1}$
$=\frac{2\sqrt{6}+2}{5}$

57. $\frac{2}{\sqrt{2}+\sqrt{3}}=\frac{2}{\sqrt{2}+\sqrt{3}}\cdot\frac{\sqrt{2}-\sqrt{3}}{\sqrt{2}-\sqrt{3}}$
$=\frac{2(\sqrt{2}-\sqrt{3})}{2-3}$
$=\frac{2\sqrt{2}-2\sqrt{3}}{-1}$
$=-2\sqrt{2}+2\sqrt{3}$

59. $\frac{8}{\sqrt{5}-\sqrt{8}}=\frac{8}{\sqrt{5}-\sqrt{8}}\cdot\frac{\sqrt{5}+\sqrt{8}}{\sqrt{5}+\sqrt{8}}$
$=\frac{8(\sqrt{5}+\sqrt{8})}{5-8}$
$=\frac{8\sqrt{5}+8\sqrt{8}}{-3}$
$=\frac{8\sqrt{5}+8\sqrt{4}\sqrt{2}}{-3}$
$=\frac{8\sqrt{5}+8\cdot 2\sqrt{2}}{-3}$
$=\frac{8\sqrt{5}+16\sqrt{2}}{-3}$
$=\frac{-8\sqrt{5}-16\sqrt{2}}{3}$

61. $\frac{5}{\sqrt{y}+3}=\frac{5}{\sqrt{y}+3}\cdot\frac{\sqrt{y}-3}{\sqrt{y}-3}$
$=\frac{5(\sqrt{y}-3)}{y-9}$
$=\frac{5\sqrt{y}-15}{y-9}$

63. $\frac{6}{4-\sqrt{y}}=\frac{6}{4-\sqrt{y}}\cdot\frac{4+\sqrt{y}}{4+\sqrt{y}}$
$=\frac{6(4+\sqrt{y})}{16-y}$
$=\frac{24+6\sqrt{y}}{16-y}$

65. $\frac{16}{\sqrt{y}+x}=\frac{16}{\sqrt{y}+x}\cdot\frac{\sqrt{y}-x}{\sqrt{y}-x}$
$=\frac{16(\sqrt{y}-x)}{y-x^2}$
$=\frac{16\sqrt{y}-16x}{y-x^2}$

67. $\frac{x}{\sqrt{x}+\sqrt{y}}=\frac{x}{\sqrt{x}+\sqrt{y}}\cdot\frac{\sqrt{x}-\sqrt{y}}{\sqrt{x}-\sqrt{y}}$
$=\frac{x(\sqrt{x}-\sqrt{y})}{x-y}$
$=\frac{x\sqrt{x}-x\sqrt{y}}{x-y}$

69. $\frac{\sqrt{x}}{\sqrt{2}-\sqrt{x}}=\frac{\sqrt{x}}{\sqrt{2}-\sqrt{x}}\cdot\frac{\sqrt{2}+\sqrt{x}}{\sqrt{2}+\sqrt{x}}$
$=\frac{\sqrt{x}(\sqrt{2}+\sqrt{x})}{2-x}$
$=\frac{\sqrt{2x}+x}{2-x}$

71. $\frac{9-3\sqrt{2}}{3}=\frac{3(3-\sqrt{2})}{3}=3-\sqrt{2}$

73. $\frac{12+24\sqrt{7}}{3}=\frac{12(1+2\sqrt{7})}{3}$
$=\frac{3\cdot 4\cdot(1+2\sqrt{7})}{3}$
$=4(1+2\sqrt{7})$

75. $\frac{4+3\sqrt{6}}{2}$
Cannot be simplified since the numerator and denominator have no common factors.

77. $\frac{6+2\sqrt{75}}{3} = \frac{6+2\sqrt{25}\sqrt{3}}{3}$
$= \frac{6+2\cdot 5\sqrt{3}}{3}$
$= \frac{2(3+5\sqrt{3})}{3}$

79. $\frac{-2+4\sqrt{80}}{10} = \frac{-2+4\sqrt{16}\sqrt{5}}{10}$
$= \frac{-2+4\cdot 4\sqrt{5}}{10}$
$= \frac{-2+16\sqrt{5}}{10}$
$= \frac{2(-1+8\sqrt{5})}{10}$
$= \frac{-1+8\sqrt{5}}{5}$

81. $\frac{16-4\sqrt{32}}{8} = \frac{4(4-\sqrt{16\cdot 2})}{8}$
$= \frac{4-\sqrt{16}\sqrt{2}}{2}$
$= \frac{4-4\sqrt{2}}{2}$
$= \frac{4(1-\sqrt{2})}{2}$
$= 2(1-\sqrt{2})$
$= 2-2\sqrt{2}$

83. $(\sqrt{x}-y)(\sqrt{x}+y)$
$= \sqrt{x}\sqrt{x} + y\sqrt{x} - y\sqrt{x} - y(y)$
$= x + y\sqrt{x} - y\sqrt{x} - y^2$
$= x - y^2$

85. $(\sqrt{x}+\sqrt{y})(\sqrt{x}-\sqrt{y})$
$= \sqrt{x}\sqrt{x} + \sqrt{x}(-\sqrt{y}) + \sqrt{y}\sqrt{x} + \sqrt{y}(-\sqrt{y})$
$= x - \sqrt{x}\sqrt{y} + \sqrt{x}\sqrt{y} - y$
$= x - y$

87. $x+2$

89. Yes

91. The missing expression is 343 since
$\sqrt{343} - \sqrt{63} = \sqrt{49\cdot 7} - \sqrt{9\cdot 7}$
$= 7\sqrt{7} - 3\sqrt{7}$
$= 4\sqrt{7}$

93. Perimeter:
$(\sqrt{2}+\sqrt{3}) + (\sqrt{2}+\sqrt{3}) + (\sqrt{2}+\sqrt{3}) + (\sqrt{2}+\sqrt{3}) = 4(\sqrt{2}+\sqrt{3})$ units
Area:
$(\sqrt{2}+\sqrt{3})(\sqrt{2}+\sqrt{3})$
$= (\sqrt{2})^2 + 2\sqrt{2}\sqrt{3} + (\sqrt{3})^2$
$= 2 + 2\sqrt{2}\sqrt{3} + 3$
$= 5 + 2\sqrt{6}$ square units

95. Perimeter:

$$5.3+(4-\sqrt{2})+(6-\sqrt{2})$$
$$=5.3+4-\sqrt{2}+6-\sqrt{2}$$
$$=5.3+4+6-2\sqrt{2}$$
$$=15.3-2\sqrt{2} \text{ units}$$

Area:

$$\frac{5.3(4-\sqrt{2})}{2}=\frac{21.2-5.3\sqrt{2}}{2}$$
$$=\frac{2(10.6-2.65\sqrt{2})}{2}$$
$$=10.6-2.65\sqrt{2} \text{ square units}$$

97. The missing expresison is 48 since

$$-5\sqrt{48}+2\sqrt{3}+3\sqrt{27}$$
$$=-5\sqrt{16\cdot 3}+2\sqrt{3}+3\sqrt{9\cdot 3}$$
$$=-5\cdot 4\sqrt{3}+2\sqrt{3}+3\cdot 3\sqrt{3}$$
$$=-20\sqrt{3}+2\sqrt{3}+9\sqrt{3}$$
$$=(-20+2+9)\sqrt{3}$$
$$=-9\sqrt{3}$$

99. **a.** $5-\sqrt{x+2}$

b. $6+\sqrt{x+3}$

101.
$$\frac{\sqrt{x}}{1-\sqrt{3}}=\frac{\sqrt{x}}{1-\sqrt{3}}\cdot\frac{1+\sqrt{3}}{1+\sqrt{3}}$$
$$=\frac{\sqrt{x}(1+\sqrt{3})}{1-3}$$
$$=\frac{\sqrt{x}(1+\sqrt{3})}{-2}$$
$$=\frac{-\sqrt{x}-\sqrt{3}}{2}$$

104.
$$\begin{array}{r} 3x-8 \\ x+4\overline{)3x^2+4x-25} \\ \underline{3x^2+12x} \\ -8x-25 \\ \underline{-8x-32} \\ 7 \end{array}$$

$$\frac{3x^2+4x-25}{x+4}=3x-8+\frac{7}{x+4}$$

105.
$$2x^2-x-36=0$$
$$(2x-9)(x+4)=0$$
$$2x-9=0 \text{ or } x+4=0$$
$$x=\frac{9}{2} \qquad x=-4$$

106
$$\frac{1}{x^2-4}-\frac{2}{x-2}$$
$$=\frac{1}{(x-2)(x+2)}-\frac{2}{x-2}$$
$$=\frac{1}{(x-2)(x+2)}-\frac{2}{x-2}\cdot\frac{x+2}{x+2}$$
$$=\frac{1}{(x-2)(x+2)}-\frac{2x+4}{(x-2)(x+2)}$$
$$=\frac{1-(2x+4)}{(x-2)(x+2)}$$
$$=\frac{-2x-3}{(x-2)(x+2)}$$

107. Mark's rate: $\frac{1}{20}$, Terry's rate: $\frac{1}{t}$

$$\frac{12}{20}+\frac{12}{t}=1$$
$$20t\left(\frac{12}{20}+\frac{12}{t}\right)=20t$$
$$12t+240=20t$$
$$240=8t$$
$$30=t$$

It would take Mrs. DeGroat 30 minutes to stack the wood by herself.

Exercise Set 9.5

1. A radical equation is an equation that contains a variable in a radicand.

3. It is necessary to check solutions because they may be extraneous.

5. Yes

7. No; $-\sqrt{64}=-8$

9. Yes

11. $\sqrt{x}=3$
$(\sqrt{x})^2=(3)^2$
$x=9$
Check: $\sqrt{x}=3$
$\sqrt{9}=3$
$3=3$ True

13. $\sqrt{x}=-5$
No solution.

15. $\sqrt{x+5}=3$
$(\sqrt{x+5})^2=3^2$
$x+5=9$
$x+5-5=9-5$
$x=4$
Check: $\sqrt{x+5}=3$
$\sqrt{4+5}=3$
$\sqrt{9}=3$
$3=3$ True

17. $\sqrt{x}+5=-7$
$\sqrt{x}=-7-5$
$\sqrt{x}=-12$
No solution.

19. $\sqrt{x}-2=7$
$\sqrt{x}=7+2$
$\sqrt{x}=9$
$(\sqrt{x})^2=9^2$
$x=81$
Check: $\sqrt{x}-2=7$
$\sqrt{81}-2=7$
$9-2=7$
$7=7$ True

21. $11=6+\sqrt{x}$
$\sqrt{x}=11-6$
$\sqrt{x}=5$
$(\sqrt{x})^2=5^2$
$x=25$
Check: $11=6+\sqrt{x}$
$11=6+\sqrt{25}$
$11=6+5$
$11=11$ True

23. $12+\sqrt{x}=3$
$\sqrt{x}=3-12$
$\sqrt{x}=-9$
No solution.

25. $\sqrt{2x-5}=x-4$
$(\sqrt{2x-5})^2=(x-4)^2$
$2x-5=x^2-8x+16$
$0=x^2-10x+21$
$0=(x-3)(x-7)$
$x-3=0$ or $x-7=0$
$x=3$ $\quad x=7$
Check:
$x=3$
$\sqrt{2x-5}=x-4$
$\sqrt{2(3)-5}=3-4$
$\sqrt{1}=-1$
$1=-1$ False

$x=7$
$\sqrt{2x-5}=x-4$
$\sqrt{2(7)-5}=7-4$
$\sqrt{9}=3$
$3=3$ True

The solution is 7; 3 is not a solution.

27. $\sqrt{3x-7}=\sqrt{x+3}$
$(\sqrt{3x-7})^2=(\sqrt{x+3})^2$
$3x-7=x+3$
$3x-x=3+7$
$2x=10$
$x=5$
Check: $\sqrt{3x-7}=\sqrt{x+3}$
$\sqrt{3(5)-7}=\sqrt{5+3}$
$\sqrt{15-7}=\sqrt{8}$
$\sqrt{8}=\sqrt{8}$ True

29. $\sqrt{4x+4}=\sqrt{6x-2}$
$(\sqrt{4x+4})^2=(\sqrt{6x-2})^2$
$4x+4=6x-2$
$4+2=6x-4x$
$6=2x$
$x=3$
Check: $\sqrt{4\cdot 3+4}=\sqrt{6\cdot 3-2}$
$\sqrt{16}=\sqrt{16}$ True

31. $\sqrt{2x+14}=3\sqrt{x}$
$(\sqrt{2x+14})^2=(3\sqrt{x})^2$
$2x+14=9x$
$14=9x-2x$
$14=7x$
$x=2$
Check: $\sqrt{2x+14}=3\sqrt{x}$
$\sqrt{2(2)+14}=3\sqrt{2}$
$\sqrt{18}=3\sqrt{2}$
$\sqrt{9\cdot 2}=3\sqrt{2}$
$\sqrt{9}\sqrt{2}=3\sqrt{2}$
$3\sqrt{2}=3\sqrt{2}$ True

33. $\sqrt{4x-5}=\sqrt{x+9}$
$(\sqrt{4x-5})^2=(\sqrt{x+9})^2$
$4x-5=x+9$
$3x-5=9$
$3x=14$
$x=\frac{14}{3}$
Check: $\sqrt{4x-5}=\sqrt{x+9}$
$\sqrt{4\left(\frac{14}{3}\right)-5}=\sqrt{\frac{14}{3}+9}$
$\sqrt{\frac{41}{3}}=\sqrt{\frac{41}{3}}$ True

35. $3\sqrt{x}=\sqrt{x+8}$
$(3\sqrt{x})^2=(\sqrt{x+8})^2$
$9x=x+8$
$8x=8$
$x=1$
Check: $3\sqrt{x}=\sqrt{x+8}$
$3\sqrt{1}=\sqrt{1+8}$
$3(1)=\sqrt{9}$
$3=3$ True

37. $4\sqrt{x}=x+3$
$(4\sqrt{x})^2=(x+3)^2$
$16x=x^2+6x+9$
$0=x^2-10x+9$
$0=(x-9)(x-1)$
$x-9=0$ or $x-1=0$
$x=9 \qquad x=1$
Check: $x=9$
$4\sqrt{x}=x+3$
$4\sqrt{9}=9+3$
$4\cdot 3=12$
$12=12$ True

$x=1$
$4\sqrt{x}=x+3$
$4\sqrt{1}=1+3$
$4\cdot 1=4$
$4=4$ True

39. $\sqrt{2x-3}=2\sqrt{3x-2}$
$(\sqrt{2x-3})^2=(2\sqrt{3x-2})^2$
$2x-3=4(3x-2)$
$2x-3=12x-8$
$-3=10x-8$
$5=10x$
$\frac{1}{2}=x$
Check: $\sqrt{2x-3}=2\sqrt{3x-2}$
$\sqrt{2\left(\frac{1}{2}\right)-3}=2\sqrt{3\left(\frac{1}{2}\right)-2}$
$\sqrt{-2}=2\sqrt{-\frac{1}{2}}$ False

$\frac{1}{2}$ is an extraneous root. There is no solution.

41. $\sqrt{x^2-4}=x+2$

$\left(\sqrt{x^2-4}\right)^2=(x+2)^2$

$x^2-4=x^2+4x+4$

$0=x^2+4x+4-x^2+4$

$0=4x+8$

$4x=-8$

$x=-2$

Check: $\sqrt{(-2)^2-4}=-2+2$

$\sqrt{4-4}=0$

$0=0$ True

43. $3+\sqrt{3x-5}=x$

$\sqrt{3x-5}=x-3$

$(\sqrt{3x-5})^2=(x-3)^2$

$3x-5=x^2-6x+9$

$0=x^2-6x+9-3x+5$

$0=x^2-9x+14$

$(x-7)(x-2)=0$

$x-7=0$ or $x-2=0$

$x=7$ or $x=2$

Check: $x=7$

$3+\sqrt{3\cdot 7-5}=7$

$3+\sqrt{16}=7$

$3+4=7$

$7=7$ True

Check: $x=2$

$3+\sqrt{3\cdot 2-5}=2$

$3+\sqrt{1}=2$

$4=2$ False

The solution is 7, 2 is an extraneous root.

45. $\sqrt{8-7x}=x-2$

$(\sqrt{8-7x})^2=(x-2)^2$

$8-7x=x^2-4x+4$

$0=x^2+3x-4$

$0=(x+4)(x-1)$

$x+4=0$ or $x-1=0$

$x=-4$ or $x=1$

Check: $x=-4$

$\sqrt{8-7x}=x-2$

$\sqrt{8-7(-4)}=-4-2$

$\sqrt{8+28}=-6$

$\sqrt{36}=-6$

$6=-6$ False

Check: $x=1$

$\sqrt{8-7x}=x-2$

$\sqrt{8-7(1)}=1-2$

$\sqrt{8-7}=-1$

$\sqrt{1}=-1$

$1=-1$ False

Both $x=-4$ and $x=-1$ are extraneous roots. There is no solution.

47. $(x+3)^{1/2}=7$

$\sqrt{x+3}=7$

$(\sqrt{x+3})^2=7^2$

$x+3=49$

$x=46$

Check: $\sqrt{46+3}=7$

$\sqrt{49}=7$

$7=7$ True

49. $(x-2)^{1/2}=(2x-9)^{1/2}$

$\sqrt{x-2}=\sqrt{2x-9}$

$(\sqrt{x-2})^2=(\sqrt{2x-9})^2$

$x-2=2x-9$

$7=x$

Check: $\sqrt{7-2}=\sqrt{2\cdot 7-9}$

$\sqrt{5}=\sqrt{14-9}$

$\sqrt{5}=\sqrt{5}$ True

51. a. $(\sqrt{x}-3)(\sqrt{x}+3)=40$

$(\sqrt{x})(\sqrt{x})+3\sqrt{x}-3\sqrt{x}-3(3)=40$

$x-9=40$

b. $x=40+9$

$x=49$

Check: $(\sqrt{49}-3)(\sqrt{49}+3)=40$

$(7-3)(7+3)=40$

$4\cdot 10=40$

$40=40$ True

53. a. $(7-\sqrt{x})(5+\sqrt{x})=35$

$7\cdot 5+7\sqrt{x}-5\sqrt{x}+(-\sqrt{x})(\sqrt{x})=35$

$35+7\sqrt{x}-5\sqrt{x}-x=35$

$35+2\sqrt{x}-x=35$

b. $35+2\sqrt{x}-x=35$

$2\sqrt{x}-x=0$

$2\sqrt{x}=x$

$(2\sqrt{x})^2=x^2$

$4x=x^2$

$0=x^2-4x$

$0=x(x-4)$

$x=0$ or $x=4$

Check $x=0$:

$(7-\sqrt{x})(5+\sqrt{x})=35$

$(7-\sqrt{0})(5+\sqrt{0})=35$

$(7)(5)=35$

$35=35$ True

Check $x=4$:

$(7-\sqrt{x})(5+\sqrt{x})=35$

$(7-\sqrt{4})(5+\sqrt{4})=35$

$(7-2)(5+2)=35$

$(5)(7)=35$

$35=35$ True

55. $n+\sqrt{n}=2$

$\sqrt{n}=2-n$

$(\sqrt{n})^2=(2-n)^2$

$n^2=4-4n+n^2$

$4n=4$

$n=1$

Check: $1+\sqrt{1}=2$

$2=2$ True

57. $\sqrt{x}+2=\sqrt{x+16}$

$(\sqrt{x}+2)^2=(\sqrt{x+16})^2$

$x+4\sqrt{x}+4=x+16$

$4\sqrt{x}=12$

$(4\sqrt{x})^2=(12)^2$

$16x=144$

$x=9$

Check: $\sqrt{x}+2=\sqrt{x+16}$

$\sqrt{9}+2=\sqrt{9+16}$

$3+2=\sqrt{25}$

$5=5$ True

59. $\sqrt{x+7}=5-\sqrt{x-8}$

$(\sqrt{x+7})^2=(5-\sqrt{x-8})^2$

$x+7=25-10\sqrt{x-8}+x-8$

$10\sqrt{x-8}=25+x-8-x-7$

$10\sqrt{x-8}=10$

$\sqrt{x-8}=1$

$(\sqrt{x-8})^2=1^2$

$x-8=1$

$x=9$

Check: $\sqrt{x+7}=5-\sqrt{x-8}$

$\sqrt{9+7}=5-\sqrt{9-8}$

$\sqrt{16}=5-\sqrt{1}$

$4=5-1$

$4=4$ True

63.

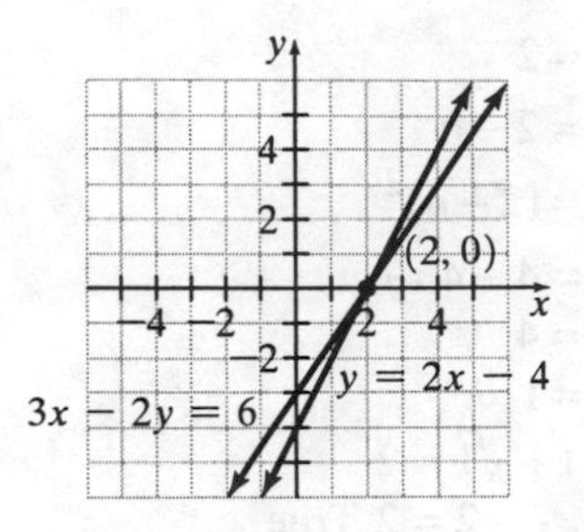

The solution is (2, 0).

64. Substitute $2x - 4$ for y in the first equation.

$$3x - 2y = 6$$
$$3x - 2(2x - 4) = 6$$
$$3x - 4x + 8 = 6$$
$$-x + 8 = 6$$
$$-x = -2$$
$$x = 2$$

Substitute 2 for x in the second equation.

$$y = 2x - 4$$
$$y = 2(2) - 4 = 4 - 4 = 0$$

The solution is (2, 0).

65. Align the x- and y-terms on the left side of the equation.

$$3x - 2y = 6$$
$$-2x + y = -4$$

Multiply the second equation by 2.

$$2[-2x + y = -4]$$

gives

$$3x - 2y = 6$$
$$\underline{-4x + 2y = -8}$$
$$-x \quad = -2$$
$$x = 2$$

Substitute 2 for x in the second equation

$$y = 2x - 4$$
$$y = 2(2) - 4 = 4 - 4 = 0$$

The solution is (2, 0).

66. Let b = the speed of the boat in still water, c = speed of the current.
Speed of boat with current = 12 mph.
Speed of boat against current = 4 mph.

$$b + c = 12$$
$$\underline{b - c = 4}$$
$$2b = 16$$
$$b = 8$$

Substitute 8 for b in the first equation.

$$b + c = 12$$
$$8 + c = 12$$
$$c = 4$$

Speed of boat in still water = 8 mph, speed of the current = 4 mph.

Exercise Set 9.6

1. A right triangle is a triangle that contains a 90° angle

3. No; only with right triangles.

5. They represent the two points in the coordinate plane that you are trying to find the distance between

7. $\sqrt{(0-0)^2 + [-7-(-2)]^2} = \sqrt{0 + (-5)^2}$
$= \sqrt{25}$
$= 5$

9.

$$a^2 + b^2 = c^2$$
$$x^2 + 4^2 = 7^2$$
$$x^2 + 16 = 49$$
$$x^2 = 33$$
$$x = \sqrt{33} \approx 5.74$$

11.

$$a^2 + b^2 = c^2$$
$$(10)^2 + 8^2 = x^2$$
$$100 + 64 = x^2$$
$$164 = x^2$$
$$x = \sqrt{164} \approx 12.81$$

13.

$$a^2 + b^2 = c^2$$
$$10^2 + (\sqrt{7})^2 = y^2$$
$$100 + 7 = y^2$$
$$107 = y^2$$
$$y = \sqrt{107} \approx 10.34$$

15. $a^2 + b^2 = c^2$
$8^2 + (\sqrt{3})^2 = x^2$
$64 + 3 = x^2$
$67 = x^2$
$x = \sqrt{67} \approx 8.19$

17. $a^2 + b^2 = c^2$
$(\sqrt{6})^2 + 14^2 = x^2$
$6 + 196 = x^2$
$202 = x^2$
$x = \sqrt{202} \approx 14.21$

19. $a^2 + b^2 = c^2$
$5^2 + x^2 = 15^2$
$25 + x^2 = 225$
$x^2 = 200$
$x = \sqrt{200} \approx 14.14$

21. $d = \sqrt{(2-5)^2 + (9-7)^2}$
$= \sqrt{9+4}$
$= \sqrt{13}$
≈ 3.61

23. $d = \sqrt{[4-(-8)]^2 + (11-4)^2}$
$= \sqrt{(12)^2 + 7^2}$
$= \sqrt{193}$
≈ 13.89

25. $a^2 + b^2 = c^2$
$(120)^2 + (53.3)^2 = x^2$
$14,400 + 2840.89 = x^2$
$x^2 = 17,240.89$
$x = \sqrt{17,240.89}$
The diagonal is $\sqrt{17,240.89} \approx 131.30$ yards.

27. $a^2 + b^2 = c^2$
$2^2 + x^2 = 8^2$
$4 + x^2 = 64$
$x^2 = 60$
$x = \sqrt{60}$
The top of the ladder will be $\sqrt{60} \approx 7.75$ meters high.

29. $A = s^2$
$s^2 = 256$
$s^2 = (16)^2$
$(s^2)^{1/2} = [(16)^2]^{1/2}$
$s = 16$
The sides are 16 feet long.

31. $A = \pi r^2$
$80 = (3.14)r^2$
$\frac{80}{3.14} = r^2$
$r^2 \approx 25.48$
$r \approx \sqrt{25.48}$
The radius of the circle is about $\sqrt{25.48} \approx 5.05$ feet

33. $a^2 + b^2 = c^2$
$(90)^2 + (90)^2 = c^2$
$8100 + 8100 = c^2$
$c^2 = 16,200$
$c = \sqrt{16,200}$
The distance is $\sqrt{16,200} \approx 127.28$ feet.

35. $a^2 + b^2 = c^2$
$(40)^2 + (20)^2 = x^2$
$1600 + 400 = x^2$
$x^2 = 2000$
$x = \sqrt{2000}$
The men are $\sqrt{2000} \approx 44.72$ feet apart.

37. $T = 2\pi\sqrt{\frac{L}{32}}$

$43 \text{ in.} = \frac{43}{12} \text{ ft}$

$T = 2 \cdot 3.14\sqrt{\frac{(\frac{43}{12})}{32}}$

$\approx 6.28\sqrt{0.11}$

$\approx 2.08 \text{ sec}$

39. $T = 2\pi\sqrt{\frac{L}{32}}$

$= 2 \cdot 3.14\sqrt{\frac{61.6}{32}}$

$= 6.28\sqrt{1.925}$

$\approx 8.71 \text{ sec}$

41. $N = 0.2(\sqrt{R})^3$

$N = 0.2(\sqrt{149.4})^3 \approx 365$ Earth days

43. $v = \sqrt{2gR}$

$v = \sqrt{2 \cdot 9.75(6,370,000)}$

$\approx 11,145.18 \text{ m / sec}$

45. $R = \sqrt{F_1^2 + F_2^2}$

$R = \sqrt{(600)^2 + (800)^2}$

$R = \sqrt{360,000 + 640,000}$

$R = \sqrt{1,000,000} = 1000$

The resulting force is 1000 pounds.

47. The gravity on the moon is

$\frac{1}{6}g = \frac{32}{6}$

$v = \sqrt{2\left(\frac{32}{6}\right) \cdot 100}$

$\approx \sqrt{1066.666}$

$\approx 32.66 \text{ ft / sec}$

49. $d = \sqrt{a^2 + b^2 + c^2}$

$= \sqrt{(37)^2 + (15)^2 + (9)^2}$

$= \sqrt{1675}$

$\approx 40.93 \text{ in.}$

51. $2(x+3) < 4x - 6$

$2x + 6 < 4x - 6$

$6 + 6 < 4x - 2x$

$12 < 2x$

$6 < x$

$x > 6$

52. $3x + 4y = 12$

$\frac{1}{2}x - 2y = 8$

Multiply the second equation by 2.

$2\left[\frac{1}{2}x - 2y = 8\right]$

gives

$$\begin{array}{r} 3x + 4y = 12 \\ + \quad x - 4y = 16 \\ \hline 4x \qquad = 28 \end{array}$$

$x = 7$

Substitute 7 for x in the first equation.

$3x + 4y = 12$

$3(7) + 4y = 12$

$21 + 4y = 12$

$4y = -9$

$y = -\frac{9}{4}$

The solution is $\left(7, -\frac{9}{4}\right)$.

53. $(4x^{-4}y^3)^{-1} = (4y^3)^{-1}(x^{-4})^{-1}$

$= (4y^3)^{-1}x^4$

$= \frac{x^4}{4y^3}$

54.

$$5+\frac{6}{x}=\frac{2}{3x}$$
$$\frac{5x+6}{x}=\frac{2}{3x}$$
$$3x(5x+6)=2x$$
$$15x^2+18x=2x$$
$$15x^2+16x=0$$
$$x(15x+16)=0$$
$$x=0 \text{ or } 15x+16=0$$
$$x=-\frac{16}{15}$$

The solution is $x=-\frac{16}{15}$ since x cannot be 0.

Exercise Set 9.7

1. a. The square root of 9

b. The cube root of 9

c. The fourth root of 9

3. Write the radicand as a product of a perfect cube and another number.

5. a. Answers will vary.

b. $\sqrt[3]{y^7}=y^{7/3}$

7. $\sqrt[3]{125}=5$ since $5^3=125$

9. $\sqrt[3]{-27}=-3$ since $(-3)^3=-27$

11. $\sqrt[4]{16}=2$ since $2^4=16$

13. $\sqrt[4]{81}=3$ since $3^4=81$

15. $\sqrt[3]{-1}=-1$ since $(-1)^3=-1$

17. $\sqrt[3]{-1000}=-10$ since $(-10)^3=-1000$

19. $\sqrt[3]{32}=\sqrt[3]{8\cdot 4}=\sqrt[3]{8}\sqrt[3]{4}=2\sqrt[3]{4}$

21. $\sqrt[3]{16}=\sqrt[3]{8\cdot 2}=\sqrt[3]{8}\sqrt[3]{2}=2\sqrt[3]{2}$

23. $\sqrt[3]{81}=\sqrt[3]{27\cdot 3}=\sqrt[3]{27}\sqrt[3]{3}=3\sqrt[3]{3}$

25. $\sqrt[4]{32}=\sqrt[4]{16\cdot 2}=\sqrt[4]{16}\sqrt[4]{2}=2\sqrt[4]{2}$

27. $\sqrt[4]{1250}=\sqrt[4]{625\cdot 2}=\sqrt[4]{625}\sqrt[4]{2}=5\sqrt[4]{2}$

29. $\sqrt[4]{x^4}=x^{4/4}=x$

31. $\sqrt[3]{y^{21}}=y^{21/3}=y^7$

33. $\sqrt[3]{x^{12}}=x^{12/3}=x^4$

35. $\sqrt[4]{x^{32}}=x^{32/4}=x^8$

37. $\sqrt[3]{x^{15}}=x^{15/3}=x^5$

39. $16^{3/4}=(\sqrt[4]{16})^3=2^3=8$

41. $125^{2/3}=(\sqrt[3]{125})^2=5^2=25$

43. $1^{2/3}=(\sqrt[3]{1})^2=1^2=1$

45. $9^{3/2}=(\sqrt[2]{9})^3=3^3=27$

47. $27^{4/3}=(\sqrt[3]{27})^4=3^4=81$

49. $125^{4/3}=(\sqrt[3]{125})^4=5^4=625$

51. $8^{-1/3}=\frac{1}{8^{1/3}}=\frac{1}{\sqrt[3]{8}}=\frac{1}{2}$

53. $27^{-2/3}=\frac{1}{27^{2/3}}=\frac{1}{(\sqrt[3]{27})^2}=\frac{1}{3^2}=\frac{1}{9}$

55. $\sqrt[3]{x^5}=x^{5/3}$

57. $\sqrt[3]{x^4}=x^{4/3}$

59. $\sqrt[4]{y^{15}}=y^{15/4}$

61. $\sqrt[3]{y^{21}}=y^{21/3}=y^7$

63. $\sqrt[4]{x}\sqrt[4]{x^3}=x^{1/4}x^{3/4}=x^{1/4+3/4}=x^{4/4}=x$

65. $\sqrt[4]{x^2}\cdot\sqrt[4]{x^2}=x^{2/4}\cdot x^{2/4}=x^{1/2}\cdot x^{1/2}=x$

67. $(\sqrt[3]{x^5})^6=(x^{5/3})^6=x^{(5/3)\cdot 6}=x^{30/3}=x^{10}$

69. $\left(\sqrt[4]{x^2}\right)^4=(x^{2/4})^4=(x^{1/2})^4=x^2$

71. For $x=8$, $(\sqrt[3]{x})^2=(\sqrt[3]{8})^2=2^2=4$

For $x=8$, $\left(\sqrt[3]{x^2}\right)=\left(\sqrt[3]{8^2}\right)=\sqrt[3]{64}=4$

73. $\sqrt[3]{5^2}$

75. $\sqrt[3]{6^2}$

77. $\sqrt[4]{5^3}$

79. The missing number is 3 since $\sqrt[3]{5^2}\cdot\sqrt[3]{5}=\sqrt[3]{5^3}=5.$

81. The missing number is 1 since $\sqrt[3]{6^1}\cdot\sqrt[3]{6^2}=\sqrt[3]{6^3}.$

83. $\sqrt[3]{xy}\cdot\sqrt[3]{x^2y^2}=\sqrt[3]{x^3y^3}=xy$

85.
$$\begin{aligned}\sqrt[4]{32}-\sqrt[4]{2}&=\sqrt[4]{16\cdot 2}-\sqrt[4]{2}\\&=\sqrt[4]{16}\cdot\sqrt[4]{2}-\sqrt[4]{2}\\&=\sqrt[4]{2}(\sqrt[4]{16}-1)\\&=\sqrt[4]{2}(2-1)\\&=\sqrt[4]{2}\end{aligned}$$

88.
$$\begin{aligned}-x^2+4xy-6&=(-2)^2+4(2)(-4)-6\\&=-4-32-6\\&=-42\end{aligned}$$

89.

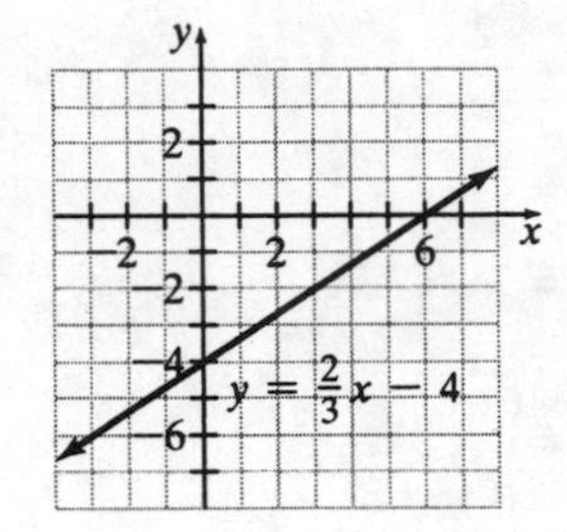

90.
$$\begin{aligned}3x^2-28x+32&=3x^2-24x-4x+32\\&=3x(x-8)-4(x-8)\\&=(3x-4)(x-8)\end{aligned}$$

91.
$$\begin{aligned}\sqrt{\frac{64x^3y^7}{2x^4}}&=\sqrt{\frac{32y^7}{x}}\\&=\frac{\sqrt{32y^7}}{\sqrt{x}}\\&=\frac{\sqrt{16y^6}\sqrt{2y}}{\sqrt{x}}\\&=\frac{4y^3\sqrt{2y}}{\sqrt{x}}\\&=\frac{4y^3\sqrt{2y}}{\sqrt{x}}\cdot\frac{\sqrt{x}}{\sqrt{x}}\\&=\frac{4y^3\sqrt{2xy}}{x}\end{aligned}$$

Review Exercises

1. $\sqrt{49}=7$ since $7^2=49$
2. $\sqrt{9}=3$ since $3^2=9$
3. $-\sqrt{64}=-8$ since $\sqrt{64}=8$ $(8^2=64)$
4. $\sqrt{6}=6^{1/2}$
5. $\sqrt{26x}=(26x)^{1/2}$
6. $\sqrt{13x^2y}=(13x^2y)^{1/2}$
7. $\sqrt{18}=\sqrt{9\cdot 2}=\sqrt{9}\sqrt{2}=3\sqrt{2}$

8. $\sqrt{44} = \sqrt{4 \cdot 11} = \sqrt{4}\sqrt{11} = 2\sqrt{11}$

9. $\sqrt{27x^7y^4} = \sqrt{9x^6y^4}\sqrt{3x} = 3x^3y^2\sqrt{3x}$

10. $\sqrt{125x^4y^6} = \sqrt{25x^4y^6}\sqrt{5} = 5x^2y^3\sqrt{5}$

11. $\sqrt{48ab^5c^4} = \sqrt{16b^4c^4}\sqrt{3ab} = 4b^2c^2\sqrt{3ab}$

12. $\sqrt{72a^2b^2c^7} = \sqrt{36a^2b^2c^6}\sqrt{2c} = 6abc^3\sqrt{2c}$

13. $\sqrt{24}\sqrt{20} = \sqrt{24 \cdot 20}$
$= \sqrt{480}$
$= \sqrt{16 \cdot 30}$
$= 4\sqrt{30}$

14. $\sqrt{7y}\sqrt{7y} = \sqrt{49y^2} = 7y$

15. $\sqrt{18x} \cdot \sqrt{2xy} = \sqrt{36x^2y}$
$= \sqrt{36x^2}\sqrt{y}$
$= 6x\sqrt{y}$

16. $\sqrt{25x^2y} \cdot \sqrt{3y} = \sqrt{75x^2y^2}$
$= \sqrt{25x^2y^2}\sqrt{3}$
$= 5xy\sqrt{3}$

17. $\sqrt{8a^3b} \cdot \sqrt{3b^4} = \sqrt{24a^3b^5}$
$= \sqrt{4a^2b^4}\sqrt{6ab}$
$= 2ab^2\sqrt{6ab}$

18. $\sqrt{5ab^3} \cdot \sqrt{20ab^4} = \sqrt{100a^2b^7}$
$= \sqrt{100a^2b^6}\sqrt{b}$
$= 10ab^3\sqrt{b}$

19. $\frac{\sqrt{50}}{\sqrt{2}} = \sqrt{\frac{50}{2}} = \sqrt{25} = 5$

20. $\sqrt{\frac{10}{490}} = \sqrt{\frac{1}{49}} = \frac{\sqrt{1}}{\sqrt{49}} = \frac{1}{7}$

21. $\sqrt{\frac{7}{28}} = \sqrt{\frac{1}{4}} = \frac{\sqrt{1}}{\sqrt{4}} = \frac{1}{2}$

22. $\frac{3}{\sqrt{5}} = \frac{3}{\sqrt{5}} \cdot \frac{\sqrt{5}}{\sqrt{5}} = \frac{3\sqrt{5}}{5}$

23. $\sqrt{\frac{a}{6}} = \frac{\sqrt{a}}{\sqrt{6}} = \frac{\sqrt{a}}{\sqrt{6}} \cdot \frac{\sqrt{6}}{\sqrt{6}} = \frac{\sqrt{6a}}{6}$

24. $\sqrt{\frac{5a}{12}} = \frac{\sqrt{5a}}{\sqrt{12}}$
$= \frac{\sqrt{5a}}{\sqrt{4}\sqrt{3}}$
$= \frac{\sqrt{5a}}{2\sqrt{3}}$
$= \frac{\sqrt{5a}}{2\sqrt{3}} \cdot \frac{\sqrt{3}}{\sqrt{3}}$
$= \frac{\sqrt{15a}}{6}$

25. $\sqrt{\frac{x^2}{3}} = \frac{\sqrt{x^2}}{\sqrt{3}} = \frac{x}{\sqrt{3}} = \frac{x}{\sqrt{3}} \cdot \frac{\sqrt{3}}{\sqrt{3}} = \frac{x\sqrt{3}}{3}$

26. $\sqrt{\frac{x^5}{8}} = \frac{\sqrt{x^5}}{\sqrt{8}}$
$= \frac{\sqrt{x^4}\sqrt{x}}{\sqrt{4}\sqrt{2}}$
$= \frac{x^2\sqrt{x}}{2\sqrt{2}}$
$= \frac{x^2\sqrt{x}}{2\sqrt{2}} \cdot \frac{\sqrt{2}}{\sqrt{2}}$
$= \frac{x^2\sqrt{2x}}{4}$

27. $\sqrt{\frac{21x^3y^7}{3x^3y^3}} = \sqrt{7y^4} = \sqrt{7}\sqrt{y^4} = y^2\sqrt{7}$

28. $\sqrt{\dfrac{30x^4y}{15x^2y^4}} = \sqrt{\dfrac{2x^2}{y^3}}$
$= \dfrac{\sqrt{2x^2}}{\sqrt{y^3}}$
$= \dfrac{\sqrt{2}\sqrt{x^2}}{\sqrt{y^2}\sqrt{y}}$
$= \dfrac{x\sqrt{2}}{y\sqrt{y}}$
$= \dfrac{x\sqrt{2}}{y\sqrt{y}} \cdot \dfrac{\sqrt{y}}{\sqrt{y}}$
$= \dfrac{x\sqrt{2y}}{y^2}$

29. $\dfrac{\sqrt{60}}{\sqrt{27a^3b^2}} = \sqrt{\dfrac{60}{27a^3b^2}}$
$= \sqrt{\dfrac{20}{9a^3b^2}}$
$= \dfrac{\sqrt{20}}{\sqrt{9a^3b^2}}$
$= \dfrac{\sqrt{4}\sqrt{5}}{\sqrt{9a^2b^2}\sqrt{a}}$
$= \dfrac{2\sqrt{5}}{3ab\sqrt{a}}$
$= \dfrac{2\sqrt{5}}{3ab\sqrt{a}} \cdot \dfrac{\sqrt{a}}{\sqrt{a}}$
$= \dfrac{2\sqrt{5a}}{3a^2b}$

30. $\dfrac{\sqrt{2a^4bc^4}}{\sqrt{7a^5bc^2}} = \sqrt{\dfrac{2a^4bc^4}{7a^5bc^2}}$
$= \sqrt{\dfrac{2c^2}{7a}}$
$= \dfrac{\sqrt{2c^2}}{\sqrt{7a}}$
$= \dfrac{c\sqrt{2}}{\sqrt{7a}}$
$= \dfrac{c\sqrt{2}}{\sqrt{7a}} \cdot \dfrac{\sqrt{7a}}{\sqrt{7a}}$
$= \dfrac{c\sqrt{14a}}{7a}$

31. $\dfrac{2}{1-\sqrt{5}} = \dfrac{2}{1-\sqrt{5}} \cdot \dfrac{1+\sqrt{5}}{1+\sqrt{5}}$
$= \dfrac{2(1+\sqrt{5})}{1-5}$
$= \dfrac{2(1+\sqrt{5})}{-4}$
$= -\dfrac{-1-\sqrt{5}}{2}$

32. $\dfrac{5}{3-\sqrt{6}} = \dfrac{5}{3-\sqrt{6}} \cdot \dfrac{3+\sqrt{6}}{3+\sqrt{6}}$
$= \dfrac{5(3+\sqrt{6})}{9-6}$
$= \dfrac{15+5\sqrt{6}}{3}$

33. $\dfrac{\sqrt{2}}{2+\sqrt{y}} = \dfrac{\sqrt{2}}{2+\sqrt{y}} \cdot \dfrac{2-\sqrt{y}}{2-\sqrt{y}}$
$= \dfrac{\sqrt{2}(2-\sqrt{y})}{4-y}$
$= \dfrac{2\sqrt{2}-\sqrt{2y}}{4-y}$

34. $\frac{2}{\sqrt{x}-5}=\frac{2}{\sqrt{x}-5}\cdot\frac{\sqrt{x}+5}{\sqrt{x}+5}$
$=\frac{2(\sqrt{x}+5)}{x-25}$
$=\frac{2\sqrt{x}+10}{x-25}$

35. $\frac{\sqrt{5}}{\sqrt{x}+\sqrt{3}}=\frac{\sqrt{5}}{\sqrt{x}+\sqrt{3}}\cdot\frac{\sqrt{x}-\sqrt{3}}{\sqrt{x}-\sqrt{3}}$
$=\frac{\sqrt{5}(\sqrt{x}-\sqrt{3})}{x-3}$
$=\frac{\sqrt{5x}-\sqrt{15}}{x-3}$

36. $7\sqrt{2}-4\sqrt{2}=(7-4)\sqrt{2}=3\sqrt{2}$

37. $4\sqrt{5}-7\sqrt{5}-3\sqrt{5}=(4-7-3)\sqrt{5}=-6\sqrt{5}$

38. $3\sqrt{x}-5\sqrt{x}=(3-5)\sqrt{x}=-2\sqrt{x}$

39. $\sqrt{x}+3\sqrt{x}-4\sqrt{x}=(1+3-4)\sqrt{x}=0\sqrt{x}=0$

40. $\sqrt{18}-\sqrt{27}=\sqrt{9\cdot 2}-\sqrt{9\cdot 3}$
$=\sqrt{9}\sqrt{2}-\sqrt{9}\sqrt{3}$
$=3\sqrt{2}-3\sqrt{3}$

41. $7\sqrt{40}-2\sqrt{10}=7\sqrt{4}\sqrt{10}-2\sqrt{10}$
$=7\cdot 2\sqrt{10}-2\sqrt{10}$
$=14\sqrt{10}-2\sqrt{10}$
$=(14-2)\sqrt{10}$
$=12\sqrt{10}$

42. $2\sqrt{98}-4\sqrt{72}=2\sqrt{49}\sqrt{2}-4\sqrt{36}\sqrt{2}$
$=2\cdot 7\sqrt{2}-4\cdot 6\sqrt{2}$
$=14\sqrt{2}-24\sqrt{2}$
$=(14-24)\sqrt{2}$
$=-10\sqrt{2}$

43. $7\sqrt{50}+2\sqrt{18}-4\sqrt{32}$
$=7\sqrt{25}\sqrt{2}+2\sqrt{9}\sqrt{2}-4\sqrt{16}\sqrt{2}$
$=7\cdot 5\sqrt{2}+2\cdot 3\sqrt{2}-4\cdot 4\sqrt{2}$
$=(35+6-16)\sqrt{2}$
$=25\sqrt{2}$

44. $4\sqrt{27}+5\sqrt{80}+2\sqrt{12}$
$=4\sqrt{9}\sqrt{3}+5\sqrt{16}\sqrt{5}+2\sqrt{4}\sqrt{3}$
$=4\cdot 3\sqrt{3}+5\cdot 4\sqrt{5}+2\cdot 2\sqrt{3}$
$=12\sqrt{3}+20\sqrt{5}+4\sqrt{3}$
$=(12+4)\sqrt{3}+20\sqrt{5}$
$=16\sqrt{3}+20\sqrt{5}$

45. $\sqrt{x}=4$
$(\sqrt{x})^2=4^2$
$x=16$
Check: $\sqrt{x}=4$
$\sqrt{16}=4$
$4=4$ True

46. $\sqrt{x}=-5$
No solution

47. $\sqrt{x-4}=3$
$(\sqrt{x-4})^2=3^2$
$x-4=9$
$x=13$
Check: $\sqrt{13-4}=3$
$\sqrt{9}=3$
$3=3$ True

48. $\sqrt{3x+1}=5$
$(\sqrt{3x+1})^2=5^2$
$3x+1=25$
$3x=24$
$x=8$
Check: $\sqrt{3x+1}=5$
$\sqrt{3(8)+1}=5$
$\sqrt{25}=5$
$5=5$ True

49.
$$\sqrt{5x+6} = \sqrt{4x+8}$$
$$(\sqrt{5x+6})^2 = (\sqrt{4x+8})^2$$
$$5x+6 = 4x+8$$
$$5x-4x = 8-6$$
$$x = 2$$
Check: $\sqrt{5x+6} = \sqrt{4x+8}$
$$\sqrt{5(2)+6} = \sqrt{4(2)+8}$$
$$\sqrt{16} = \sqrt{16} \text{ True}$$

50. $4\sqrt{x} - x = 4$
$$4\sqrt{x} = x+4$$
$$(4\sqrt{x})^2 = (x+4)^2$$
$$16x = x^2+8x+16$$
$$0 = x^2-8x+16$$
$$0 = (x-4)^2$$
$$x-4 = 0$$
$$x = 4$$
Check: $4\sqrt{x} - x = 4$
$$4\sqrt{4} - 4 = 4$$
$$4 \cdot 2 - 4 = 4$$
$$4 = 4 \text{ True}$$

51.
$$\sqrt{x^2-3} = x-1$$
$$\left(\sqrt{x^2-3}\right)^2 = (x-1)^2$$
$$x^2-3 = x^2-2x+1$$
$$2x-1-3 = 0$$
$$2x-4 = 0$$
$$2x = 4$$
$$x = 2$$
Check: $\sqrt{x^2-3} = x-1$
$$\sqrt{2^2-3} = 2-1$$
$$\sqrt{1} = 1$$
$$1 = 1 \text{ True}$$

52. $\sqrt{4x+8} - \sqrt{7x-13} = 0$
$$\sqrt{4x+8} = \sqrt{7x-13}$$
$$(\sqrt{4x+8})^2 = (\sqrt{7x-13})^2$$
$$4x+8 = 7x-13$$
$$8+13 = 7x-4x$$
$$21 = 3x$$
$$x = \frac{21}{3} = 7$$
Check: $\sqrt{4x+8} - \sqrt{7x-13} = 0$
$$\sqrt{4(7)+8} - \sqrt{7(7)-13} = 0$$
$$\sqrt{36} - \sqrt{36} = 0$$
$$0 = 0 \text{ True}$$

53.
$$4\sqrt{5x-2} = 8$$
$$\sqrt{5x-2} = 2$$
$$(\sqrt{5x-2})^2 = 2^2$$
$$5x-2 = 4$$
$$5x = 6$$
$$x = \frac{6}{5}$$
Check: $4\sqrt{5x-2} = 8$
$$4\sqrt{5\left(\frac{6}{5}\right)-2} = 8$$
$$4\sqrt{6-2} = 8$$
$$4\sqrt{4} = 8$$
$$4(2) = 8$$
$$8 = 8 \text{ True}$$

54.
$$a^2+b^2 = c^2$$
$$10^2+24^2 = x^2$$
$$100+576 = x^2$$
$$x^2 = 676$$
$$x = \sqrt{676} = 26$$

55.
$$a^2+b^2 = c^2$$
$$10^2+x^2 = 15^2$$
$$100+x^2 = 225$$
$$x^2 = 225-100$$
$$x^2 = 125$$
$$x = \sqrt{125} \approx 11.18$$

56. $a^2 + b^2 = c^2$
$x^2 + (\sqrt{3})^2 = (\sqrt{15})^2$
$x^2 + 3 = 15$
$x^2 = 12$
$x = \sqrt{12} \approx 3.46$

57. $a^2 + b^2 = c^2$
$6^2 + 5^2 = x^2$
$36 + 25 = x^2$
$61 = x^2$
$x = \sqrt{61} \approx 7.81$

58. $a^2 + b^2 = c^2$
$h^2 + 3^2 = 12^2$
$h^2 + 9 = 144$
$h^2 = 135$
$h = \sqrt{135}$
The height of the ladder on the house is $\sqrt{135} \approx 11.62$ feet.

59. $a^2 + b^2 = c^2$
$15^2 + 6^2 = d^2$
$225 + 36 = d^2$
$261 = d^2$
$d = \sqrt{261}$
The length of the diagonal is $\sqrt{261} \approx 16.16$ inches.

60. $d = \sqrt{(x_2 - x_1)^2 + (y_2 - y_1)^2}$
$= \sqrt{(1-4)^2 + [7-(-3)]^2}$
$= \sqrt{(-3)^2 + (10)^2}$
$= \sqrt{9 + 100}$
$= \sqrt{109}$
≈ 10.44

61. $d = \sqrt{(x_2 - x_1)^2 + (y_2 - y_1)^2}$
$= \sqrt{(-6-6)^2 + (8-5)^2}$
$= \sqrt{(-12)^2 + (3)^2}$
$= \sqrt{144 + 9}$
$= \sqrt{153}$
≈ 12.37

62. $A = \frac{s^2\sqrt{3}}{4}$
$A = \frac{36^2\sqrt{3}}{4}$
$= \frac{1296\sqrt{3}}{4}$
$= 324\sqrt{3}$
$\approx 324(1.732)$
≈ 561.18 square inches

63. $d = \sqrt{(3/2)h}$
$= \sqrt{(3/2)40}$
$= \sqrt{60}$
≈ 7.75
He can see about 7.5 miles.

64. $s = \sqrt{(3V)/h}$
$= \sqrt{3(85,503,750)/450}$
$= \sqrt{570,025}$
$= 755$
The sides are 755 feet long.

65. $\sqrt[3]{64} = 4$ since $4^3 = 64$

66. $\sqrt[3]{-64} = -4$ since $(-4)^3 = -64$

67. $\sqrt[4]{16} = 2$ since $2^4 = 16$

68. $\sqrt[4]{81} = 3$ since $3^4 = 81$

69. $\sqrt[3]{27} = 3$ since $3^3 = 27$

70. $\sqrt[3]{-8} = -2$ since $(-2)^3 = -8$

71. $\sqrt[4]{32} = \sqrt[4]{16 \cdot 2} = \sqrt[4]{16}\sqrt[4]{2} = 2\sqrt[4]{2}$

72. $\sqrt[3]{48} = \sqrt[3]{8 \cdot 6} = \sqrt[3]{8}\sqrt[3]{6} = 2\sqrt[3]{6}$

73. $\sqrt[3]{54} = \sqrt[3]{27 \cdot 2} = \sqrt[3]{27}\sqrt[3]{2} = 3\sqrt[3]{2}$

74. $\sqrt[4]{96} = \sqrt[4]{16 \cdot 6} = \sqrt[4]{16}\sqrt[4]{6} = 2\sqrt[4]{6}$

75. $\sqrt[3]{x^{21}} = x^{21/3} = x^7$

76. $\sqrt[3]{x^{24}} = x^{24/3} = x^8$

77. $27^{2/3} = \left(\sqrt[3]{27}\right)^2 = 3^2 = 9$

78. $25^{1/2} = \sqrt{25} = 5$

79. $27^{-2/3} = (\sqrt[3]{27})^{-2} = 3^{-2} = \frac{1}{3^2} = \frac{1}{9}$

80. $64^{2/3} = (\sqrt[3]{64})^2 = 4^2 = 16$

81. $8^{-4/3} = (\sqrt[3]{8})^{-4} = 2^{-4} = \frac{1}{2^4} = \frac{1}{16}$

82. $49^{3/2} = (\sqrt{49})^3 = 7^3 = 343$

83. $\sqrt[3]{x^7} = x^{7/3}$

84. $\sqrt[3]{x^8} = x^{8/3}$

85. $\sqrt[4]{y^9} = y^{9/4}$

86. $\sqrt{x^5} = x^{5/2}$

87. $\sqrt{y^3} = y^{3/2}$

88. $\sqrt[4]{y^7} = y^{7/4}$

89. $\sqrt[3]{x}\sqrt[3]{x^2} = x^{1/3}x^{2/3} = x^{1/3+2/3} = x^{3/3} = x$

90. $\sqrt[3]{x} \cdot \sqrt[3]{x} = x^{1/3} \cdot x^{1/3}$
$= x^{1/3+1/3}$
$= x^{2/3}$
$= \sqrt[3]{x^2}$

91. $\sqrt[3]{x^2} \cdot \sqrt[3]{x^7} = x^{2/3} \cdot x^{7/3}$
$= x^{2/3+7/3}$
$= x^{9/3}$
$= x^3$

92. $\sqrt[4]{x^2} \cdot \sqrt[4]{x^6} = x^{2/4} \cdot x^{6/4}$
$= x^{2/4+6/4}$
$= x^{8/4}$
$= x^2$

93. $\left(\sqrt[3]{x^3}\right)^4 = (x^{3/3})^4 = x^4$

94. $\left(\sqrt[4]{x}\right)^4 = (x^{1/4})^4 = x$

95. $\left(\sqrt[4]{x^8}\right)^3 = (4^{8/4})^3 = (x^2)^3 = x^{3(2)} = x^6$

96. $\left(\sqrt[4]{x^3}\right)^8 = (x^{3/4})^8 = x^{(3/4)8} = x^6$

Practice Test

1. $\sqrt{3x} = (3x)^{1/2}$

2. $x^{2/3} = \sqrt[3]{x^2}$

3. $\sqrt{(x-5)^2} = x - 5$

4. $\sqrt{80} = \sqrt{16(5)} = \sqrt{16}\sqrt{5} = 4\sqrt{5}$

5. $\sqrt{12x^2} = \sqrt{4x^2}\sqrt{3} = 2x\sqrt{3}$

6. $\sqrt{50x^7y^3} = \sqrt{25x^6y^2}\sqrt{2xy} = 5x^3y\sqrt{2xy}$

7. $\sqrt{8x^2y}\cdot\sqrt{6xy}=\sqrt{48x^3y^2}$
$=\sqrt{16x^2y^2}\sqrt{3x}$
$=4xy\sqrt{3x}$

8. $\sqrt{15xy^2}\cdot\sqrt{5x^3y^3}=\sqrt{75x^4y^5}$
$=\sqrt{25x^4y^4}\sqrt{3y}$
$=5x^2y^2\sqrt{3y}$

9. $\sqrt{\frac{5}{125}}=\sqrt{\frac{1}{25}}=\frac{\sqrt{1}}{\sqrt{25}}=\frac{1}{5}$

10. $\frac{\sqrt{7x^2y}}{\sqrt{7y^3}}=\sqrt{\frac{7x^2y}{7y^3}}=\sqrt{\frac{x^2}{y^2}}=\frac{\sqrt{x^2}}{\sqrt{y^2}}=\frac{x}{y}$

11. $\frac{1}{\sqrt{6}}=\frac{1}{\sqrt{6}}\cdot\frac{\sqrt{6}}{\sqrt{6}}=\frac{\sqrt{6}}{6}$

12. $\sqrt{\frac{4x}{5}}=\frac{\sqrt{4x}}{\sqrt{5}}$
$=\frac{\sqrt{4}\sqrt{x}}{\sqrt{5}}$
$=\frac{2\sqrt{x}}{\sqrt{5}}$
$=\frac{2\sqrt{x}}{\sqrt{5}}\cdot\frac{\sqrt{5}}{\sqrt{5}}$
$=\frac{2\sqrt{5x}}{5}$

13. $\sqrt{\frac{40x^2y^5}{6x^3y^7}}=\sqrt{\frac{20}{3xy^2}}$
$=\frac{\sqrt{20}}{\sqrt{3xy^2}}$
$=\frac{\sqrt{4(5)}}{\sqrt{y^2}\sqrt{3x}}$
$=\frac{\sqrt{4}\sqrt{5}}{y\sqrt{3x}}$
$=\frac{2\sqrt{5}}{y\sqrt{3x}}$
$=\frac{2\sqrt{5}}{y\sqrt{3x}}\cdot\frac{\sqrt{3x}}{\sqrt{3x}}$
$=\frac{2\sqrt{15x}}{3xy}$

14. $\frac{3}{2-\sqrt{7}}=\frac{3}{2-\sqrt{7}}\cdot\frac{2+\sqrt{7}}{2+\sqrt{7}}$
$=\frac{3(2+\sqrt{7})}{4-7}$
$=\frac{3(2+\sqrt{7})}{-3}$
$=-2-\sqrt{7}$

15. $\frac{6}{\sqrt{x}-3}=\frac{6}{\sqrt{x}-3}\cdot\frac{\sqrt{x}+3}{\sqrt{x}+3}$
$=\frac{6(\sqrt{x}+3)}{x-9}$
$=\frac{6\sqrt{x}+18}{x-9}$

16. $\sqrt{48}+\sqrt{75}+2\sqrt{3}=\sqrt{16\cdot 3}+\sqrt{25\cdot 3}+2\sqrt{3}$
$=4\sqrt{3}+5\sqrt{3}+2\sqrt{3}$
$=11\sqrt{3}$

17. $5\sqrt{y}+7\sqrt{y}-\sqrt{y}=(5+7-1)\sqrt{y}=11\sqrt{y}$

18. $\sqrt{x-8}=4$
$(\sqrt{x-8})^2=4^2$
$x-8=16$
$x=24$
Check: $\sqrt{24-8}=4$
$\sqrt{16}=4$
$4=4$ True

19. $2\sqrt{x-4}+4=x$
$2\sqrt{x-4}=x-4$
$(2\sqrt{x-4})^2=(x-4)^2$
$4(x-4)=x^2-8x+16$
$4x-16=x^2-8x+16$
$0=x^2-12x+32$
$0=(x-4)(x-8)$
$x-4=0$ or $x-8=0$
$x=4$ or $x=8$
Check: $x=4$
$2\sqrt{x-4}+4=x$
$2\sqrt{4-4}+4=4$
$2\sqrt{0}+4=4$
$4=4$ True
Check: $x=8$
$2\sqrt{x-4}+4=x$
$2\sqrt{8-4}+4=8$
$2\sqrt{4}+4=8$
$2\cdot 2+4=8$
$8=8$ True
The solutions are 4 and 8.

20. $a^2+b^2=c^2$
$9^2+5^2=x^2$
$81+25=x^2$
$106=x^2$
$x=\sqrt{106}\approx 10.30$

21. $\sqrt{(-4-3)^2+[-5-(-2)]^2}=\sqrt{(-7)^2+(-3)^2}$
$=\sqrt{49+9}$
$=\sqrt{58}$
≈ 7.62

22. $27^{-4/3}=\frac{1}{27^{4/3}}=\frac{1}{(\sqrt[3]{27})^4}=\frac{1}{3^4}=\frac{1}{81}$

23. $\sqrt[4]{x^5}\cdot\sqrt[4]{x^7}=x^{5/4}x^{7/4}$
$=x^{(5/4)+(7/4)}$
$=x^{12/4}$
$=x^3$

24. $s^2=121$
$s=\sqrt{121}=11$
The side is 11 meters.

25. $v=\sqrt{2gh}=\sqrt{2(32)10}=\sqrt{640}\approx 25.30$ ft / sec

Cumulative Review Test

1. a. –9, 735, and 4 are integers.

b. 4 and 735 are whole numbers.

c. –9, 735, 0.5, 4, and $\frac{1}{2}$ are rational numbers.

d. $\sqrt{13}$ is an irrational number.

e. All of the numbers are real numbers.

2. $6x-5y^2+xy$
$=6(-5)-5(2)^2+(-5)(2)$
$=-30-5(4)-10$
$=-30-20-10$
$=-60$

3. $-7(3-x)=4(x+2)-3x$
$-21+7x=4x+8-3x$
$-21+7x=x+8$
$7x-x=8+21$
$6x=29$
$x=\frac{29}{6}$

4. $3(x+2) > 5-4(2x-7)$
$3x+6 > 5-8x+28$
$3x+8x > 5+28-6$
$11x > 27$
$x > \frac{27}{11}$

$\frac{27}{11}$

5. Use the slope-intercept form with $m = \frac{2}{5}$ and $(x_1, y_1) = (-3, 1)$.
$y - y_1 = m(x - x_1)$
$y - 1 = \frac{2}{5}[x-(-3)]$
$y - 1 = \frac{2}{5}(x+3)$
$y - 1 = \frac{2}{5}x + \frac{6}{5}$
$y = \frac{2}{5}x + \frac{11}{5}$

6. The y-intercept is $(0, -2)$ and the slope is
$m = \frac{1-(-2)}{1-0} = \frac{1+2}{1} = 3.$
$y = 3x - 2$

7.

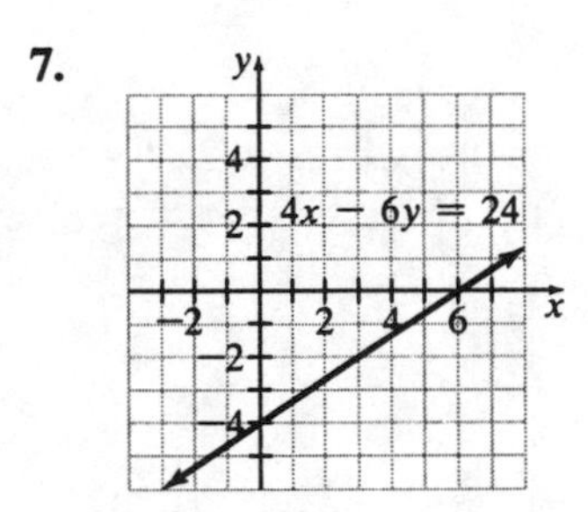

8. $-2x+3y = 6$
$x - 2y = -5$
Solve the second equation for x.
$x = -5 + 2y$
Substitute $-5 + 2y$ for x in the first equation.
$-2(-5+2y)+3y = 6$
$10 - 4y + 3y = 6$
$-y = 6 - 10$
$-y = -4$
$y = 4$

Substitute 4 for y in the equation
$x = -5 + 2y$.
$x = -5 + 2(4) = -5 + 8 = 3$
The solution is $(3, 4)$.

9. $\frac{4a^3b^{-5}}{28a^8b} = \frac{1}{7a^5b^6}$

10. $3x^3 + x^2 + 6x + 2 = x^2(3x+1) + 2(3x+1)$
$= (x^2+2)(3x+1)$

11. $x^2 + 12x - 28 = (x+14)(x-2)$

12. $x^2 - 3x = 0$
$x(x-3) = 0$
$x = 0$ or $x - 3 = 0$
$x = 0$ or $x = 3$

13. $\frac{y+5}{8} + \frac{2y-7}{8} = \frac{y+5+2y-7}{8} = \frac{3y-2}{8}$

14. $\frac{5}{y+2} = \frac{3}{2y-7}$
$5(2y-7) = 3(y+2)$
$10y - 35 = 3y + 6$
$10y - 3y = 6 + 35$
$7y = 41$
$y = \frac{41}{7}$

15. $3\sqrt{11} - 4\sqrt{11} = (3-4)\sqrt{11} = -\sqrt{11}$

16. $\sqrt{\frac{3z}{28y}} = \frac{\sqrt{3z}}{\sqrt{28y}} \cdot \frac{\sqrt{28y}}{\sqrt{28y}}$
$= \frac{\sqrt{84yz}}{28y}$
$= \frac{\sqrt{4}\sqrt{21yz}}{28y}$
$= \frac{2\sqrt{21yz}}{28y}$
$= \frac{\sqrt{21yz}}{14y}$

17. $\sqrt{x+5}=6$
$(\sqrt{x+5})^2=6^2$
$x+5=36$
$x=36-5$
$x=31$
Check: $\sqrt{31+5}=6$
$\sqrt{36}=6$
$6=6$ True

18. $\frac{3}{11}=\frac{x}{10}$
$11x=3(10)$
$11x=30$
$x=\frac{30}{11}=2\frac{8}{11}$

She must use $2\frac{8}{11}$ cups of flour.

19. Let x = the amount raised from January 1, 1995 to June 30, 1996.
$x+0.08x=484.3$
$1.08x=484.3$
$x=\frac{484.3}{1.08}\approx 448.4$
About $448.4 million was raised.

20. Let x = the distance. Since $d=rt$, $t=\frac{d}{r}$.
$\frac{x}{100}+\frac{x}{125}=2$
$500\left(\frac{x}{100}+\frac{x}{125}\right)=500(2)$
$5x+4x=1000$
$9x=1000$
$x=\frac{1000}{9}\approx 111.1$
The distance is about 111.1 miles.

Chapter 10

Exercise Set 10.1

1. If $x^2 = a$, then $x = \sqrt{a}$ or $x = -\sqrt{a}$.

3. In any golden rectangle, the length is about 1.62 times its width.

5. a. $x^2 = 6$ has 2 solutions

b. $x^2 = 0$ has 1 solution

c. $(x-2)^2 = 1$ has 2 solutions

7. $x^2 = 100$
$x = \pm\sqrt{100}$
$x = 10, -10$

9. $x^2 = 144$
$x = \pm\sqrt{144}$
$x = 12, -12$

11. $y^2 = 169$
$y = \pm\sqrt{169}$
$y = 13, -13$

13. $x^2 = 121$
$x = \pm\sqrt{121}$
$x = 11, -11$

15. $x^2 = 12$
$x = \pm\sqrt{12}$
$x = \pm\sqrt{4}\sqrt{3}$
$x = \pm 2\sqrt{3}$
$x = 2\sqrt{3}, -2\sqrt{3}$

17. $3x^2 = 12$
$x^2 = 4$
$x = \pm\sqrt{4}$
$x = 2, -2$

19. $2w^2 = 34$
$w^2 = 17$
$w = \pm\sqrt{17}$
$w = \sqrt{17}, -\sqrt{17}$

21. $2x^2 + 1 = 19$
$2x^2 = 18$
$x^2 = 9$
$x = \pm\sqrt{9}$
$x = 3, -3$

23. $9w^2 + 5 = 20$
$9w^2 = 15$
$w^2 = \frac{15}{9}$
$w = \pm\sqrt{\frac{15}{9}}$
$w = \pm\frac{\sqrt{15}}{3}$
$w = \frac{\sqrt{15}}{3}, -\frac{\sqrt{15}}{3}$

25. $16x^2 - 17 = 56$
$16x^2 = 73$
$x^2 = \frac{73}{16}$
$x = \pm\sqrt{\frac{73}{16}}$
$x = \pm\frac{\sqrt{73}}{\sqrt{16}}$
$x = \pm\frac{\sqrt{73}}{4}$
$x = \frac{\sqrt{73}}{4}, -\frac{\sqrt{73}}{4}$

27. $(x-3)^2 = 25$
$x-3 = \pm\sqrt{25}$
$x-3 = \pm 5$
$x = 3 \pm 5$
$x = 3+5$ or $x = 3-5$
$x = 8$ or $x = -2$
The solutions are 8 and −2.

29. $(x+3)^2 = 81$
$x+3 = \pm\sqrt{81}$
$x+3 = \pm 9$
$x = -3 \pm 9$
$x = -3+9$ or $x = -3-9$
$x = 6$ or $x = -12$
The solutions are 6 and −12.

31. $(x+4)^2 = 64$
$x+4 = \pm\sqrt{64}$
$x+4 = \pm 8$
$x = -4 \pm 8$
$x = -4+8$ or $x = -4-8$
$x = 4$ or $x = -12$
The solutions are 4 and −12.

33. $(x+7)^2 = 32$
$x+7 = \pm\sqrt{32}$
$x+7 = \pm\sqrt{16}\sqrt{2}$
$x+7 = \pm 4\sqrt{2}$
$x = -7 \pm 4\sqrt{2}$
The solutions are $-7+4\sqrt{2}$ and $-7-4\sqrt{2}$.

35. $(x+6)^2 = 20$
$x+6 = \pm\sqrt{20}$
$x+6 = \pm\sqrt{4}\sqrt{5}$
$x+6 = \pm 2\sqrt{5}$
$x = -6 \pm 2\sqrt{5}$
The solutions are $-6+2\sqrt{5}$ and $-6-2\sqrt{5}$.

37. $(x+2)^2 = 25$
$x+2 = \pm\sqrt{25}$
$x+2 = \pm 5$
$x = -2 \pm 5$
$x = -2+5$ or $x = -2-5$
$x = 3$ or $x = -7$
The solutions are 3 and −7.

39. $(x-9)^2 = 100$
$x-9 = \pm\sqrt{100}$
$x-9 = \pm 10$
$x = 9 \pm 10$
$x = 9+10$ or $x = 9-10$
$x = 19$ or $x = -1$
The solutions are 19 and −1.

41. $(2x+3)^2 = 18$
$2x+3 = \pm\sqrt{18}$
$2x+3 = \pm\sqrt{9}\sqrt{2}$
$2x+3 = \pm 3\sqrt{2}$
$2x = -3 \pm 3\sqrt{2}$
$x = \frac{-3 \pm 3\sqrt{2}}{2}$
The solutions are $\frac{-3+3\sqrt{2}}{2}$ and $\frac{-3-3\sqrt{2}}{2}$.

43. $(4x+1)^2 = 20$
$4x+1 = \pm\sqrt{20}$
$4x+1 = \pm\sqrt{4}\sqrt{5}$
$4x+1 = \pm 2\sqrt{5}$
$4x = -1 \pm 2\sqrt{5}$
$x = \frac{-1 \pm 2\sqrt{5}}{4}$
The solutions are $\frac{-1+2\sqrt{5}}{4}$ and $\frac{-1-2\sqrt{5}}{4}$.

45. $(2x-6)^2 = 18$

$$2x-6 = \pm\sqrt{18}$$
$$2x-6 = \pm\sqrt{9}\sqrt{2}$$
$$2x-6 = \pm 3\sqrt{2}$$
$$2x = 6 \pm 3\sqrt{2}$$
$$x = \frac{6 \pm 3\sqrt{2}}{2}$$

The solutions are $\frac{6+3\sqrt{2}}{2}$ and $\frac{6-3\sqrt{2}}{2}$.

47. $x = \pm 6$

$x^2 = (\pm 6)^2$

$x^2 = 36$

Other answers are possible.

49. $x^2 - 9 = 27$; need an equation equivalent to $x^2 = 36$.

51. a. $-3x^2 + 9x - 6 = 0$

Multiply both sides by -1

$$-1(-3x^2 + 9x - 6) = (-1)(0)$$
$$3x^2 - 9x + 6 = 0$$

b. $-3x^2 + 9x - 6 = 0$

Multiply both sides by $-\frac{1}{3}$

$$-\frac{1}{3}(3x^2 + 9x - 6) = \left(-\frac{1}{3}\right)(0)$$
$$x^2 - 3x + 2 = 0$$

53. Let x = width of rectangle,
then $1.62x$ = length of rectangle.

Area = length · width

$$2000 = (1.62x)x$$
$$2000 = 1.62x^2$$
$$\frac{2000}{1.62} = x^2$$
$$x = \pm\sqrt{\frac{2000}{1.62}} \approx \pm 35.14$$

Since the width is positive $x \approx 35.14$ feet.
The length is $1.62 \cdot 35.14 \approx 56.93$ feet.

55. a. Left x^2, right $(x+3)^2$

b. $x^2 = 36$

$$x = \pm\sqrt{36}.$$
$$x = \pm 6$$

Since the length cannot be negative, the length of each side of the square is = 6 inches.

c. $x^2 = 50$

$$x = \pm\sqrt{50}$$
$$x = \pm\sqrt{25}\sqrt{2}$$
$$x = \pm 5\sqrt{2}$$

Since the length cannot be negative, the length of each side of the square is $5\sqrt{2} \approx 7.07$ inches.

d. $(x+3)^2 = 81$

$$x+3 = \pm\sqrt{81}$$
$$x+3 = \pm 9$$

Since the length cannot be negative, the length of each side of the square is = 9 inches.

e. $(x+3)^2 = 92$

$$x+3 = \pm\sqrt{92}$$

Since the length cannot be negative, the length of each side of the square is $\sqrt{92} \approx 9.59$ inches.

57. $I = p^2 r$

$$\frac{I}{r} = p^2$$
$$\sqrt{\frac{I}{r}} = p$$
$$p = \sqrt{\frac{I}{r}}$$

59. $a^2 + b^2 = c^2$

$$b^2 = c^2 - a^2$$
$$\sqrt{b^2} = \sqrt{c^2 - a^2}$$
$$b = \sqrt{c^2 - a^2}$$

61.
$$A = p(1+r)^2$$
$$\frac{A}{p} = (1+r)^2$$
$$\sqrt{\frac{A}{p}} = \sqrt{(1+r)^2}$$
$$\sqrt{\frac{A}{p}} = 1+r$$
$$\sqrt{\frac{A}{p}} - 1 = r$$
$$r = \sqrt{\frac{A}{p}} - 1$$

62. $m = \frac{y_2 - y_1}{x_2 - x_1} = \frac{3-(-1)}{1-0} = \frac{3+1}{1} = 4$

The y-intercept is $(0, -1)$. Thus, the equation of the line is $y = 4x - 1$.

63.
$$\begin{aligned} 4x^2 - 10x - 24 &= 2(2x^2 - 5x - 12) \\ &= 2(2x^2 - 8x + 3x - 12) \\ &= 2[2x(x-4) + 3(x-4)] \\ &= 2(2x+3)(x-4) \end{aligned}$$

64. $\frac{3-\frac{1}{y}}{6-\frac{1}{y}} = \frac{\left(3-\frac{1}{y}\right)y}{\left(6-\frac{1}{y}\right)y} = \frac{3y-1}{6y-1}$

65.
$$\begin{aligned} \frac{\sqrt{135a^4b}}{\sqrt{3a^5b^5}} &= \sqrt{\frac{135a^4b}{3a^5b^5}} \\ &= \sqrt{\frac{45}{ab^4}} \\ &= \frac{\sqrt{45}}{\sqrt{ab^4}} \\ &= \frac{\sqrt{9}\sqrt{5}}{\sqrt{b^4}\sqrt{a}} \\ &= \frac{3\sqrt{5}}{b^2\sqrt{a}} \\ &= \frac{3\sqrt{5}}{b^2\sqrt{a}} \cdot \frac{\sqrt{a}}{\sqrt{a}} \\ &= \frac{3\sqrt{5a}}{ab^2} \end{aligned}$$

Exercise Set 10.2

1. a. A perfect square trinomial is a trinomial that can be expressed as the square of a binomial.

b. $x^2 + 8x + 16$; the constant is the square of half the coefficient of the x-term.

3. The constant is the square of half the coefficient of the x-term.

5. $\left(\frac{-12}{2}\right)^2 = (-6)^2 = 36$

7.

$$x^2+3x-4=0$$
$$x^2+3x=4$$
$$x^2+3x+\frac{9}{4}=4+\frac{9}{4}$$
$$\left(x+\frac{3}{2}\right)^2=\frac{16}{4}+\frac{9}{4}$$
$$\left(x+\frac{3}{2}\right)^2=\frac{25}{4}$$
$$x+\frac{3}{2}=\pm\sqrt{\frac{25}{4}}$$
$$x+\frac{3}{2}=\pm\frac{5}{2}$$
$$x=-\frac{3}{2}\pm\frac{5}{2}$$
$$x=-\frac{3}{2}+\frac{5}{2} \text{ or } x=-\frac{3}{2}-\frac{5}{2}$$
$$x=1 \quad \text{or } x=-4$$

The solutions are 1 and –4.

9.

$$x^2-8x+7=0$$
$$x^2-8x=-7$$
$$x^2-8x+16=-7+16$$
$$(x-4)^2=9$$
$$x-4=\pm\sqrt{9}$$
$$x-4=\pm3$$
$$x=4\pm3$$
$$x=4+3 \text{ or } x=4-3$$
$$x=7 \quad \text{or } x=1$$

The solutions are 7 and 1.

11.

$$x^2+3x+2=0$$
$$x^2+3x=-2$$
$$x^2+3x+\frac{9}{4}=-2+\frac{9}{4}$$
$$\left(x+\frac{3}{2}\right)^2=-\frac{8}{4}+\frac{9}{4}$$
$$\left(x+\frac{3}{2}\right)^2=\frac{1}{4}$$
$$x+\frac{3}{2}=\pm\sqrt{\frac{1}{4}}$$
$$x+\frac{3}{2}=\pm\frac{1}{2}$$
$$x=-\frac{3}{2}\pm\frac{1}{2}$$
$$x=-\frac{3}{2}+\frac{1}{2} \text{ or } x=-\frac{3}{2}-\frac{1}{2}$$
$$x=-1 \quad \text{or } x=-2$$

The solutions are –1 and –2.

13.

$$z^2-2z-8=0$$
$$x^2-2z=8$$
$$z^2-2z+1=8+1$$
$$(z-1)^2=9$$
$$z-1=\pm\sqrt{9}$$
$$z-1=\pm3$$
$$z=1\pm3$$
$$z=1+3 \text{ or } z=1-3$$
$$z=4 \quad \text{or } z=-2$$

The solutions are 4 and –2.

15.

$$x^2=-6x-9$$
$$x^2+6x=-9$$
$$x^2+6x+9=-9+9$$
$$(x+3)^2=0$$
$$x+3=\pm\sqrt{0}$$
$$x+3=\pm0$$
$$x=-3\pm0$$
$$x=-3$$

The solution is –3.

17.

$$\begin{aligned} x^2 &= 2x+15 \\ x^2-2x &= 15 \\ x^2-2x+1 &= 15+1 \\ (x-1)^2 &= 16 \\ x-1 &= \pm\sqrt{16} \\ x-1 &= \pm 4 \\ x &= 1\pm 4 \end{aligned}$$

$x = 1+4$ or $x = 1-4$
$x = 5$ or $x = -3$
The solutions are 5 and –3.

19.

$$\begin{aligned} x^2+10x+24 &= 0 \\ x^2+10x &= -24 \\ x^2+10x+25 &= -24+25 \\ (x+5)^2 &= 1 \\ x+5 &= \pm\sqrt{1} \\ x+5 &= \pm 1 \\ x &= -5\pm 1 \end{aligned}$$

$x = -5+1$ or $x = -5-1$
$x = -4$ or $x = -6$
The solutions are –4 and –6.

21.

$$\begin{aligned} x^2 &= 15x-56 \\ x^2-15x &= -56 \\ x^2-15+\frac{225}{4} &= -56+\frac{225}{4} \\ \left(x-\frac{15}{2}\right)^2 &= -\frac{224}{4}+\frac{225}{4} \\ \left(x-\frac{15}{2}\right)^2 &= \frac{1}{4} \\ x-\frac{15}{2} &= \pm\sqrt{\frac{1}{4}} \\ x-\frac{15}{2} &= \pm\frac{1}{2} \\ x &= \frac{15}{2}\pm\frac{1}{2} \end{aligned}$$

$x = \frac{15}{2}+\frac{1}{2}$ or $x = \frac{15}{2}-\frac{1}{2}$
$x = 8$ or $x = 7$

The solutions are 8 and 7.

23.

$$\begin{aligned} -4x &= -x^2+12 \\ x^2-4x &= 12 \\ x^2-4x+4 &= 12+4 \\ (x-2)^2 &= 16 \\ x-2 &= \pm\sqrt{16} \\ x-2 &= \pm 4 \\ x &= 2\pm 4 \end{aligned}$$

$x = 2+4$ or $x = 2-4$
$x = 6$ or $x = -2$
The solutions are 6 and –2.

25.

$$\begin{aligned} z^2-4z &= -2 \\ z^2-4z+4 &= -2+4 \\ (z-2)^2 &= 2 \\ z-2 &= \pm\sqrt{2} \\ z &= 2\pm\sqrt{2} \end{aligned}$$

The solutions are $2+\sqrt{2}$ and $2-\sqrt{2}$.

27.

$$\begin{aligned} 8x+3 &= -x^2 \\ x^2+8x+3 &= 0 \\ x^2+8x &= -3 \\ x^2+8x+16 &= -3+16 \\ (x+4)^2 &= 13 \\ x+4 &= \pm\sqrt{13} \\ x &= -4\pm\sqrt{13} \end{aligned}$$

The solutions are $-4+\sqrt{13}$ and $-4-\sqrt{13}$.

29.

$$
\begin{aligned}
m^2 + 7m + 2 &= 0 \\
m^2 + 7m &= -2 \\
m^2 + 7m + \frac{49}{4} &= -2 + \frac{49}{4} \\
\left(m + \frac{7}{2}\right)^2 &= \frac{41}{4} \\
m + \frac{7}{2} &= \pm\sqrt{\frac{41}{4}} \\
m + \frac{7}{2} &= \pm\frac{\sqrt{41}}{2} \\
m &= -\frac{7}{2} \pm \frac{\sqrt{41}}{2} \\
m &= \frac{-7 \pm \sqrt{41}}{2}
\end{aligned}
$$

The solutions are $\frac{-7+\sqrt{41}}{2}$ and $\frac{-7-\sqrt{41}}{2}$.

31.

$$
\begin{aligned}
2x^2 + 4x - 6 &= 0 \\
\frac{1}{2}(2x^2 + 4x - 6) &= \frac{1}{2}(0) \\
x^2 + 2x - 3 &= 0 \\
x^2 + 2x &= 3 \\
x^2 + 2x + 1 &= 3 + 1 \\
(x+1)^2 &= 4 \\
x + 1 &= \pm\sqrt{4} \\
x + 1 &= \pm 2 \\
x &= -1 \pm 2
\end{aligned}
$$

$x = -1 + 2$ or $x = -1 - 2$
$x = 1$ or $x = -3$

The solutions are 1 and –3.

33.

$$
\begin{aligned}
2x^2 + 18x + 4 &= 0 \\
\frac{1}{2}(2x^2 + 18x + 4) &= \frac{1}{2}(0) \\
x^2 + 9x + 2 &= 0 \\
x^2 + 9x &= -2 \\
x^2 + 9x + \frac{81}{4} &= -2 + \frac{81}{4} \\
\left(x + \frac{9}{2}\right)^2 &= \frac{73}{4} \\
x + \frac{9}{2} &= \pm\sqrt{\frac{73}{4}} \\
x + \frac{9}{2} &= \pm\frac{\sqrt{73}}{2} \\
x &= -\frac{9}{2} \pm \frac{\sqrt{73}}{2} \\
x &= \frac{-9 \pm \sqrt{73}}{2}
\end{aligned}
$$

The solutions are
$\frac{-9+\sqrt{73}}{2}$ and $\frac{-9-\sqrt{73}}{2}$.

35.
$$3x^2 - 15x - 18 = 0$$
$$\frac{1}{3}(3x^2 - 15x - 18) = \frac{1}{3}(0)$$
$$x^2 - 5x - 6 = 0$$
$$x^2 - 5x = 6$$
$$x^2 - 5x + \frac{25}{4} = 6 + \frac{25}{4}$$
$$\left(x - \frac{5}{2}\right)^2 = \frac{24}{4} + \frac{25}{4}$$
$$\left(x - \frac{5}{2}\right)^2 = \frac{49}{4}$$
$$x - \frac{5}{2} = \pm\sqrt{\frac{49}{4}}$$
$$x - \frac{5}{2} = \pm\frac{7}{2}$$
$$x = \frac{5}{2} \pm \frac{7}{2}$$
$$x = \frac{5}{2} + \frac{7}{2} \text{ or } x = \frac{5}{2} - \frac{7}{2}$$
$$x = 6 \quad \text{ or } x = -1$$

The solutions are 6 and −1.

37.
$$3x^2 - 11x - 4 = 0$$
$$\frac{1}{3}(3x^2 - 11x - 4) = \frac{1}{3}(0)$$
$$x^2 - \frac{11}{3}x - \frac{4}{3} = 0$$
$$x^2 - \frac{11}{3}x = \frac{4}{3}$$
$$x^2 - \frac{11}{3}x + \frac{121}{36} = \frac{4}{3} + \frac{121}{36}$$
$$\left(x - \frac{11}{6}\right)^2 = \frac{48}{36} + \frac{121}{36}$$
$$\left(x - \frac{11}{6}\right)^2 = \frac{169}{36}$$
$$x - \frac{11}{6} = \pm\sqrt{\frac{169}{36}}$$
$$x - \frac{11}{6} = \pm\frac{13}{6}$$
$$x = \frac{11}{6} \pm \frac{13}{6}$$
$$x = \frac{11}{6} + \frac{13}{6} \text{ or } x = \frac{11}{6} - \frac{13}{6}$$
$$x = 4 \quad \text{ or } x = -\frac{1}{3}$$

The solutions are 4 and $-\frac{1}{3}$.

39.
$$3x^2 + 6x = 6$$
$$\frac{1}{3}(3x^2 + 6x) = \frac{1}{3}(6)$$
$$x^2 + 2x = 2$$
$$x^2 + 2x + 1 = 2 + 1$$
$$(x+1)^2 = 3$$
$$x + 1 = \pm\sqrt{3}$$
$$x = -1 \pm \sqrt{3}$$

The solutions are $-1 + \sqrt{3}$ and $-1 - \sqrt{3}$.

41.
$$x^2 - 5x = 0$$
$$x^2 - 5x + \frac{25}{4} = 0 + \frac{25}{4}$$
$$\left(x - \frac{5}{2}\right)^2 = \frac{25}{4}$$
$$x - \frac{5}{2} = \pm\sqrt{\frac{25}{4}}$$
$$x - \frac{5}{2} = \pm\frac{5}{2}$$
$$x = \frac{5}{2} \pm \frac{5}{2}$$
$$x = \frac{5}{2} + \frac{5}{2} \text{ or } x = \frac{5}{2} - \frac{5}{2}$$
$$x = 5 \quad \text{ or } x = 0$$

The solutions are 5 and 0.

43.
$$2x^2 = 12x$$
$$2x^2 - 12x = 0$$
$$\frac{1}{2}(2x^2 - 12x) = \frac{1}{2}(0)$$
$$x^2 - 6x = 0$$
$$x^2 - 6x + 9 = 0 + 9$$
$$(x - 3)^2 = 9$$
$$x - 3 = \pm\sqrt{9}$$
$$x - 3 = \pm 3$$
$$x = 3 \pm 3$$
$$x = 3 + 3 \text{ or } x = 3 - 3$$
$$x = 6 \quad \text{ or } x = 0$$

The solutions are 6 and 0.

45. a. $x^2 + 20x + 100$

b. The constant is the square of half the coefficient of the x-term.
$$\left(\frac{20}{2}\right)^2 = 10^2 = 100$$

47. Let x be the number.
$$x^2 + 3x = 4$$
$$x^2 + 3x + \frac{9}{4} = 4 + \frac{9}{4}$$
$$\left(x + \frac{3}{2}\right)^2 = \frac{16}{4} + \frac{9}{4}$$
$$\left(x + \frac{3}{2}\right)^2 = \frac{25}{4}$$
$$x + \frac{3}{2} = \pm\sqrt{\frac{25}{4}}$$
$$x + \frac{3}{2} = \pm\frac{5}{2}$$
$$x = -\frac{3}{2} \pm \frac{5}{2}$$
$$x = -\frac{3}{2} + \frac{5}{2} \text{ or } x = -\frac{3}{2} - \frac{5}{2}$$
$$x = 1 \quad \text{ or } x = -4$$
The numbers are 1 and –4.

49. Let x be the number.
$$(x + 3)^2 = 9$$
$$x + 3 = \pm\sqrt{9}$$
$$x + 3 = \pm 3$$
$$x = -3 \pm 3$$
$$x = -3 + 3 \text{ or } x = -3 - 3$$
$$x = 0 \quad \text{ or } x = -6$$
The numbers are 0 and –6.

51. Let x and y be the numbers.
$$xy = 21$$
$$y = x + 4$$
Substitute $x + 4$ for y in the first equation.
$$xy = 21$$
$$x(x + 4) = 21$$
$$x^2 + 4x = 21$$
$$x^2 + 4x + 4 = 21 + 4$$
$$(x + 2)^2 = 25$$
$$x + 2 = \pm\sqrt{25}$$
$$x + 2 = \pm 5$$
$$x = -2 \pm 5$$
$$x = -2 + 5 \text{ or } x = -2 - 5$$
$$x = 3 \quad \text{ or } x = -7$$
Since the numbers must be positive, $x = 3$; $y = 3 + 4 = 7$. The numbers are 3 and 7.

53. $+\frac{3}{5}x$ or $-\frac{3}{5}x$

The coefficient of the x-term is plus or minus twice the square root of the constant.

$$b = \pm 2\sqrt{c} = \pm 2\sqrt{\frac{9}{100}} = \pm 2\left(\frac{3}{10}\right) = \pm\frac{3}{5}$$

55. a.

$$\begin{aligned} x^2 - 14x - 1 &= 0 \\ x^2 - 14x &= 1 \\ x^2 - 14x + 49 &= 1 + 49 \\ (x-7)^2 &= 50 \\ x - 7 &= \pm\sqrt{50} \\ x - 7 &= \pm 5\sqrt{2} \\ x &= 7 \pm 5\sqrt{2} \end{aligned}$$

b. $x = 7 + 5\sqrt{2}$

$$\begin{aligned} x^2 - 14x - 1 &= 0 \\ (7+5\sqrt{2})^2 - 14(7+5\sqrt{2}) - 1 &= 0 \\ 49 + 70\sqrt{2} + 50 - 98 - 70\sqrt{2} - 1 &= 0 \\ 49 + 50 - 98 - 1 + 70\sqrt{2} - 70\sqrt{2} &= 0 \\ 0 &= 0 \text{ True} \end{aligned}$$

$x = 7 - 5\sqrt{2}$

$$\begin{aligned} x - 14x - 1 &= 0 \\ (7-5\sqrt{2})^2 - 14(7-5\sqrt{2}) - 1 &= 0 \\ 49 - 70\sqrt{2} + 50 - 98 + 70\sqrt{2} - 1 &= 0 \\ 49 + 50 - 98 - 1 - 70\sqrt{2} + 70\sqrt{2} &= 0 \\ 0 &= 0 \text{ True} \end{aligned}$$

57.

$$\begin{aligned} x^2 - \frac{2}{3}x - \frac{1}{5} &= 0 \\ x^2 - \frac{2}{3}x &= \frac{1}{5} \\ x^2 - \frac{2}{3}x + \frac{1}{9} &= \frac{1}{5} + \frac{1}{9} \\ \left(x - \frac{1}{3}\right)^2 &= \frac{14}{45} \\ x - \frac{1}{3} &= \pm\sqrt{\frac{14}{45}} \\ x &= \frac{1}{3} \pm \sqrt{\frac{14}{45}} \\ x &= \frac{1}{3} \pm \frac{1}{3}\sqrt{\frac{14}{5}} \\ x &= \frac{1}{3} \pm \frac{1}{3}\frac{\sqrt{14}}{\sqrt{5}} \\ x &= \frac{1}{3} \pm \frac{1}{3}\frac{\sqrt{14}}{\sqrt{5}} \cdot \frac{\sqrt{5}}{\sqrt{5}} \\ x &= \frac{1}{3} \pm \frac{1}{3}\frac{\sqrt{70}}{5} \\ x &= \frac{1}{3} \pm \frac{1}{15}\sqrt{70} \\ x &= \frac{5 \pm \sqrt{70}}{15} \end{aligned}$$

The solutions are: $\frac{5+\sqrt{70}}{15}$ and $\frac{5-\sqrt{70}}{15}$.

59.

$$\begin{aligned} 0.1x^2 + 0.2x - 0.54 &= 0 \\ 10(0.1x^2 + 0.2x - 0.54) &= 10(0) \\ x^2 + 2x - 5.4 &= 0 \\ x^2 + 2x &= 5.4 \\ x^2 + 2x + 1 &= 5.4 + 1 \\ (x+1)^2 &= 6.4 \\ x + 1 &= \pm\sqrt{6.4} \\ x &= -1 \pm \sqrt{6.4} \end{aligned}$$

The solutions are: $-1 + \sqrt{6.4}$ and $-1 - \sqrt{6.4}$.

61. If the slopes are the same and the y-intercepts are different, the equations represent parallel lines.

62. $3x - 4y = 6$
$2x + y = 8$
Multiply the second equation by 4.
$4[2x + y = 8]$
gives

$$\begin{aligned} 3x - 4y &= 6 \\ 8x + 4y &= 32 \\ \hline 11x \quad &= 38 \\ x \quad &= \frac{38}{11} \end{aligned}$$

Substitute $\frac{38}{11}$ for x in the first equation.

$$\begin{aligned} 3x - 4y &= 6 \\ 3\left(\frac{38}{11}\right) - 4y &= 6 \\ \frac{114}{11} - 4y &= 6 \\ -4y &= -\frac{48}{11} \\ y &= \frac{12}{11} \end{aligned}$$

The solution is $\left(\frac{38}{11}, \frac{12}{11}\right)$.

63.

$$\begin{aligned} &\frac{x^2}{x^2 - x - 6} - \frac{x-2}{x-3} \\ &= \frac{x^2}{(x-3)(x+2)} - \frac{x-2}{x-3} \\ &= \frac{x^2}{(x-3)(x+2)} - \frac{(x-2)}{(x-3)} \cdot \frac{(x+2)}{(x+2)} \\ &= \frac{x^2}{(x-3)(x+2)} - \frac{x^2 - 4}{(x-3)(x+2)} \\ &= \frac{x^2 - (x^2 - 4)}{(x-3)(x+2)} \\ &= \frac{4}{(x-3)(x+2)} \end{aligned}$$

64.

$$\begin{aligned} \sqrt{2x+3} &= 2x - 3 \\ (\sqrt{2x+3})^2 &= (2x-3)^2 \\ 2x + 3 &= 4x^2 - 12x + 9 \\ 0 &= 4x^2 - 14x + 6 \\ 0 &= 2(2x-1)(x-3) \\ 2x - 1 = 0 &\text{ or } x - 3 = 0 \\ x = \frac{1}{2} \quad &\text{ or } x = 3 \end{aligned}$$

Check: $x = \frac{1}{2}$

$$\begin{aligned} \sqrt{2x+3} &= 2x - 3 \\ \sqrt{2\left(\frac{1}{2}\right) + 3} &= 2\left(\frac{1}{2}\right) - 3 \\ \sqrt{4} &= -2 \\ 2 &= -2 \text{ False} \end{aligned}$$

Check: $x = 3$

$$\begin{aligned} \sqrt{2x+3} &= 2x - 3 \\ \sqrt{2(3)+3} &= 2(3) - 3 \\ \sqrt{9} &= 3 \\ 3 &= 3 \text{ True} \end{aligned}$$

$\frac{1}{2}$ is an extraneous root. The solution is 3.

Exercise Set 10.3

1. a. $b^2 - 4ac$

b. If the discriminant is:
greater than 0 there are two solutions;
equal to 0 there is one solution;
less than 0 there is no real solution.

3. $x = \dfrac{-b \pm \sqrt{b^2 - 4ac}}{2a}$

5. The first step to take when solving a quadratic equation is to write the equation in standard form.

7. The values used for b and c are incorrect because the equation was not first put in standard form.

9. $b^2 - 4ac = (4)^2 - 4(1)(-6) = 16 + 24 = 40$
Since the discriminant is positive, this equation has two distinct real number solutions.

11. $b^2 - 4ac = (1)^2 - 4(2)(1) = 1 - 8 = -7$
Since the discriminant is negative, this equation has no real number solution.

13. $b^2 - 4ac = (3)^2 - 4(5)(-7) = 9 + 140 = 149$
Since the discriminant is positive, this equation has two distinct real number solutions.

15. $2x^2 = 16x - 32$
$2x^2 - 16x + 32 = 0$
$b^2 - 4ac = (-16)^2 - 4(2)(32) = 256 - 256 = 0$
Since the discriminant is zero, this equation has one real number solution.

17. $b^2 - 4ac = (-8)^2 - 4(1)(5) = 64 - 20 = 44$
Since the discriminant is positive, this equation has two distinct real number solutions.

19. $4x = 8 + x^2$
$0 = x^2 - 4x + 8$
$b^2 - 4ac = (-4)^2 - 4(1)(8) = 16 - 32 = -16$
Since the discriminant is negative, this equation

21. $b^2 - 4ac = (7)^2 - 4(1)(-3) = 49 + 12 = 61$
Since the discriminant is positive, this equation has two distinct real number solutions.

23. $b^2 - 4ac = (0)^2 - 4(4)(-9) = 0 + 144 = 144$
Since the discriminant is positive, this equation has two distinct real number solutions.

25. $a = 1, b = -1, c = -6$

$$x = \frac{-b \pm \sqrt{b^2 - 4ac}}{2a} = \frac{-(-1) \pm \sqrt{(-1)^2 - 4(1)(-6)}}{2(1)} = \frac{1 \pm \sqrt{1 + 24}}{2} = \frac{1 \pm \sqrt{25}}{2} = \frac{1 \pm 5}{2}$$

$x = \frac{1+5}{2}$ or $x = \frac{1-5}{2}$
$x = 3$ or $x = -2$

27. $a = 1, b = 9, c = 18$

$$x = \frac{-b \pm \sqrt{b^2 - 4ac}}{2a} = \frac{-9 \pm \sqrt{(9)^2 - 4(1)(18)}}{2(1)} = \frac{-9 \pm \sqrt{81 - 72}}{2} = \frac{-9 \pm \sqrt{9}}{2} = \frac{-9 \pm 3}{2}$$

$x = \frac{-9+3}{2}$ or $x = \frac{-9-3}{2}$
$x = -3$ or $x = -6$

29. Write in standard form

$x^2 - 6x + 5 = 0$

$a = 1, b = -6, c = 5$

$$x = \frac{-b \pm \sqrt{b^2 - 4ac}}{2a} = \frac{-(-6) \pm \sqrt{(-6)^2 - 4(1)(5)}}{2(1)} = \frac{6 \pm \sqrt{36 - 20}}{2} = \frac{6 \pm \sqrt{16}}{2} = \frac{6 \pm 4}{2}$$

$x = \frac{6+4}{2}$ or $x = \frac{6-4}{2}$

$x = 5$ or $x = 1$

31. Write in standard form

$x^2 - 10x + 24 = 0$

$a = 1, b = -10, c = 24$

$$x = \frac{-b \pm \sqrt{b^2 - 4ac}}{2a} = \frac{-(-10) \pm \sqrt{(-10)^2 - 4(1)(24)}}{2(1)} = \frac{10 \pm \sqrt{100 - 96}}{2} = \frac{10 \pm \sqrt{4}}{2} = \frac{10 \pm 2}{2}$$

$x = \frac{10+2}{2}$ or $x = \frac{10-2}{2}$

$x = 6$ or $x = 4$

33. Write in standard form

$x^2 - 100 = 0$

$a = 1, b = 0, c = -100$

$$x = \frac{-b \pm \sqrt{b^2 - 4ac}}{2a} = \frac{-0 \pm \sqrt{(0)^2 - 4(1)(-100)}}{2(1)} = \frac{\pm\sqrt{400}}{2} = \frac{\pm 20}{2} = \pm 10$$

$x = 10$ or $x = -10$

35. $a = 1, b = -3, c = 0$

$$x = \frac{-b \pm \sqrt{b^2 - 4ac}}{2a} = \frac{-(-3) \pm \sqrt{(-3)^2 - 4(1)(0)}}{2(1)} = \frac{3 \pm \sqrt{9}}{2} = \frac{3 \pm 3}{2}$$

$x = \frac{3+3}{2}$ or $x = \frac{3-3}{2}$

$x = 3$ or $x = 0$

37. $a = 1, b = -7, c = 10$

$$n = \frac{-b \pm \sqrt{b^2 - 4ac}}{2a} = \frac{-(-7) \pm \sqrt{(-7)^2 - 4(1)(10)}}{2(1)} = \frac{7 \pm \sqrt{49 - 40}}{2} = \frac{7 \pm \sqrt{9}}{2} = \frac{7 \pm 3}{2}$$

$n = \frac{7+3}{2}$ or $n = \frac{7-3}{2}$

$n = 5$ or $n = 2$

39. $a = 2, b = -7, c = 4$

$$y = \frac{-b \pm \sqrt{b^2 - 4ac}}{2a}$$
$$= \frac{-(-7) \pm \sqrt{(-7)^2 - 4(2)(4)}}{2(2)}$$
$$= \frac{7 \pm \sqrt{49 - 32}}{4}$$
$$= \frac{7 \pm \sqrt{17}}{4}$$
$$x = \frac{7 + \sqrt{17}}{4} \text{ or } x = \frac{7 - \sqrt{17}}{4}$$

41. Write in standard form

$6x^2 + x - 1 = 0$

$a = 6, b = 1, c = -1$

$$x = \frac{-b \pm \sqrt{b^2 - 4ac}}{2a}$$
$$= \frac{-1 \pm \sqrt{(1)^2 - 4(6)(-1)}}{2(6)}$$
$$= \frac{-1 \pm \sqrt{1 + 24}}{12}$$
$$= \frac{-1 \pm \sqrt{25}}{12}$$
$$= \frac{-1 \pm 5}{12}$$
$$x = \frac{-1 + 5}{12} \text{ or } x = \frac{-1 - 5}{12}$$
$$x = \frac{1}{3} \quad \text{or } x = -\frac{1}{2}$$

43. Write in standard form

$2x^2 - 5x - 7 = 0$

$a = 2, b = -5, c = -7$

$$x = \frac{-b \pm \sqrt{b^2 - 4ac}}{2a}$$
$$= \frac{-(-5) \pm \sqrt{(-5)^2 - 4(2)(-7)}}{2(2)}$$
$$= \frac{5 \pm \sqrt{25 + 56}}{4}$$
$$= \frac{5 \pm \sqrt{81}}{4}$$
$$= \frac{5 \pm 9}{4}$$
$$x = \frac{5 + 9}{4} \text{ or } x = \frac{5 - 9}{4}$$
$$x = \frac{7}{2} \quad \text{or } x = -1$$

45. $a = 2, b = -4, c = 3$

$$s = \frac{-b \pm \sqrt{b^2 - 4ac}}{2a}$$
$$= \frac{-(-4) \pm \sqrt{(-4)^2 - 4(2)(3)}}{2(2)}$$
$$= \frac{4 \pm \sqrt{16 - 24}}{4}$$
$$= \frac{4 \pm \sqrt{-8}}{4}$$

Since $\sqrt{-8}$ is not a real number, this equation has no real number solution.

47. Write in standard form

$4x^2 - x - 5 = 0$

$a = 4, b = -1, c = -5$

$$x = \frac{-b \pm \sqrt{b^2 - 4ac}}{2a}$$

$$= \frac{-(-1) \pm \sqrt{(-1)^2 - 4(4)(-5)}}{2(4)}$$

$$= \frac{1 \pm \sqrt{1+80}}{8}$$

$$= \frac{1 \pm \sqrt{81}}{8}$$

$$= \frac{1 \pm 9}{8}$$

$$x = \frac{1+9}{8} \text{ or } x = \frac{1-9}{8}$$

$$x = \frac{5}{4} \qquad \text{or } x = -1$$

49. Write in standard form

$2x^2 - 7x - 9 = 0$

$a = 2, b = -7, c = -9$

$$x = \frac{-b \pm \sqrt{b^2 - 4ac}}{2a}$$

$$= \frac{-(-7) \pm \sqrt{(-7)^2 - 4(2)(-9)}}{2(2)}$$

$$= \frac{7 \pm \sqrt{49+72}}{4}$$

$$= \frac{7 \pm \sqrt{121}}{4}$$

$$= \frac{7 \pm 11}{4}$$

$$x = \frac{7+11}{4} \text{ or } x = \frac{7-11}{4}$$

$$x = \frac{9}{2} \qquad \text{or } x = -1$$

51. $a = -2, b = 11, c = -15$

$$x = \frac{-b \pm \sqrt{b^2 - 4ac}}{2a}$$

$$= \frac{-11 \pm \sqrt{(11)^2 - 4(-2)(-15)}}{2(-2)}$$

$$= \frac{-11 \pm \sqrt{121-120}}{-4}$$

$$= \frac{-11 \pm \sqrt{1}}{-4}$$

$$= \frac{-11 \pm 1}{-4}$$

$$x = \frac{-11+1}{-4} \text{ or } x = \frac{-11-1}{-4}$$

$$x = \frac{5}{2} \qquad \text{or } x = 3$$

53. Let x = the smaller integer,
then $x + 1$ = the larger integer.

$$x(x+1) = 42$$

$$x^2 + x = 42$$

$$x^2 + x - 42 = 0$$

$a = 1, b = 1, c = -42$

$$x = \frac{-b \pm \sqrt{b^2 - 4ac}}{2a}$$

$$= \frac{-1 \pm \sqrt{(1)^2 - 4(1)(-42)}}{2(1)}$$

$$= \frac{-1 \pm \sqrt{1+168}}{2}$$

$$= \frac{-1 \pm \sqrt{169}}{2}$$

$$= \frac{-1 \pm 13}{2}$$

$$x = \frac{-1+13}{2} \text{ or } x = \frac{-1-13}{2}$$

$$x = 6 \qquad \text{or } x = -7$$

Since the numbers are positive, $x = 6$. The numbers are 6 and $6 + 1 = 7$.

55. Let w = width of rectangle,
then $2w - 3$ = length of rectangle.
Area = length · width

$$20 = (2w-3)w$$
$$20 = 2w^2 - 3w$$
$$0 = 2w^2 - 3w - 20$$

$a = 2$, $b = -3$, $c = -20$

$$x = \frac{-b \pm \sqrt{b^2 - 4ac}}{2a}$$
$$= \frac{-(-3) \pm \sqrt{(-3)^2 - 4(2)(-20)}}{2(2)}$$
$$= \frac{3 \pm \sqrt{9+160}}{4}$$
$$= \frac{3 \pm \sqrt{169}}{4}$$
$$= \frac{3 \pm 13}{4}$$
$$x = \frac{3+13}{4} \text{ or } x = \frac{3-13}{4}$$
$$x = 4 \qquad \text{or } x = -\frac{5}{2}$$

Since the width is positive, $w = 4$. The width is 4 feet and the length is $2(4) - 3 = 5$ feet.

57. Let x = the width of the tile border.
Area of the pool
$= (30)(40) = 1200$ square feet
Area of the pool plus border
$= (2x+30)(2x+40) = 4x^2 + 140x + 1200$
Area of the border
$= 4x^2 + 140x + 1200 - 1200$
$= 4x^2 + 140x$

$$4x^2 + 140x = 296$$
$$4x^2 + 140x - 296 = 0$$
$$x^2 + 35x - 74 = 0$$

$a = 1$, $b = 35$, $c = -74$

$$x = \frac{-b \pm \sqrt{b^2 - 4ac}}{2a}$$
$$= \frac{-35 \pm \sqrt{(35)^2 - 4(1)(-74)}}{2(1)}$$
$$= \frac{-35 \pm \sqrt{1225+296}}{2}$$
$$= \frac{-35 \pm \sqrt{1521}}{2}$$
$$= \frac{-35 \pm 39}{2}$$
$$x = \frac{-35+39}{2} \text{ or } x = \frac{-35-39}{2}$$
$$x = 2 \qquad \text{or } x = -37$$

Since the width must be positive, the border can be 2 feet wide.

59. $2x^2 + 3x + c = 0$
$a = 2$, $b = 3$
The discriminant is
$b^2 - 4ac = 3^2 - 4(2)c = 9 - 8c.$

a. The equation has two real number solutions when $b^2 - 4ac > 0$.

$$9 - 8c > 0$$
$$9 > 8c$$
$$\frac{9}{8} > c$$
$$c < \frac{9}{8}$$

b. The equation has one real number solution when $b^2 - 4ac = 0$.

$$9 - 8c = 0$$
$$9 = 8c$$
$$\frac{9}{8} = c$$
$$c = \frac{9}{8}$$

c. The equation has no real number solution when $b^2 - 4ac < 0$.

$$9 - 8c < 0$$
$$9 < 8c$$
$$\frac{9}{8} < c$$
$$c > \frac{9}{8}$$

61. Let l = the length (along the river) and w = the width of the region, then the length of fencing is $l + 2w$.

Since $l + 2w = 400$, $l = 400 - 2w$.

$$A = l \cdot w$$
$$15,000 = (400 - 2w)w$$
$$15,000 = 400w - 2w^2$$
$$2w^2 - 400w + 15,000 = 0$$
$$w^2 - 200w + 7500 = 0$$

$a = 1$, $b = -200$, $c = 7500$

$$w = \frac{-b \pm \sqrt{b^2 - 4ac}}{2a}$$
$$= \frac{-(-200) \pm \sqrt{(-200)^2 - 4(1)(7500)}}{2(1)}$$
$$= \frac{200 \pm \sqrt{40,000 - 30,000}}{2}$$
$$= \frac{200 \pm \sqrt{10,000}}{2}$$
$$= \frac{200 \pm 100}{2}$$
$$w = \frac{200 + 100}{2} \text{ or } w = \frac{200 - 100}{2}$$
$$w = 150 \qquad \text{or } w = 50$$

$l = 400 - 2w \qquad l = 400 - 2w$
$= 400 - 2(150) \qquad = 400 - 2(50)$
$= 400 - 300 \qquad = 400 - 100$
$= 100 \qquad = 300$

The region should be 100 feet long and 150 feet wide or 300 feet long and 50 feet wide.

63. a.

$$x^2 - 13x + 42 = 0$$
$$(x - 7)(x - 6) = 0$$
$$x - 7 = 0 \text{ or } x - 6 = 0$$
$$x = 7 \qquad \text{or } x = 6$$

b.

$$x^2 - 13x + 42 = 0$$
$$x^2 - 13x = -42$$
$$x^2 - 13x + \frac{169}{4} = -42 + \frac{169}{4}$$
$$\left(x - \frac{13}{2}\right)^2 = \frac{1}{4}$$
$$x - \frac{13}{2} = \pm\sqrt{\frac{1}{4}}$$
$$x - \frac{13}{2} = \pm\frac{1}{2}$$
$$x = \frac{13}{2} \pm \frac{1}{2}$$
$$x = \frac{13}{2} + \frac{1}{2} \text{ or } x = \frac{13}{2} - \frac{1}{2}$$
$$x = 7 \qquad \text{or } x = 6$$

c. $a = 1$, $b = -13$, $c = 42$

$$x = \frac{-b \pm \sqrt{b^2 - 4ac}}{2a}$$
$$= \frac{-(-13) \pm \sqrt{(-13)^2 - 4(1)(42)}}{2(1)}$$
$$= \frac{13 \pm \sqrt{169 - 168}}{2}$$
$$= \frac{13 \pm \sqrt{1}}{2}$$
$$= \frac{13 \pm 1}{2}$$
$$x = \frac{13 + 1}{2} \text{ or } x = \frac{13 - 1}{2}$$
$$x = 7 \qquad \text{or } x = 6$$

64. a.

$$6x^2 + 11x - 35 = 0$$
$$6x^2 + 21x - 10x - 35 = 0$$
$$3x(2x + 7) - 5(2x + 7) = 0$$
$$(3x - 5)(2x + 7) = 0$$
$$3x - 5 = 0 \text{ or } 2x + 7 = 0$$
$$x = \frac{5}{3} \text{ or } \qquad x = -\frac{7}{2}$$

b.
$$6x^2+11x-35=0$$
$$x^2+\frac{11}{6}x-\frac{35}{6}=0$$
$$x^2+\frac{11}{6}x=\frac{35}{6}$$
$$x^2+\frac{11}{6}x+\frac{121}{144}=\frac{35}{6}+\frac{121}{144}$$
$$\left(x+\frac{11}{12}\right)^2=\frac{961}{144}$$
$$x+\frac{11}{12}=\pm\sqrt{\frac{961}{144}}$$
$$x+\frac{11}{12}=\pm\frac{31}{12}$$
$$x=-\frac{11}{12}\pm\frac{31}{12}$$
$$x=-\frac{11}{12}+\frac{31}{12} \text{ or } x=-\frac{11}{12}-\frac{31}{12}$$
$$x=\frac{20}{12}=\frac{5}{3} \text{ or } x=-\frac{42}{12}=-\frac{7}{2}$$

c. $a=6$, $b=11$, $c=-35$
$$x=\frac{-b\pm\sqrt{b^2-4ac}}{2a}$$
$$=\frac{-11\pm\sqrt{11^2-4(6)(-35)}}{2(6)}$$
$$=\frac{-11\pm\sqrt{121+840}}{12}$$
$$=\frac{-11\pm\sqrt{961}}{12}$$
$$=\frac{-11\pm 31}{12}$$
$$x=\frac{-11+31}{12} \text{ or } x=\frac{-11-31}{12}$$
$$x=\frac{20}{12}=\frac{5}{3} \text{ or } x=\frac{-42}{12}=-\frac{7}{2}$$

65. a. $2x^2+3x-4=0$
Since there are no integers whose product is –8 and whose sum is 3, this equation cannot be solved by factoring.

b.
$$2x^2+3x-4=0$$
$$x^2+\frac{3}{2}x-2=0$$
$$x^2+\frac{3}{2}x=2$$
$$x^2+\frac{3}{2}x+\frac{9}{16}=2+\frac{9}{16}$$
$$\left(x+\frac{3}{4}\right)^2=\frac{41}{16}$$
$$x+\frac{3}{4}=\pm\sqrt{\frac{41}{16}}$$
$$x+\frac{3}{4}=\pm\frac{\sqrt{41}}{4}$$
$$x=-\frac{3}{4}\pm\frac{\sqrt{41}}{4}$$
$$x=\frac{-3\pm\sqrt{41}}{4}$$
$$x=\frac{-3+\sqrt{41}}{4} \text{ or } x=\frac{-3-\sqrt{41}}{4}$$

c. $a=2$, $b=3$, $c=-4$
$$x=\frac{-b\pm\sqrt{b^2-4ac}}{2(a)}$$
$$=\frac{-3\pm\sqrt{3^2-4(2)(-4)}}{2(2)}$$
$$=\frac{-3\pm\sqrt{9+32}}{4}$$
$$=\frac{-3\pm\sqrt{41}}{4}$$
$$x=\frac{-3+\sqrt{41}}{4} \text{ or } x=\frac{-3-\sqrt{41}}{4}$$

66. a.
$$6x^2=54$$
$$6x^2-54=0$$
$$6(x^2-9)=0$$
$$6(x-3)(x+3)=0$$
$$x-3=0 \text{ or } x+3=0$$
$$x=3 \text{ or } x=-3$$

b. $6x^2 = 54$
$x^2 = 9$
$x = \pm\sqrt{9}$
$x = \pm 3$
$x = 3$ or $x = -3$

c. $6x^2 - 54 = 0$
$a = 6,\ b = 0,\ c = -54$

$$x = \frac{-b \pm \sqrt{b^2 - 4ac}}{2(a)} = \frac{-0 \pm \sqrt{0^2 - 4(6)(-54)}}{2(6)} = \frac{\pm\sqrt{1296}}{12} = \frac{\pm 36}{12} = \pm 3$$

$x = 3$ or $x = -3$

Exercise Set 10.4

1. The graph of a quadratic equation of the form $y = ax^2 + bx + c, a \neq 0$ is called a parabola.

3. Answers will vary.

5. **a.** Where the graph crosses the x-axis.

   b. The x-intercepts are found by setting $y = 0$ and solving for x.

7. **a.** $x = -\dfrac{b}{2a}$

   b. This line is called the axis of symmetry.

9. $a = 1,\ b = 6,\ c = -3$
$$x = -\frac{b}{2a} = -\frac{6}{2(1)} = -3$$
The axis of symmetry is $x = -3$.
Find the y-coordinate of the vertex:
$y = x^2 + 6x - 3$
$y = (-3)^2 + 6(-3) - 3$
$= 9 - 18 - 3$
$= -12$
The vertex is $(-3, -12)$
Since $a > 0$, the parabola opens upward.

11. $a = -1,\ b = 3,\ c = -4$
$$x = -\frac{b}{2a} = -\frac{3}{2(-1)} = \frac{3}{2}$$
The axis of symmetry is $x = \dfrac{3}{2}$
Find the y-coordinate of the vertex:
$y = -x^2 + 3x - 4$
$$y = -\left(\frac{3}{2}\right)^2 + 3\left(\frac{3}{2}\right) - 4 = -\frac{9}{4} + \frac{9}{2} - 4 = -\frac{7}{4}$$
The vertex is $\left(\dfrac{3}{2}, -\dfrac{7}{4}\right)$
Since $a < 0$, the parabola opens downward.

13. $a = -3,\ b = 5,\ c = 8$
$$x = -\frac{b}{2a} = -\frac{5}{2(-3)} = \frac{5}{6}$$
The axis of symmetry is $x = \dfrac{5}{6}$
Find the y-coordinate of the vertex:
$y = -3x^2 + 5x + 8$
$$y = -3\left(\frac{5}{6}\right)^2 + 5\left(\frac{5}{6}\right) + 8 = -\frac{25}{12} + \frac{25}{6} + 8 = \frac{121}{12}$$
The vertex is $\left(\dfrac{5}{6}, \dfrac{121}{12}\right)$.
Since $a < 0$, the parabola opens downward.

15. $a = 4,\ b = 8,\ c = 3$
$$x = -\frac{b}{2a} = -\frac{8}{2(4)} = -1$$
The axis of symmetry is $x = -1$.
Find the y-coordinate of the vertex:
$y = 4x^2 + 8x + 3$
$y = 4(-1)^2 + 8(-1) + 3$
$= 4 - 8 + 3$
$= -1$
The vertex is $(-1, -1)$.
Since $a > 0$, the parabola opens upward.

17. $a = 2, b = 3, c = 8$

$x = -\frac{b}{2a} = -\frac{3}{2(2)} = -\frac{3}{4}$

The axis of symmetry is $x = -\frac{3}{4}$.

Find the y-coordinate of the vertex:

$y = 2x^2 + 3x + 8$

$y = 2\left(-\frac{3}{4}\right)^2 + 3\left(-\frac{3}{4}\right) + 8$

$= \frac{9}{8} - \frac{9}{4} + 8$

$= \frac{55}{8}$

The vertex is $\left(-\frac{3}{4}, \frac{55}{8}\right)$.

Since $a > 0$, the parabola opens upward.

19. $a = -5, b = 6, c = -1$

$x = -\frac{b}{2a} = -\frac{6}{2(-5)} = \frac{3}{5}$

The axis of symmetry is $x = \frac{3}{5}$.

Find the y-coordinate of the vertex:

$y = -5x^2 + 6x - 1$

$y = -5\left(\frac{3}{5}\right)^2 + 6\left(\frac{3}{5}\right) - 1$

$= -\frac{9}{5} + \frac{18}{5} - 1$

$= \frac{4}{5}$

The vertex is $\left(\frac{3}{5}, \frac{4}{5}\right)$.

Since $a < 0$, the parabola opens downward.

21. $a = 1, b = 0, c = -4$

Since $a > 0$, the parabola opens upward.

Axis of symmetry is $x = -\frac{b}{2a} = -\frac{0}{2(1)} = 0$.

y-coordinate of the vertex:

$y = x^2 - 4$

$y = 0^2 - 4 = -4$

The vertex is (0, –4)

$y = x^2 - 4$

Let $x = -2$ $y = (-2)^2 - 4 = 0$

Let $x = 2$ $y = 2^2 - 4 = 0$

x	y
–2	0
2	0

$0 = x^2 - 4$

$0 = (x+2)(x-2)$

$x = -2$ or $x = 2$

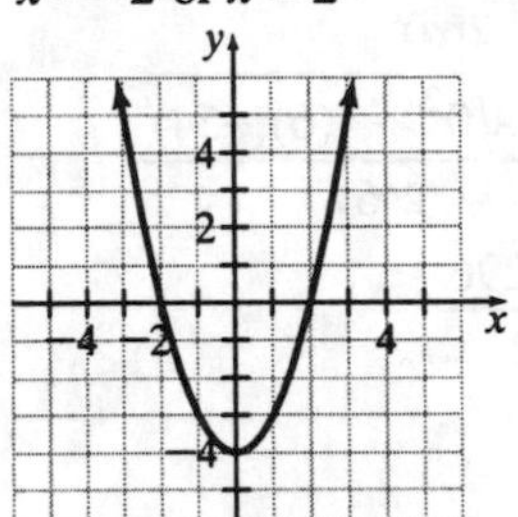

23. $a = -1, b = 0, c = 5$

Since $a < 0$, the parabola opens downward.

Axis of symmetry is $x = -\frac{b}{2a} = -\frac{0}{2(-1)} = 0$

y-coordinate of the vertex:

$y = -x^2 + 5$

$y = -0^2 + 5 = 5$

The vertex is (0, 5).

$y = -x^2 + 5$

Let $x = -1$ $y = -(-1)^2 + 5 = 4$

Let $x = 1$ $y = -(1)^2 + 5 = 4$

x	y
–1	4
1	4

$0 = -x^2 + 5$
$x^2 = 5$
$x = \pm\sqrt{5}$

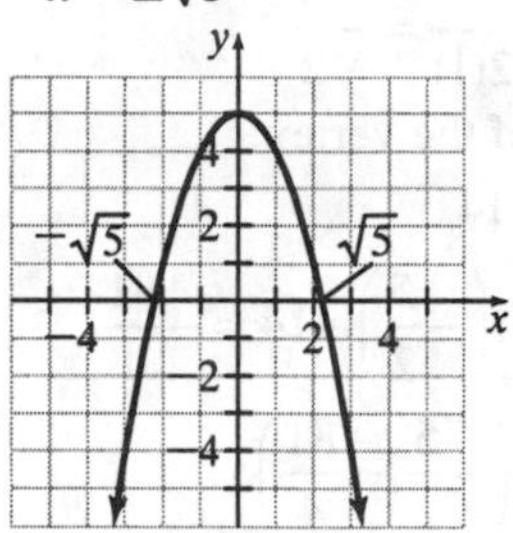

25. $a = 1, b = 4, c = 3$
Since $a > 0$, the parabola opens upward.
Axis of symmetry is $x = -\frac{b}{2a} = -\frac{4}{2(1)} = -2$
y-coordinate of the vertex:
$y = x^2 + 4x + 3$
$y = (-2)^2 + 4(-2) + 3 = 1$
The vertex is (–2, –1).

$y = x^2 + 4x + 3$

Let $x = -1$ $\quad y = (-1)^2 + 4(-1) + 3 = 0$
Let $x = 0$ $\quad y = 0^2 + 4(0) + 3 = 3$

x	y
–1	0
0	3

$0 = x^2 + 4x + 3$
$0 = (x + 3)(x + 1)$
$x = -3$ or $x = -1$

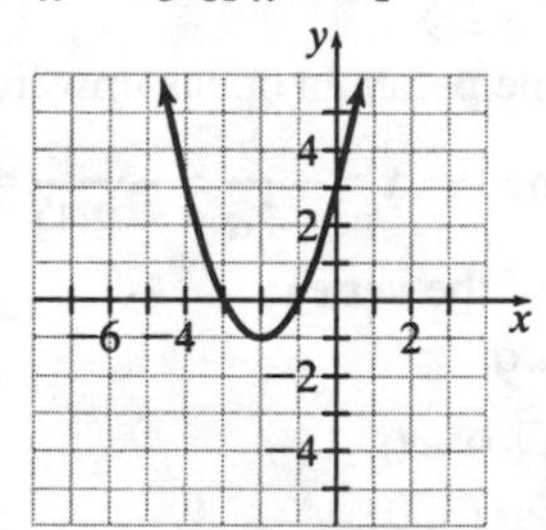

27. $a = 1, b = 4, c = 4$
Since $a > 0$, the parabola opens upward.
Axis of symmetry is $x = -\frac{b}{2a} = -\frac{4}{2(1)} = -2$
y-coordinate of the vertex:
$y = x^2 + 4x + 4$
$y = (-2)^2 + 4(-2) + 4 = 0$
The vertex is (–2, 0)

$y = x^2 + 4x + 4$

Let $x = -3$ $\quad y = (-3)^2 + 4(-3) + 4 = 1$
Let $x = -1$ $\quad y = (-1)^2 + 4(-1) + 4 = 1$

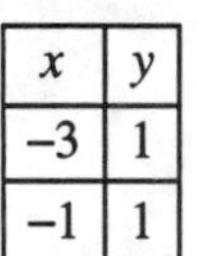

x	y
–3	1
–1	1

$0 = x^2 + 4x + 4$
$0 = (x + 2)^2$
$x = -2$

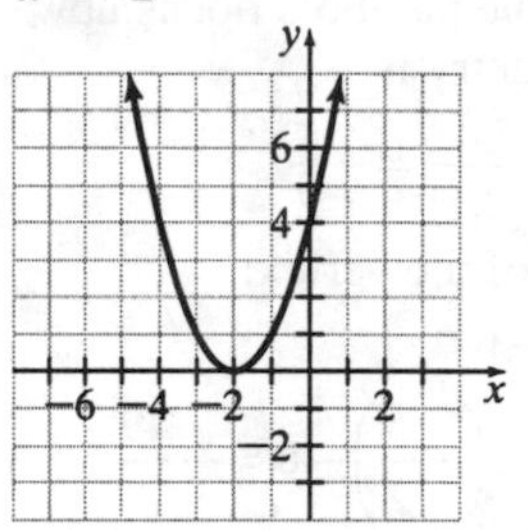

29. $a = -1, b = -2, c = 3$
Since $a < 0$, the parabola opens downward.
Axis of symmetry is
$x = -\frac{b}{2a} = -\frac{-2}{2(-1)} = -1$
y-coordinate of the vertex:
$y = -x^2 - 2x + 3$
$y = -(-1)^2 - 2(-1) + 3 = 4$
The vertex is (–1, 4)

$y = -x^2 - 2x + 3$

Let $x = 1$ $\quad y = -(1)^2 - 2(1) + 3 = 0$
Let $x = -3$ $\quad y = -(-3)^2 - 2(-3) + 3 = 0$

x	y
–3	0
1	0

$0 = -x^2 - 2x + 3$
$0 = x^2 + 2x - 3$
$0 = (x+3)(x-1)$
$x = -3$ or $x = 1$

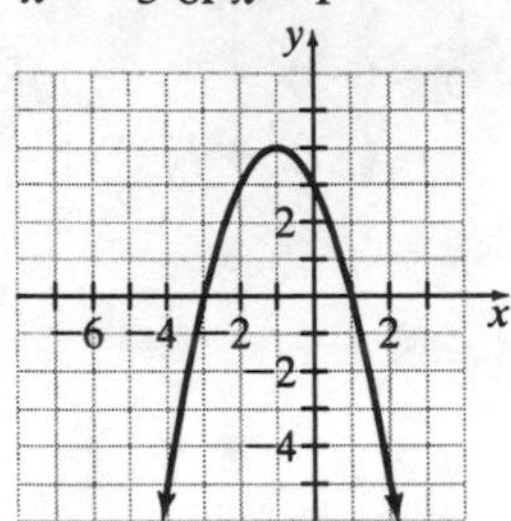

31. $a = 1, b = 5, c = -6$
Since $a > 0$, the parabola opens upward.
Axis of symmetry is

$$x = -\frac{b}{2a} = -\frac{5}{2(1)} = -\frac{5}{2}$$

y-coordinate of the vertex:

$$y = x^2 + 5x - 6$$

$$y = \left(-\frac{5}{2}\right)^2 + 5\left(-\frac{5}{2}\right) - 6 = -\frac{49}{4}$$

The vertex is $\left(-\frac{5}{2}, -\frac{49}{4}\right)$.

$y = x^2 + 5x - 6$

Let $x = -6$ $\quad y = (-6)^2 + 5(-6) - 6 = 0$

Let $x = 1$ $\quad y = 1^2 + 5(1) - 6 = 0$

x	y
−6	0
1	0

$0 = x^2 + 5x - 6$
$0 = (x+6)(x-1)$
$x = -6$ or $x = 1$

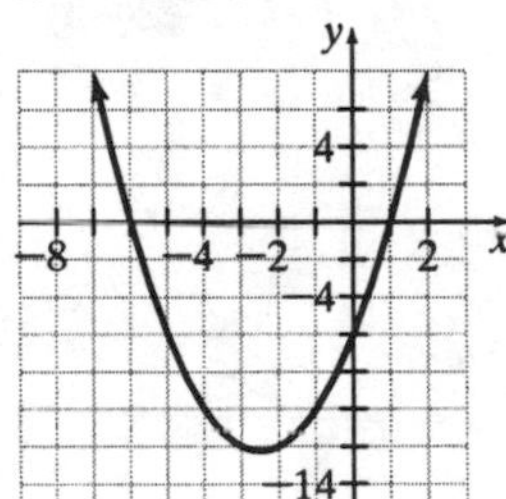

33. $a = 1, b = 5, c = -14$
Since $a > 0$, the parabola opens upward.
Axis of symmetry is

$$x = -\frac{b}{2a} = -\frac{5}{2(1)} = -\frac{5}{2}$$

y-coordinate of the vertex:

$$y = x^2 + 5x - 14$$

$$y = \left(-\frac{5}{2}\right)^2 + 5\left(-\frac{5}{2}\right) - 14 = -\frac{81}{4}$$

The vertex is $\left(-\frac{5}{2}, -\frac{81}{4}\right)$

$y = x^2 + 5x - 14$

Let $x = -7$ $\quad y = (-7)^2 + 5(-7) - 14 = 0$

Let $x = 2$ $\quad y = 2^2 + 5(2) - 14 = 0$

x	y
−7	0
2	0

$0 = x^2 + 5x - 14$
$0 = (x+7)(x-2)$
$x = -7$ or $x = 2$

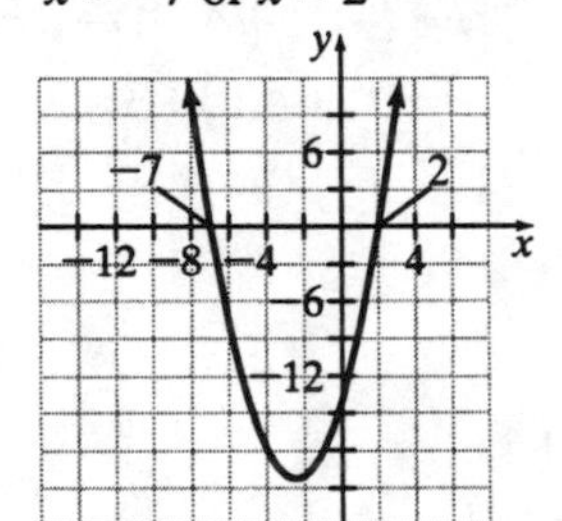

35. $a = 1, b = -6, c = 9$
Since $a > 0$, the parabola opens upward.
Axis of symmetry: $x = -\frac{b}{2a} = -\frac{-6}{2(1)} = 3$
y-coordinate of the vertex:
$y = x^2 - 6x + 9$
$y = 3^2 - 6(3) + 9 = 0$
The vertex is at (3, 0)

$y = x^2 - 6x + 9$

Let $x = 4$ $\quad y = 4^2 - 6(4) + 9 = 1$

Let $x = 5$ $\quad y = 5^2 - 6(5) + 9 = 4$

Let $x = 6$ $\quad y = 6^2 - 6(6) + 9 = 9$

x	y
4	1
5	4
6	9

$0 = x^2 - 6x + 9$
$0 = (x-3)^2$
$x = 3$

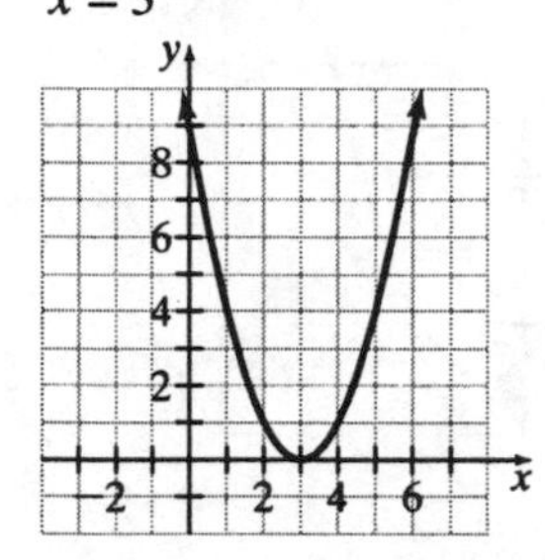

37. $a = 1, b = -6, c = 0$
Since $a > 0$, the parabola opens upward.
axis of symmetry is $x = -\frac{b}{2a} = -\frac{-6}{2(1)} = 3$
y-coordinate of the vertex:
$y = x^2 - 6x$
$y = 3^2 - 6(3) = -9$
The vertex is $(3, -9)$.

$y = x^2 - 6x$
Let $x = 0$ $\quad y = 0^2 - 6(0) = 0$
Let $x = 6$ $\quad y = 6^2 - 6(6) = 0$

x	y
0	0
6	0

$0 = x^2 - 6x$
$0 = x(x-6)$
$x = 0$ or $x = 6$

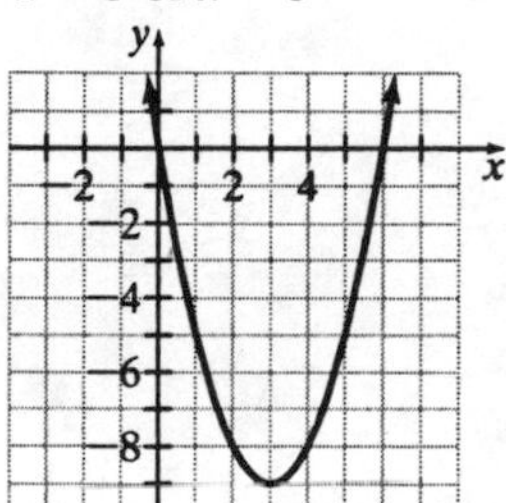

39. $a = 4, b = 12, c = 9$
Since $a > 0$, the parabola opens upward.
Axis of symmetry is
$x = -\frac{b}{2a} = -\frac{12}{2(4)} = -\frac{3}{2}$
y-coordinate of the vertex:
$y = 4x^2 + 12x + 9$
$y = 4\left(-\frac{3}{2}\right)^2 + 12\left(-\frac{3}{2}\right) + 9 = 0$
The vertex is $\left(-\frac{3}{2}, 0\right)$.

$y = 4x^2 + 12x + 9$
Let $x = -1$ $\quad y = 4(-1)^2 + 12(-1) + 9 = 1$
Let $x = 0$ $\quad y = 4(0)^2 + 12(0) + 9 = 9$

x	y
−1	1
0	9

$0 = 4x^2 + 12x + 9$
$0 = (2x+3)^2$
$x = -\frac{3}{2}$

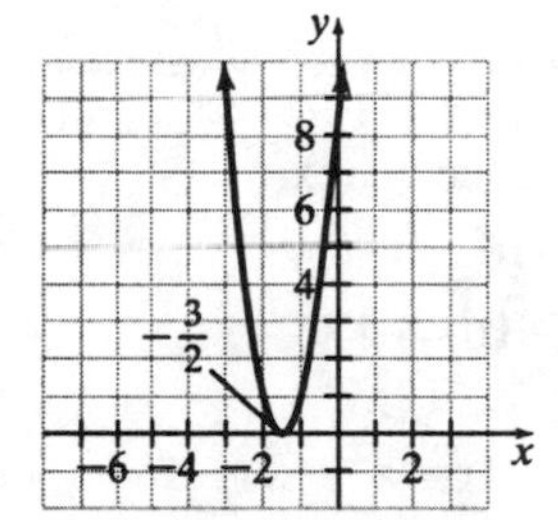

41. $a = -1, b = 11, c = -28$
Since $a < 0$, the parabola opens downward.
Axis of symmetry: $x = -\frac{b}{2a} = -\frac{11}{2(-1)} = \frac{11}{2}$
y-coordinate of the vertex:
$y = -x^2 + 11x - 28$
$y = -\left(\frac{11}{2}\right)^2 + 11\left(\frac{11}{2}\right) - 28 = \frac{9}{4}$
The vertex is at $\left(\frac{11}{2}, \frac{9}{4}\right)$.

$y = -x^2 + 11x - 28$

Let $x = 6$ $\quad y = -6^2 + 11(6) - 28 = 2$

Let $x = 7$ $\quad y = -7^2 + 11(7) - 28 = 0$

Let $x = 8$ $\quad y = -8^2 + 11(8) - 28 = -4$

x	y
6	2
7	0
8	–4

$0 = -x^2 + 11x - 28$
$0 = -(x - 4)(x - 7)$
$x = 4$ or $x = 7$

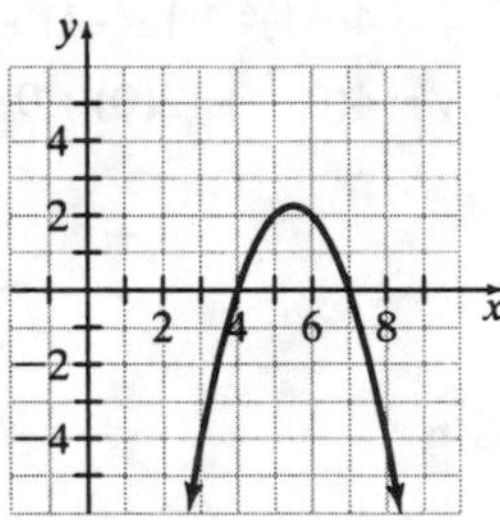

43. $a = 1, b = -2, c = -16$
Since $a > 0$, the parabola opens upward.

Axis of symmetry is $x = -\frac{b}{2a} = -\frac{-2}{2(1)} = 1$

y-coordinate of the vertex:

$y = x^2 - 2x - 16$
$y = 1^2 - 2(1) - 16 = -17$

The vertex is $(1, -17)$.

$y = x^2 - 2x - 16$

Let $x = -2$ $\quad y = (-2)^2 - 2(-2) - 16 = -8$

Let $x = 0$ $\quad y = 0^2 - 2(0) - 16 = -16$

x	y
–2	–8
0	–16

$0 = x^2 - 2x - 16$

$$x = \frac{-(-2) \pm \sqrt{(-2)^2 - 4(1)(-16)}}{2(1)}$$
$$= \frac{2 \pm \sqrt{68}}{2}$$
$$= 1 \pm \sqrt{17}$$

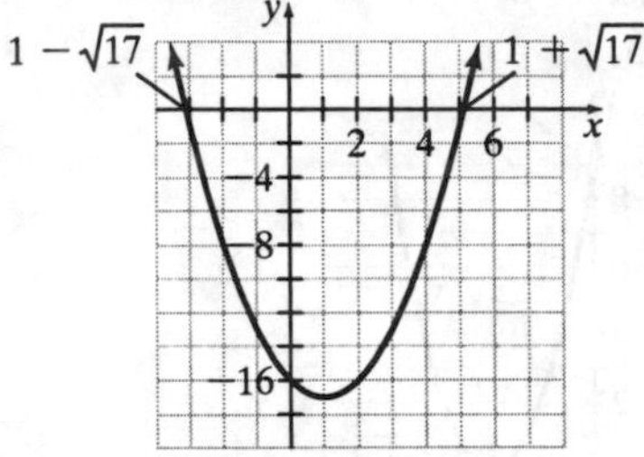

45. $a = -2, b = 3, c = -2$
Since $a < 0$, the parabola opens downward.

Axis of symmetry: $x = -\frac{b}{2a} = -\frac{3}{2(-2)} = \frac{3}{4}$

y-coordinate of the vertex:

$y = -2x^2 + 3x - 2$

$y = -2\left(\frac{3}{4}\right)^2 + 3\left(\frac{3}{4}\right) - 2 = -\frac{7}{8}$

The vertex is $\left(\frac{3}{4}, -\frac{7}{8}\right)$.

$y = -2x^2 + 3x - 2$

Let $x = -1$ $\quad y = -2(-1)^2 + 3(-1) - 2 = -7$

Let $x = 0$ $\quad y = -2(0)^2 + 3(0) - 2 = -2$

Let $x = 1$ $\quad y = -2(1)^2 + 3(1) - 2 = -1$

Let $x = 2$ $\quad y = -2(2)^2 + 3(2) - 2 = -4$

x	y
–1	–7
0	–2
1	–1
2	–4

$0 = -2x^2 + 3x - 2$

$x = \dfrac{-3 \pm \sqrt{3^2 - 4(-2)(-2)}}{2(-2)}$

$= \dfrac{-3 \pm \sqrt{-7}}{-4}$

No real number solution

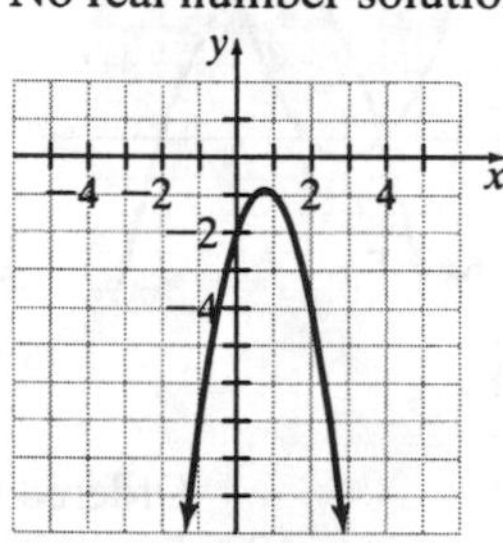

No x-intercepts

47. $a = 2, b = -1, c = -15$
Since $a > 0$, the parabola opens upward.
Axis of symmetry is $x = -\dfrac{b}{2a} = -\dfrac{-1}{2(2)} = \dfrac{1}{4}$

y-coordinate of the vertex:

$y = 2x^2 - x - 15$

$y = 2\left(\dfrac{1}{4}\right)^2 - \dfrac{1}{4} - 15 = -\dfrac{121}{8}$

The vertex is $\left(\dfrac{1}{4}, -\dfrac{121}{8}\right)$

$y = 2x^2 - x - 15$

Let $x = -2$ $\quad y = 2(-2)^2 - (-2) - 15 = -5$

Let $x = 2$ $\quad y = 2(2)^2 - 2 - 15 = -9$

x	y
–2	–5
2	–9

$0 = 2x^2 - x - 15$

$0 = (2x + 5)(x - 3)$

$x = -\dfrac{5}{2}$ or $x = 3$

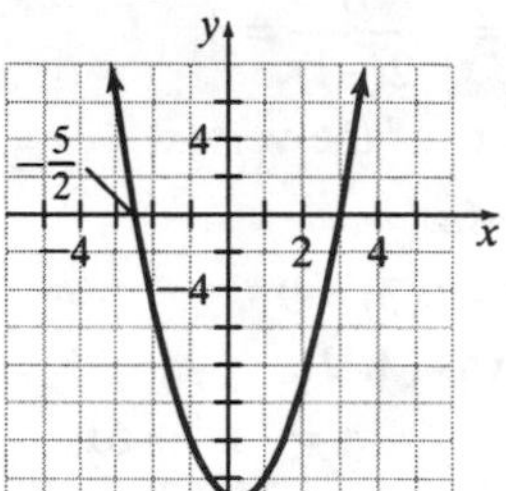

49. $b^2 - 4ac = (-2)^2 - 4(3)(-16)$
$= 4 + 192$
$= 192$
Since the discriminant is positive, there will be two x-intercepts.

51. $b^2 - 4ac = (-6)^2 - 4(4)(-7)$
$= 36 + 112$
$= 148$
Since the discriminant is positive, there will be two x-intercepts.

53. $b^2 - 4ac = (-20)^2 - 4(1)(100)$
$= 400 - 400$
$= 0$
Since the discriminant is zero, there will be one x-intercept.

55. $b^2 - 4ac = 2^2 - 4(1.6)(-3.9)$
$= 4 + 24.96$
$= 28.96$
Since the discriminant is positive, there will be two x-intercepts.

57. None; the vertex is below the x-axis and the parabola opens downward.

59. One; the vertex of the parabola is on the x-axis.

61. Yes; if y is set to 0, both equations have the same solutions, 4 and –2.

63. a. $a = -1$, $b = 6$, $c = 0$
Since $a < 0$, the parabola opens downward.
Axis of symmetry is

$$x = -\frac{b}{2a} = -\frac{6}{2(-1)} = 3$$

y-coordinate of the vertex:

$$y = -x^2 + 6x$$

$$y = -(3)^2 + 6(3) = 9$$

The vertex is (3, 9).

$$y = -x^2 + 6x$$

Let $x = 0$ $y = -(0)^2 + 6(0) = 0$

Let $x = 6$ $y = -(6)^2 + 6(6) = 6$

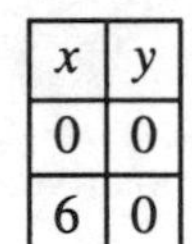

x	y
0	0
6	0

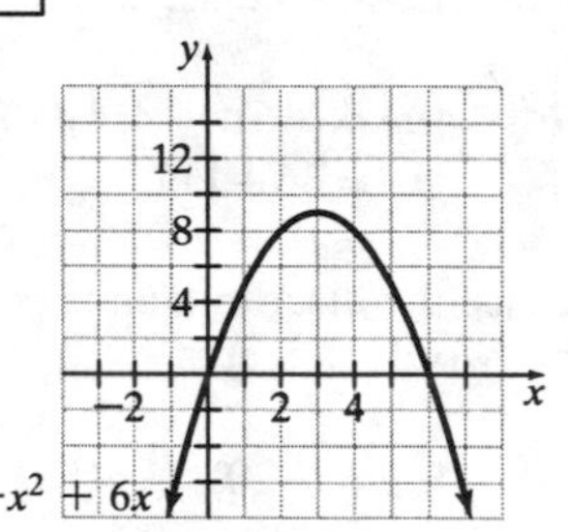

b. For $y = x^2 - 2x$
$a = 1$, $b = -2$, $c = 0$
Since $a > 0$, the parabola opens upward.
Axis of symmetry is

$$x = -\frac{b}{2a} = -\frac{-2}{2(1)} = 1$$

y-coordinate of the vertex:

$$y = x^2 - 2x$$

$$y = (1)^2 - 2(1) = -1$$

The vertex is (1, –1).

$$y = x^2 - 2x$$

Let $x = 2$ $y = 2^2 - 2(2) = 0$

Let $x = -1$ $y = (-1)^2 - 2(-1) = 3$

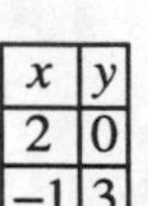

x	y
2	0
−1	3

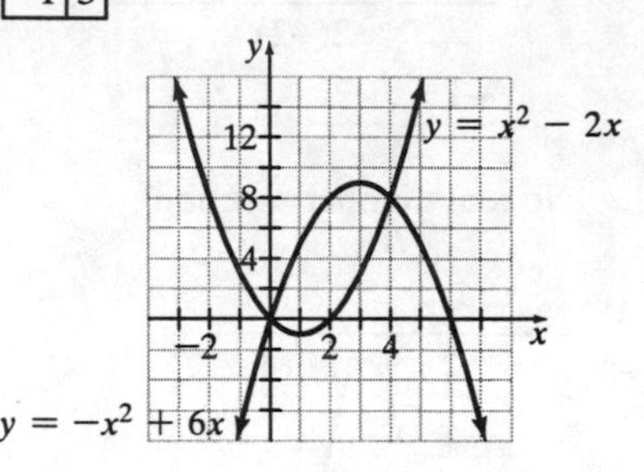

c. (0, 0), (4, 8)

65. $4x - 6y = 20$ Ordered Pair

Let $x = 0$, then $y = -\frac{10}{3}$ $\left(0, -\frac{10}{3}\right)$

Let $y = 0$, then $x = 5$ $(5, 0)$

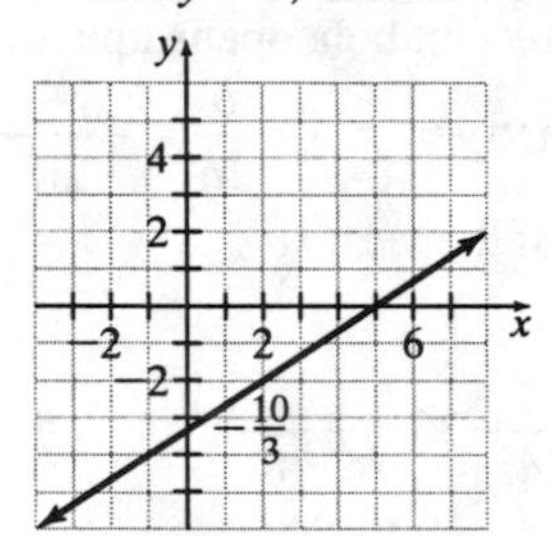

66. Draw the line $y = 2$ (horizontal line). Since the inequality symbol is <, draw a dashed line.
Check: (0, 0)

$$y < 2$$

$$0 < 2 \text{ True}$$

Since the origin does satisfy the inequality, all the points on the same side of the line as the origin satisfy the inequality.

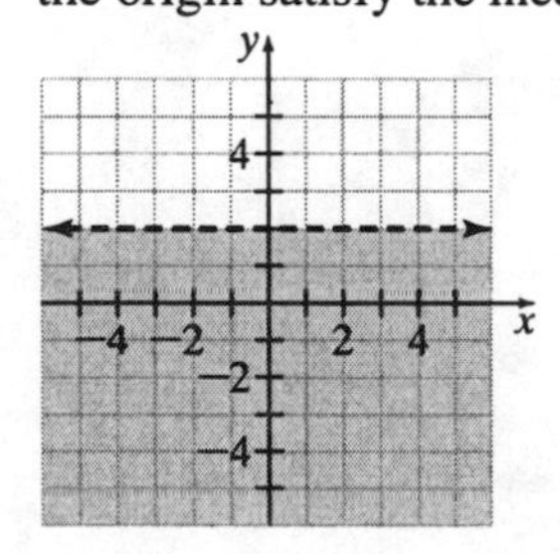

67. $\frac{3}{x+3}-\frac{x-2}{x-4}=\frac{3}{x+3}\cdot\frac{x-4}{x-4}-\frac{x-2}{x-4}\cdot\frac{x+3}{x+3}$
$$=\frac{3x-12}{(x+3)(x-4)}-\frac{x^2+x-6}{(x+3)(x-4)}$$
$$=\frac{3x-12-(x^2+x-6)}{(x+3)(x-4)}$$
$$=\frac{-x^2+2x-6}{(x+3)(x-4)}$$

68. $\frac{1}{3}(x+6)=3-\frac{1}{4}(x-5)$
$$12\left[\frac{1}{3}(x+6)\right]=12\left[3-\frac{1}{4}(x-5)\right]$$
$$4(x+6)=36-3(x-5)$$
$$4x+24=36-3x+15$$
$$7x=27$$
$$x=\frac{27}{7}$$

Review Exercises

1. $x^2=100$
$x=\pm\sqrt{100}$
$x=\pm 10$
The solutions are 10 and -10.

2. $x^2=12$
$x=\pm\sqrt{12}$
$x=\pm 2\sqrt{3}$
The solutions are $2\sqrt{3}$ and $-2\sqrt{3}$.

3. $2x^2=12$
$\frac{1}{2}(2x^2)=\frac{1}{2}(12)$
$x^2=6$
$x=\pm\sqrt{6}$
The solutions are $\sqrt{6}$ and $-\sqrt{6}$.

4. $x^2+3=9$
$x^2=6$
$x=\pm\sqrt{6}$
The solutions are $\sqrt{6}$ and $-\sqrt{6}$.

5. $x^2-4=16$
$x^2=20$
$x=\pm\sqrt{20}$
$x=\pm 2\sqrt{5}$
The solutions are $2\sqrt{5}$ and $-2\sqrt{5}$.

6. $2x^2-4=10$
$2x^2=14$
$x^2=7$
$x=\pm\sqrt{7}$
The solutions are $\sqrt{7}$ and $-\sqrt{7}$.

7. $4x^2-30=2$
$4x^2=32$
$x^2=8$
$x=\pm\sqrt{8}$
$x=\pm 2\sqrt{2}$
The solutions are $2\sqrt{2}$ and $-2\sqrt{2}$.

8. $(x-3)^2=12$
$x-3=\pm\sqrt{12}$
$x-3=\pm 2\sqrt{3}$
$x=3\pm 2\sqrt{3}$
The solutions are $3+2\sqrt{3}$ and $3-2\sqrt{3}$.

9. $(3x-5)^2=50$
$3x-5=\pm\sqrt{50}$
$3x-5=\pm 5\sqrt{2}$
$3x=5\pm 5\sqrt{2}$
$x=\frac{5\pm 5\sqrt{2}}{3}$
The solutions are $\frac{5+5\sqrt{2}}{3}$ and $\frac{5-5\sqrt{2}}{3}$.

10. $(2x+4)^2 = 30$

$$2x+4 = \pm\sqrt{30}$$
$$2x = -4 \pm \sqrt{30}$$
$$x = \frac{-4 \pm \sqrt{30}}{2}$$

The solutions are $\frac{-4+\sqrt{30}}{2}$ and $\frac{-4-\sqrt{30}}{2}$.

11.

$$x^2 - 9x + 18 = 0$$
$$x^2 - 9x = -18$$
$$x^2 - 9x + \frac{81}{4} = -18 + \frac{81}{4}$$
$$\left(x - \frac{9}{2}\right)^2 = \frac{9}{4}$$
$$x - \frac{9}{2} = \pm\sqrt{\frac{9}{4}}$$
$$x - \frac{9}{2} = \pm\frac{3}{2}$$
$$x = \frac{9}{2} \pm \frac{3}{2}$$
$$x = \frac{9}{2} + \frac{3}{2} \text{ or } x = \frac{9}{2} - \frac{3}{2}$$
$$x = 6 \quad \text{or } x = 3$$

The solutions are 6 and 3.

12.

$$x^2 - 11x + 28 = 0$$
$$x^2 - 11x = -28$$
$$x^2 - 11x + \frac{121}{4} = -28 + \frac{121}{4}$$
$$\left(x - \frac{11}{2}\right)^2 = \frac{9}{4}$$
$$x - \frac{11}{2} = \pm\sqrt{\frac{9}{4}}$$
$$x - \frac{11}{2} = \pm\frac{3}{2}$$
$$x = \frac{11}{2} \pm \frac{3}{2}$$
$$x = \frac{11}{2} + \frac{3}{2} \text{ or } x = \frac{11}{2} - \frac{3}{2}$$
$$x = 7 \quad \text{or } x = 4$$

The solutions are 7 and 4.

13.

$$x^2 - 18x + 17 = 0$$
$$x^2 - 18x = -17$$
$$x^2 - 18x + 81 = -17 + 81$$
$$(x-9)^2 = 64$$
$$x - 9 = \pm\sqrt{64}$$
$$x - 9 = \pm 8$$
$$x = 9 \pm 8$$
$$x = 9 + 8 \text{ or } x = 9 - 8$$
$$x = 17 \quad \text{or } x = 1$$

The solutions are 17 and 1.

14.

$$x^2 + x - 6 = 0$$
$$x^2 + x = 6$$
$$x^2 + x + \frac{1}{4} = 6 + \frac{1}{4}$$
$$\left(x + \frac{1}{2}\right)^2 = \frac{25}{4}$$
$$x + \frac{1}{2} = \pm\sqrt{\frac{25}{4}}$$
$$x + \frac{1}{2} = \pm\frac{5}{2}$$
$$x = -\frac{1}{2} \pm \frac{5}{2}$$
$$x = -\frac{1}{2} + \frac{5}{2} \text{ or } x = -\frac{1}{2} - \frac{5}{2}$$
$$x = 2 \quad \text{or } x = -3$$

The solutions are 2 and –3.

15.

$$x^2 - 3x - 54 = 0$$
$$x^2 - 3x = 54$$
$$x^2 - 3x + \frac{9}{4} = 54 + \frac{9}{4}$$
$$\left(x - \frac{3}{2}\right)^2 = \frac{225}{4}$$
$$x - \frac{3}{2} = \pm\sqrt{\frac{225}{4}}$$
$$x - \frac{3}{2} = \pm\frac{15}{2}$$
$$x = \frac{3}{2} \pm \frac{15}{2}$$
$$x = \frac{3}{2} + \frac{15}{2} \text{ or } x = \frac{3}{2} - \frac{15}{2}$$
$$x = 9 \quad \text{or } x = -6$$

The solutions are 9 and –6.

16.

$$x^2 = -5x + 6$$
$$x^2 + 5x = 6$$
$$x^2 + 5x + \frac{25}{4} = 6 + \frac{25}{4}$$
$$\left(x + \frac{5}{2}\right)^2 = \frac{49}{4}$$
$$x + \frac{5}{2} = \pm\sqrt{\frac{49}{4}}$$
$$x + \frac{5}{2} = \pm\frac{7}{2}$$
$$x- = -\frac{5}{2} \pm \frac{7}{2}$$
$$x = -\frac{5}{2} + \frac{7}{2} \text{ or } x = -\frac{5}{2} - \frac{7}{2}$$
$$x = 1 \quad \text{or } x = -6$$

The solutions are 1 and –6.

17.

$$x^2 - 3x - 8 = 0$$
$$x^2 - 3x = 8$$
$$x^2 - 3x + \frac{9}{4} = 8 + \frac{9}{4}$$
$$\left(x - \frac{3}{2}\right)^2 = \frac{41}{4}$$
$$x - \frac{3}{2} = \pm\sqrt{\frac{41}{4}}$$
$$x - \frac{3}{2} = \pm\frac{\sqrt{41}}{2}$$
$$x = \frac{3}{2} \pm \frac{\sqrt{41}}{2}$$
$$x = \frac{3 \pm \sqrt{41}}{2}$$
$$x = \frac{3 + \sqrt{41}}{2} \text{ or } x = \frac{3 - \sqrt{41}}{2}$$

The solutions are $\frac{3 + \sqrt{41}}{2}$ and $\frac{3 - \sqrt{41}}{2}$.

18.

$$x^2 + 2x - 5 = 0$$
$$x^2 + 2x = 5$$
$$x^2 + 2x + 1 = 5 + 1$$
$$(x + 1)^2 = 6$$
$$x + 1 = \pm\sqrt{6}$$
$$x = -1 \pm \sqrt{6}$$
$$x = -1 + \sqrt{6} \text{ or } x = -1 - \sqrt{6}$$

The solutions are $-1 + \sqrt{6}$ and $-1 - \sqrt{6}$.

19.
$$2x^2 - 8x = 64$$
$$\frac{1}{2}(2x^2 - 8x) = \frac{1}{2}(64)$$
$$x^2 - 4x = 32$$
$$x^2 - 4x + 4 = 32 + 4$$
$$(x-2)^2 = 36$$
$$x - 2 = \pm\sqrt{36}$$
$$x - 2 = \pm 6$$
$$x = 2 \pm 6$$
$$x = 2 + 6 \text{ or } x = 2 - 6$$
$$x = 8 \quad \text{or } x = -4$$

The solutions are 8 and –4.

20.
$$2x^2 - 4x = 30$$
$$\frac{1}{2}(2x^2 - 4x) = \frac{1}{2}(30)$$
$$x^2 - 2x = 15$$
$$x^2 - 2x + 1 = 15 + 1$$
$$(x-1)^2 = 16$$
$$x - 1 = \pm\sqrt{16}$$
$$x - 1 = \pm 4$$
$$x = 1 \pm 4$$
$$x = 1 + 4 \text{ or } x = 1 - 4$$
$$x = 5 \quad \text{or } x = -3$$

The solutions are 5 and –3.

21.
$$3x^2 + 2x - 8 = 0$$
$$\frac{1}{3}(3x^2 + 2x - 8) = \frac{1}{3}(0)$$
$$x^2 + \frac{2}{3}x - \frac{8}{3} = 0$$
$$x^2 + \frac{2}{3}x = \frac{8}{3}$$
$$x^2 + \frac{2}{3}x + \frac{1}{9} = \frac{8}{3} + \frac{1}{9}$$
$$\left(x + \frac{1}{3}\right)^2 = \frac{25}{9}$$
$$x + \frac{1}{3} = \pm\sqrt{\frac{25}{9}}$$
$$x + \frac{1}{3} = \pm\frac{5}{3}$$
$$x = -\frac{1}{3} \pm \frac{5}{3}$$
$$x = -\frac{1}{3} + \frac{5}{3} \text{ or } x = -\frac{1}{3} - \frac{5}{3}$$
$$x = \frac{4}{3} \quad \text{or } x = -2$$

The solutions are $\frac{4}{3}$ and -2.

22.
$$6x^2 - 19x + 15 = 0$$
$$\frac{1}{6}(6x^2 - 19x + 15) = \frac{1}{6}(0)$$
$$x^2 - \frac{19}{6}x + \frac{5}{2} = 0$$
$$x^2 - \frac{19}{6}x = -\frac{5}{2}$$
$$x^2 - \frac{19}{6}x + \frac{361}{144} = -\frac{5}{2} + \frac{361}{144}$$
$$\left(x - \frac{19}{12}\right)^2 = \frac{1}{144}$$
$$x - \frac{19}{12} = \pm\sqrt{\frac{1}{144}}$$
$$x - \frac{19}{12} = \pm\frac{1}{12}$$
$$x = \frac{19}{12} \pm \frac{1}{12}$$
$$x = \frac{19}{12} + \frac{1}{12} \quad \text{or} \quad x = \frac{19}{12} - \frac{1}{12}$$
$$x = \frac{5}{3} \quad \text{or} \quad x = \frac{3}{2}$$

The solutions are $\frac{5}{3}$ and $\frac{3}{2}$.

23. $b^2 - 4ac = 4^2 - 4(-2)(6) = 16 + 48 = 64$
Since the discriminant is positive, there are two distinct real number solutions.

24. Write in standard form.
$-3x^2 + 4x - 9 = 0$
$b^2 - 4ac = (4)^2 - 4(-3)(-9) = 16 - 108 = -92$

Since the discriminant is negative, there is no real number solution.

25. $b^2 - 4ac = (-10)^2 - 4(1)(25) = 100 - 100 = 0$
Since the discriminant is zero, there is one real number solution.

26. $b^2 - 4ac = (-1)^2 - 4(1)(8) = 1 - 32 = -31$
Since the discriminant is negative, there is no real number solution.

27. Write in standard form.
$2x^2 - 3x - 6 = 0$
$b^2 - 4ac = (-3)^2 - 4(2)(-6) = 9 + 48 = 57$
Since the discriminant is positive, there are two real number solutions.

28. $b^2 - 4ac = (-4)^2 - 4(3)(5) = 16 - 60 = -44$
Since the discriminant is negative, there is no real number solution.

29. $b^2 - 4ac = (-4)^2 - 4(-3)(8) = 16 + 96 = 112$
Since the discriminant is positive, there are two real number solutions.

30. $b^2 - 4ac = (-9)^2 - 4(1)(6) = 81 - 24 = 57$
Since the discriminant is positive, there are two real number solutions.

31. $a = 1, b = -11, c = 18$
$$x = \frac{-b \pm \sqrt{b^2 - 4ac}}{2a}$$
$$= \frac{-(-11) \pm \sqrt{(-11)^2 - 4(1)(18)}}{2(1)}$$
$$= \frac{11 \pm \sqrt{121 - 72}}{2}$$
$$= \frac{11 \pm \sqrt{49}}{2}$$
$$= \frac{11 \pm 7}{2}$$
$$x = \frac{11 + 7}{2} \quad \text{or} \quad x = \frac{11 - 7}{2}$$
$$x = 9 \quad \text{or} \quad x = 2$$
The solutions are 9 and 2.

32. $a = 1, b = -7, c = -44$

$$x = \frac{-b \pm \sqrt{b^2 - 4ac}}{2a}$$
$$= \frac{-(-7) \pm \sqrt{(-7)^2 - 4(1)(-44)}}{2(1)}$$
$$= \frac{7 \pm \sqrt{49 + 176}}{2}$$
$$= \frac{7 \pm \sqrt{225}}{2}$$
$$= \frac{7 \pm 15}{2}$$
$$x = \frac{7 + 15}{2} \text{ or } x = \frac{7 - 15}{2}$$
$$x = 11 \quad \text{or } x = -4$$

The solutions are 11 and –4.

33. Write in standard form.

$x^2 - 10x + 9 = 0$

$a = 1, b = -10, c = 9$

$$x = \frac{-b \pm \sqrt{b^2 - 4ac}}{2a}$$
$$= \frac{-(-10) \pm \sqrt{(-10)^2 - 4(1)(9)}}{2(1)}$$
$$= \frac{10 \pm \sqrt{100 - 36}}{2}$$
$$= \frac{10 \pm \sqrt{64}}{2}$$
$$= \frac{10 \pm 8}{2}$$
$$x = \frac{10 + 8}{2} \text{ or } x = \frac{10 - 8}{2}$$
$$x = 9 \quad \text{or } x = 1$$

The solutions are 9 and 1.

34. Write in standard form.

$5x^2 - 7x - 6 = 0$

$a = 5, b = -7, c = -6$

$$x = \frac{-b \pm \sqrt{b^2 - 4ac}}{2a}$$
$$= \frac{-(-7) \pm \sqrt{(-7)^2 - 4(5)(-6)}}{2(5)}$$
$$= \frac{7 \pm \sqrt{49 + 120}}{10}$$
$$= \frac{7 \pm \sqrt{169}}{10}$$
$$= \frac{7 \pm 13}{10}$$
$$x = \frac{7 + 13}{10} \text{ or } x = \frac{7 - 13}{10}$$
$$x = 2 \quad \text{or } x = -\frac{3}{5}$$

The solutions are 2 and $-\frac{3}{5}$.

35. Write in standard form.

$x^2 - 7x - 18 = 0$

$a = 1, b = -7, c = -18$

$$x = \frac{-b \pm \sqrt{b^2 - 4ac}}{2a}$$
$$= \frac{-(-7) \pm \sqrt{(-7)^2 - 4(1)(-18)}}{2(1)}$$
$$= \frac{7 \pm \sqrt{49 + 72}}{2}$$
$$= \frac{7 \pm \sqrt{121}}{2}$$
$$= \frac{7 \pm 11}{2}$$
$$x = \frac{7 + 11}{2} \text{ or } x = \frac{7 - 11}{2}$$
$$x = 9 \quad \text{or } x = -2$$

The solutions are 9 and –2.

36. $a = 1, b = -1, c = -30$

$$x = \frac{-b \pm \sqrt{b^2 - 4ac}}{2a}$$
$$= \frac{-(-1) \pm \sqrt{(-1)^2 - 4(1)(-30)}}{2(1)}$$
$$= \frac{1 \pm \sqrt{1+120}}{2}$$
$$= \frac{1 \pm \sqrt{121}}{2}$$
$$= \frac{1 \pm 11}{2}$$
$$x = \frac{1+11}{2} \text{ or } x = \frac{1-11}{2}$$
$$x = 6 \quad \text{ or } x = -5$$

The solutions are 6 and –5.

37. $a = 6, b = 1, c = -15$

$$x = \frac{-b \pm \sqrt{b^2 - 4ac}}{2a}$$
$$= \frac{-1 \pm \sqrt{1^2 - 4(6)(-15)}}{2(6)}$$
$$= \frac{-1 \pm \sqrt{1+360}}{12}$$
$$= \frac{-1 \pm \sqrt{361}}{12}$$
$$= \frac{-1 \pm 19}{12}$$
$$x = \frac{-1+19}{12} \text{ or } x = \frac{-1-19}{12}$$
$$x = \frac{3}{2} \quad \text{ or } x = -\frac{5}{3}$$

The solutions are $\frac{3}{2}$ and $-\frac{5}{3}$.

38. $a = -2, b = 3, c = 6$

$$x = \frac{-b \pm \sqrt{b^2 - 4ac}}{2a}$$
$$= \frac{-3 \pm \sqrt{3^2 - 4(-2)(6)}}{2(-2)}$$
$$= \frac{-3 \pm \sqrt{9+48}}{-4}$$
$$= \frac{-3 \pm \sqrt{57}}{-4}$$
$$= \frac{3 \pm \sqrt{57}}{4}$$
$$x = \frac{3+\sqrt{57}}{4} \text{ or } x = \frac{3-\sqrt{57}}{4}$$

The solutions are $\frac{3+\sqrt{57}}{4}$ and $\frac{3-\sqrt{57}}{4}$.

39. $a = 2, b = 4, c = -3$

$$x = \frac{-b \pm \sqrt{b^2 - 4ac}}{2a}$$
$$= \frac{-4 \pm \sqrt{4^2 - 4(2)(-3)}}{2(2)}$$
$$= \frac{-4 \pm \sqrt{16+24}}{4}$$
$$= \frac{-4 \pm \sqrt{40}}{4}$$
$$= \frac{-4 \pm 2\sqrt{10}}{4}$$
$$= \frac{2(-2 \pm \sqrt{10})}{4}$$
$$= \frac{-2 \pm \sqrt{10}}{2}$$
$$x = \frac{-2+\sqrt{10}}{2} \text{ or } x = \frac{-2-\sqrt{10}}{2}$$

The solutions are $\frac{-2+\sqrt{10}}{2}$ and $\frac{-2-\sqrt{10}}{2}$.

40. $a = 1, b = -6, c = 7$

$$x = \frac{-b \pm \sqrt{b^2 - 4ac}}{2a}$$
$$= \frac{-(-6) \pm \sqrt{(-6)^2 - 4(1)(7)}}{2(1)}$$
$$= \frac{6 \pm \sqrt{36 - 28}}{2}$$
$$= \frac{6 \pm \sqrt{8}}{2}$$
$$= \frac{6 \pm 2\sqrt{2}}{2}$$
$$= \frac{2(3 \pm \sqrt{2})}{2}$$
$$= 3 \pm \sqrt{2}$$

$x = 3 + \sqrt{2}$ or $x = 3 - \sqrt{2}$

The solutions are $3 + \sqrt{2}$ and $3 - \sqrt{2}$.

41. $a = 3, b = -4, c = 6$

$$x = \frac{-b \pm \sqrt{b^2 - 4ac}}{2a}$$
$$= \frac{-(-4) \pm \sqrt{(-4)^2 - 4(3)(6)}}{2(3)}$$
$$= \frac{4 \pm \sqrt{16 - 72}}{6}$$
$$= \frac{4 \pm \sqrt{-56}}{6}$$

Since $\sqrt{-56}$ is not a real number, there is no real number solution.

42. $a = 3, b = -6, c = -8$

$$x = \frac{-b \pm \sqrt{b^2 - 4ac}}{2a}$$
$$= \frac{-(-6) \pm \sqrt{(-6)^2 - 4(3)(-8)}}{2(3)}$$
$$= \frac{6 \pm \sqrt{36 + 96}}{6}$$
$$= \frac{6 \pm \sqrt{132}}{6}$$
$$= \frac{6 \pm 2\sqrt{33}}{6}$$
$$= \frac{2(3 \pm \sqrt{33})}{6}$$
$$= \frac{3 \pm \sqrt{33}}{3}$$

$x = \frac{3 + \sqrt{33}}{3}$ or $x = \frac{3 - \sqrt{33}}{3}$

The solutions are $\frac{3 + \sqrt{33}}{3}$ and $\frac{3 - \sqrt{33}}{3}$.

43. $a = 7, b = -3, c = 0$

$$x = \frac{-b \pm \sqrt{b^2 - 4ac}}{2a}$$
$$= \frac{-(-3) \pm \sqrt{(-3)^2 - 4(7)(0)}}{2(7)}$$
$$= \frac{3 \pm \sqrt{9}}{14}$$
$$= \frac{3 \pm 3}{14}$$

$x = \frac{3 + 3}{14}$ or $x = \frac{3 - 3}{14}$

$x = \frac{3}{7}$ or $x = 0$

The solutions are $\frac{3}{7}$ and 0.

44. $a = 2, b = -5, c = 0$

$$x = \frac{-b \pm \sqrt{b^2 - 4ac}}{2a}$$
$$= \frac{-(-5) \pm \sqrt{(-5)^2 - 4(2)(0)}}{2(2)}$$
$$= \frac{5 \pm \sqrt{25}}{4}$$
$$= \frac{5 \pm 5}{4}$$
$$x = \frac{5+5}{4} \text{ or } x = \frac{5-5}{4}$$
$$x = \frac{5}{2} \quad \text{or } x = 0$$

The solutions are $\frac{5}{2}$ and 0.

45.
$$x^2 - 13x + 42 = 0$$
$$(x-6)(x-7) = 0$$
$$x - 6 = 0 \text{ or } x - 7 = 0$$
$$x = 6 \text{ or } \quad x = 7$$

46.
$$x^2 + 15x + 56 = 0$$
$$(x+7)(x+8) = 0$$
$$x + 7 = 0 \text{ or } x + 8 = 0$$
$$x = -7 \text{ or } \quad x = -8$$

47.
$$x^2 - 3x - 70 = 0$$
$$(x-10)(x+7) = 0$$
$$x - 10 = 0 \text{ or } x + 7 = 0$$
$$x = 10 \text{ or } \quad x = -7$$

48.
$$x^2 + 6x = 27$$
$$x^2 + 6x - 27 = 0$$
$$(x-3)(x+9) = 0$$
$$x - 3 = 0 \text{ or } x + 9 = 0$$
$$x = 3 \text{ or } \quad x = -9$$

49.
$$x^2 - 4x - 60 = 0$$
$$(x+6)(x-10) = 0$$
$$x + 6 = 0 \text{ or } x - 10 = 0$$
$$x = -6 \text{ or } \quad x = 10$$

50.
$$x^2 - x - 42 = 0$$
$$(x-7)(x+6) = 0$$
$$x - 7 = 0 \text{ or } x + 6 = 0$$
$$x = 7 \text{ or } \quad x = -6$$

51.
$$x^2 + 11x - 12 = 0$$
$$(x+12)(x-1) = 0$$
$$x + 12 = 0 \text{ or } x - 1 = 0$$
$$x = -12 \quad \text{or } x = 1$$

52.
$$x^2 + 6x = 0$$
$$x(x+6) = 0$$
$$x = 0 \text{ or } x + 6 = 0$$
$$x = 0 \text{ or } \quad x = -6$$

53.
$$x^2 = 81$$
$$x = \pm\sqrt{81}$$
$$x = \pm 9$$
$$x = 9 \text{ or } x = -9$$

54.
$$2x^2 + 5x = 3$$
$$2x^2 + 5x - 3 = 0$$
$$(2x-1)(x+3) = 0$$
$$2x - 1 = 0 \text{ or } x + 3 = 0$$
$$x = \frac{1}{2} \text{ or } \quad x = -3$$

55.
$$2x^2 = 9x - 10$$
$$2x^2 - 9x + 10 = 0$$
$$(2x-5)(x-2) = 0$$
$$2x - 5 = 0 \text{ or } x - 2 = 0$$
$$x = \frac{5}{2} \text{ or } \quad x = 2$$

56.
$$6x^2 + 5x = 6$$
$$6x^2 + 5x - 6 = 0$$
$$(2x+3)(3x-2) = 0$$
$$2x + 3 = 0 \text{ or } 3x - 2 = 0$$
$$x = -\frac{3}{2} \text{ or } \quad x = \frac{2}{3}$$

57. $a = -2, b = 6, c = 9$

$$x = \frac{-b \pm \sqrt{b^2 - 4ac}}{2a}$$
$$= \frac{-6 \pm \sqrt{6^2 - 4(-2)(9)}}{2(-2)}$$
$$= \frac{-6 \pm \sqrt{36 + 72}}{-4}$$
$$= \frac{-6 + \sqrt{108}}{-4}$$
$$= \frac{-6 \pm 6\sqrt{3}}{-4}$$
$$= \frac{-2(3 \pm 3\sqrt{2})}{-2(2)}$$
$$= \frac{3 \pm 3\sqrt{2}}{2}$$
$$x = \frac{3 + 3\sqrt{3}}{2} \text{ or } x = \frac{3 - 3\sqrt{3}}{2}$$

58. $3x^2 - 11x + 10 = 0$
$(3x - 5)(x - 2) = 0$
$3x - 5 = 0$ or $x - 2 = 0$
$x = \frac{5}{3}$ or $x = 2$

59. $-3x^2 - 5x + 8 = 0$
$-1(-3x^2 - 5x + 8) = (-1)(0)$
$3x^2 + 5x - 8 = 0$
$(3x + 8)(x - 1) = 0$
$3x + 8 = 0$ or $x - 1 = 0$
$x = -\frac{8}{3}$ or $x = 1$

60. Write in standard form.
$x^2 + 3x - 6 = 0$
$a = 1, b = 3, c = -6$

$$x = \frac{-b \pm \sqrt{b^2 - 4ac}}{2a}$$
$$= \frac{-3 \pm \sqrt{3^2 - 4(1)(-6)}}{2(1)}$$
$$= \frac{-3 \pm \sqrt{9 + 24}}{2}$$
$$= \frac{-3 \pm \sqrt{33}}{2}$$
$$x = \frac{-3 + \sqrt{33}}{2} \text{ or } x = \frac{-3 - \sqrt{33}}{2}$$

61. $a = 4, b = -9, c = 0$

$$x = \frac{-b \pm \sqrt{b^2 - 4ac}}{2a}$$
$$= \frac{-(-9) \pm \sqrt{(-9)^2 - 4(4)(0)}}{2(4)}$$
$$= \frac{9 \pm \sqrt{81}}{8}$$
$$= \frac{9 \pm 9}{8}$$
$$x = \frac{9 + 9}{8} \text{ or } x = \frac{9 - 9}{8}$$
$$x = \frac{9}{4} \text{ or } x = 0$$

62. $3x^2 + 5x = 0$
$x(3x + 5) = 0$
$x = 0$ or $3x + 5 = 0$
$x = 0$ or $x = -\frac{5}{3}$

63. $a = 1, b = -4, c = -5$

$$x = -\frac{b}{2a} = -\frac{-4}{2(1)} = 2$$

The axis of symmetry is $x = 2$.
y-coordinate of the vertex:
$y = x^2 - 4x - 5$
$y = (2)^2 - 4(2) - 5$
$= 4 - 8 - 5$
$= -9$
The vertex is $(2, -9)$
Since $a > 0$, the parabola opens upward.

64. $a = 1, b = -12, c = 6$

$x = -\frac{b}{2a} = -\frac{-12}{2(1)} = 6$

The axis of symmetry is $x = 6$

y-coordinate of the vertex:

$y = x^2 - 12x + 6$

$y = 6^2 - 12(6) + 6$

$= 36 - 72 + 6$

$= -30$

The vertex is $(6, -30)$

Since $a > 0$, the parabola opens upward.

65. $a = 1, b = -3, c = 7$

$x = -\frac{b}{2a} = -\frac{-3}{2(1)} = \frac{3}{2}$

The axis of symmetry is $x = \frac{3}{2}$.

y-coordinate of the vertex:

$y = x^2 - 3x + 7$

$y = \left(\frac{3}{2}\right)^2 - 3\left(\frac{3}{2}\right) + 7$

$= \frac{9}{4} - \frac{9}{2} + 7$

$= \frac{19}{4}$

The vertex is $\left(\frac{3}{2}, \frac{19}{4}\right)$.

Since $a > 0$, the parabola opens upward.

66. $a = -1, b = -2, c = 15$

$x = -\frac{b}{2a} = -\frac{-2}{2(-1)} = -1$

The axis of symmetry is $x = -1$.

y-coordinate of the vertex:

$y = -x^2 - 2x + 15$

$y = -(-1)^2 - 2(-1) + 15$

$= -1 + 2 + 15$

$= 16$

The vertex is $(-1, 16)$.

Since $a < 0$, the parabola opens downward.

67. $a = 2, b = 7, c = 3$

$x = -\frac{b}{2a} = -\frac{7}{2(2)} = -\frac{7}{4}$

The axis of symmetry is $x = -\frac{7}{4}$.

y-coordinate of the vertex:

$y = 2x^2 + 7x + 3$

$y = 2\left(-\frac{7}{4}\right)^2 + 7\left(-\frac{7}{4}\right) + 3$

$= \frac{49}{8} - \frac{49}{4} + 3$

$= -\frac{25}{8}$

The vertex is $\left(-\frac{7}{4}, -\frac{25}{8}\right)$.

Since $a > 0$, the parabola opens upward.

68. $a = -1, b = -5, c = 0$

$x = -\frac{b}{2a} = -\frac{-5}{2(-1)} = \frac{-5}{2}$

The axis of symmetry is $x = -\frac{5}{2}$.

y-coordinate of the vertex:

$y = -x^2 - 5x$

$y = -\left(-\frac{5}{2}\right)^2 - 5\left(-\frac{5}{2}\right)$

$= -\frac{25}{4} + \frac{25}{2}$

$= \frac{25}{4}$

The vertex is $\left(-\frac{5}{2}, \frac{25}{4}\right)$.

Since $a < 0$, the parabola opens downward.

69. $a = -1, b = 0, c = -8$

$x = -\frac{b}{2a} = -\frac{0}{2(-1)} = 0$

The axis of symmetry is $x = 0$.

y-coordinate of the vertex:

$y = -x^2 - 8$

$y = -0^2 - 8 = -8$

The vertex is $(0, -8)$.

Since $a < 0$, the parabola opens downward.

70. $a=-1, b=-1, c=20$

$x=-\frac{b}{2a}=-\frac{-1}{2(-1)}=-\frac{1}{2}$

The axis of symmetry is $x=-\frac{1}{2}$.

y-coordinate of the vertex:

$y=-x^2-x+20$

$y=-\left(-\frac{1}{2}\right)^2-\left(-\frac{1}{2}\right)+20$

$=-\frac{1}{4}+\frac{1}{2}+20$

$=\frac{81}{4}$

The vertex is $\left(-\frac{1}{2},\frac{81}{4}\right)$.

Since $a<0$, the parabola opens downward.

71. $a=-4, b=8, c=5$

$x=-\frac{b}{2a}=-\frac{8}{2(-4)}=1$

The axis of symmetry is $x=1$.

y-coordinate of the vertex:

$y=-4x^2+8x+5$

$y=-4(1)^2+8(1)+5$

$=-4+8+5$

$=9$

The vertex is (1, 9).

Since $a<0$, the parabola opens downward.

72. $a=3, b=5, c=-8$

$x=-\frac{b}{2a}=-\frac{5}{2(3)}=-\frac{5}{6}$

The axis of symmetry is $x=-\frac{5}{6}$.

y-coordinate of the vertex:

$y=3x^2+5x-8$

$y=3\left(-\frac{5}{6}\right)^2+5\left(-\frac{5}{6}\right)-8$

$=\frac{25}{12}-\frac{25}{6}-8$

$=-\frac{121}{12}$

The vertex is $\left(-\frac{5}{6},-\frac{121}{12}\right)$.

Since $a>0$, the parabola opens upward.

73. $a=1, b=-2, c=0$

Since $a>0$, the parabola opens upward.

The axis of symmetry is

$y=-\frac{b}{2a}=-\frac{-2}{2(1)}=1$

y-coordinate of the vertex:

$y=x^2-2x$

$y=(1)^2-2(1)=-1$

The vertex is (1, −1).

$y=x^2-2x$

Let $x=2$ $\quad y=2^2-2(2)=0$

Let $x=0$ $\quad y=0^2-2(0)=0$

x	y
2	0
0	0

$0=x^2-2x$

$0=x(x-2)$

$x=0$ or $x=2$

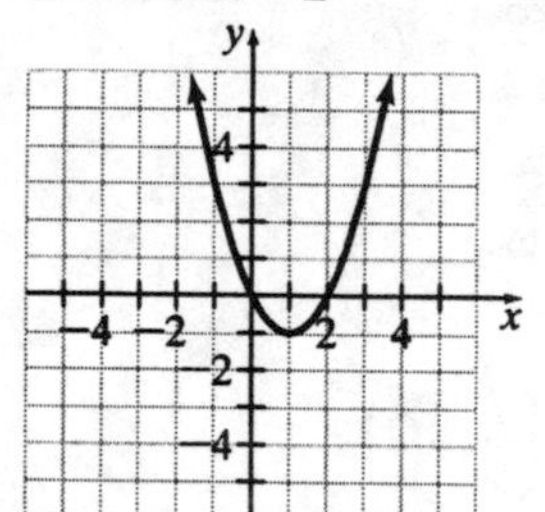

74. $a=-3, b=0, c=6$

Since $a<0$, the parabola opens downward.

The axis of symmetry is

$x=-\frac{b}{2a}=-\frac{0}{2(-3)}=0$

y-coordinate of the vertex:

$y=-3x^2+6$

$y=-3(0)^2+6=6$

The vertex is (0, 6).

$y=-3x^2+6$

Let $x=-1$ $\quad y=-3(-1)^2+6=3$

Let $x=1$ $\quad y=-3(1)^2+6=3$

x	y
−1	3
1	3

$0 = -3x^2 + 6$
$3x^2 = 6$
$x^2 = 2$
$x = \pm\sqrt{2}$

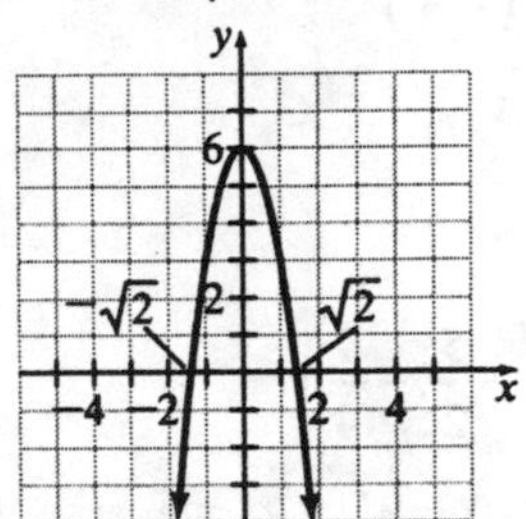

75. $a = 1$, $b = -2$, $c = -15$
Since $a > 0$, the parabola opens upward.
The axis of symmetry is

$x = -\frac{b}{2a} = -\frac{-2}{2(1)} = 1$

y-coordinate of the vertex:

$y = x^2 - 2x - 15$
$y = (1)^2 - 2(1) - 15$
$= 1 - 2 - 15$
$= -16$

The vertex is $(1, -16)$

$y = x^2 - 2x - 15$
Let $x = -3$ $\quad y = (-3)^2 - 2(-3) - 15 = 0$
Let $x = 5$ $\quad y = (5)^2 - 2(5) - 15 = 0$

x	y
−3	0
5	0

$0 = x^2 - 2x - 15$
$0 = (x + 3)(x - 5)$
$x = -3$ or $x = 5$

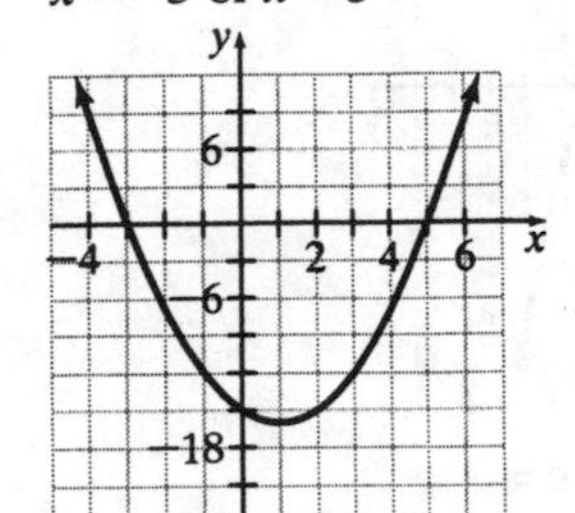

76. $a = -1$, $b = 5$, $c = -6$
Since $a < 0$, the parabola opens downward.
The axis of symmetry is

$x = -\frac{b}{2a} = -\frac{5}{2(-1)} = \frac{5}{2}$

y-coordinate of the vertex:

$y = -x^2 + 5x - 6$
$y = -\left(\frac{5}{2}\right)^2 + 5\left(\frac{5}{2}\right) - 6$
$= -\frac{25}{4} + \frac{25}{2} - 6$
$= \frac{1}{4}$

The vertex is $\left(\frac{5}{2}, \frac{1}{4}\right)$

$y = -x^2 + 5x - 6$
Let $x = 0$ $\quad y = -0^2 + 5(0) - 6 = -6$
Let $x = 3$ $\quad y = -(3)^2 + 5(3) - 6 = 0$

x	y
0	−6
3	0

$0 = -x^2 + 5x - 6$
$x^2 - 5x + 6 = 0$
$(x - 2)(x - 3) = 0$

$x = 2$ or $x = 3$

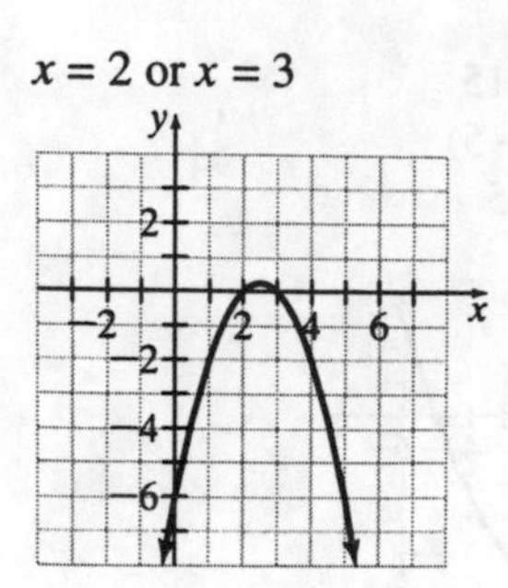

77. $a = 1, b = -1, c = 1$
Since $a > 0$, the parabola opens upward.
The axis of symmetry is

$$x = -\frac{b}{2a} = -\frac{-1}{2(1)} = \frac{1}{2}$$

y-coordinate of the vertex:

$$y = x^2 - x + 1$$

$$y = \left(\frac{1}{2}\right)^2 - \frac{1}{2} + 1 = \frac{3}{4}$$

The vertex is $\left(\frac{1}{2}, \frac{3}{4}\right)$.

$y = x^2 - x + 1$
Let $x = 0$ $\quad y = 0^2 - 0 + 1 = 1$
Let $x = 2$ $\quad y = 2^2 - (2) + 1 = 3$

x	y
0	1
2	3

$$0 = x^2 - x + 1$$

$$x = \frac{-(-1) \pm \sqrt{(-1)^2 - 4(1)(1)}}{2(1)}$$

$$= \frac{1 \pm \sqrt{-3}}{2}$$

No real number solution

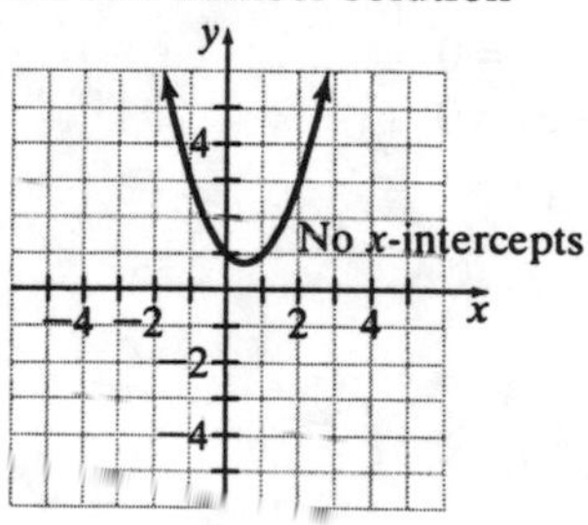

78. $a = 1, b = 5, c = 4$
Since $a > 0$, the parabola opens upward.
The axis of symmetry is

$$x = -\frac{b}{2a} = -\frac{5}{2(1)} = -\frac{5}{2}$$

y-coordinate of the vertex:

$$y = x^2 + 5x + 4$$

$$y = \left(-\frac{5}{2}\right)^2 + 5\left(-\frac{5}{2}\right) + 4$$

$$= \frac{25}{4} - \frac{25}{2} + 4$$

$$= -\frac{9}{4}$$

The vertex is $\left(-\frac{5}{2}, -\frac{9}{4}\right)$.

$y = x^2 + 5x + 4$
Let $x = -2$ $\quad y = (-2)^2 + 5(-2) + 4 = -2$
Let $x = -1$ $\quad y = (-1)^2 + 5(-1) + 4 = 0$
Let $x = 0$ $\quad y = (0)^2 + 5(0) + 4 = 4$

x	y
−2	−2
−1	0
0	4

$$0 = x^2 + 5x + 4$$
$$0 = (x + 1)(x + 4)$$
$$x + 1 = 0 \text{ or } x + 4 = 0$$
$$x = -1 \quad \text{ or } x = -4$$

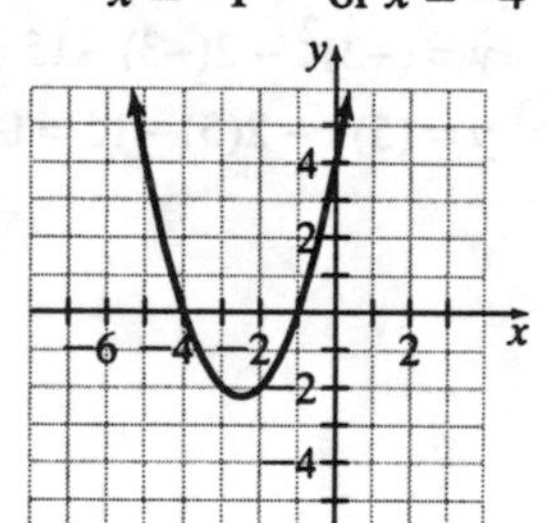

79. $a = -1, b = -4, c = 0$
Since $a < 0$, the parabola opens downward.
The axis of symmetry is
$$x = -\frac{b}{2a} = -\frac{-4}{2(-1)} = -2$$
y-coordinate of the vertex:
$$y = -x^2 - 4x$$
$$y = -(-2)^2 - 4(-2) = -4 + 8 = 4$$
The vertex is (–2, 4).
$$y = -x^2 - 4x$$
Let $x = 0$ $\quad y = -0^2 - 4(0) = 0$
Let $x = -4$ $\quad y = -(-4)^2 - 4(-4) = 0$

x	y
0	0
–4	0

$$0 = -x^2 - 4x$$
$$0 = -x(x + 4)$$
$$x = 0 \text{ or } x = -4$$

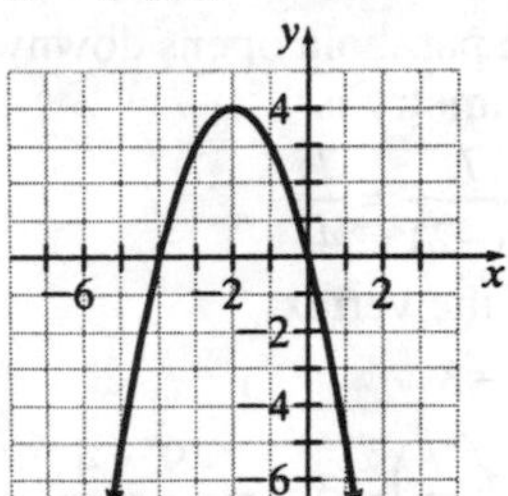

80. $a = 1, b = 4, c = 3$
Since $a > 0$, the parabola opens upward.
The axis of symmetry is
$$x = -\frac{b}{2a} = -\frac{4}{2(1)} = -2$$
y-coordinate of the vertex:
$$y = x^2 + 4x + 3$$
$$y = (-2)^2 + 4(-2) + 3$$
$$= 4 - 8 + 3$$
$$= -1$$
The vertex is (–2, –1).
$$y = x^2 + 4x + 3$$
Let $x = -1$ $\quad y = (-1)^2 + 4(-1) + 3 = 0$
Let $x = 0$ $\quad y = (0)^2 + 4(0) + 3 = 3$

x	y
–1	0
0	3

$$0 = x^2 + 4x + 3$$
$$0 = (x + 1)(x + 3)$$
$$x + 1 = 0 \text{ or } x + 3 = 0$$
$$x = -1 \text{ or } \quad x = -3$$

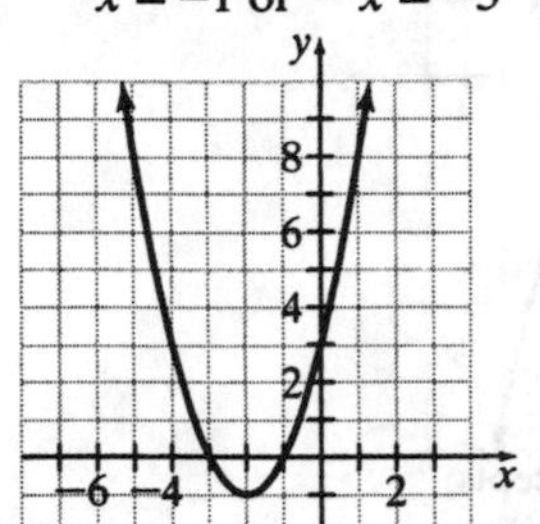

81. $a = -1, b = 1, c = -3$
Since $a < 0$, the parabola opens downward.
The axis of symmetry is
$$x = -\frac{b}{2a} = -\frac{1}{2(-1)} = \frac{1}{2}$$
y-coordinate of the vertex:
$$y = -x^2 + x - 3$$
$$y = -\left(\frac{1}{2}\right)^2 + \frac{1}{2} - 3$$
$$= -\frac{1}{4} + \frac{1}{2} - 3$$
$$= -\frac{11}{4}$$
The vertex is $\left(\frac{1}{2}, -\frac{11}{4}\right)$.
$$y = -x^2 + x - 3$$
Let $x = 0$ $\quad y = -0^2 + 0 - 3 = -3$
Let $x = 2$ $\quad y = -(2)^2 + 2 - 3 = -5$

x	y
0	-3
2	-5

$$0 = -x^2 + x - 3$$

$$x = \frac{-1 \pm \sqrt{1^2 - 4(-1)(-3)}}{2(-1)}$$

$$= \frac{-1 \pm \sqrt{-11}}{-2}$$

No real number solution

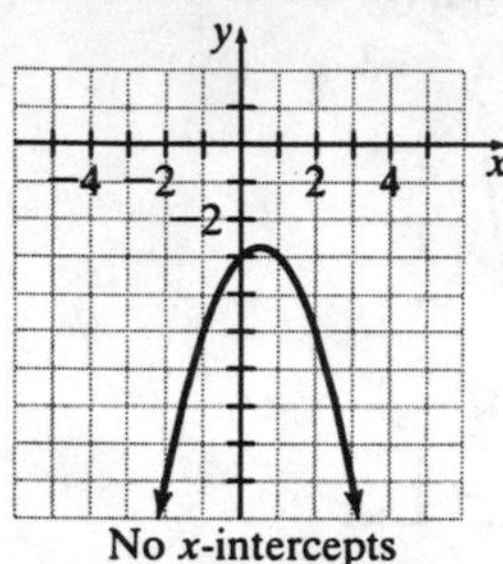

No x-intercepts

82. $a = 3, b = -4, c = -8$
Since $a > 0$, the parabola opens upward.
The axis of symmetry is

$$x = -\frac{b}{2a} = -\frac{-4}{2(3)} = \frac{2}{3}$$

y-coordinate of the vertex:

$$y = 3x^2 - 4x - 8$$

$$y = 3\left(\frac{2}{3}\right)^2 - 4\left(\frac{2}{3}\right) - 8$$

$$= \frac{4}{3} - \frac{8}{3} - 8$$

$$= -\frac{28}{3}$$

The vertex is $\left(\frac{2}{3}, -\frac{28}{3}\right)$.

$y = 3x^2 - 4x - 8$

Let $x = 0$ $y = 3(0)^2 - 4(0) - 8 = -8$

Let $x = 2$ $y = 3(2)^2 - 4(2) - 8 = -4$

x	y
0	−8
2	−4

$$0 = 3x^2 - 4x - 8$$

$$x = \frac{-(-4) \pm \sqrt{(-4)^2 - 4(3)(-8)}}{2(3)}$$

$$= \frac{4 \pm \sqrt{112}}{6}$$

$$= \frac{2(2 \pm 2\sqrt{7})}{2(3)}$$

$$= \frac{2 \pm 2\sqrt{7}}{3}$$

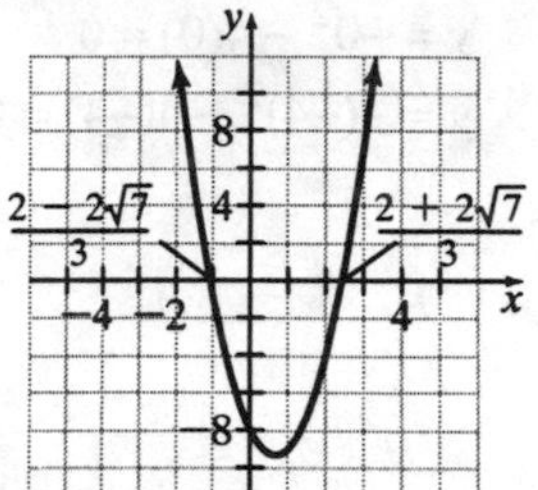

83. $a = -2, b = 7, c = -3$
Since $a < 0$, the parabola opens downward.
The axis of symmetry is

$$x = -\frac{b}{2a} = -\frac{7}{2(-2)} = \frac{7}{4}$$

y-coordinate of the vertex:

$$y = -2x^2 + 7x - 3$$

$$y = -2\left(\frac{7}{4}\right)^2 + 7\left(\frac{7}{4}\right) - 3 = -\frac{49}{8} + \frac{49}{4} - 3 = \frac{25}{8}$$

The vertex is $\left(\frac{7}{4}, \frac{25}{8}\right)$.

$y = -2x^2 + 7x - 3$

Let $x = 0$ $y = -2(0)^2 + 7(0) - 3 = -3$

Let $x = 3$ $y = -2(3)^2 + 7(3) - 3 = 0$

x	y
0	−3
3	0

$$0 = -2x^2 + 7x - 3$$

$$0 = -(2x - 1)(x - 3)$$

$x = \frac{1}{2}$ or $x = 3$

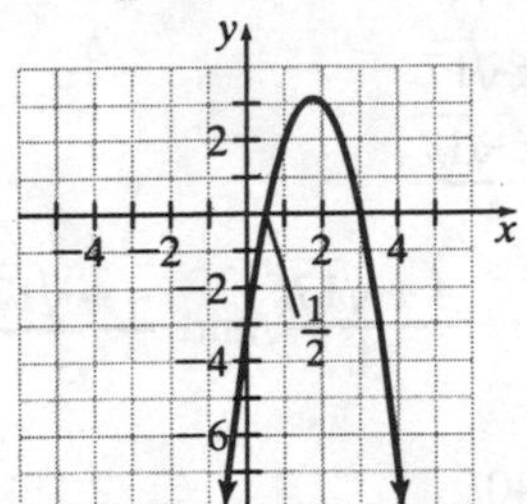

84. $a = 1, b = -5, c = 4$
Since $a > 0$, the parabola opens upward.
The axis of symmetry is

$x = -\frac{b}{2a} = -\frac{-5}{2(1)} = \frac{5}{2}$

y-coordinate of the vertex:

$y = x^2 - 5x + 4$

$y = \left(\frac{5}{2}\right)^2 - 5\left(\frac{5}{2}\right) + 4 = \frac{25}{4} - \frac{25}{2} + 4 = -\frac{9}{4}$

The vertex is $\left(\frac{5}{2}, -\frac{9}{4}\right)$.

$y = x^2 - 5x + 4$

Let $x = 4$ $\quad y = 4^2 - 5(4) + 4 = 0$

Let $x = 1$ $\quad y = 1^2 - 5(1) + 4 = 0$

x	y
4	0
1	0

$0 = x^2 - 5x + 4$

$0 = (x - 1)(x - 4)$

$x = 1$ or $x = 4$

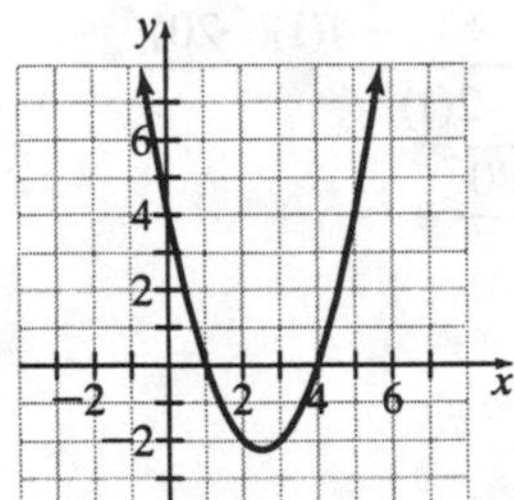

85. Let x = the smaller integer, then $x + 2$ = the larger integer.

$x(x + 2) = 48$

$x^2 + 2x = 48$

$x^2 + 2x + 1 = 48 + 1$

$(x + 1)^2 = 49$

$x + 1 = \pm\sqrt{49}$

$x + 1 = \pm 7$

$x = -1 \pm 7$

$x = -1 + 7$ or $x = -1 - 7$

$x = 6$ or $x = -8$

Since the numbers are positive, $x = 6$ and $x + 2 = 6 + 2 = 8$.

86. Let x = the smaller integer, then $x + 3$ = the larger integer.

$x(x + 3) = 88$

$x^2 + 3x = 88$

$x^2 + 3x + \frac{9}{4} = 88 + \frac{9}{4}$

$\left(x + \frac{3}{2}\right)^2 = \frac{361}{4}$

$x + \frac{3}{2} = \pm\sqrt{\frac{361}{4}}$

$x + \frac{3}{2} = \pm\frac{19}{2}$

$x = -\frac{3}{2} \pm \frac{19}{2}$

$x = -\frac{3}{2} + \frac{19}{2}$ or $x = -\frac{3}{2} - \frac{19}{2}$

$x = 8$ or $x = -11$

Since the numbers are positive, $x = 8$ and $x + 3 = 8 + 3 = 11$.

87. Let w = width of table,
then $2w + 6$ = length of table.

$$\text{Area} = \text{length} \cdot \text{width}$$
$$920 = w(2w+6)$$
$$2w^2 + 6w = 920$$
$$\frac{1}{2}(2w^2 + 6w) = \frac{1}{2}(920)$$
$$w^2 + 3w = 460$$
$$w^2 + 3w + \frac{9}{4} = 460 + \frac{9}{4}$$
$$\left(w + \frac{3}{2}\right)^2 = \frac{1849}{4}$$
$$w + \frac{3}{2} = \pm\sqrt{\frac{1849}{4}}$$
$$w + \frac{3}{2} = \pm\frac{43}{2}$$
$$w = -\frac{3}{2} \pm \frac{43}{2}$$
$$w = -\frac{3}{2} + \frac{43}{2} \text{ or } w = -\frac{3}{2} - \frac{43}{2}$$
$$w = 20 \quad \text{or } w = -23$$

Since w is positive, the width is 20 inches and the length is 2(20) + 6 = 46 inches.

88. Let w = width of the desktop,
then $w + 20$ = length of the desktop.

$$\text{Area} = \text{length} \cdot \text{width}$$
$$1344 = w(w+20)$$
$$1344 = w^2 + 20w$$
$$w^2 + 20w - 1344 = 0$$
$$(w-28)(w+48) = 0$$
$$w - 28 = 0 \text{ or } w + 48 = 0$$
$$w = 28 \text{ or } \quad w = -48$$

Since w is positive, the width is 28 inches and the length is 28 + 20 = 48 inches.

Practice Test

1. $x^2 - 2 = 26$

$$x^2 = 28$$
$$x = \pm\sqrt{28}$$
$$x = \pm 2\sqrt{7}$$

The solutions are $2\sqrt{7}$ and $-2\sqrt{7}$.

2. $(3x-4)^2 = 17$

$$3x - 4 = \pm\sqrt{17}$$
$$3x = 4 \pm \sqrt{17}$$
$$x = \frac{4 \pm \sqrt{17}}{3}$$

The solutions are $\frac{4+\sqrt{17}}{3}$ and $\frac{4-\sqrt{17}}{3}$.

3. $x^2 - 6x = 40$

$$x^2 - 6x + 9 = 40 + 9$$
$$(x-3)^2 = 49$$
$$x - 3 = \pm\sqrt{49}$$
$$x - 3 = \pm 7$$
$$x = 3 \pm 7$$
$$x = 3 + 7 \text{ or } x = 3 - 7$$
$$x = 10 \quad \text{or } x = -4$$

The solutions are 10 and –4.

4. $x^2 + 4x = 60$

$$x^2 + 4x + 4 = 60 + 4$$
$$(x+2)^2 = 64$$
$$x + 2 = \pm\sqrt{64}$$
$$x + 2 = \pm 8$$
$$x = -2 \pm 8$$
$$x = -2 + 8 \text{ or } x = -2 - 8$$
$$x = 6 \quad \text{or } x = -10$$

The solutions are 6 and –10.

5. $a = 1, b = -8, c = -20$

$$x = \frac{-b \pm \sqrt{b^2 - 4ac}}{2a}$$
$$= \frac{-(-8) \pm \sqrt{(-8)^2 - 4(1)(-20)}}{2(1)}$$
$$= \frac{8 \pm \sqrt{64 + 80}}{2}$$
$$= \frac{8 \pm \sqrt{144}}{2}$$
$$= \frac{8 \pm 12}{2}$$
$$x = \frac{8+12}{2} \text{ or } x = \frac{8-12}{2}$$
$$x = 10 \quad \text{or } x = -2$$

The solutions are 10 and –2.

6. $2x^2 + 5 = -8x$

$2x^2 + 8x + 5 = 0$

$a = 2, b = 8, c = 5$

$$x = \frac{-b \pm \sqrt{b^2 - 4ac}}{2a}$$
$$= \frac{-8 \pm \sqrt{8^2 - 4(2)(5)}}{2(2)}$$
$$= \frac{-8 \pm \sqrt{64 - 40}}{4}$$
$$= \frac{-8 \pm \sqrt{24}}{4}$$
$$= \frac{-8 \pm 2\sqrt{6}}{4}$$
$$= \frac{-4 \pm \sqrt{6}}{2}$$
$$x = \frac{-4 + \sqrt{6}}{2} \text{ or } x = \frac{-4 - \sqrt{6}}{2}$$

The solutions are $\frac{-4 + \sqrt{6}}{2}$ and $\frac{-4 - \sqrt{6}}{2}$.

7. $16x^2 = 49$

$$x^2 = \frac{49}{16}$$
$$x = \pm\sqrt{\frac{49}{16}}$$
$$x = \pm\frac{7}{4}$$

The solutions are $\frac{7}{4}$ and $-\frac{7}{4}$.

8. $2x^2 + 9x = 5$

$2x^2 + 9x - 5 = 0$

$(2x - 1)(x + 5) = 0$

$2x - 1 = 0$ or $x + 5 = 0$

$x = \frac{1}{2}$ or $x = -5$

The solutions are $\frac{1}{2}$ and -5.

9. $x = \frac{-b \pm \sqrt{b^2 - 4ac}}{2a}$

10. Answers will vary.

11. $b^2 - 4ac = (-4)^2 - 4(3)(2) = 16 - 24 = -8$
Since the discriminant is negative, the equation has no real solution.

12. $b^2 - 4ac = (8)^2 - 4(1)(-3) = 64 + 12 = 76$
Since the discriminant is positive, the equation has two distinct real solutions.

13. $a = -1, b = -6, c = 7$

Axis of symmetry: $x = -\frac{b}{2a} = -\frac{-6}{2(-1)} = -3$

The axis of symmetry is $x = -3$.

14. $a = 4, b = -8, c = 9$

Axis of symmetry: $x = -\frac{b}{2a} = -\frac{-8}{2(4)} = 1$

The axis of symmetry is $x = 1$.

15. $a = -1, b = -6, c = 7$
Since $a < 0$, the graph opens downward.

16. $a = 4, b = -8, c = 9$
Since $a > 0$, the graph opens upward.

17. The vertex of the graph of a parabola is the lowest point on a parabola that opens upward or the highest point on a parabola that opens downward.

18. $a = -1, b = -10, c = -16$

Axis of symmetry: $x = -\frac{b}{2a} = -\frac{-10}{2(-1)} = -5$

The axis of symmetry is $x = -5$.

y-coordinate of vertex:

$$y = -x^2 - 10x - 16$$
$$y = -(-5)^2 - 10(-5) - 16$$
$$= -25 + 50 - 16$$
$$= 9$$

The vertex is $(-5, 9)$.

19. $a = 3, b = -8, c = 9$

Axis of symmetry: $x = -\dfrac{b}{2a} = -\dfrac{-8}{2(3)} = \dfrac{4}{3}$

The axis of symmetry is $x = \dfrac{4}{3}$.

y-coordinate of vertex:

$y = 3x^2 - 8x + 9$

$y = 3\left(\dfrac{4}{3}\right)^2 - 8\left(\dfrac{4}{3}\right) + 9$

$= \dfrac{16}{3} - \dfrac{32}{3} + 9$

$= \dfrac{11}{3}$

The vertex is $\left(\dfrac{4}{3}, \dfrac{11}{3}\right)$.

20. $a = 1, b = 2, c = -8$

Since $a > 0$, the parabola opens upward.

Axis of symmetry: $x = -\dfrac{b}{2a} = -\dfrac{2}{2(1)} = -1$

y-coordinate of vertex:

$y = x^2 + 2x - 8$

$y = (-1)^2 + 2(-1) - 8 = 1 - 2 - 8 = -9$

The vertex is $(-1, -9)$

$y = x^2 + 2x - 8$

Let $x = 0$ $y = 0^2 + 2(0) - 8 = -8$

Let $x = 1$ $y = 1^2 + 2(1) - 8 = -5$

Let $x = 2$ $y = 2^2 + 2(2) - 8 = 0$

Let $x = 3$ $y = 3^2 + 2(3) - 8 = 7$

x	y
0	−8
1	−5
2	0
3	7

$0 = x^2 + 2x - 8$

$0 = (x + 4)(x - 2)$

$x = -4$ or $x = 2$

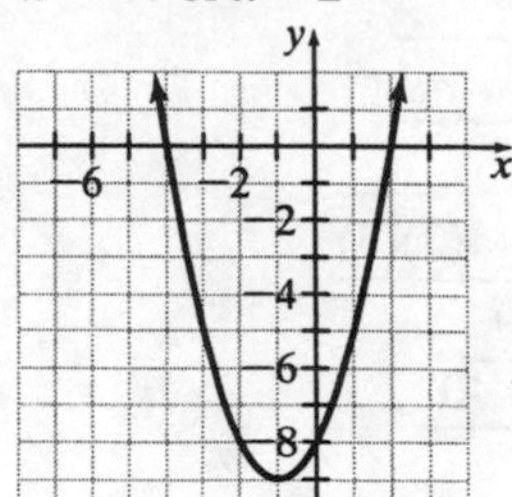

21. $a = -1, b = 6, c = -9$

Since $a < 0$, the parabola opens downward.

Axis of symmetry: $x = -\dfrac{b}{2a} = -\dfrac{6}{2(-1)} = 3$

y-coordinate of vertex:

$y = -x^2 + 6x - 9$

$y = -3^2 + 6(3) - 9 = -9 + 18 - 9 = 0$

The vertex is $(3, 0)$.

$y = -x^2 + 6x - 9$

Let $x = 0$ $y = -0^2 + 6(0) - 9 = -9$

Let $x = 1$ $y = -1^2 + 6(1) - 9 = -4$

Let $x = 2$ $y = -2^2 + 6(2) - 9 = -1$

x	y
0	−9
1	−4
2	−1

$0 = -x^2 + 6x - 9$

$0 = -(x - 3)^2$

$x = 3$

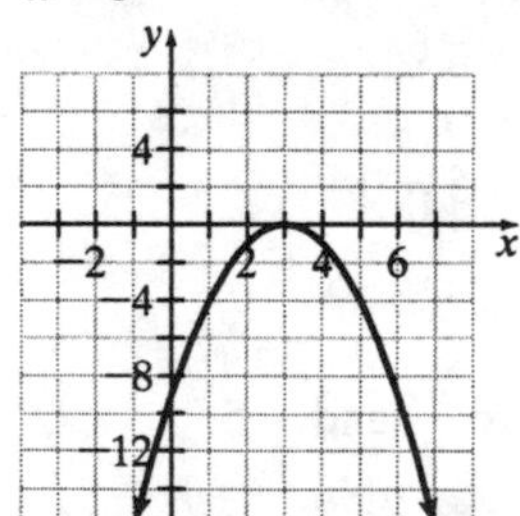

22. $a = 2$, $b = -6$, $c = 0$
Since $a > 0$, the parabola opens upward.
Axis of symmetry: $x = -\frac{b}{2a} = -\frac{-6}{2(2)} = \frac{3}{2}$
y-coordinate of vertex:
$y = 2x^2 - 6x$
$y = 2\left(\frac{3}{2}\right)^2 - 6\left(\frac{3}{2}\right) = \frac{9}{2} - \frac{18}{2} = -\frac{9}{2}$
The vertex is $\left(\frac{3}{2}, -\frac{9}{2}\right)$.

$y = 2x^2 - 6x$
Let $x = 1$ $\quad y = 2(1)^2 - 6(1) = -4$
Let $x = 2$ $\quad y = 2(2)^2 - 6(2) = -4$

x	y
1	–4
2	–4

$0 = 2x^2 - 6x$
$0 = 2x(x - 3)$
$x = 0$ or $x = 3$

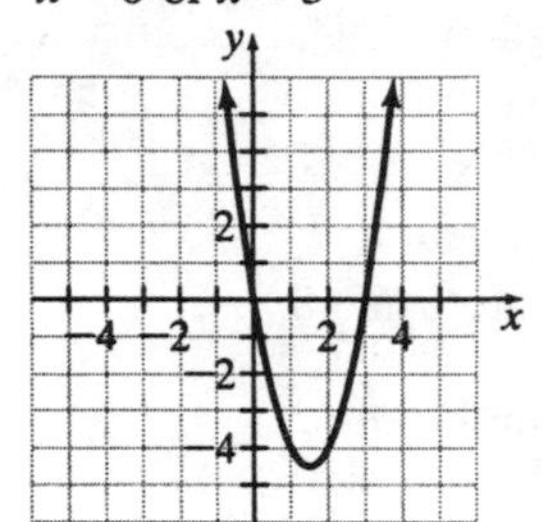

23. Let w = width of the mural,
then $3w + 1$ = length of the mural.
$$\text{Area} = \text{length} \cdot \text{width}$$
$$30 = (3w + 1)w$$
$$3w^2 + w = 30$$
$$3w^2 + w - 30 = 0$$
$$(3w + 10)(w - 3) = 0$$
$$3w + 10 = 0 \text{ or } w - 3 = 0$$
$$w = -\frac{10}{3} \text{ or } w = 3$$
Since w is positive, the width is 3 feet and the length is $3(3) + 1 = 10$ feet.

24. Let x = the larger integer,
then $x - 2$ = the smaller integer.
$$x(x - 2) = 99$$
$$x^2 - 2x = 99$$
$$x^2 - 2x + 1 = 99 + 1$$
$$(x - 1)^2 = 100$$
$$x - 1 = \pm\sqrt{100}$$
$$x - 1 = \pm 10$$
$$x = 1 \pm 10$$
$$x = 1 + 10 \text{ or } x = 1 - 10$$
$$x = 11 \quad \text{or } x = -9$$
Since the integer is positive, it is 11.

25. Let x = Shawn's age,
then $x - 4$ = Aaron's age.
$$x(x - 4) = 45$$
$$x^2 - 4x = 45$$
$$x^2 - 4x + 4 = 45 + 4$$
$$(x - 2)^2 = 49$$
$$x - 2 = \pm\sqrt{49}$$
$$x - 2 = \pm 7$$
$$x = 2 \pm 7$$
$$x = 2 + 7 \text{ or } x = 2 - 7$$
$$x = 9 \quad \text{or } x = -5$$
Since the age must be positive, Shawn is 9 years old.

Cumulative Review Test

1. $-3x^2y + 2y^2 - xy$
$= -3(2)^2(-3) + 2(-3)^2 - 2(-3)$
$= 36 + 18 + 6$
$= 60$

2.
$$\frac{1}{2}z - \frac{2}{7}z = \frac{1}{5}(3z - 1)$$
$$70\left[\frac{1}{2}z - \frac{2}{7}z\right] = 70\left[\frac{1}{5}(3z - 1)\right]$$
$$35z - 20z = 14(3z - 1)$$
$$15z = 42z - 14$$
$$-27z = -14$$
$$z = \frac{14}{27}$$

3. $\frac{x}{8}=\frac{2}{3}$
$3x=(8)(2)$
$3x=16$
$x=\frac{16}{3}$ or $5\frac{1}{3}$
The length of side x is $5\frac{1}{3}$ inches.

4. $2(x-3)\le 6x-5$
$2x-6\le 6x-5$
$2x-1\le 6x$
$-1\le 4x$
$-\frac{1}{4}\le x$
$x\ge -\frac{1}{4}$

$-\frac{1}{4}$

5. $A=\frac{m+n+P}{3}$
$3A=m+n+P$
$3A-m-n=P$
$P=3A-m-n$

6. $(2a^4b^3)^3(3a^2b^5)^2$
$=2^3a^{4(3)}b^{3(3)}\cdot 3^2a^{2(2)}b^{5(2)}$
$=8a^{12}b^9\cdot 9a^4b^{10}$
$=72a^{12+4}b^{9+10}$
$=72a^{16}b^{19}$

7. $x+2\overline{)x^2+6x+5}$ quotient $x+4$
x^2+2x
$4x+5$
$4x+8$
-3

$\frac{x^2+6x+5}{x+2}=x+4-\frac{3}{x+2}$

8. $2x^2-3xy-4xy+6y^2$
$=x(2x-3y)-2y(2x-3y)$
$=(x-2y)(2x-3y)$

9. $9x^2-48x+15$
$=3(3x^2-16x+5)$
$=3(3x-1)(x-5)$

10. $\frac{4}{a^2-16}+\frac{2}{(a-4)^2}$
$=\frac{4}{(a-4)(a+4)}+\frac{2}{(a-4)^2}$
$=\frac{4}{(a-4)(a+4)}\cdot\frac{a-4}{a-4}+\frac{2}{(a-4)^2}\cdot\frac{a+4}{a+4}$
$=\frac{4a-16}{(a+4)(a-4)^2}+\frac{2a+8}{(a+4)(a-4)^2}$
$=\frac{6a-8}{(a+4)(a-4)^2}$

11. $x+\frac{48}{x}=14$
$x\left[x+\frac{48}{x}\right]=14\cdot x$
$x^2+48=14x$
$x^2-14x+48=0$
$(x-6)(x-8)=0$
$x-6=0$ or $x-8=0$
$x=6$ or $x=8$
The solutions are 6 and 8.

12. Write in standard form.
$y=4x-8$

	Ordered Pair
Let $x=0$, then $y=-8$	$(0,-8)$
Let $x=2$, then $y=0$	$(2,0)$

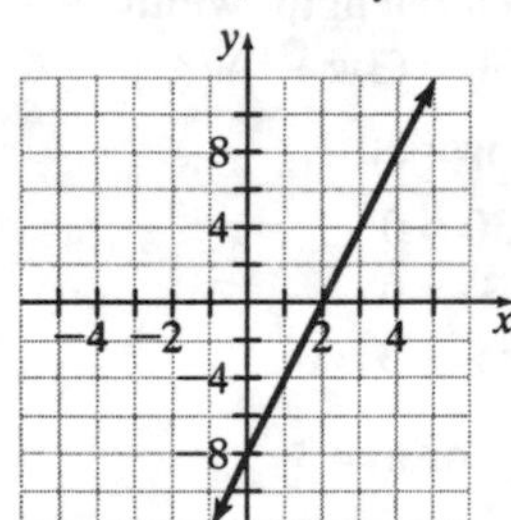

13. $3x - 4y = 12$
$4x - 3y = 6$
Multiply the first equation by 4 and the second equation by -3.
$4[3x - 4y = 12]$
$-3[4x - 3y = 6]$
gives

$$\begin{array}{r} 12x - 16y = 48 \\ -12x + 9y = -18 \\ \hline -7y = 30 \\ y = -\frac{30}{7} \end{array}$$

Substitute $-\frac{30}{7}$ for y in the first equation.

$$\begin{aligned} 3x - 4y &= 12 \\ 3x - 4\left(-\frac{30}{7}\right) &= 12 \\ 3x + \frac{120}{7} &= 12 \\ 3x &= -\frac{36}{7} \\ x &= -\frac{12}{7} \end{aligned}$$

The solution is $\left(-\frac{12}{7}, -\frac{30}{7}\right)$.

14.

$$\begin{aligned} \sqrt{\frac{3x^2y^3}{54x}} &= \sqrt{\frac{xy^3}{18}} \\ &= \frac{\sqrt{xy^3}}{\sqrt{18}} \\ &= \frac{\sqrt{y^2}\sqrt{xy}}{\sqrt{9}\sqrt{2}} \\ &= \frac{y\sqrt{xy}}{3\sqrt{2}} \\ &= \frac{y\sqrt{xy}}{3\sqrt{2}} \cdot \frac{\sqrt{2}}{\sqrt{2}} \\ &= \frac{y\sqrt{2xy}}{6} \end{aligned}$$

15.

$$\begin{aligned} 2\sqrt{28} - 3\sqrt{7} + \sqrt{63} &= 2\sqrt{4}\sqrt{7} - 3\sqrt{7} + \sqrt{9}\sqrt{7} \\ &= 2 \cdot 2\sqrt{7} - 3\sqrt{7} + 3\sqrt{7} \\ &= 4\sqrt{7} - 3\sqrt{7} + 3\sqrt{7} \\ &= 4\sqrt{7} \end{aligned}$$

16.

$$\begin{aligned} x - 2 &= \sqrt{x^2 - 12} \\ (x-2)^2 &= \left(\sqrt{x^2 - 12}\right)^2 \\ x^2 - 4x + 4 &= x^2 - 12 \\ -4x + 4 &= -12 \\ -4x &= -16 \\ x &= 4 \end{aligned}$$

Check:

$$\begin{aligned} x - 2 &= \sqrt{x^2 - 12} \\ 4 - 2 &= \sqrt{4^2 - 12} \\ 2 &= \sqrt{16 - 12} \\ 2 &= \sqrt{4} \\ 2 &= 2 \text{ True} \end{aligned}$$

17. $2x^2 + x - 8 = 0$
$a = 2$, $b = 1$, $c = -8$

$$\begin{aligned} x &= \frac{-b \pm \sqrt{b^2 - 4ac}}{2a} \\ &= \frac{-1 \pm \sqrt{(1)^2 - 4(2)(-8)}}{2(2)} \\ &= \frac{-1 \pm \sqrt{1 + 64}}{4} \\ &= \frac{-1 \pm \sqrt{65}}{4} \end{aligned}$$

$x = \frac{-1 + \sqrt{65}}{4}$ or $x = \frac{-1 - \sqrt{65}}{4}$

The solutions are $\frac{-1 + \sqrt{65}}{4}$ and $\frac{-1 - \sqrt{65}}{4}$.

18.

$$\begin{aligned} \frac{500 \text{ square feet}}{4 \text{ pounds fertilizer}} &= \frac{3200 \text{ square feet}}{x \text{ pounds fertilizer}} \\ \frac{500}{4} &= \frac{3200}{x} \\ 500x &= 4 \cdot 3200 \\ 500x &= 12{,}800 \\ x &= 25.6 \end{aligned}$$

25.6 pounds of fertilizer are needed for 3200 square feet of lawn.

19. Let w = width of garden
Then $3w - 3$ = length of garden

$$P = 2l + 2w$$
$$74 = 2(3w - 3) + 2w$$
$$74 = 6w - 6 + 2w$$
$$80 = 8w$$
$$10 = w$$

The width is 10 feet and the length is $3(10) - 3 = 27$ feet.

20. Let w = walking speed
Then $w + 3$ = jogging speed

Time to walk 2 miles = $\frac{2}{w}$

Time to jog 2 miles = $\frac{2}{w+3}$

Total time was 1 hour.

$$\frac{2}{w} + \frac{2}{w+3} = 1$$
$$w(w+3)\left[\frac{2}{w} + \frac{2}{w+3}\right] = 1(w+3)w$$
$$2(w+3) + 2w = w^2 + 3w$$
$$2w + 6 + 2w = w^2 + 3w$$
$$4w + 6 = w^2 + 3w$$
$$0 = w^2 - w - 6$$
$$0 = (w-3)(w+2)$$
$$(w-3)(w+2) = 0$$
$$w - 3 = 0 \text{ or } w + 2 = 0$$
$$w = 3 \text{ or } \quad w = -2$$

Since w must be positive, his walking speed is 3 mph and his jogging speed is $3 + 3 = 6$ mph.